International Conference on the Utilization of Tidal Power, Nova Scotia Technical College, 1970
Tidal power

TIDAL POWER

TIDAL POWER

Proceedings of an International Conference on the Utilization of Tidal Power held May 24-29, 1970, at the Atlantic Industrial Research Institute, Nova Scotia Technical College, Halifax, Nova Scotia

Edited by
T. J. Gray
Director, Atlantic Industrial Research Institute
Nova Scotia Technical College
Halifax, Nova Scotia

and

O. K. Gashus
Head, Electrical Engineering Department
Nova Scotia Technical College
Halifax, Nova Scotia

PLENUM PRESS • NEW YORK–LONDON • 1972

First Printing – February 1972
Second Printing – December 1972

Library of Congress Catalog Card Number 70-179031
ISBN 0-306-30559-3

A Division of Plenum Publishing Corporation
227 West 17th Street, New York, N.Y. 10011

United Kingdom edition published by Plenum Press, London
A Division of Plenum Publishing Company, Ltd.
Davis House (4th Floor), 8 Scrubs Lane, Harlesden, NW10 6SE, London, England

Printed in the United States of America

FOREWORD

At a time when public attention is focused on the environment, while simultaneously society is increasing at an ever-accelerating rate its demand for electrical power, the possibility of utilizing the power of the oceans by pollution free tidal power generation is most attractive. Tidal power has been used to a limited extent over several centuries but only recently has any significant effort been dedicated to realizing some of the vast potential.

The first pilot project at La Rance has now been operating successfully for several years and the second experimental station using up-dated construction techniques has been in operation at Kislaya Guba since 1969. These projects have contributed valuable experience and establish the technical feasibility of this important source of electrical power, while providing guidance in those areas requiring further development to realize economic viability.

More than fifty sites can be readily identified around the world where tidal power schemes could realistically be developed. With improvements in technology, this number might well be extended by utilization of a large number of river estuaries. Such developments must be considered not only on the basis of the production of electrical power but also in respect of associated benefits. Considerable bodies of water would be partially confined, thereby improving recreational facilities as has already been experienced at La Rance. The reduction in tidal extremes can improve land utilization in the vicinity by the elimination of flooding, while the tidal basin itself enhances the possibility of fish farming. Transportation systems can use with advantage highways constructed above the tidal barrage and secondary industry can be confidently anticipated as developing both during the development and construction of the tidal power installation and subsequently.

These many advantages can result from a power generating system essentially free of environmental hazard. Tidal power has significantly less effect on the ecology than even conventional hydro-electric systems, where the flooding of large areas has a profound effect. Inherently tidal power schemes leave the tidal regimen controlled but essentially unchanged. The penalty for this in the past has been higher cost but this has now been entirely changed by improvements in the technology, rising cost of conventional systems, sharply rising cost and projected shortages of fuel and the cost of pollution control.

This International Conference has brought together world experts in an overall consideration of the many complexities involved in the realization of large tidal power projects. The Conference provided a forum for extended discussions, which have been embodied in the revised articles from the participating authors. It has been effective in no small measure in stimulating renewed interest in Tidal Power in general and the Bay of Fundy scheme in particular.

The Editors wish to express their sincere appreciation for a supporting grant from the Federal Department of Energy, Mines and Resources, Ottawa, the use of the facilities of Nova Scotia Technical College and of the Atlantic Industrial Research Institute, and for the assistance of many members of their staff. Particular acknowledgment is made to Mr. C. MacLennan (Harza Engineering), Mr. L. Kirkpatrick (Nova Scotia Power Commission), and to the Board of Trustees of the Atlantic Industrial Research Institute for their sustaining interest and assistance during the planning period.

The sincere appreciation of the Editors and Conferees is extended to the Province of Nova Scotia, to the Voluntary Economic Planning Board of Nova Scotia, to the Royal Bank of Canada, and to the Halifax Herald Limited, for entertainment during the Conference, and to the numerous organizations who contributed in many ways to its success.

A very special acknowledgment is due to Linda Flanders, who prepared the preprint manuscripts and acted in so many important ways as general secretary for the Conference. The preprints, which contributed notably to the success of the Conference, were produced with the personal assistance of Mr. George Baker of the Kentville Advertiser, many at the eleventh hour. A final acknowledgment is due to Shirley Tutt, who completed the arduous preparation of the manuscript of these Proceedings.

T. J. Gray
O. K. Gashus

Halifax, Nova Scotia
June, 1971

CONTRIBUTORS

L. B. Bernshtein, Institute Hydroproject, The Ministry of Power and Electrification of the U.S.S.R., Moscow, U.S.S.R.

M. Braikevitch, English Electric-AEI Turbine Generators, Ltd., G. E. C. Power Engineering Ltd., Liverpool, England

H. A. Erith, The Shawinigan Engineering Company, Ltd., Montreal, Quebec, Canada

H. E. Fentzloff, Hochtief AG, Essen, West Germany

J. D. Gwynn, Engineering and Power Development Consultants, Kent, England

R. V. L. Hall, Olin Corporation, Metals Research Laboratories, New Haven, Connecticut

N. S. Heaps, Institute of Coastal Oceanography and Tides, Bidston Observatory, Birkenhead, Cheshire, England

F. L. Lawton, Atlantic Tidal Power Programming Board, Halifax, Nova Scotia, Canada

Georges Mauboussin, ancien Directeur de la Région d'Equipement Marémotrice, d'Electricité de France

J. F. McGurn, The International Nickel Company of Canada Limited, Toronto, Ontario, Canada

R. M. McMullen, Atlantic Oceanographic Laboratory, Bedford Institute, Dartmouth, Nova Scotia, Canada

V. M. Odd, Ministry of Technology, Hydraulics Research Station, Wallingford, England

M. W. Owen, Ministry of Technology, Hydraulics Research Station, Wallingford, England

F. E. Parkinson, LaSalle Hydraulic Laboratory, LaSalle, Quebec, Canada

B. R. Pelletier, Atlantic Oceanographic Laboratory, Bedford Institute, Dartmouth, Nova Scotia, Canada

M. J. Pryor, Olin Corporation, Metals Research Laboratories, New Haven, Connecticut

E. Ruus, Department of Civil Engineering, The University of British Columbia, Vancouver, British Colombia, Canada

B. Severn, Balfour, Beatty and Company, Limited, Croydon, England

T. L. Shaw, Department of Civil Engineering, University of Bristol, Bristol, England

T. J. Sluymer, H. G. Acres Limited, Niagara Falls, Ontario, Canada

K. E. Sorenson, Harza Engineering Company, Chicago, Illinois

F. Spaargaren, Waterloopkundige Afdeling van de Deltadienst, The Hague, Netherlands

M. C. Swales, Montreal Engineering Company, Montreal, Quebec, Canada

P. R. Tozer, H. G. Acres Limited, Niagara Falls, Ontario, Canada

A. N. T. Varzeliotis, Engineering Division, Inland Waters Branch, Department of Energy, Mines, and Resources, Vancouver, British Colombia, Canada

D. H. Waller, Atlantic Industrial Research Institute, Halifax, Nova Scotia, Canada

J. G. Warnock, Acres Limited, Toronto, Ontario, Canada

C. R. Wilder, Portland Cement Association, Public Works and Transportation Section, Skokie, Illinois

E. M. Wilson, Department of Civil Engineering, University of Salford, Lancashire, England

J. A. M. Wilson, H. G. Acres Limited, Niagara Falls, Ontario, Canada

A. J. Woestenenk, Bitumarin, Zaltbommel, Netherlands

CONTENTS

TIDAL POWER IN THE BAY OF FUNDY

F. L. Lawton*

INTRODUCTION

It is a distinct honour to have the opportunity of discussing with you a few of the many engineering and economic aspects involved in the utilization of tidal power, as exemplified by the Bay of Fundy investigations.

It is singularly appropriate that this International Conference on the Utilization of Tidal Power should be held under the auspices of the Atlantic Industrial Research Institute and Nova Scotia Technical College for the latter is not only the inspiration of the Institute but, more significantly, the heart of engineering in the Atlantic Provinces of Canada.

It is unusually appropriate that this Conference should be held in this historic seaport and City of Halifax which has seen so much of engineering and economic change since the days of the water, wood and wind technology. Now we are seeing the advent of a closely integrated high-speed air-water-land containerized transport. Both technologies rely on the utilization of energy, originally wood, wind and waterpower, later coal and petroleum derivatives. This Conference deals with another major source of energy.

It is most appropriate that this Conference should be held in Halifax which, with its sister city across the Bedford Basin and the harbour, is the centre of a shining galaxy of research, development and educational centres in the form of the Atlantic Oceanographic Laboratory of the Bedford Institute of Oceanography, the National Research Council of Canada, the Nova Scotia Research Foundation, the Atlantic Industrial Research Institute, the Nova Scotia Technical College, Dalhousie University and several others.

The recently completed investigation of the technical and economic feasibility of development of the large tidal power resources in the Bay of Fundy was not a matter of accident. It was based on the deliberate conviction that the resources existed. Could they be put to use economically?

To answer this question, it is worth recalling a few facts. Canada stems

---

*formerly Study Director, Atlantic Tidal Power Programming Board, Halifax, Nova Scotia.

from two founding peoples, French and English, both maritime by heritage. New Brunswick and Nova Scotia, a century ago, were in the forefront of ship-building.

Some of the largest and finest four-masted schooners and other notable sea-going vessels were built in Minas Basin and elsewhere in the Maritimes. These vessels were fabricated at such places as Maitland, Summerville, Newport Landing, Hantsport, Windsor, Parrsboro and Port Greville. In fact, some 100 vessels were built at Newport Landing. The "D.W. Lawrence", the largest ship afloat in her time, was built in Maitland.

The success achieved in ship-building in the nineteenth century was due to the skills displayed in interweaving a multiplicity of disciplines based on the use of local resources, timber, know-how, good labour, and financial perspicacity.

Many skills are called for in the investigation of tidal power development. To name but a few: a comprehensive knowledge is required of the tides; the oceanographic arts and sciences; modern surveying methods using geophysical means and aerial photography; mathematics; computer usage; civil, mechanical and electrical engineering; corrosion; and many others. These are being discussed at this Conference.

In brief, the resources exist. The know-how is available. Do we have the financial acumen to make tidal power development in the Bay of Fundy economically viable?

The National Energy Board estimates that demand for electrical energy in Canada will reach some 620×10^9 kWh in 1990 or about 3.9 times the 159×10^9 kWh usage in 1966. Concurrently, the hydro-electric energy will decrease from the 1966 position of 82% of total generation to only 45% with nuclear and fossil-fuel-fired, mostly coal, generation picking up in relative importance. In fact nuclear generation capacity is forecast to increase from 2,500 MW in 1975 to 31,500 MW in 1990.

In the Maritime Power Pool comprising New Brunswick, Nova Scotia and Prince Edward Island, 1968 peak demand of 1,161 MW and energy consumption of 6,445 million kWh is forecast to increase to 5,571 MW and 31,455 million kWh by 1990.

In contrast, United States usage from 1929 to 1965 shows a growth rate of 7.1% per year, electrical energy usage doubling every 10 years.

The forecast usage of electrical energy demonstrates the need of long-range planning in the electric power field if regional and national needs of Canada are to be met.

Such was undoubtedly in the minds of those who drew up, on behalf of the governments of Canada, New Brunswick and Nova Scotia, the intergovernmental agreement of August, 1966, under which the Atlantic Tidal Power Programming Board and the Atlantic Tidal Power Engineering and Management Committee have carried out the investigation of tidal power possibilities in the Bay of Fundy which forms the basis for much of the discussion in this paper.

The investigation of the engineering and construction feasibility and of the economic viability of tidal power possibilities in the Bay of Fundy, which constitutes the basis of this paper, was so extensive that space and time permit dealing with the highlights only. Moreover, as the economic aspect is being covered in a companion paper at this Conference but brief reference is made to this aspect herein.

HISTORICAL BACKGROUND

In his search for useful sources of energy, mankind has always been interested in the potential energy of the restless tides in those areas of the world where the tides reach a substantial range. It should not be overlooked, despite the great interest currently being shown in the exploitation of tidal energy, that this is one of the older forms of energy utilized by man.

Tide Mills

Early records indicate that tide mills were being worked along the Atlantic Coast of Europe, notably in Great Britain, France and Spain by the 11th century. One such installation in the Deben Estuary, in Great Britain, was mentioned as early as 1170 in the records of the Parish of Woodbridge. This is believed to be still in operation. Tidal energy was widely used in coastal areas where the tides attained a sufficient range to the middle of the 19th century. Part of the water supply of London in 1824 was provided by 20 ft. diameter waterwheels installed in 1580 under the arches of London Bridge. A tidal power installation for pumping sewage was still in use in Hamburg in 1880. Other installations have been reported throughout this era in Russia, North America, and Italy. Some of the old structures were of impressive size. A tide mill in Rhode Island built in the 18th century used 20 ton wheels 11 ft. in diameter and 26 ft. in width.

Many methods of utilization of tidal energy have been tried in the past based on the potential energy of the tides or the kinetic energy of the tidal currents or combinations of both. Devices employed have included waterwheels, lifted platforms or weights, air compressors, water pressurization and many others. Abell has reported that between 1856 and 1939 some 280 patents dealing with utilization of tidal energy were registered.

Early tide mills were designed to extract a relatively small proportion of the total energy potential producing small amounts of mechanical energy, about 30 to 100 kW, used at the immediate site. Such amounts of power served needs before the advent of the electric motor and long-distance power transmission. The disappearance of tidal power generation towards the end of the 19th century has been attributed by Pierre Ailleret, a noted French power engineer, to power economics. With the advent of power generation on an industrial scale, utilizing for the most part the hydroelectric possibilities of rivers and the thermal energy of fossil-fuel-fired power plants, the price of energy began to decline. Today it is roughly 1/20th, in terms of man hours, what it was at the beginning of the 18th century. Small tide mills of the past could not meet this fall in price.

Electricité de France, the French national power utility, has carried out extensive tidal power investigations in L'Aber Vrach on the northwestern coast of Brittany and in the vicinity of Mont St. Michel near Saint Malo.

In addition large tidal power plants utilizing the estuary of the Severn River in England, Carlingford and Strangford estuaries in Northern Ireland, the Gulf of San Jose in Argentina, various sites in western Australia and in Russia, notably near Murmansk, have been investigated. Less extensive studies have been made into the possibility of developing tidal power at Cook Inlet in Alaska.

The USSR has slowly and methodically pursued work leading to the ultimate development of tidal power generation, the principal proponent being Bernstein, a distinguished Russian engineer. They have in mind the possible development of a 320,000 kW tidal power plant at Lumbovskaya, where a bay with an area of 70 km^2 can be cut off by a relatively short dam. Other tidal power schemes, all in embayments of the White Sea, and in estuaries of rivers flowing into it, would utilize flood tides running up to about 30 ft. in range.

However, new concepts of construction and marked advances in very large, better-adapted generating units have awakened interest in tidal power as possibly competitive with other forms of energy.

Bay of Fundy Investigations

A number of studies into the possibility of large scale tidal power developments at various sites in the Bay of Fundy have been made during the present century. During 1910-20, Dr. O. Turnbull, a Canadian inventor from Rothesay, New Brunswick, gave consideration to utilization of the tidal power resources. He published a proposal for an installation on the Petitcodiac River in 1919 and continued to work on the possibilities for the next two decades. In 1928 the Petitcodiac Tidal Power Co. was formed in Moncton to develop a site on the Petitcodiac River, and made several representations to governmental bodies for assistance. Nothing came of this proposal. In 1915, the Cape Split Development Co. of Nova Scotia carried out investigations into the possibilities of Minas Passage.

In 1945, a study and design for an installation on the Petitcodiac and Memramcook Rivers in New Brunswick to develop a power output of 76,000 hp at 100% load factor was carried out for the Government of Canada. In the 1950-60 period, proposals were devised for developments in Canadian waters of the Bay of Fundy by a number of consulting engineering firms, of which one related to the Chignecto Basin area. In 1962, the ACRES, FENCO and LASALLE firms were retained by the New Brunswick government to carry out further investigations into the possibility of economic tidal power development in the embranchments of Chignecto Bay.

The Passamaquoddy Bay area, involving both U.S.A. and Canadian waters, has received intensive study by various organizations. In 1919, Dexter P. Cooper proposed to the United States government that Passamaquoddy Bay, straddling the New Brunswick - Maine border, be developed. Eventually Cooper's project was taken over by the U.S.A., which allocated $45 million to a project there in 1934. Work was stopped in 1935 after $7 million had been spent on acquiring rights and initial construction work. In 1950, a detailed report was prepared for the International Joint Commission on the Passamaquoddy tidal power possibilities. This study was reviewed again in 1952 by the Corps of Engineers. The last comprehensive investigation was carried out jointly by the U.S.A. and Canada under the International Joint Commission in 1956-59 with later supplementary studies by the U.S.A. Department of the Interior.

The International Passamaquoddy Engineering Board, in its report of October, 1959, to the International Joint Commission, based its double-basin project, selected for specific design and costing, on the utilization of Passamaquoddy Bay, as the high pool, with Cobscook Bay in Maine and Friar Roads in New Brunswick as the low pool. The high pool was planned with an area of 101 sq. mi. and the low pool 41 sq. mi. The power plant, located at Carryingplace Cove, would have thirty 320 in. diameter propeller-type, vertical-axis turbines driving 10 MW, 13.8 kV, 60 cycle generators at 40 rpm, operating under an average head of 11 ft.

The investigations contemplated the tidal power plant alone as well as in combination with a hydro-electric plant at Rankin Rapids on the Upper Saint John River in Maine, about 175 mi. from the tidal power plant. With Rankin Rapids having 460 MW dependable capacity and an annual generation of 1220 million kWh, the combination would provide a total dependable capacity of 555 MW and generation of 3063 million kWh. If the same Rankin Rapids hydro-electric plant were built primarily to serve the normal power load in Maine, with an additional 260 MW installed to firm up the variable output of the tidal plant, the combination would have a dependable capacity of 355 MW and an annual generation of 1843 million kWh. A third alternative involved a pumped storage plant on the Digdeguash River emptying into Passamaquoddy Bay east of St. Andrews, New Brunswick, with a usable storage of 204,000 acre ft. The investigation showed the dependable capacity of the combination would amount to 323 MW with a net annual generation of 1759 million kWh.

As different interest rates then prevailed in the United States and Canada, the studies indicated that, assuming power output and project first cost divided equally between the two countries, power costs for the favourable combination of a tidal power plant and all of the Rankin Rapids hydroelectric development would amount to 8.4 mills per kWh in the United States and 11.5 in Canada on the assumption of a 50 year amortization period. With a 75 year amortization period corresponding figures would be 7.2 and 10.6 mills per kWh.

In April, 1961, the International Joint Commission reported to the United States and Canada on its investigation of the International Passamaquoddy tidal project. In this study, in addition to other concepts previously noted, the 300 MW installed capacity in the tidal power plant was assumed combined with 220 MW of steam electric capacity. The two would provide a total dependable capacity of 300 MW and an average annual generation of about 2143 million kWh. The Commission reported an at-site energy cost, in this case, of 13.5 mills per kWh in U.S.A. and 18.6 in Canada, as against the corresponding cost from alternative steam electric plants of 10.6 and 7.3 mills per kWh.

Existing Tidal Power Plants

The only large modern tidal power development is that across the estuary of the Rance River emptying into the Atlantic Ocean between the old walled city of Saint-Malo and Dinard. Here the tides have an average range of 27 ft. which is utilized in a power station containing 24 units each rated at 10 MW providing an annual production of 544 million kWh.

Officially inaugurated in November, 1966, this power plant is notable for the development and utilization of the so-called bulb-type turbine unit, essentially a horizontal-axis propeller turbine with variable pitch runner blades. The turbine is connected to a generator enclosed in a nacelle or bulb upstream from the turbine in the water passage by which water is conveyed to the turbine. The runner is so designed that it can operate as a turbine with flow from the basin to the sea or from the sea to the basin as well as pumping in the two directions. It can also serve as an orifice passing about 50% of its normal flow.

The USSR has recently completed a small (400 kW) tidal power plant in Kislaya Bay, a small, deep basin with an area of one square kilometer connected with the sea by a narrow estuary about 100 ft. in width. The tide is relatively low, the range being somewhat under 11 ft. Conceived as an experimental plant involving minimal expenditures, the basic concept entailed the use of prefabricated units or caissons literally built under factory conditions at a suitable location, floated and towed to site in the tidal power development and there sunk onto prepared foundations.

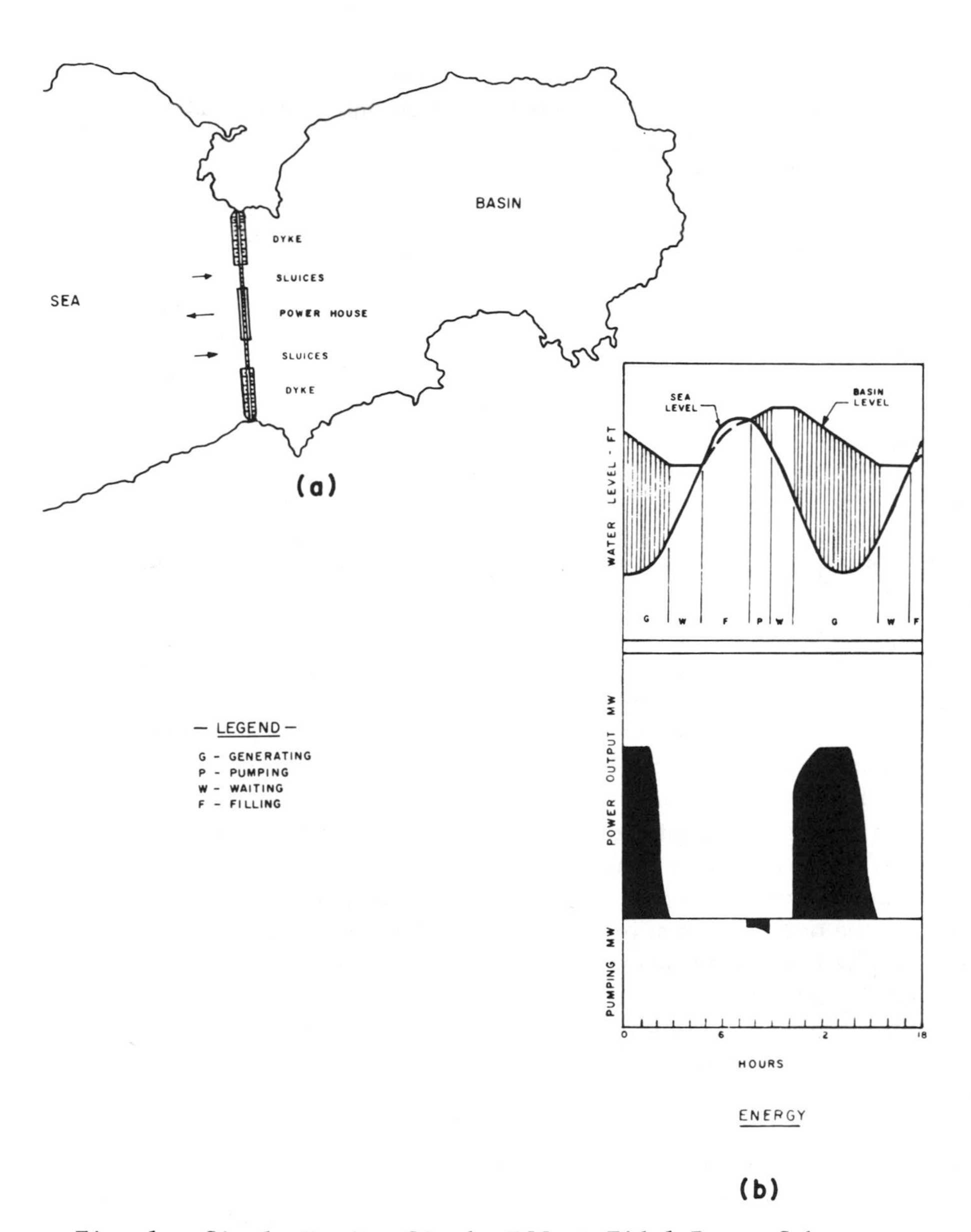

Fig. 1. Single-Basin, Single-Effect Tidal-Power Scheme.

BASIC ASPECTS OF TIDAL POWER

Schemes receiving detailed consideration in connection with possible development of Bay of Fundy tidal power sites were as follows:

(1) single-effect single-basin generation;
(2) double-effect single-basin generation and pumping;
(3) linked-basin; and
(4) paired-basin schemes.

In addition, external pumped storage and the utilization of existing hydro-electric storage developments have been given close attention.

The several practical schemes, from an economic point of view, were selected from numerous concepts proposed at one time or another, many extremely complex and correspondingly costly. The more complex schemes were conceived to overcome the basic weakness of tidal power in the past due to its characteristic -- a variable production not necessarily in phase with human needs.

The single-effect single-basin scheme is exemplified by Fig. 1. The oldest form of tidal power generation, it was the basis of many tidal mills which came into existence in the 10th and 11th centuries in Western Europe. Fig. 2 indicates a more modern version, the double-effect single-basin scheme.

The double-effect single-basin generation and pumping scheme, exemplified by the Rance development, utilizes minimum civil works since it involves a single basin with reliance on double-effect generation and pumping as well as utilization of the generating units as orifices to supplement sluiceway capacity to overcome the basic weakness of tidal power.

The linked-basin concept, exemplified by Passamaquoddy, entails the use of two more or less contiguous basins of suitable proportions. Such conditions exist at the mouths of Shepody Bay and Cumberland Basin. One basin can be operated as a high pool and the other as low pool with generation always in the one sense from the high pool to the low pool. Fig. 3(a) indicates the physical arrangements for such a concept and Fig. 3(b) shows the nature of the production.

The paired-basin scheme consists of two single-effect single-basin schemes which are interconnected electrically. Such an arrangement affords somewhat more flexibility in operation of the plants to meet market demands and may, in certain cases where there is a difference in tidal phase, permit deriving even greater benefit.

Because of the importance of basing the Board's work on all concepts offering promise, our investigations have been primarily related to the four concepts enumerated. In addition, an evaluation was made of the significance of external pumped storage and of the utilization of existing hydro-electric power developments with major storages in producing or firming up the dependable-peakoutput of tidal power schemes studied.

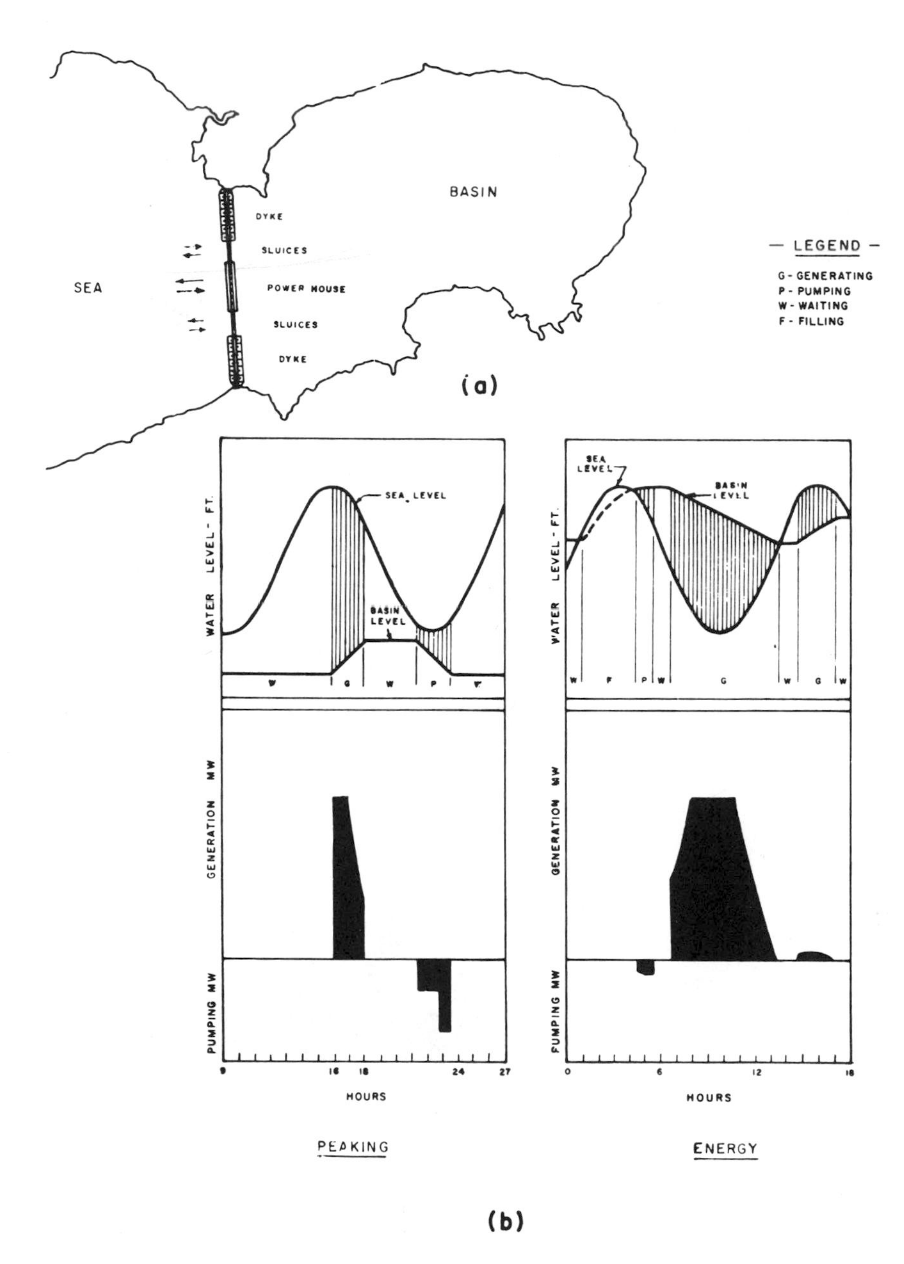

Fig. 2. Single-Basin, Double-Effect Tidal-Power Scheme.

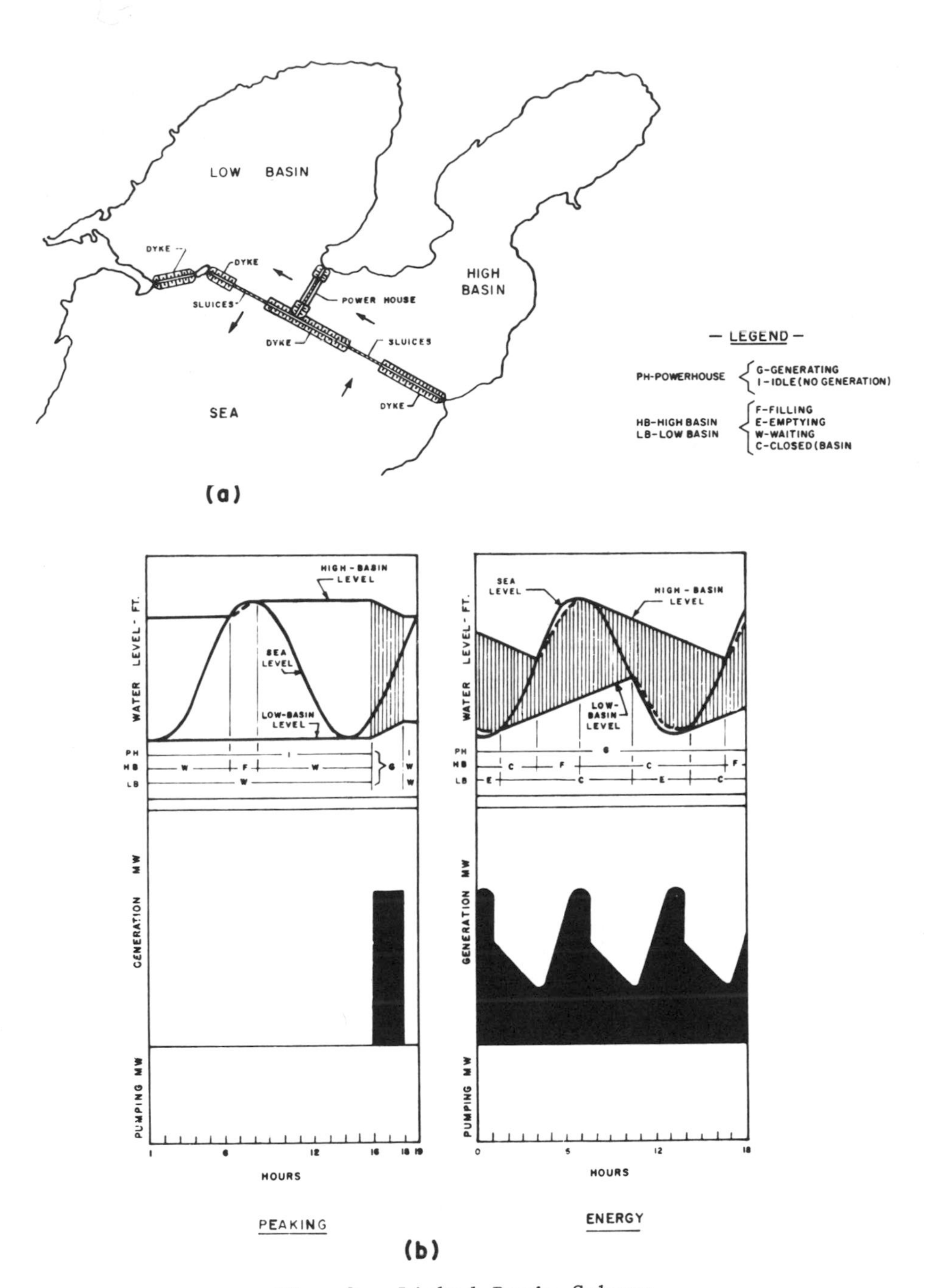

Fig. 3. Linked-Basin Scheme.

MAGNITUDE OF THE RESOURCE

An indication of the magnitude of tidal power resources in the Bay of Fundy is provided by Table 1. The locations are shown by Fig. 4.

Table 1
Some Bay of Fundy Tidal Power Potentials

Site	Average Natural Energy MW	Site	Average Natural Energy MW
7.1	2590	8.1	5140
7.2	1630	8.2	18910
7.4	920	9.1	800

The "natural energy" is computed according to the Bernstein formula adapted for variable basin area and an average of rms tidal amplitude.

ORGANIZATION FOR INVESTIGATION

Under the Intergovernmental Agreement of August, 1966, the sponsoring parties created the Atlantic Tidal Power Programming Board with overall responsibility for the investigation and the Atlantic Tidal Power Engineering and Management Committee, the latter charged with engineering and management aspects. The Board and the Committee each consisted of five members with provision for alternatives when necessary.

A Study Director was appointed with responsibility for the development and direction of the studies under the Committee, itself reporting to the Board. A small staff, 19 at the peak of activities, was organized in Halifax. This comprised several engineers, stenographic personnel, an administrative clerk and draughtsmen.

In the early part of the work, three general consulting engineering groups were retained for area studies. The services of a number of specialist consulting firms were utilized for geophysical surveys; marine surveys of various types including current measurements, sediment surveys and sediment coring; soils mechanics tests; diamond drilling for armourstone and rockfill sources; aerial photography; advanced development concepts; optimal powerhouse designs; and the extremely important tidal regime and power productivity investigations. In addition, a number of staff consultants were retained to assist the Study Office, from time to time, on specific problems such as geology, generating units, sedimentation, and power-system studies, both from the expansion and stability points of view.

The investigational work began on November 1, 1966, with the assump-

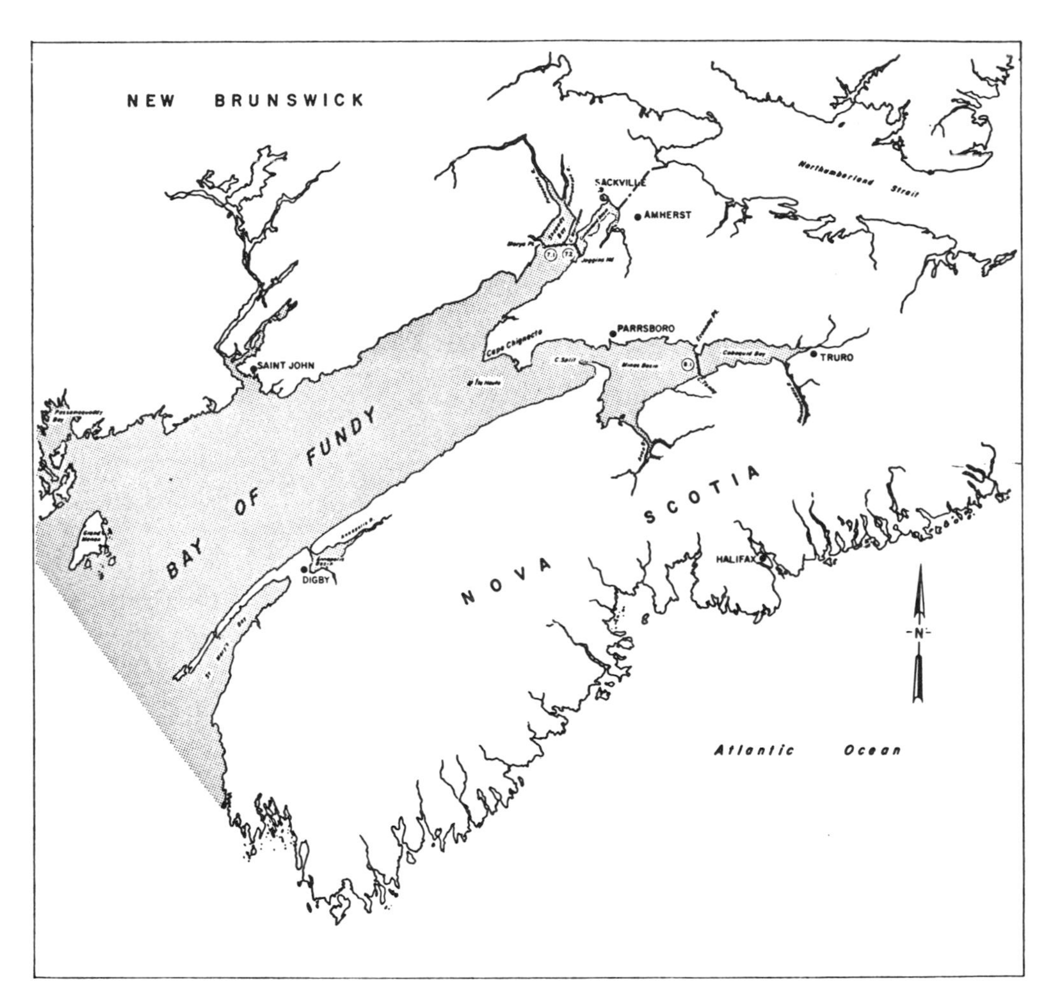

Fig. 4. Locations of Tidal-Power Schemes.

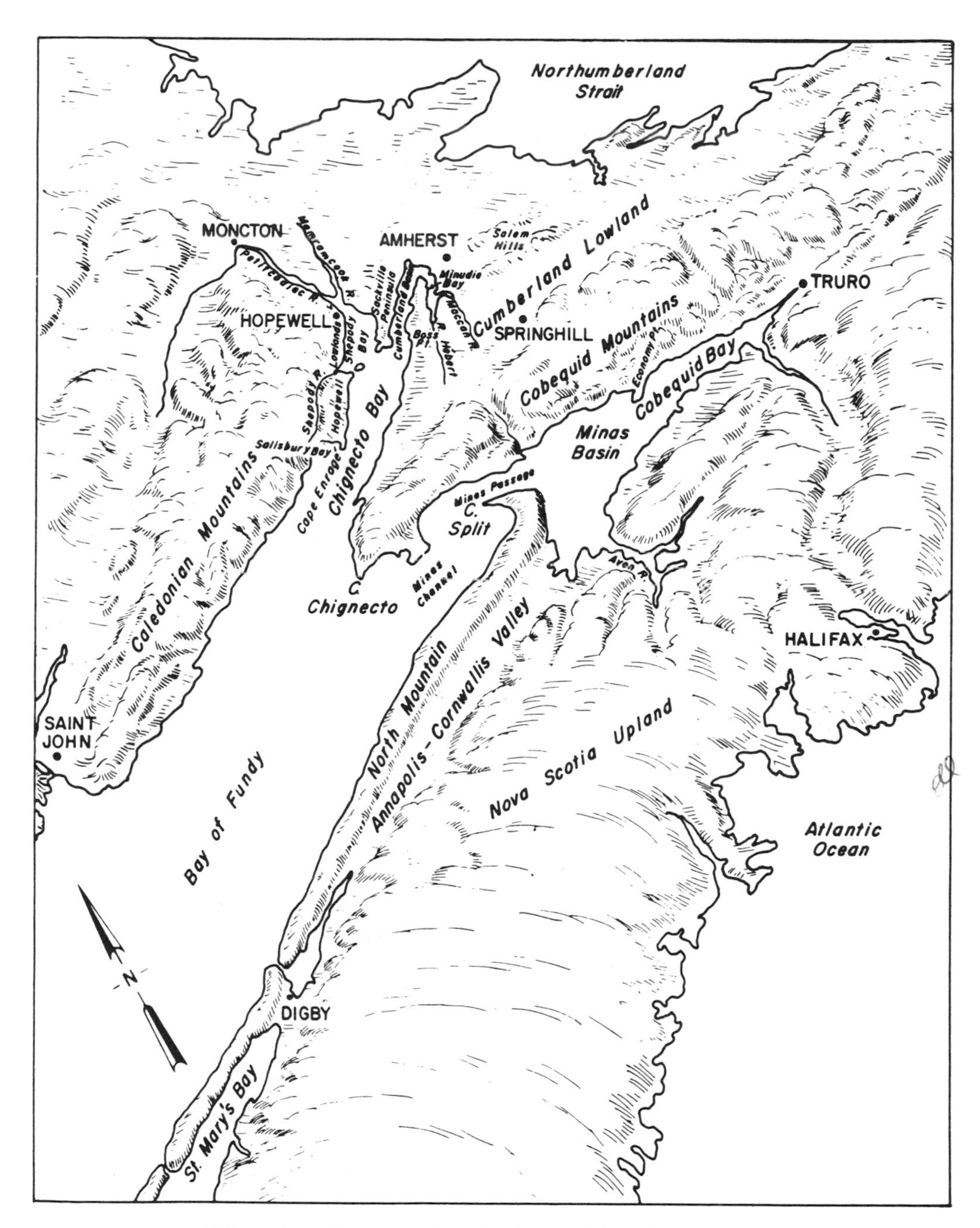

Fig. 5. Topography in Bay of Fundy Region.

tion of responsibility by the Study Director. All studies and reports were completed 41 months later by March 31, 1970. During this period no fewer than 23 sites for development were appraised on a qualitative basis; three were studied in detail. The detailed studies comprised all requisite work for determination of the technical feasibility and economic viability of tidal power development. These studies covered investigation of development of each of the three sites as a single-effect single-basin scheme; each as a double-effect single-basin scheme; two as a linked-basin scheme; and two as a paired-basin scheme.

During the course of the studies the Board met 11 times and the Committee 47 times.

PHYSICAL CONDITIONS IN THE BAY OF FUNDY

Geology

The Bay of Fundy region has been one of great geological instability producing uplifting, tilting and folding of the earth's crust. These tectonic movements, combined with several periods of general subsidence and elevation with respect to sea level, account for its present topographical features as indicated by Fig. 5. Although the broad geological aspects of this area have been understood for some time, detailed information is still scanty in many areas.

The Bay occupies a zone in which considerable faulting has occurred. It has been formed by the partial inundation of lowlands which extend on a series of soft and easily eroded sedimentary rocks into central New Brunswick and northern Nova Scotia. In the south, these lowlands are bounded by a uniform upland region which covers much of southwest and central Nova Scotia. This upland is formed on an extensive sequence of hard metamorphic and granitic rocks.

Well defined, isolated ranges of hills composed of harder rocks divide the lowlands into a number of sub-areas. The Caledonian Mountains extend along the New Brunswick shore from Saint John to Moncton and separate a central New Brunswick lowland region from the Bay of Fundy. The Cobequid Mountains, extending from Cape Chignecto in the west to New Glasgow in the east, separate the Cumberland Lowland from the Minas Basin Lowland. A unique, asymmetrical ridge, North Mountain, extends the length of the Nova Scotia shore from Brier Island in the southwest to the hook at Cape Blomidon -- Cape Split. Caused by a tilted layer of resistant lava (basalt), and almost unbroken throughout its length, it separates St. Mary's Bay, Annapolis Basin and the Annapolis-Cornwallis Valley from the Bay of Fundy.

Changes in sea level which occurred during the advances and retreats of the continental ice sheet during the last Ice Age caused first the carving and later the flooding of river systems in the soft lowland sediments. This is most readily apparent in the shapes of Shepody Bay and the Cumberland Basin. Minas Basin and Cobequid Bay are also basically a drowned river system, perhaps enlarged somewhat by shoreline features such as headlands, islands, and bays. These can be explained by evaluating the twin conditions of exposure to erosion and rock resistance.

As a result of glaciation, varying depths and types of glacial deposits now occur in the area. Upland areas generally have a thin but persistent covering of sandy glacial till. Lowland areas are mostly covered with thicker deposits of till. However, in certain areas, extensive deposits of sand and gravel outwash are found as terraces. The largest terraces are found along the north shore of Minas Basin within and south of the Cobequid Mountains. Similar but smaller deposits are found adjacent to the Caledonian Mountains and near North Mountain.

The 1953 seismic zoning map of Canada, as published in the National Building Code, places this region in Zone 2, a zone where moderate earthquake damage may be expected. The Division of Seismology, Dominion Observatory, suggests that the area may be subject to seismic intensities as great as six on the Modified Mercalli Scale on the average of once every 60 years, and as great as seven every 130 years. These figures correspond to magnitudes 5.0 and 5.7 on the Gutenberg-Richter Scale, the former representing the threshold of damage. Relatively minor to moderate damage might reasonably be expected from earthquake shocks in this area; tidal power structures should be designed so as to minimize such damage. On the other hand, no allowance has been made in the design of structures for tidal waves generated by earthquakes beneath the ocean.

Sediment

Sediment ranging from clay to gravel is found in the Bay of Fundy. Suspended sediments are predominantly silt with varying percentages of clay and sand depending on location. Bottom sediments are reworked sand and gravel, intact glacial sediments having a wide range of grain size, and recently deposited muds, silts and sands.

In many embranchments of the Bay, large quantities of sediments occur. These are extremely mobile under the influence of the prevailing strong tidal currents. Extensive sand deposits and large streams of sediment-laden water can be readily identified on aerial photographs. Large permanent sand bodies are clearly marked on navigation charts. Marshland reclamation has caused major deposits on the seaward side of tidal barriers which have radically modified the regime of the local estuaries. Evidence from these developments and other observations indicate that careful consideration must be given to problems of sediment movement in the design of tidal power plants in the Bay of Fundy.

The effect of accumulation of river sediments in reservoirs associated with some river type hydro-electric developments is well known to engineers. In the case of tidal power developments in the Bay of Fundy, the rate of production of new sediments appears to be very small compared to the storage volumes of the basins or reservoirs that might be cut off from the open Bay by a barrage. It follows that the useful life of such basins need not be limited by loss of storage capacity resulting from sedimentation provided the tidal power plants are properly designed to avoid undesirable redistribution and excess accumulation of the existing coarse sediments and to pass the finer suspended sediments.

Ice

Ice develops in the upper reaches of the Bay of Fundy during the colder winter months. Surveys have shown that ice occurs in thicknesses up to 15 ft. or more, the maximum thickness relating to shore-fast ice and cake ice which grows in size after grounding by rafting and by accretion of layers of frozen muddy water and sediments. Ridging and overlapping layers of sheet and cake ice in headland areas give rise to shore-fast ice.

Winds and tidal currents are effective in shifting the ice pack back and forth, closing and opening bays, basins and river areas. Ice forming along shore lines and in river areas breaks off and feeds pack ice developing in the central portions of Minas Basin and Chignecto Bay. Ice is continually bled from these basins into the Bay of Fundy. Accumulation of ice pack varies with the season and the climatic conditions of wind, temperature and snow. Periods of extremely cold weather give rise to 100% ice coverage in some areas.

Climate

The Bay of Fundy region, lying in the belt of westerly winds, is largely influenced by air moving from land to sea. Consequently, its climate is not as maritime in character as that of the Atlantic Coast of Nova Scotia. The path of the most frequently occurring cyclonic storms passes south of the Bay. Storms result from the movement of cold polar air striking moisture-laden air currents moving up from the south. Analysis of tropical storms and hurricanes since 1886 indicates that most severe storms occur during June to November, develop over the Gulf Stream and then move rapidly towards or close to Nova Scotia. Winter storms often produce violent gales and rain changing to snow. On the other hand, influxes of moist Atlantic air result in mild spells in winter time and cool weather during the summer time.

January, the coldest month, shows temperatures ranging from 20° to 24°F along the southerly shore and somewhat lower at 18° to 20°F along the northerly shore. Minimum temperatures as low as -52°F have been recorded at Chipman, New Brunswick, and -42°F at Stewiacke in the upper regions of Minas Basin in Nova Scotia. Summer temperatures are coolish, ranging from 60° to 65°F with an occasional maximum of 80°F.

The heaviest precipitation occurs along the north shore of the Bay. Snowfall ranges from 70 to 90 in. over the area with mean annual precipitation ranging from 29 in. at Sackville to 54 in. at Saint John, New Brunswick. Fog occurs fairly frequently with some coastal stations reporting nearly 100 days of fog annually. The greatest frequency of occurrence is in July. In 1967, fog occurred during almost the entire month of July in the morning if not all day.

Prevailing winds are westerly. Complete reversals during a tidal cycle are not unusual.

Tides

The great tides of the Bay of Fundy make it of unusual interest from the point of view of tidal power development. At the head of Minas Basin, for instance, the spring tides attain a range of 52.9 ft. between lower low and higher high waters. In Chignecto Bay and its embranchments the tidal range is 46 ft. but, at the southerly end, in St. Mary's Bay, the range is only about 22 ft.

Bay of Fundy tides are semi-diurnal with two high waters and two low waters each with approximately the same height. The interval between the transit of the moon and the occurrence of high water is nearly constant. As a result, the tides are extremely regular with two tides of nearly the same magnitude and pattern each 24 hr. 50 min. lunar day. Table 2 shows the range of tides encountered at several sites of interest. The tide is the so-called anomalistic type, the variation in range with the distance of the moon from perigee to apogee being the greatest variation. The spring tides each month vary by a relatively small amount.

Table 2

Tidal Ranges at Locations of Potential Tidal Power Developments

Site	Location	Tidal Range, Ft. Maximum	Minimum	RMS*
7.1	Mary's Pt. to Grindstone Is. to Cape Maringouin	43.7	19.1	32.4
7.2	Ward Pt. to Joggins Head	44.4	21.6	32.0
8.1	Economy Pt. to Cape Tenny	52.9	23.9	37.3
8.2	Clark Hd. to Cape Blomidon	46.4	19.8	32.1
9.1	Entrance to Digby Gut	29.5	11.7	20.3

*RMS tides are derived on the basis that the "root-mean-square" average of N values of a variable A is defined as the square root of the sum of the squares of the variable divided by the number N. Precise values of RMS tidal ranges are not easily developed because of the many influences which must be considered.

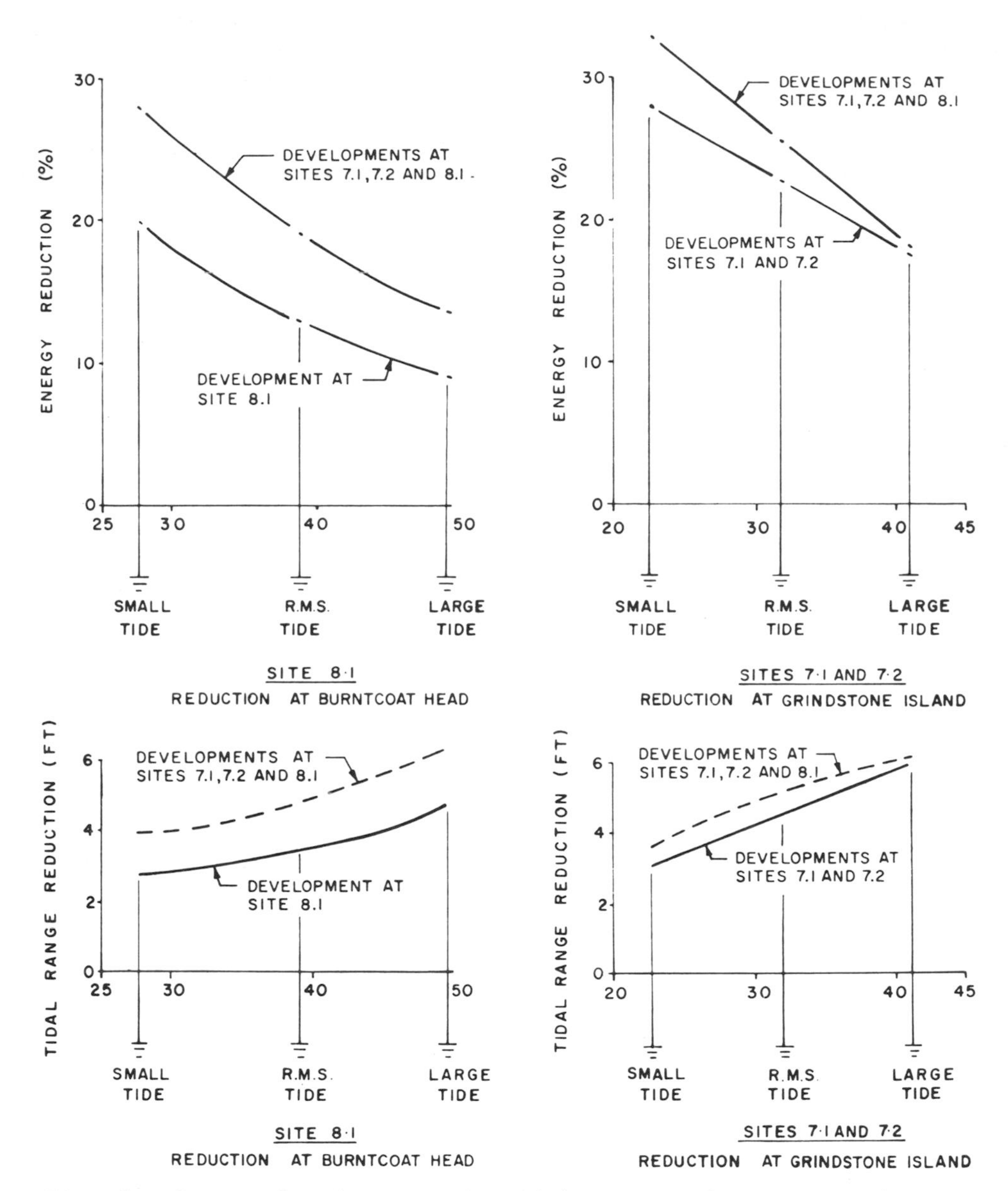

Fig. 6. Computed reductions in tidal range and energy production with operation of tidal-power plants.

The physiographic features of the Bay play a very important role in the amplification of its tides, the dual lateral and vertical convergence as the heads of the Bay are approached resulting in an amplification of some 30 to 40% with the predominant factor affecting amplification being the approximation of the effective length of the Bay to the resonance or quarter-wave length of 185 mi.

Work done in 1965 by Canadian researchers had suggested that complete barriers in Chignecto Bay and Minas Channel would reduce the tidal amplitude at such barriers by about 20%. Realizing the importance of the potential effects on the tidal regime, arrangements were made to develop a mathematical model which, in conjunction with a computer programme, would enable close determination of changes in tidal range and the corresponding change in energy production which would result from the implantation of tidal power plants. Fig. 6 presents some preliminary data.

It will be noted that the reduction in tidal range increases with the addition of generating units because the power plant discharges tend to be out-of-phase with the natural flow and, accordingly, function quite positively as major sources of attenuation or weakening of the forces causing build-up or resonance in the tides.

Similar effects have been shown by Bonnefille and Chabert d'Hières. Vantroys has also dealt with this general subject in a notable paper.

In addition to the effect on tidal range on the sea side of a tidal power plant, there will be effects inside the basin caused by the particular mode of operation of the plant. For instance, the changes with single-effect generation would be different from those associated with double-effect generation. Pumping would naturally change the high water level in the tidal power plant basin beyond the higher high water of the natural tide, and, depending on the mode selected, decrease the lower low water below natural elevations. The extent of this is not likely to be great; the full effect is difficult to anticipate at this stage.

In general, our investigations showed that the effect of tidal power plants on the tidal regime diminishes the closer these are located with respect to the heads of the principal arm or arms.

Waves and Water Levels

An orientation of tidal power dams on the upper embranchments of the Bay of Fundy, which ensures economy of construction and maximum productivity from the plants, may expose the structures to large waves under certain conditions of wind velocity, duration and direction. The crest elevations to which structures would have to be built would therefore depend on the most severe waves likely to occur, the simultaneous increase in water level due to storm or hurricane surge induced by the wind, return period of these conditions, and the runup of large waves on the exposed structures. These events could occur at either high or low astronomical tides, but only the high tide need be considered in the selection of crest elevations. Since all factors would be additive under the worst conditions, the final selection of crest elevations for design would depend on the degree and frequency of overtopping considered acceptable.

To obtain the best estimate of the most severe wind and waves likely to occur during a certain period, without resorting to a prolonged period of wind and wave measurements, a study was made of wind records for a number of years from existing land-based stations. Since the velocities of winds blowing from sea to land are reduced overland, the recorded velocities were increased by 20% to simulate over-water velocities. The records indicate that winds up to 70 mph have occurred from the critical direction (the direction measured from each site in which the wind of interest would generate the highest wave) at a frequency estimated at 20 yr. This velocity increased by 20%, i.e. to 84 mph, with a return period of about 20 yr. was used to estimate the most severe wave conditions likely to occur at the sites. No attempt was made to analyze wind climate over a range of lesser velocities or to estimate the corresponding wave spectra other than that presumed to be the maximum.

The study was divided into two parts. Firstly, a fetch was chosen within each embranchment where the maximum wave could be generated without influence from the broader and longer Bay of Fundy; secondly, critical fetches on the Bay of Fundy directed towards the entrances to Chignecto Bay and Minas Basin were chosen such that waves could be propagated into the embranchments and undergo further generation as they progressed towards the sites. The latter analysis indicated that waves with periods in the six to 11 sec. range in the Bay of Fundy spectrum would be refracted into the embranchments and undergo further generation (from the reduced height after refraction) resulting in the most severe spectrum at the sites. At Sites 7.1 and 7.2 the characteristics of these waves are estimated as follows:

Significant deep-water wave height	=	18.0 ft.
Significant deep-water wave period	=	8.8 sec.
One percent deep-water wave height	=	24.3 ft.
One percent deep-water wave period	=	10.7 sec.
Duration of wind for maximum wave growth	=	1.8 hr.

For Site 8.1 the wave characteristics are estimated to be:

Significant deep-water wave height	=	19.7 ft.
Significant deep-water wave period	=	7.7 sec.
One percent deep-water wave height	=	29.7 ft.
One percent deep-water wave period	=	12.3 sec.
Duration of wind for maximum wave growth	=	2.0 hr.

The highest elevation to which the water level may rise on the seaward side of a tidal power plant, neglecting long-term changes, would be due to the simultaneous occurrence of high spring tides due to astronomical causes, wave height, run-up of the highest wave and surge. Other factors such as the transport of water by wave action, slope of the sea bed in front of the development, shoreline configuration and direction of tidal streams may contribute in small measure. These factors, however, were neglected in this study.

The vertical height to which water from a wave will run-up on a given structure depends on the characteristics of the structure itself, such as shape, slope, roughness and permeability; the depth of water at and seaward of the structure; and the characteristics of the incident wave. For rubble-mound slopes of the type proposed for the dykes, the results of the model tests performed by Hudson are considered most applicable to prototype conditions in cases where the ratio of water depth at the structure to wave height is greater than three. Run-up of the significant wave on the seaward side of deep-water dykes predicted from the Hudson data for side slopes of 1:1.75 is 14.8 ft. at Sites 7.1 and 7.2. It is 13.8 ft. at Site 8.1. These values were arbitrarily increased by 20% to compensate for scale effect inherent in small scale model tests.

Increase in water level due to surge results primarily from surface currents induced through the interaction of the tangential force of the wind and the water surface. This results in a flow of water and a consequent buildup or increase in level in the downwind direction. Surge due to hurricanes is caused by pressure reduction from normal in addition to wind induced surface currents. To differentiate between the primary forces causing surge on the open coast, the term "storm surge" is used to describe the result of a force due to wind only and the term "hurricane surge" to the result originating from a wind accompanied by an atmospheric pressure reduction from normal. On enclosed lakes and reservoirs where an increase in water level in the downwind direction is accompanied by a decrease in the upwind direction, the term "wind setup" is used.

The prediction of storm and hurricane surge on the embranchments of interest in the Bay of Fundy was based on the published results of investigations. Using standard project hurricane parameters compiled by the U.S. Weather Bureau for the Gulf of Maine, it was estimated that hurricane surge at Sites 7.1, 7.2 and 8.1 at low astronomical tide would approach 5.6 ft., i.e. 3.9 ft. due to the wind and 1.7 ft. due to the atmospheric pressure reduction from normal. The corresponding value at high tide level is 5.0 ft. The only measurements of surge which may be used for comparison are those due to the Saxby Gales which occurred on October 5, 1869. The accuracy of these measurements should be considered doubtful due to methods used to compile them some time after the event. W. Bell Dawson, in his 1917 publication entitled "Tides at the Head of the Bay of Fundy", refers to measurement of water marks on piers, wharves and buildings, and the transfer of these measurements to distant bench marks, in order to relate them to predicted astronomical tide levels. Nevertheless, these measurements resulted in estimates of surge at Noel River on Minas Basin of 3.05 ft., in Cumberland

Basin 3.17 ft. and 6.45 ft. at Moncton on the Petitcodiac River.

The magnitude of storm surge as distinct from hurricane surge on Chignecto Bay and Minas Basin was derived for straight-line wind velocities of 84 mph on critical fetches within each embranchment. At Sites 7.1 and 7.2, the storm surge was estimated to be 1.9 ft. at high and 3.2 ft. at low astronomical tides. The corresponding values at Site 8.1 were 1.9 ft. and 3.1 ft., respectively.

The prediction of a design maximum water level at the sites, without a thorough investigation of all the factors involved, is not amenable to easy or accurate resolution. In a feasibility study of this nature, the best estimates which can be obtained are necessarily approximate, largely a matter of engineering judgment concerning that combination of factors likely to result in the highest water level.

The particular characteristics of hurricanes and the effects produced by them are of greatest concern due to their frequent occurrence. The Bay of Fundy is so oriented that many hurricanes of tropic origin moving up the Atlantic Coast pass on routes parallel to its axis or at acute angles thereto. These hurricanes lose some of their energy on approaching the land mass of Nova Scotia or otherwise veer eastward south of Newfoundland. A characteristic of hurricanes in the northern hemisphere is that their winds circulate in a counter-clockwise direction. Consequently, the direction of surface winds blowing over the Bay of Fundy from a hurricane centre south of Nova Scotia would be expected to vary approximately between East and North with the passage of the hurricane. Winds of this direction would not generate waves of any appreciable height on the embranchments.

It is generally accepted that the larger the diameter of full hurricane winds the less severe are the effects, although they may be prominent several hundreds of miles from the centre. The average diameter of full hurricane winds on the Atlantic ranges from 75 to 100 mi. with a maximum of about 500 mi. In view of this, and in view of the loss of energy on nearing a land mass, it appears unlikely that the "circular" hurricane winds at reduced velocities as opposed to winds of a straight-line direction would generate larger waves on the narrow embranchments of Chignecto Bay and Minas Basin. However, full hurricane surge should be anticipated.

For the purposes of comparison using the simplification that all waves have crests and troughs equal distances above and below still water level, a hurricane wave 12 ft. high and with an eight second period, occurring at high astronomical tide level and at times of maximum surge, would have a run-up value of about 11 ft. on a 1:1.5 slope. The height above still water level reached would be

$$22.0\text{ ft.} \quad \left(\frac{\text{wave height}}{2} + \text{hurricane surge} + \text{wave run-up}\right).$$

The significant wave developed for Site 8.1 generated by 84 mph straight-line winds is 19.7 ft. high and has a 7.7 second period. The run-up of this wave is 14.5 ft. on a 1:1.5 slope. With the addition of half the wave height and storm surge at high tide, the level reached above still water would be 26.2 ft. These examples indicate that hurricane conditions are not likely to give rise to water levels as high as those predicted for a straight-linewind direction accompanied by local storm surge.

FIELD INVESTIGATIONS

Field investigations included a number of examinations, reconnaissance and other surveys in the field for the following purposes:

1. Examination of the existing conditions at 23 sites for the purpose of selecting one or more sites warranting more detailed study.
2. Geophysical surveys were carried out at the same sites to determine underwater foundation conditions.
3. A general geological examination was made for the purpose of appraising probable structural geological conditions at the sites of interest.
4. More detailed geological examinations were made at the three sites finally selected.
5. Temporary tide gauges were installed to measure the rise and fall of the tide during the course of surveys at the three sites finally selected.
6. Continuously recording current meters were installed to measure the direction and velocity of tidal currents at selected locations and depths.
7. Seismic-reflection profiles were carried out to accurately reproduce the depth of water reduced to GSCD*, indicate the quality of sediments, the quality and characteristics of the bedrock, and any geological discontinuities along and parallel to the axis of the Sites 7.1, 7.2 and 8.1.
8. Borings were obtained by which cores of sediments overlying bedrock at the three sites studied in detail could be classified and analyzed by laboratory testing.
9. Water-sediment samples were secured to determine total solids content and establish instantaneous current velocities and directions, temperature and salinity. These data were obtained at four different elevations at ebb tide, mid-range on the flood, flood, and mid-range on the ebb tide.
10. Sufficient geophysical and sediment boring data were developed to assess distribution, thickness and classification of selected submarine sediment deposits.
11. A limited amount of diamond drilling was carried out in the inter-tidal zone to determine bedrock conditions.

---

*GSCD is the Geodetic Survey of Canada Datum, based on the value of mean sea level prior to 1910 as determined from a period of observation at Halifax and Yarmouth, N.S., and Father Point, Que., on the east coast.

12. Diamond drilling was undertaken to determine possible sources and the probable quality of armourstone and large quarry-run rockfill for dyke construction.
13. A survey of the inter-tidal zone, utilizing aerial photography and ground control, was made to complete gaps in information necessary for determination of basin areas and volumes at various levels.
14. Observations were made on sediment in existing estuaries and at tidal dams (aboiteaux) in order to appraise the effects demonstrated by these natural-scale sedimentation models.
15. Ice surveys were carried out during two winter seasons to determine the magnitude of the ice problems which would be encountered by tidal power plants constructed at the preferred locations.

Geophysical Surveys

Geophysical surveys carried out in previous investigations and prior to the studies forming the basis of this paper have added materially to the knowledge of geological structures under the waters of the Bay of Fundy. These geophysical surveys have provided a fair amount of information on the thickness and general nature of the overlying sediments.

In 1965, surveys were carried out for the Atlantic Development Board to determine geological conditions in Minas Channel, Passage and Basin. The surveys consisted of over-water and land-based geophysical surveys of several types including continuous seismic profiling, land-based geological surveys, and a limited sampling programme to determine characteristics of the bottom sediments. The surveys were concentrated in Minas Channel and Minas Passage (the narrow entrance to Minas Basin) but included limited reconnaissance surveys within Minas Basin. These surveys located a deep scour trench in Minas Passage where the water was up to 425 ft. deep. Unconsolidated materials, inferred to be mostly sands and gravels, were found to be widespread and in some areas to be as much as 300 ft. thick.

These surveys served to discourage further tidal power studies within Minas Passage and Channel. The very limited data collected within Minas Basin indicated that other sites might be feasible. However, the survey profile lines did not follow closely any of the power development sites considered in the present study so that further surveys were required.

To aid in the preliminary feasibility analyses of tidal power, continuous seismic reflection profiles were carried out along planned survey lines in several areas within Chignecto Bay, Minas Basin and Cobequid Bay, and the Annapolis Basin and St. Mary's Bay. Some 125 mi. of profiles were run. Interpretation of the geophysical records permitted the drawing of profiles showing the inferred geological conditions along and adjacent to the axes of proposed tidal power developments.

Foundation conditions were interpreted to be markedly less favourable than had been anticipated in the Annapolis Basin area, and they were not particularly favourable in St. Mary's Bay.

Conditions in Chignecto Bay at Sites 7.1, 7.2 and 7.4 appeared to be favourable, although bedrock was generally overlain by considerable thicknesses of unconsolidated sediments, some of which were believed to be very soft. Bedrock structures were not well defined. A sudden change in bedding-plane attitudes was observed in the middle of Cumberland Basin.

The Minas Basin profiles showed that conditions at the west end of Minas Basin were not particularly favourable. Extremely thick deposits of unconsolidated materials, in places up to 400 ft., were interpreted as overlying bedrock in this area. The weight of tidal power development structures would probably cause major settlements in such materials. In the vicinity of Economy Point at Site 8.1 conditions were generally better. Along profiles from Economy Point to Cape Tenny, bedrock was inferred to be covered with thin discontinuous deposits of unconsolidated materials over most of the distance. A sediment-filled trough, about 0.5 mi. wide, near the central portion of the basin was interpreted.

More detailed marine surveys were required to evaluate properly the three sites selected for preliminary engineering feasibility studies. Consequently, in 1968, continuous-seismic reflection profiling was conducted along the axes of the proposed sites as part of a larger programme which included measurements of tidal currents, sampling of waterborne suspended sediment, vibrocore coring of the unconsolidated sediments at 36 locations, and drilling into bedrock at three intertidal zone locations. Seismic surveys were also run over two submerged sand bodies in order to determine their sizes, shapes and compositions, making a total of almost 90 mi. of seismic profiles obtained in 1968. As a result, a fairly complete picture of bottom conditions was obtained.

At Sites 7.1 and 7.2, conditions were inferred to be much as they had been from previous surveys. These 1968 surveys tended to establish the presence of fault zones under both Cumberland Basin and Shepody Bay. Unconsolidated sediment thicknesses along Site 7.2 were interpreted to be somewhat less than by earlier surveys.

At Site 8.1, the 1968 surveys confirmed the earlier surveys except in the vicinity of the central buried-valley zone. This feature was deemed to be a complexly faulted zone rather than a valley. The 1968 seismic records contain some information which tends to support this view. However, the present data are somewhat ambiguous; either interpretation may prove correct. In any case, this is a zone of potential weakness requiring special treatment.

Sediment Coring

Previous work had revealed that all sites providing interesting possibilities for tidal power developments were undoubtedly overlain by substantial thicknesses of unconsolidated sediments. Consequently, it was important that an adequately comprehensive knowledge of these unconsolidated sediments be obtained in order that the cost of constructing tidal power developments could be determined.

Much thought was given to the most effective method of determining the thickness of unconsolidated sediment and, also, the nature of the underlying bedrock. Relatively undisturbed samples of the unconsolidated sediments between sea bed and bedrock could be recovered together with cores from the drill penetration of the underlying bedrock by the use of various types of drill rigs. However, the use of such equipment in the deep tidal waters, at Sites 7.1, 7.2 and 8.1, to evaluate site conditions would be expensive. Consequently, consideration was given to the use of the vibrocoring technique since this method would enable recovering relatively undisturbed samples from the full profile of unconsolidated sediments between sea bed and underlying bedrock, and do it at a reasonable cost.

Work during the summer of 1968 comprised ten vibrocore borings along the alignment at Site 7.1. Thirteen were carried out along the alignment of Site 7.2, and nine borings were located along the axis of Site 8.1.

Core recoveries varied considerably with the type of material being penetrated. Most cores averaged 15 to 20 ft. in length, but a few were as long as 30 ft. Although some compaction and remoulding resulted from the sampling procedure, the sediment cores generally were adequate for standard laboratory strength and consolidation tests.

Bottom-sampling operations were severely restricted by the tidal currents. The entire coring operation, including the final location of the ship on the drilling site, was limited to approximately one hour during each slack tide period. The rapid 180° change in current direction between flood and ebb tides made it difficult for the drilling boat to maintain position during coring operations.

Laboratory examination of the vibrocore samples taken at Sites 7.1 and 7.2 showed the sediments to contain considerable quantities of weak, compressible, organic silts, particularly in the central parts of the channels. Examination of the vibrocore samples obtained at Site 8.1 revealed them to be predominantly compact sands and silts.

Site Drilling

In order to provide necessary data on the bedrock structure for correlation with geophysical profiles run along the respective centre lines at Sites 7.1, 7.2 and 8.1, as well as 200 ft. upbay and downbay from the centre lines, diamond drilling was carried out to the extent of placing one hole on the alignment of Site 7.2 and two on the alignment of Site 8.1, in both cases in the intertidal zone. Bore-hole data from previous surveys in the vicinity of Sites 7.1 and 7.2 were also used.

It was necessary to locate each drill hole in the intertidal zone as far from the high water mark as possible in order to overlap with the ends of the geophysical profiles. Moreover, an important consideration was the positioning of each hole with respect to obtaining a reasonable period of time during which the drilling could be carried out. In brief, each drill hole was so located as to allow for about two hours drilling time at low tide.

In the case of Site 7.2 and the southerly end of Site 8.1, it was possible to locate the drill holes essentially at the ends of the geophysical profiles. However, at Economy Point, foreshore conditions required that the hole be located at a distance of about 500 ft. from the end of the geophysical run. This did not, however, hinder geological interpretation since the sedimentary rockbeds exposed in the vicinity of Economy Point are essentially horizontal and very consistent throughout the area so that the drill-hole data could be projected offshore to tie in with the geophysical profile.

At the easterly end of Site 7.2, the drill hole was situated just north of the town of Joggins. Here, at the borehole location 500 ft. offshore from the high tide level, the bedrock was found to consist of interbedded shales and sandstones dipping about 30° towards the west. This structure is well exposed along the shoreline and offshore beyond the low tide mark. The drill hole encountered rocks identical to those exposed at the surface; i.e. shales with interbedded siltstone and sandstone layers dipping at 20 to 30° from the horizontal.

At the Economy Point abutment of Site 8.1, the drill hole was located 300 ft. offshore from the high tide mark. Here bedrock outcrops along the shoreline and at isolated spots up to two miles offshore consist of nearly horizontally-lying, interbedded shales, siltstones and sandstones. The drill hole penetrated the same type of rock at depth, specifically horizontally-lying, red siltstone with some interbedded shales. The relatively low core recovery was due to the poorly cemented nature of the rocks.

At the Cape Tenny abutment of Site 8.1, the sub-surface rock conditions were explored by a vertical drill hole located 600 ft. offshore from the high tide mark. Bedrock is exposed for the full width of the intertidal zone. It was found to consist of cross-bedded, reddish-brown, fine to coarse grained sandstone with an overall dip at approximately five degrees towards the north. Low core recovery indicated the rock to be poorly cemented. Considerable material was reported by the geologist to have been lost due to erosion by the drilling water. The most notable feature at this location was the occurrence of a fresh water artesian flow at a rate of approximately 10 gallons per minute (gpm) which was encountered at a depth of three feet.

Tidal Currents

Knowledge of tidal currents at tidal power development sites is important because of their bearing on construction operations in deep tidal waters and on the movement of sediment.

Not much work had actually been done, previously, on the establishment of tidal currents at the sites of potential tidal power developments of interest. However, a good general knowledge of surface currents had been provided by the work of various researchers in the greater part of the Bay of Fundy.

In 1960, high-altitude aerial photography was used as a technique of water-current measurement. This work enabled the preparation of reasonably complete surface current data for Chignecto Bay as far upbay as Sites 7.1 and 7.2 and into Minas Basin, although not as far as Site 8.1.

In 1965, a survey was carried out to acquire data for an investigation into the changes in the tides which could be brought about by the construction of barriers at various sites across Chignecto Bay, Shepody Bay, Cumberland Basin and Minas Channel, all at the head of the Bay of Fundy.

Additional information with regard to the development of the tidal oscillation within the Bay was obtained from 13 tide gauges. Two of these gauges were installed at either end of the cross-section which was regarded as the seaward boundary of the Bay for the purpose of investigation. The remainder were installed near and in the approaches to the sites at which power developments might be constructed. Observations of the tidal streams and residual currents were taken at 16 locations within the Bay. Three of these lay on the boundary cross-section and the remainder in the upper part of the Bay.

To provide information necessary for the calculations of volume transport by the tidal streams and currents, a number of profiles across the channels were delineated by echo sounder and the depths were referred to datums corresponding to the lower low water levels of large tides.

The observed rates and directions of flow were resolved into rates along two component directions. The direction of the major component at each site conforms approximately to the general direction of the channel while that of the minor component runs transversely across the channel at right angles to the direction of the major component. The rates in the two component directions were analyzed separately.

A study of ice surveys carried out during the winters of 1966-67 and 1967-68 provided useful information on the general nature of the tidal currents during the cold season.

Leads and areas of open water suggested tidal currents are more predominant along the southeast shore of New Brunswick from Cape Maringouin to Cape Enragé on the ebb tide. At low water the ice is moved down-Bay, leads developing in all directions; evident patterns suggest a down-Bay trend.

In Minas Basin on the ebb tide shortly after high water, leads in the ice cover were more predominant along the northern shore of Cobequid Bay and Minas Basin and along the western shore of Windsor Bay. A ribbon ice floe 0.75 mi. wide moving through Minas Passage, approximately 0.5 mi. offshore from Cape Blomidon, was dispersed into open patterns by tidal currents after moving around Cape Split and Cape D'Or into Minas Channel. At low water, there was a greater concentration of leads from Five Islands to Cape D'Or, in the Windsor Bay area, and along the southern shore of Minas Basin from Walton to Cape Tenny, thus indicating a generally westerly movement of the ice pack under the influence of ebb tide currents.

Tidal currents were measured in the vicinity of Sites 7.1, 7.2 and 8.1 during the course of the investigation which forms the basis for this paper. Recording current meters were operated at five stations for periods of one full month or more in conjunction with recording tide gauges. The current meters were self-contained, digital instruments, measuring current direction and speed simultaneously. The tidal current directions and velocities together with corresponding tidal curves were recorded for ebb, mid-flood, flood and mid-ebb tide.

From these data the following observations on current velocities were found to apply to the Chignecto Bay area:

1. A very consistent pattern with respect to observed current speeds and direction of flow was obtained from one tide cycle to the next. A gradual increase in current speed was observed as the cycle changed from spring tide to mean tide.
2. Maximum speeds at the current meter station in Shepody Bay for ebbing tides ranged from 1.7 to 2.7 fps; the observed flood tide maximums were slightly less. A complete 180° change in direction from flood to ebb or ebb to flood took approximately 1.5 hr. with a relatively gradual decrease in current velocity before slack periods.
3. At the current-meter station in Chignecto Bay, maximum current velocities ranged from 1.7 to 2.4 fps. In this area, the time required to change direction from flood to ebb was relatively rapid, generally only 20 to 30 min. in duration.
4. At the current-meter station in Cumberland Basin, the maximum current velocities recorded for flooding and ebbing tides ranged from 1.7 to 2.5 fps. A complete change in water direction took approximately 45 min. to 1 hr.

For the Minas Basin area, the following can be noted:-

1. Maximum current velocities recorded for flooding and ebbing tides at the current meter station at the entrance to Cobequid Bay ranged from 2.9 to 3.9 fps. The current direction data indicated a very irregular pattern during the flood cycle at this station.
2. At a current-meter station between Cape Blomidon and Partridge Island, the current velocity appeared to be more consistent from one cycle to the next. The maximum velocity ranged from 2.2 to 4.2 fps for flooding and ebbing tides. In both areas the time required to make a complete 180° change in direction of flow was approximately 30 min.
3. During the course of the survey it was noted that the flow of water in the centre of the Basin changed directions at times different from the water

flowing closer to the shore.

Sediment Observations

Measurements of the suspended sediment load were made on five rivers tributary to the Bay of Fundy with watersheds in New Brunswick and Nova Scotia during the period of December, 1967 to June, 1968. Data were obtained for the Salmon River at Murray, Kennebecasis River at Apohaqui, Annapolis River at Wilmot Station, Palmers Creek near Dorchester, and Petitcodiac River near Petitcodiac.

The maximum daily mean discharge at the five stations ranged from a minimum of 584 cfs to a maximum of 4880 cfs with a minimum discharge ranging between 1.6 and 214 cfs. The maximum suspended sediment load measured ranged from a minimum of 2.115 to a maximum of 5.485 tons/day/sq. mi. of watershed while the minimum dropped almost to the vanishing point, it ranging from nil (at three locations) to a maximum of 0.006 tons/day/sq. mi. of watershed.

In the Bay of Fundy and its embranchments, water-sediment samples were taken with a point-integrating, suspended-sediment sampler at four different elevations at ebb tide, mid-range on the flood, flood, and mid-range on the ebb tide in order to determine the total solids content and measure the instantaneous current speed, direction, temperature, and salinity at several locations. The elevations were 1.5 ft., 4.0 ft., mid-depth and 2.0 ft. below the water surface.

Samples were taken at five stations along the axis of Site 7.1 and five stations along a line between New Horton and Shulie across Chignecto Bay.

The total solids content at stations at Site 7.1 ranged from 0.044 to 0.435 grams/liter. The values for total solids content at stations in Chignecto Bay, although consistently higher than at Site 7.1, still did not exceed 1.645 grams/liter, a relatively small weight per total volume of samples recovered.

For Site 7.2, water-sediment samples were taken at five stations. The results of the measurements indicated that the total solids content ranged from 0.176 to 3.240 grams/liter. No hydrometer analyses were possible due to the small amount of sediment content in the water.

Two major sand-body complexes occur in Windsor and Cobequid Bays in the form of plano-convex or tabular sand masses running to thicknesses of some 100 ft. and lengths of about 20 mi. These sand deposits appear to exhibit features showing they are the result of two very different tidal regimes.

Evidence from geophysical and other studies indicate that the Cobequid Bay sand-body complex is much more intertidal than that in Windsor Bay. It is reported that the period in which the tidal channels are independent of the tide in the Bay is as much as five hours on the ebb and two hours on the flood, as a result of which the discharge, during these periods, is adjusted to the characteristics of the channel. Current velocity and, to a lesser degree, water depth is constant. Basically, the tidal regime can be described as a low water-channel phase and a high-water, sheet-flood phase.

A consultant retained to advise on sediment erosion, transport and deposition in Minas Basin concluded:

1. The major agents of sediment erosion, transport and deposition are waves, wave-generated currents, tidal currents, along-shore drift and open channel flow.
2. Most of the sediments are derived from two sources. These sources are sea-cliff bedrock along-shore and Pleistocene sediments on the floor of the Minas Basin.
3. The major agent of erosion is wave activity on shore. Such erosion yields mud, gravel and sand.
4. Gravels occur along shore as lag concentrates produced by erosion of sea cliffs. Such marginal gravels migrate by along-shore drift an average linear distance of 4.0 ft./yr.
5. Sandy gravel bars are formed by transport and deposition of gravels by open-channel flow at low tide.
6. Sand is eroded, transported and deposited by tidal currents and occurs primarily in asymmetrical sand bars.
7. The major factor controlling sand transport and accumulation is the shear strength of the sands.
8. Mud is eroded by wave activity, transported in suspension and deposited on high tidal flats and in estuaries by settling during slack water at high tide.
9. Assuming a life span of 150 yr., it is estimated that the reservoir behind a potential tidal power development in Minas Basin could silt up to one-third of the total volume of basin space landward from the development.

At Site 8.1, water-sediment samples were taken at five stations between Economy Point and Cape Tenny. Laboratory measurements indicated that the total solids content for all stations ranged from nil to 0.79 grams/liter, a relatively negligible amount. As a result of the small amount of solids recovered per total volume of samples taken, no hydrometer analyses were possible.

During the course of the marine surveys, it was found that deposits of medium to coarse sand and gravel are present at the outer edges of the inter-tidal zone at both Cape Tenny and Economy Point. A field examination of two en-echelon* sand bars at Cape Tenny was made during the low tide to determine the sediment quality and bedding characteristics. Each deposit was approximately 700 to 800 ft. in length, 500 to 600 ft. in width, and

---

*En-echelon means parallel structural features offset like the edges of shingles on a roof when viewed from the side.

varied in thickness from a few feet at the edges up to 20 ft. or more near the centre. A series of pronounced sand waves up to five feet high were formed along the top of the sand deposits. The trend of the sand wave ridges was roughly north-south with a longer more gently dipping surface facing easterly and the steeper slope to the west indicating a general migration pattern from east to west with the ebbing tide. Numerous smaller scale ripple marks formed in the area tend to support these observations.

These shore-line sediments consist generally of reddish-brown to grayish-brown and buff, fine to coarse sand with several layers of gravel and cobbles one to three feet thick. In some instances, particularly near the edges of the sand deposits, large boulders up to three feet in diameter are present. The sand is very loose and water saturated in most areas; however, in some of the troughs between ridges where the slightly finer material seemed to form a matrix for the coarse material, the sediment is decidedly more firm. Sediments are resting directly upon Triassic bedrock which is exposed around the entire periphery of the deposit.

Indigenous Construction Materials

The power houses and sluiceways of the tidal power developments would require very substantial volumes of concrete either in reinforced or mass forms, whether these structures be built in the dry behind cofferdams or prefabricated in on-shore construction facilities, towed out to location, and sunk onto prepared foundations. The dykes require large volumes of rockfill in various sizes together with,in some designs, materials suitable for impervious cores and filters.

During the summer of 1968, preliminary exploration by diamond drilling was carried out at potential rock-quarry sites in New Brunswick and Nova Scotia for the purpose of determining the suitability of the outcrops for the production of large-size armourstone. This assessment included a geological interpretation of the jointing patterns and other structural features of the rock as revealed by drilling, the probable behavior of the rock drilled under normal quarrying techniques, and the percentage and size of the largest blocks of rock which a quarry would be likely to yield.

Two sources of rock suitable for primary armourstone were located in Nova Scotia, both in areas of granitic rock and with good access by road and railroad. However, they are a considerable haul distance from any of the selected sites; one is located near Lawrencetown (40 mi. northeast of Digby), and the other is near Meaghers Grant (25 mi. northeast of Halifax). Sandstone quarries near Cape Dorchester, New Brunswick, appear to be capable of producing large blocks of rock, but this stone is believed to be insufficiently durable to be used as primary armour although suitable for secondary armour.

With respect to granular materials, large volumes of sand and gravel are found along the north shore of the Minas Basin. These deposits appear to be quite consistent in character. Test results indicate they would be suitable for use in concrete. Elsewhere in the Bay of Fundy region, granular deposits are generally lacking or of poor quality. Near Chignecto Bay the only suitable aggregate supplies so far located occur near Alma, New Brunswick.

Basin Areas and Volumes

Basin surface areas and volumes as a function of elevation at Sites 7.1, 7.2 and 8.1 were determined from a synthesis of existing charts and maps, in part, and from a combination of aerial photography and ground control for the remainder. This latter was accomplished by carrying out successive aerial photography of the intertidal zone over one half of one tidal cycle at predetermined intervals of time. The line of contact between the foreshore and the water surface indicated on the aerial photographs with appropriate adjustments for variations in tidal levels enabled the establishment of basin areas at any desired levels referred to GSCD.

Having carried out surveys around the peripheries of the basins which would be associated with tidal power developments at Sites 7.1, 7.2 and 8.1, relative to the elevations of the crests of existing dykes, elevations at low points in adjacent highways, roads and railways; elevations at the inverts of sewerage systems; elevations of bridge seats for such structures crossing bays and streams entering the reservoir; and relevant data on the controlling elevations for other man-made structures: the practical upper and lower limiting levels for the basins could be determined. These limiting levels were selected so as to have little effect on existing structures or, at most, give rise to circumstances which could be readily corrected at nominal cost.

The practical limiting water levels established for Sites 7.1, 7.2 and 8.1, determined as above, were:-

Site	Maximum limiting level*, ft.	Minimum limiting level*, ft.	Volume between limiting levels, million cu. ft.
7.1	24.5	-26.5	55,000
7.2	25.0	-25.0	44,400
8.1	30.0	-29.0	154,000

*with respect to GSCD

FEASIBILITY OF BUILDING TIDAL POWER PLANTS

A controlling factor in the design and construction of a tidal power development is the nature of the unconsolidated sediments lying between the sea bed and the bedrock. There is little problem where bedrock, even of relatively poor quality, exists in the foundation areas at the right elevation. However, unconsolidated silty-sandy sediments may pose substantial problems related to its removal in founding structures on sound rock or to designs and construction adequately suited to such an unconsolidated foundation.

There is a substantial body of laboratory and actual construction experience attesting to the feasibility of depositing rock and other materials in flowing water moving at high velocities for the purpose of achieving river and tidal-estuary closures whether for cofferdams or permanent structures such as causeways or dykes in a tidal power plant. This ability has been confirmed by actual construction experience in a number of instances.

Closure of the eastern entrances to Scapa Flow in Scotland during World War II by causeways entailed the successful placement of rockfilled gabions containing five tons of quarry-run rock in water depths of up to 60 ft. and in water velocities up to 1.7 fps. Primary armour of 10-ton random placed rock and secondary armour of five-ton rock were used. This closure was made by the friction control method in which the gap being closed is gradually built up, more or less, uniformly from the bottom.

The Canso Causeway in Nova Scotia was successfully built across the Strait of Canso in a depth of 187 ft. of water, over 25 ft. of light silt. Work started in July, 1952, was completed in May, 1955. About 10,100,000 cu. yd. of rock was used. The crest width of this causeway is 80 ft.; bottom width is about 650 ft. at approximately 200 ft. depth, and its length is some 3500 ft. The slope of both westerly and easterly faces is one to 1.25 horizontal. Construction was carried out by the pinch-off method involving closure by end dumping of fill material.

The Arrow Dam, one of the three major storage developments on the Canadian reach of the Columbia River in British Columbia, is located in the Castlegar Narrows downstream from Arrow Lake about 20 mi. from the Canada-United States border. At the site, the river valley was eroded to great depth during or prior to the last Ice Age, then backfilled with material of fluvio-glacial origin probably, at least, 7,000 yr. ago. Drilling carried out to a depth of about 500 ft. did not encounter bedrock. It did show that the river-bedmaterials consisted largely of sands and gravels with a permeability of between 0.1 and 0.01 cm./sec. with random occurrences of open type gravels. Exploratory work revealed that a sound rock spur on the left bank provided an excellent foundation for the concrete control structures, but would necessitate a large cofferdam. The investigation indicated it would be impractical to dewater the river channel for the major portion of the dam.

The dam has a maximum height of 190 ft. and contains approximately

8.5 million cu. yd. of material obtained from a nearby site on the left bank. The lower portion of the dam was built in the wet to approximately normal low-water level due to the depth of 70 to 100 ft. of water at the site between the rock spur and the right bank. The bed of the river was initially filled with sands and gravels to a suitable intermediate level some 50 ft. lower than normal low-water level following which an upstream blanket was placed and the downstream supporting zone then raised to divert flow through the control structure; the remaining underwater fill was then completed.

The Zuyder Zee was closed by a dyke across the mouth. Started in 1923, this earthern dyke was completed in 1932. Twenty miles long, the closure involved the deposition of material in water in depths up to 16 ft. deep below mean sea level with relatively open exposure to the stormy North Sea. Closure of the Zuyder Zee in the final gap was affected by the dumping of boulder clay. This boulder clay could withstand velocities of 12 to 13 fps without serious scour.

In general, closures of rivers or tidal estuaries by the dumping of rock or other materials in flowing water have been achieved by two methods:-

1. The end-dump method in which the fill is built progressively into the channel from one or both banks at its full crest height gradually restricting the flow to a narrow, high velocity section which is finally closed by the use of larger rocks or precast concrete masses. This has been termed the "pinch-off" type closure.
2. The friction-control method in which the fill is built up from the bottom in incremental layers, each of which extends the full width of the channel. The depth of flow thus remains uniform across the river width, and is gradually reduced as the fill height is increased. In this method, the fill is deposited by barge, by overhead cableway, or by trucks or railway cars from a trestle.

ELEMENTS OF TIDAL POWER PLANTS

Power Houses

Tidal power plants consist of three basic elements: the power house or setting for the generating units; the sluiceways with their gates for the filling or emptying of the controlled basins; and the dykes, usually rockfill, constituting the closures between power houses and sluiceways and between either and the abutments of the development.

In view of its importance to the economic development of tidal power, a number of specific investigations of optimum power house design were carried out. These studies considered power house design from the following points of view:

1. Functional requirements as reflected in simplicity of design, maximum

economy of construction, and minimum operating and maintenance costs.
2. Efficiency of water passages.
3. Provisions for inspection, maintenance and replacement of equipment.
4. Requirements for both single-effect and double-effect generating units.
5. Adaptability to construction in the dry or in the wet.

The designations in the tabulation hereunder are used throughout the subsequent discussion. The four types are compared hereunder:-

Type	Characteristic Feature	Shown by	Method of Construction
A	Built in situ	Fig. 7	In the dry behind coffer-dams
B	Caisson principle	Fig. 8	In the wet
C	Caisson principle	Fig. 9	In the wet
D	Caisson principle	Fig. 10	In the wet

Generally speaking, Type A is representative of all designs involving construction in situ behind temporary cofferdams. However, the width - at right angles to the longitudinal axis of the overall development - is substantially less with the single-effect generating unit as compared with the double-effect unit. Type A may be considered, broadly, of conventional design.

Types B, C and D, all involving the principle of a prefabricated element or caisson intended for construction in the wet, reflect studies seeking that design providing lowest possible cost of construction. These three designs and many others more preliminary in nature involved different approaches to the objective.

Type B (Fig. 8) arose from a concept avoiding the necessity for unwatering power house sections to gain access to the turbine generators for maintenance purposes as well as installation. This entailed consideration of a prefabricated element or caisson built in an on-shore facility which could be floated into position and sunk onto prepared foundations at which time temporary bulkheads over the water inlets and outlets would be removed to permit the water levels to equalize inside and outside the caisson. The idea behind this concept was to reduce the overall foundation pressures to a minimum by a decrease in ballast requirements.

Further development of Type B provided that the turbine generator unit be contained in a module consisting of a section of water passage. This module would be raised or lowered within the confines of the caisson for major servicing, etc., by using suitable synchronized machinery mounted on tracks on top of the deck. The generating unit would be locked in position during operation. Access to the interior of the unit for minor maintenance and service attention could be achieved, preferably at low tide, while it is still in operating position.

A further refinement of Type B is possible in which the caisson length

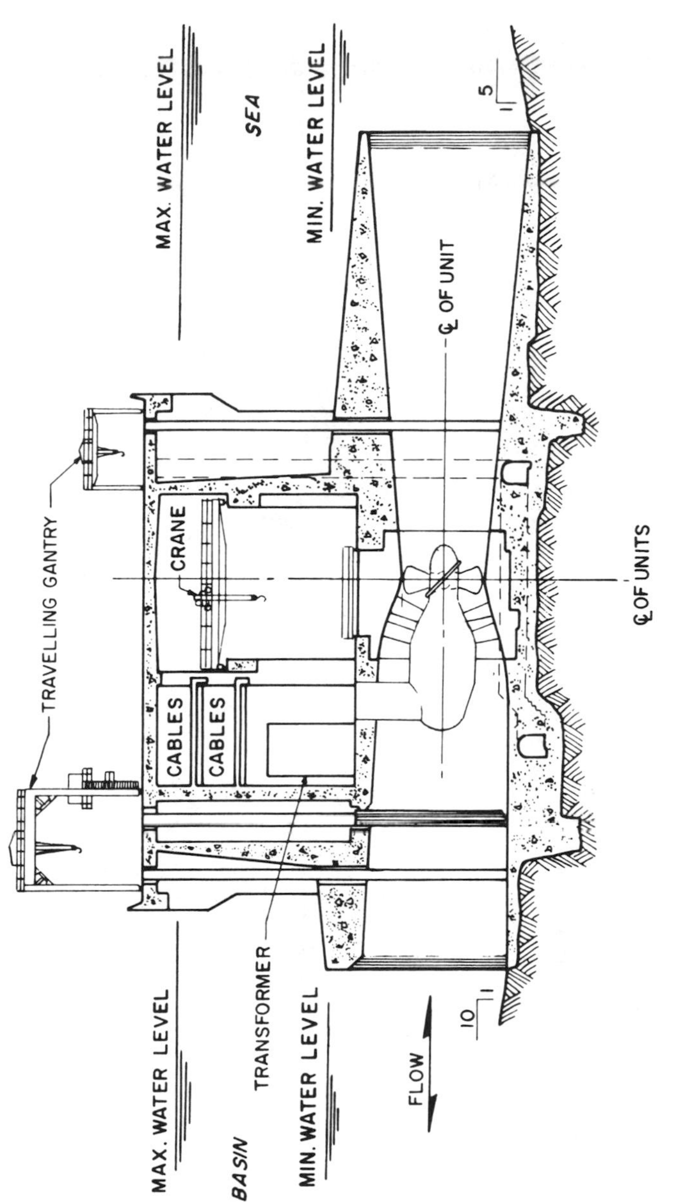

Fig. 7. Type A power house.

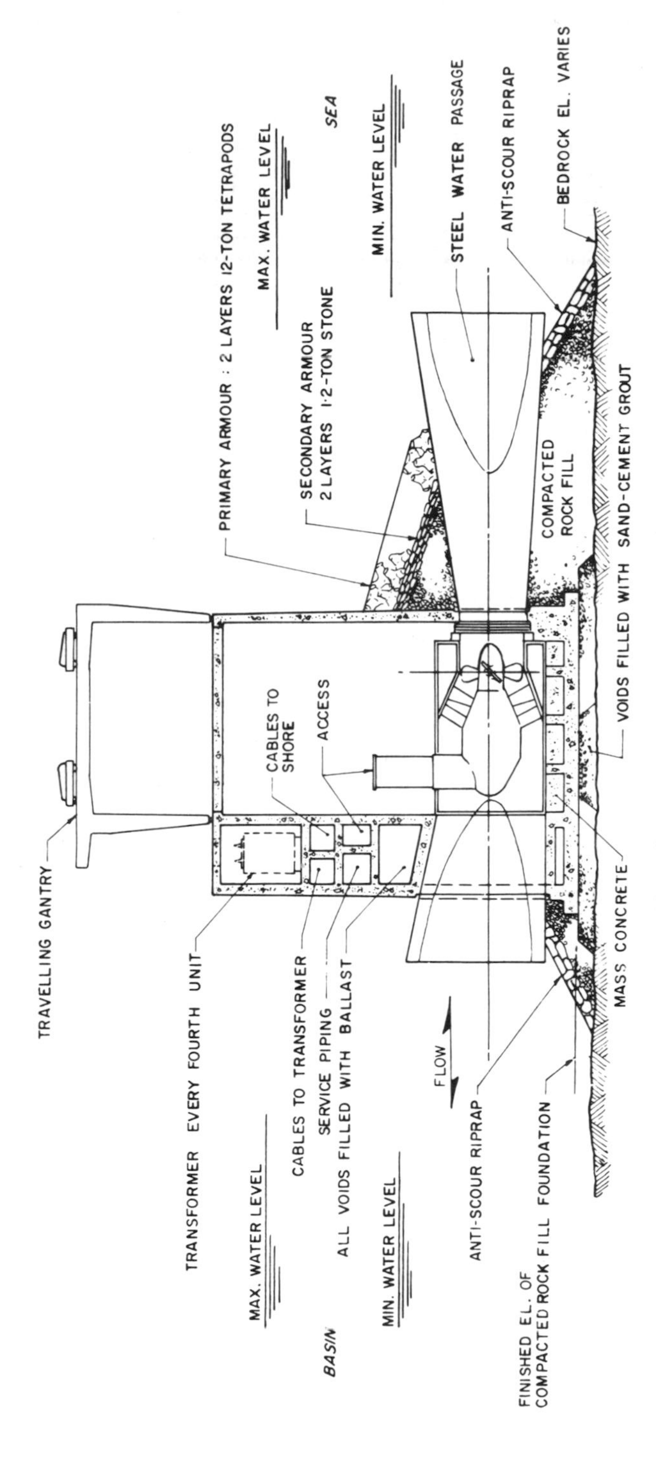

Fig. 8. Type B power house.

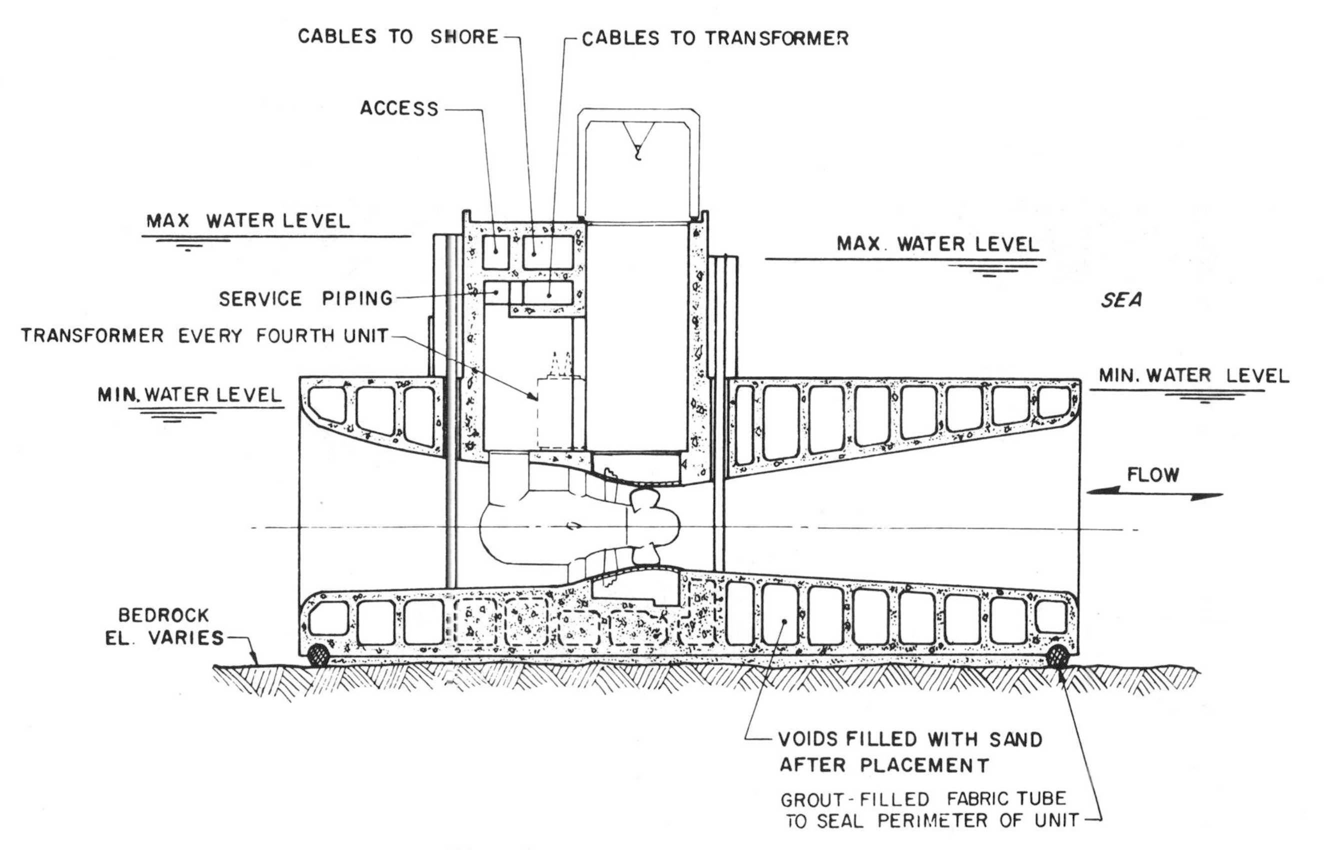

Fig. 9. Type C power house.

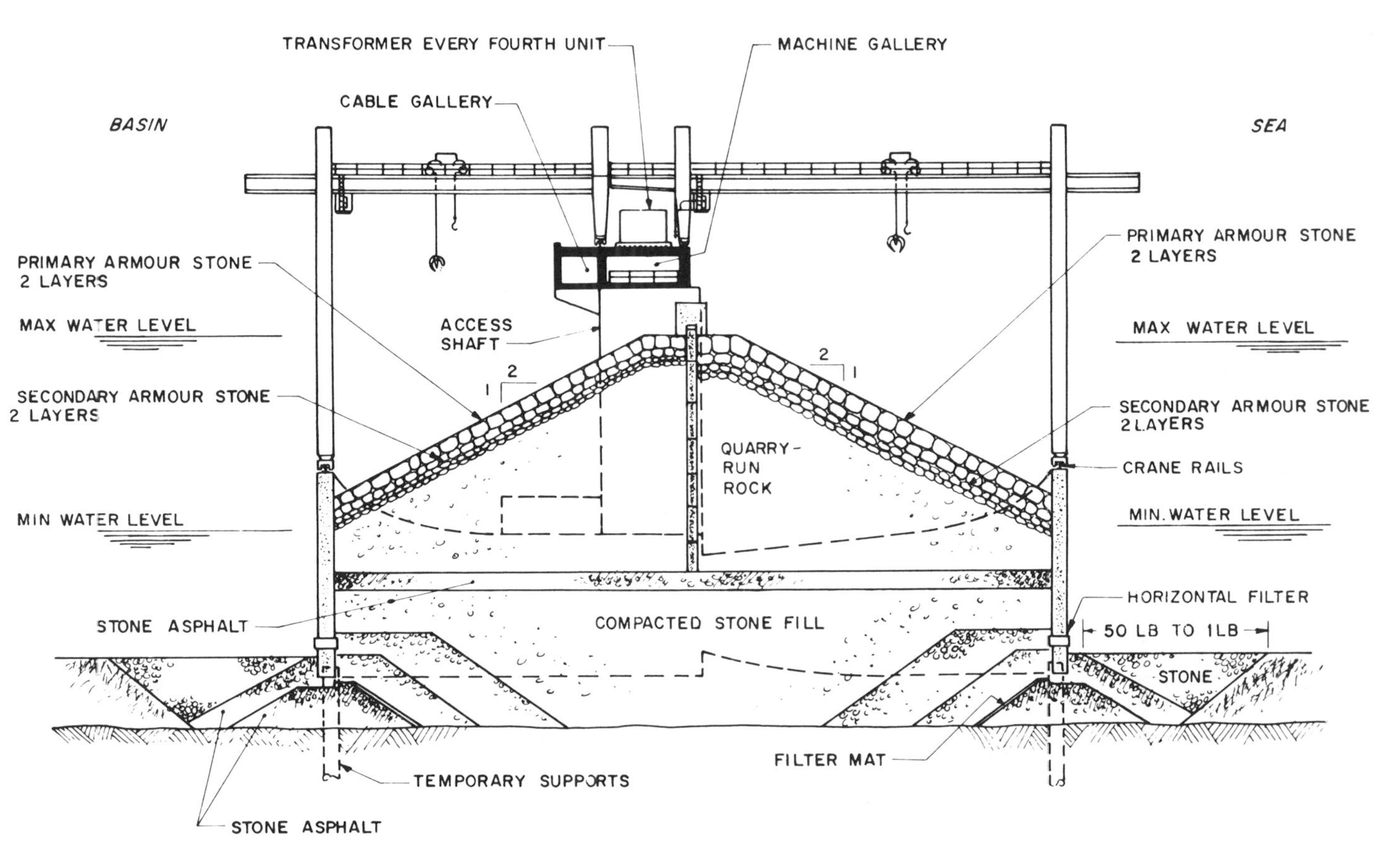

Fig. 10. Type D power house.

would be limited to that necessary to provide for the generating unit module only, in the interest of economy. The water-passage segments outside the central caisson may be visualized as separate modules such as concrete box sections placed by any convenient means or prefabricated steel sections suitably reinforced both circumferentially and longitudinally with designed structural shapes to withstand any in-service forces, i.e. ice load, wave impact, etc.

Type C, shown by Fig. 9, is associated with three variations of the general design. These are:

1. Individual power house units separated by a stop-log system;
2. Twin power house units separated by a stop-log system;
3. Individual power house units placed adjacent to one another.

Variants (1) and (2) have the advantage that the need for accuracy in placing the units is not as high as with (3) as the stop-logs may be cast to suit the space between individual units. In addition, the stop-logs eliminate the need for an equivalent length of dyke at the ends of the power house and sluice structures which could result in an overall cost saving for the project.

Type C represents a more conventional approach than Type B, adapting a conventional power house design primarily conceived for construction in the dry to the caisson concept of construction in the wet.

Type D, Fig. 10, provides for realization of a concept consisting of power house elements buried in the base of a rockfill dam. It is based on the most radical design of power house module with the emphasis on two factors. The first of these is the important factor of towage requirements of specialized marine equipment, the related aspect of sophisticated and advanced construction procedures, and highly competent specialized marine labour. The second factor involves maximum use of indigenous construction material in the form of rockfill and armourstone.

Each element would consist of two parallel water passages jointed together by means of two end bulkheads. Once in place, these end bulkheads would serve primarily to retain the sandstone fill placed around the water passages. On top of each water passage would be a concrete shaft, providing access to a generating unit, placed within the water passage.

Fig. 11 shows a model of a Type D power house module. It indicates two twin-water passage elements in the oblique view of the ocean side. Fig. 12, from the ocean side, indicates on the left the basic structural nature of the water-passage element while on the right it depicts a complete power house module with gantry cranes and the rockfill completing the power house module. Fig. 13 shows a completed twin-water passage element carried by the construction pontoon.

Basic design requirements indicate that the water-retaining height of the power house dam should be El 30, GSCD. The overall height must be such that no appreciable overtopping will occur under the most extreme combinations of high water and waves.

Fig. 11. Type D power-house model, showing the cross-section of a complete module and, on the left, the ocean end of a twin water-passage element.

Fig. 12. Type D power-house model. On the left may be seen a water-passage element housing two generating units; on the right the complete module.

Fig. 13. Type D power-house model. (Above) construction pontoon for twin water-passage element. (Below) completed twin water-passage element carried by construction pontoon.

Of the various designs based on the caisson principle, B, C and D, it should be noted that Type D embodies more innovations in structural design than do Types B and C, and hence, inherently involved more potential uncertainties at this time.

All four designs of power house may be described as open-type involving gantry cranes for provision of cranage services for erection and maintenance. However, the designation is somewhat of a misnomer as the power house designs all embody adequate provision for water tightness. All designs also incorporate provision for auxiliary equipment and services, both mechanical and electrical. All contemplate the use of high voltage cables for transmission of the electrical output to an on-shore terminal facility with control room. All designs contemplate the same basic grouping of generating units and related features such as governors, voltage regulators, etc.

A preliminary study of the water-passage geometry indicated that shortening the water passage length by 25 ft. without altering the angle of flare would result in a loss of gross head of approximately 10%. A detailed review of the water-passage geometry can only be made when the turbine characteristics and mode of operation are fixed.

Sluiceways

Tidal power plants operate on the continuously varying difference in level between the water in the basin on the landward side of the development and the water in the sea. The basin must be filled from the sea or emptied to the sea as required by the operating regime of the power plant so that production can be coordinated with the load curve of the power network with which it is interconnected.

This requires suitable sluiceways equipped with gates which can be operated quickly and reliably. These gates must be as free from maintenance as feasible despite the impact of storm-engendered waves, masses of ice carried by the flow, freezing of operating mechanisms and the coating thereof with ice as well as damage due to the corrosive nature of the marine environment.

Sluiceways with crest gates are much more subject to damage from the causes stated than those of submerged, venturi-type design. Exemplification of this is afforded by the relative freedom from ice problems of submerged intake structures in conventional hydro-electric power plants in northern climates.

Consideration of the type of sluiceway structure, gate setting, gate hoisting system, gate structure, gate heating and operating conditions led to the conclusion that single-leaf, fixed-wheel, vertical-lift gates, equipped with hydraulic hoists set in a venturi-shaped water passage with a throat opening 40 ft. sq. would be about the optimum for Bay of Fundy operating conditions.

Generally the largest sluiceway gates which can be fabricated and transported or assembled on site from sub-elements prove to be the most economical. However, this is not necessarily true for tidal power plants for various reasons. There is an optimum cost relationship between gate size and the structure constituting the sluiceway. This is particularly true if the sluiceway structure is built from caissons fabricated at an on-shore construction facility, towed to site, and sunk onto prepared foundations.

The following should be noted:-

1. The gates are required to open and close each tidal cycle resulting in 705 complete operations annually. This exceedingly high service requirement is in sharp contrast with normal hydro-electric practice in which gates are only operated a few times per annum.
2. The gates, their embedded parts, their hoisting systems, and their superstructures would be exposed to the severely corrosive environment of sea water and spray.
3. To some extent even submerged gates associated with venturi-type sluiceways would be subjected to severe dynamic loads from wave action.
4. The climatic conditions, with cold spells resulting in maximum daily air temperatures below the freezing point of sea water persisting for five to 10 days at a time over a three month winter period, indicate the possibility of severe ice buildup on exposed surfaces of the gates and the hoisting equipment due to freezing of windborne spray.
5. The combination of low temperatures and tidal variations tends to the formation of ice floes so that during periods of heavy wave action severe buffeting of the gate on either side of the development might occur.
6. Gate and hoist superstructures exposed above deck level would be subjected to wind loads of considerable magnitude due to the location of the tidal power developments in the Bay of Fundy.

One concept of construction is based on the conventional mode of construction in the dry behind cofferdams. An alternative concept of construction involves assembling the sluiceway structure from unit sections fabricated in a construction facility on shore which are then towed out or otherwise moved to a site and sunk onto a prepared foundation.

In either case, the sediments occurring at the Bay of Fundy tidal power sites are of such a nature that the sluiceways would have to be built on bedrock. This would necessitate dredging in the case of the caisson concept being utilized.

Dykes

A number of possible dyke designs were considered, taking into account the many relevant engineering and economic aspects. These designs are discussed hereunder.

Studies by the International Passamaquoddy Engineering Board led to a design embodying the construction procedure indicated by the cross-section shown on Fig. 14(a). In the maximum water depths which would be encountered, up to 125 ft. and more below mean sea level (msl), rockfill would be placed by bottom dump scows to form a ridge on each side of the axis of the dam. In the valley between the rockfills, gravel and sand would be placed by lowering the materials in large bottom-dump buckets. Intermingling of the materials would occur as indicated by the lenticular contact zones. Between depths of 125 to 25 ft. below msl, a clay core would be added by placing large masses of clay from bottom-dump scows. Concurrent placement of rockfill, clay, and of gravel and sand transition zones would be necessary. Above 25 ft. below msl, flows would be cut off by end dumping, a rockfill along one side of the dam. Completion of the crest construction would be accomplished by end-dump and bucket-placement method.

The design shown by Fig. 14(b) consists essentially of a run-of-quarry rockfill embankment. This section with the indicated slope protection added to the seaward side constitutes a completely stable dyke for the purposes of the tidal power development. The filters shown to the left of the run-of-quarry embankment section are provided for the purpose of reducing seepage through the dyke.

A third dyke design (Fig. 15(a) consisting of prefabricated crest caissons towed out to location and sunk onto prepared foundation fills was studied. To ensure adequate stability of the embankment under all conditions, it would be necessary to load the seaward toe of the deeper embankments supporting crest caissons as shown. This would constitute a local thickening of the anti-scour rock blanket.

In this design the fill would be topped out at about 10 ft. below extreme low water level. A concrete crest structure extending up to approximately high water level would then be placed. This crest structure would consist of caissons in the form of twin culverts below the roadway. During construction the open culverts could be used to pass part of the tidal flows making closure easier. After closure the culverts would be closed by stop logs, the caissons filled with sand and gravel ballast, and protected by suitable paving against erosion. This would permit the uninterrupted passage of storm waves at high tide through the space between the tops of the caissons and the roadway. Containment of storm waves would otherwise require a heavier design. The access roadway across the dyke would be 24 ft. wide but greater width could be accommodated if required.

A fourth design was considered in which the dyke cross-section would consist of a quarry-run rockfill with voids filled with silty sandy sediment. This filling could be accomplished in a number of ways; for example, by placing sediments on the dyke slopes and relying on hydraulic action to transport the material into the body of the dyke or by placing sediments hydraulically concurrently with the dumping of the rockfill. The dyke crest and slopes would be armoured to protect against wave and ice action.

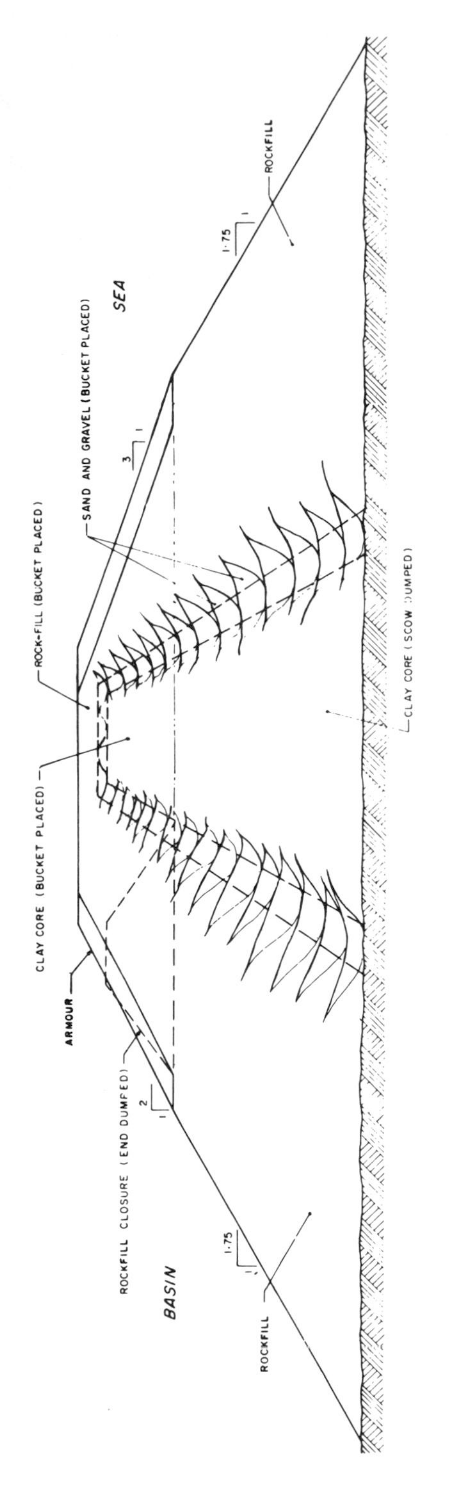

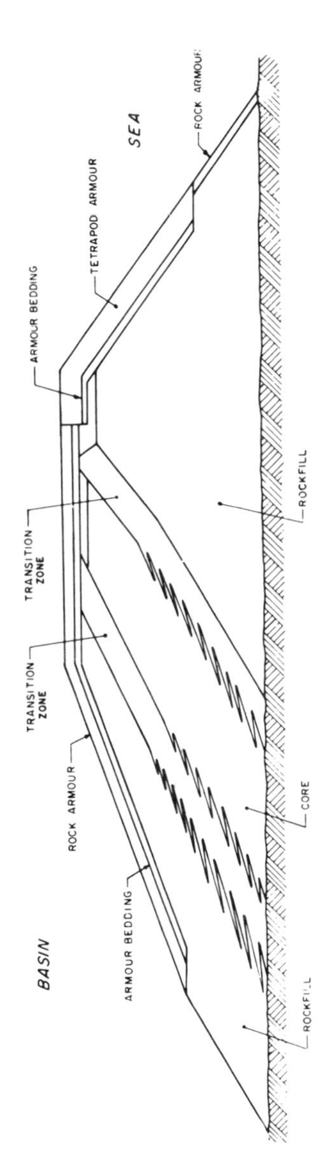

Fig. 14. Central-core (a) and sloping-core (b) dykes.

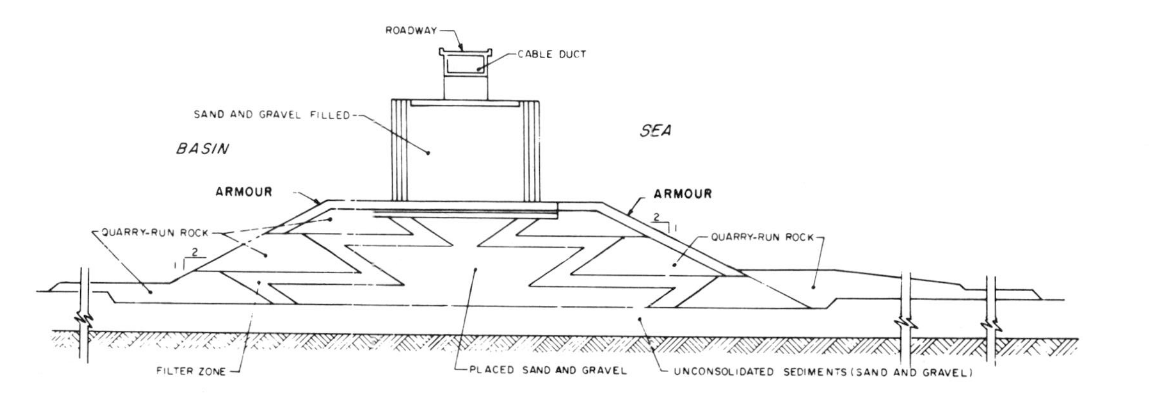

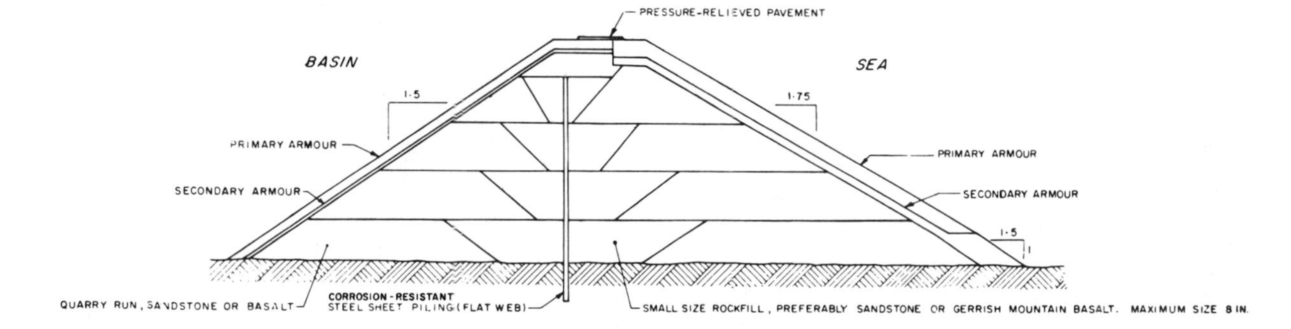

Fig. 15. Crest-caisson (c) and steel-sheet (d) piling membrane deep-water dykes.

Various alternatives to the four types mentioned were examined. All of these involved the use of rockfill as the basic dam material but attempted to provide alternative arrangements for the reduction of seepage. These included consideration of the use of a cast-in-situ concrete barrier placed through a selected rock core in the dyke following closure. In all of these designs and alternatives, it was considered that technical difficulties and uncertainties in placing, along with higher costs and the necessity for higher contingency allowances made it desirable to develop a more economic design.

Such a design, shown as Fig. 15(b), took into account the availability of an abundance of sandstone bedrock or basalt common to Sites 7.1, 7.2 and 8.1. This material does not meet requirements for armourstone and much is unsuitable for quarry-run core material. However, in this design, steel-sheet piling driven through this material would provide impermeability.

In the design of the Burlington Beach Wharf in Hamilton Harbour, rockfill enclosing berms were constructed by end dumping in water through which interlocking steel-sheet piling was later driven to form bulkhead walls. The maximum size of rock in the fill was limited to eight inches to ensure that steel-sheet piling could penetrate the material. Driving was accomplished without problems which demonstrated that steel-sheet piling can be driven through rockfill having the above classification. The same technique could be applied to dykes for Bay of Fundy tidal power plants with the following advantages:

1. The sandstone could easily be extracted in the desirable size range; most probably by ripping.
2. Resistance to driving steel-sheet piling would be minimal as the sandstone rock in the path of the driven steel could be easily broken under the driving impact.
3. Resistance to driving due to cohesion would not exist.
4. Placement of the material in water would be simplified, i.e. excessive loss of fines would not be a matter for concern.
5. During construction, the dyke would be less vulnerable to attack from the elements.
6. Concern over the imperviousness of the dyke with a steel core would be non-existent. Also, the possibility of blowouts under a rapid draw-down of head would be removed as pore water pressures would be minimized.
7. The steel-sheet piling could probably be extended into the in-situ materials, that is bedrock, thus providing an effective barrier to possible seepage of water. Other possible solutions of this problem would be difficult to assess or even resolve using cores of finer materials.
8. The stability of the dyke would be improved due to the structural strength imparted to it by the steel-sheet piling.
9. The crest width of the dyke would not be determined by the thickness of transition zones, impervious cores and the like, and could therefore be reduced to the absolute minimum necessary for stability, transportation

and construction purposes. In this way, very large quantities of fill materials could be saved with corresponding savings in cost.

The major disadvantage of a steel sheet-piling membrane is associated with the problem of corrosion of the steel. Manufacturers of steel sheet-piling have long recognized the need of some "rolled in" anti-corrosion provision if the material was to compete with other materials for permanent installation in corrosive environments. Applied coatings were found to be only partially effective. Cathodic protection failed in large measure due to the intermittent nature of the process in the intertidal zone; it was found to be almost totally ineffective in the splash zone.

A most important development has been the production of a steel composition highly resistant to sea water corrosion without loss of strength. Specimens of this steel have been tested by others with very encouraging results, so much so that its useful life may be considered at least twice that of ordinary mild steel.

Quite apart from the resistance of the steel to corrosion in a sea water environment, in the case of a dyke, the steel will be permanently buried and will not be required to carry any load except small hydraulic pressures. It may be allowed to deform or even fail structurally without impairing its usefulness. For this reason, it may be considered an effective seepage barrier as long as a vestige of its cross-section remains. As a matter of fact, it may be reasonably assumed that the natural breakdown of the sandstone into finer segments, will have long since built up a dyke cross-section providing, in itself, an effective seepage barrier.

It was recognized that stability requirements and settlement limitations are much the same whether a core of clayey material or one of granular material is used. The high shear strength and good drainage characteristics of the construction materials in granular and quarry-run rock cores permitted drawing the conclusion that settlement within the embankment would be substantially completed upon termination of construction provided adequate foundations are available. It was assumed the dykes would be founded on stable till or sand and gravel with the top 10 ft. of overburden comprising silt or clay removed by either dredging or displacement.

On the basis of various tests by others as well as experience on the dumping of materials in deep water, exterior slopes for rockfill were taken at 1.75 to one.

It should be noted that the crest of the dyke would be armoured to allow for overtopping by waves developed under the most severe storm conditions. The access roadway and other facilities required for the tidal power development would be installed over the armour protection.

The significant wave height at each site has been used in the calculation of runup and slope protection requirements for the dam embankments. The stillwater level has been taken as the elevation of higher high water large tides. The top width has been chosen to provide access for purposes of construction as well as for operational maintenance and inspection.

Not all of the dykes would be designed for construction in accordance with Fig. 15(b). For instance, at Site 8.1 the surface of the intertidal zone becomes dry for a distance of about 3,700 ft. from Economy Point and 2,700 ft. from Cape Tenny, measured horizontally at low tide. In these sections the dykes can be constructed entirely in the dry using land based equipment and an earth-fill core technique. These portions are referred to as "shallow water dykes" to differentiate them from the portions where steel-sheet piling is proposed, i.e. deep water dykes.

GENERATING UNITS

The energy potential in a tidal power development is exploited under low to very low heads. The only turbine types which are adaptable to such plants are the axial flow, high specific speed types including the Kaplan, bulb, straight flow and tube designs. (See Fig. 16)

The Kaplan turbine, the classical type of high specific speed turbine adaptable to low heads, was developed in the 1920's. Applied in a head range from about eight to 260 ft., it is characterized by a short intake terminating in a scroll case, a vertical shaft prime mover, with fixed stay vanes, adjustable wicket gates and adjustable runner blades, and an elbow-type draft tube. The most advantageous coordination of the adjustable wicket gate and runner blade positions is normally determined from model and prototype tests. These prime movers show a high efficiency over a large portion of their operating range but are adaptable to only one operating mode, direct generation.

The bulb-type turbine, of which the outstanding installation in tidal power application is at the Rance development in France, made its first appearance in the mid-30's with two units in a small run-of-river plant. Major developments took place in the post-war years in France, being promoted to a large extent by a desire to create a unit suitable for application in tidal power development. A significant advance in the field of low-head turbines, it has enabled exploitation of heads between about four and 60 ft. The bulb turbine has a relatively straight and horizontal, or near horizontal, waterway. It is essentially a Kaplan-type turbine with stay vanes, adjustable wicket gates and adjustable runner blades, thereby providing so-called double regulation and a horizontal shaft. The stay vanes provide the structural support from which the unit is cantilevered. The generator is contained in a roughly torpedo-shaped housing, generally at the upstream end of the unit.

The bulb-type generating unit is characterized by improved hydraulic efficiency, greater discharge capability, and greater power output for the same runner size or, conversely, smaller turbine dimensions for the same efficiency and discharge. It reaches higher specific speeds than the Kaplan

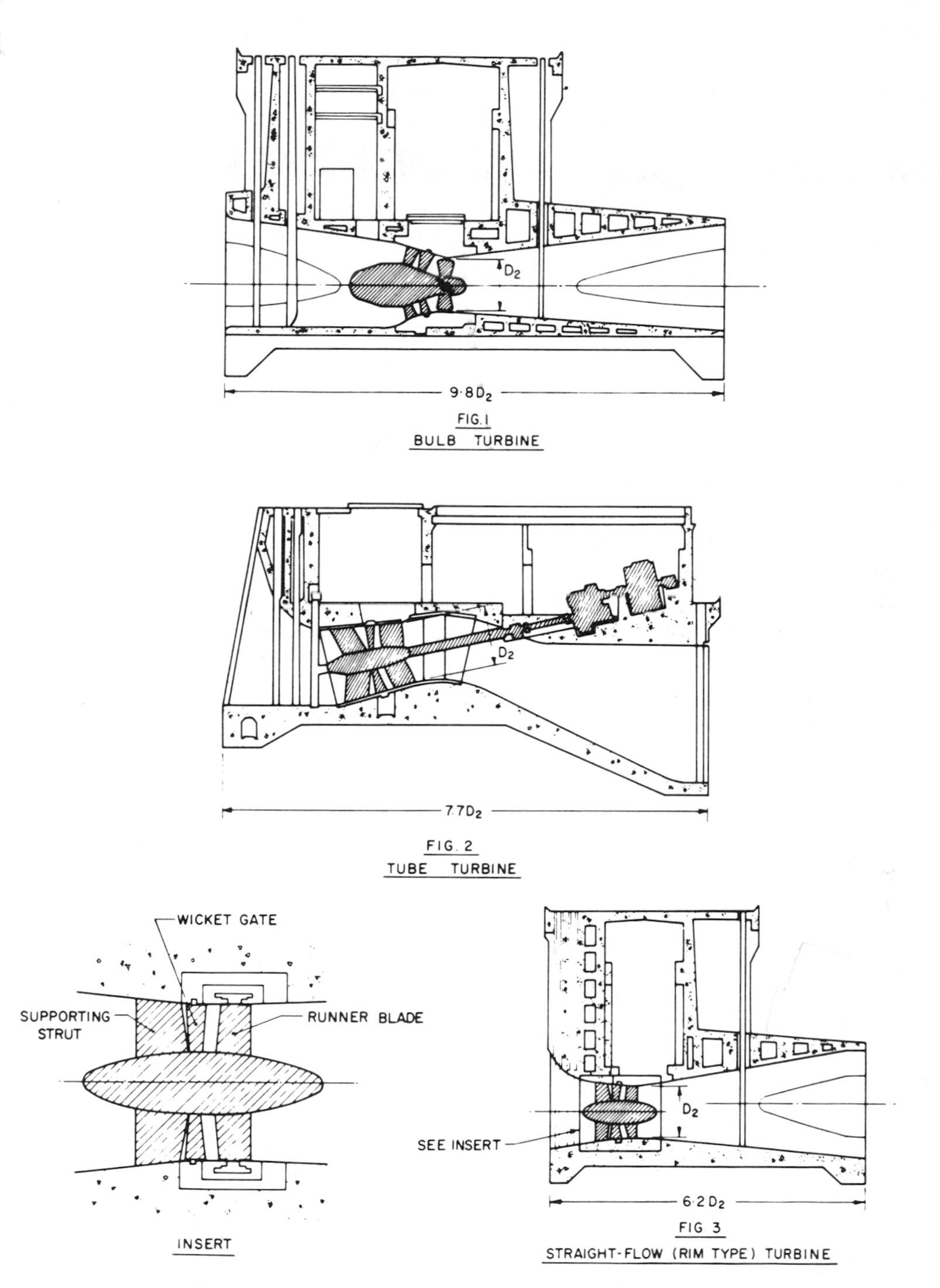

Fig. 16. General arrangements for generating units.

and thus can be used for even lower heads. The essentially horizontal draft tube plays a very important part in the recapture of kinetic energy at the exit from the runner, which often accounts for 60 to 70% of the potential corresponding to the head. The operation and efficiency of the draft tube is significant in the efficiency and performance of the generating unit. Because the waterway provides for straight flow and is essentially horizontal, attractive savings can be achieved in the volume and cost of civil works associated with the power houses. The horizontal distance between the center lines of units is reduced by the absence of a scroll case. Excavation need not be carried out to the same depth due to the elimination of a vertical draft tube with elbow.

The bulb-type turbine is adaptable to operation in five possible modes, as a turbine and as a pump in both directions, and as an orifice, the latter at up to about 50% of the discharge capacity.

The straight-flow turbine, first suggested by Harza in 1919, saw its initial major application during the late 30's in a number of small, central European run-of-river plants. Adaptable to the same head range as a bulb unit, and after considerable difficulty with later war-time installations in Bavaria, the outlook for prompt development of this type of unit in large runner sizes and its applicability to tidal power has been advanced materially by a development programme in England.

The waterway is horizontal, or near horizontal, and is extremely short. The unit is characterized by the generator design; the rotor mounted as a rim on the peripheral ends of the runner blades forming an integral part of the runner. The rotor turns in a sealed annular recess in the water passage. The stator is mounted in the dry surrounding the rotor recess. Two sets of suitably shaped struts support the short runner shaft with wicket gates and runner blades.

The straight-flow generating unit is very similar to the bulb type in so far as hydraulic performance is concerned. It possesses the major advantage of further reduction in volume and cost of civil engineering works in the power house setting, but it is primarily adaptable to generation in one direction only, such as in single-effect single-basin and linked-basin schemes.

The tube-type turbine competes with the bulb and straight-flow types in the low-head field. A number of small run-of-river units have been built with others having runner diameters up to 26 ft. or more specified.

The unit is characterized by a water passage, more or less horizontal, with a gentle S-shaped double bend. The runner is essentially a double-regulated Kaplan-type wheel mounted on a shaft with an inclination of up to about 24°. The shaft runs from the water passage through one of the S-bends into the power house where the generator is located in the dry.

The tube turbine may not have quite as high a hydraulic efficiency as the types with which it competes, but the difference is not critical. It does permit substantial savings to be achieved in civil works and has the advantage of relative simplicity of design.

CORROSION

Because tidal power plants are located in a marine environment, corrosion assumes a far greater importance than in the case of conventional hydroelectric power plants operating on run-of-river and stored water. These corrosion aspects extend to every feature of the development: structural elements, sluiceways and their gates, turbines and other parts of the generating unit (as in the case of both bulb-type and straight-flow units); embedded parts; and contacts of control and relay equipment. Corrosion is particularly serious as a maintenance problem in connection with the embedded parts of the sluiceway gates, trashracks if used, emergency gates, and the turbines. Ordinarily special steels are used, such as stainless steels. However, the Rance tidal power plant near St. Malo across the Rance estuary is equipped with 24 turbines, 12 of which have stainless steel runners and the remainder aluminim-bronze runners. Operating experience with the two types is about on a par.

Many research workers have carried out long-term and extensive work designed to provide the best possible materials for various marine applications. The many aspects involved in corrosion have been carefully investigated, these ranging from the effects of temperature, humidity, orientation, shelter, elevation, variation and extent of exposure due to location in the splash zone or in the intertidal zone, the velocity of movement of sea water and the significance of pitting corrosion versus crevice corrosion. In addition, such aspects as selective corrosion, stress corrosion, cracking and biological effects have been investigated.

Before finalizing the design for the generating units used in the Rance tidal power development, a very extensive and intensive study was made of various alloys utilizing a test circuit containing two so-called material test units representing the bulb units on a scale of 1:18. One of the units was operated essentially as a turbine, the other as a pump.

As a result of this test program which was started in October, 1956, and carried out for several tens of thousands of hours, three promising alloys were selected for all further tests utilizing a 9,000 kW bulb-type unit installed in an unused lock in St. Malo Harbour. The full-scale tests on the St. Malo unit extended over a year and, indeed, partly over two years. These tests proved to be particularly valuable.

Much other valuable experience on sea-water corrosion has been gained from experience with pumps used for circulating cooling water from the sea through the condensers of thermal-electric power plants. Wide experience has also been gained with ships' propellers.

Many long-term corrosion investigations carried out by various organizations have produced extremely worthwhile basic data readily available.

In addition to the utilization of various alloys, other forms of protection are used for various purposes. One form consists of various types of protective coatings, normally used on carbon-steel surfaces for the purpose of reducing cathodic areas to a minimum. Electricité de France, for the Rance units, used special coatings on the thrust and guide bearings, governor oil tanks and other components in contact with hot oil; an epoxy paint able to withstand sea water attack and sustain the necessary polymerisation temperature. A final coat of anti-fouling paint with six undercoats of a special vinyl paint for surfaces in contact with sea water was found to be desirable.

The Rance generating-unit components on the basin side up to and including the stay ring were made of carbon-steel while all other components, except for the seaward side of the throat ring, were made of a stainless steel where in contact with sea water. A cathodic-protection system working at a density of 100 millamperes per square metre was utilized. The negative side of an external direct-current supply was connected to the metal components to be protected with the positive side to point anodes distributed over the unit. These point anodes were located:

1. Twelve on the sea side of the throat ring near the anchor ring.
2. Twelve on the outer distributor ring near the throat ring.
3. Eight on the tie-rod ring.
4. Eight on the access shaft.

The system operated with the metal elements to be protected at a potential of -800 millivolts measurable at two points on the stay ring, one at the tip of the upstream or seaward part of the throat ring, and one on the nose cap.

CONCRETE DETERIORATION

Many investigations have been made into the proper procedures to be followed in the design, production and placing of concrete in structures exposed to a severe marine environment involving continual wetting and drying, periodic freezing and thawing, and the battering action of the waves. Experience indicates that good aggregates, the proper cement in respect of type, quality and quantity, and the proper reinforcing steel must be used with a design providing for adequate cover of concrete over steel.

Much information is available in this area from actual Canadian experience in the Maritimes, in the United Kingdom, the U.S.A., and elsewhere.

The U.S. Army Engineers Waterways Experimental Station at Vicksburg, Miss., has carried out long-term concrete exposure tests at Treat Island, Maine, where tidal waters and weathering conditions are essentially analogous to those which would prevail at tidal power developments in the Bay of Fundy. Research work at Treat Island began when the concrete laboratory for the Passamaquoddy tidal power project was set up in 1936. Exposure tests have been carried out since that time providing some 32 yr. of continuing testing.

SITE SELECTION

An initial step in the overall investigation was the identification of sites offering an interesting potential for development as only one or two of the most favourable of the numerous sites which seemed to be feasible could be investigated in depth. No fewer than 23 sites appeared favourable, these being grouped in three main geographical areas; Chignecto Bay; Minas Basin; and Annapolis Basin-Saint Mary's Bay. Preliminary studies of sites within each area were undertaken by three general consulting-engineering firms retained for the purpose. Their work comprised the development of data relative to the best sites in their area of study for the purpose of further investigation, the preparation of preliminary designs and cost estimates, and consideration of advanced concepts.

Site evaluation was carried out in two stages. The initial selection was based on a parametrical analysis of each site considering some or all of the following factors: the length of dam required; the average depth of water at the site alignment; the approximate length of the power house; the size of the tidal basin; tidal ranges at the site; and energy capability at the site. Exposure to severe wave climates, adequacy of foundations, and sediment problems were also taken into account in the evaluation.

On the basis of the foregoing, 17 of the initial 23 sites were eliminated. The remaining six were subjected to further study and analysis during 1967, and provided a basis for the selection of three sites, 7.1, 7.2 and 8.1 for preliminary engineering design and production studies.

The majority of the 23 sites initially considered were eliminated on the basis of parametric analyses as observed. Many simplifications and approximations related to this type of analysis were made as a matter of necessity. Consequently, it was deemed desirable to review the validity of results deriving from the application of imperfect criteria for selection. This review confirmed the validity of the preliminary parametric studies.

COSTS OF TIDAL POWER PLANTS

Determination of the costs of tidal power plants involving complete feasibility-type estimates of the many aspects for no fewer than three sites warrants treatment in a paper by itself. The highlights have been given in a companion paper being presented at this Conference under the title "Economics of Tidal Power".

PRODUCTIVITY OF TIDAL POWER PLANTS

Scope

Productivity studies were carried out in depth for the following conditions:

1. Maximum-energy production, single-effect and double-effect, single-basin schemes.
2. Dependable-peak production during a specified period and maximum-energy production during the rest of the year for double-effect schemes; effect of changing values for day-time energy on production of dependable peak.

Less detailed studies were made for the following purposes:

3. Dependable-peak production during a specified period and maximum-energy production during the rest of the year for paired-basin schemes.
4. Dependable-peak production during a specified period and maximum-energy production during the rest of the year for a linked-basin scheme.
5. Effect of using single-regulated versus double-regulated turbines on maximum-energy production.
6. Dependable-peak production during a specified period from a single-basin single-effectscheme in combination with a pumped-storage facility.

The specified period determined after examination of possible market outlets was fixed at two hours per day, five days per week, for the three winter months of December, January and February.

The development of the programmes necessary for the computation of the productivity studies involved a very comprehensive background in hydraulics, power and higher mathematics. Consequently, only the highlights can be touched on in this paper.

Considerations Influencing Power Production

The computations necessary to determine the power production for a given tidal site and power plant are very complex. The complexity arises from such aspects as the continuously varying tides in a lunar day and throughout the lunar month, the operating mode, and many others. As a result, it is necessary to resort to powerful digital computers. There are many specific aspects which must be considered in power production studies and consequently programmed into the digital computers.

These aspects are quite unlike those associated with run-of-river hydroelectric developments where it is usually not difficult to determine the head, the discharge and the required spillway capacity. In the case of a tidal power development, the head varies continuously and depends upon the tidal amplitude, the basin area, the capacity of the sluiceway used to fill or empty the basin, the capacity of the generating units, and the cycle of operation followed.

The installed capacity is not fixed by the available flow but by economic considerations. The output of the generating units varies continuously from nil to a maximum obtained for a relatively short time in a year. An economic study is needed to choose the generator rating. A similar situation applies to the sluiceway capacity. This is chosen to allow proper filling and emptying of the controlled basin or basins for optimization of energy production rather than the capacity necessary to handle a given maximum discharge.

Effective consideration of the production aspects associated with a tidal power plant involves a systematic variation of all pertinent parameters in order that the optimum generating and sluiceway capacities and the cycle of operation can be determined.

Two groups of parameters must be considered. The first comprises the relationship between basin area and water level, tidal range, and period. The second relates to the energy drawn from the generating unit consisting of the turbine and the generator, the characteristics of the turbine being normally shown on a hill chart depicting the characteristics of a specific design. A typical hill chart is illustrated by Fig. 17.

A generating unit is defined by the following characteristics:-

(a) Turbine runner diameter D
(b) Speed of rotation n
(c) Generator limiting capacity (maximum permissible output) N

A tidal power plant is defined by the generating-unit dimensions and the number, K, of units installed. The four variables are thus:

$$D, n, N, K.$$

In addition, the capacity, V, of the sluiceway has to be adjusted to the capacity of the turbines to allow proper filling and emptying of the basin. Various values of these parameters will yield different values of energy production for a given site and a given tidal range. The period of the tide, being much the same from one site to another in the Bay of Fundy, does not affect comparisons of production.

The field of variation of the five parameters listed determines this optimum value. Three values of each characteristic are chosen. The minimum and maximum values are assumed such as to encompass the optimum value.

For the purpose of productivity calculations, the four parameters above can be combined and reduced to three, namely:

(a) KD^2 proportional to the total turbine throat area and thus to the turbine discharge capacity;
(b) nD proportional to the peripheral speed;
(c) N/D^2 proportional to the unit limiting capacity.

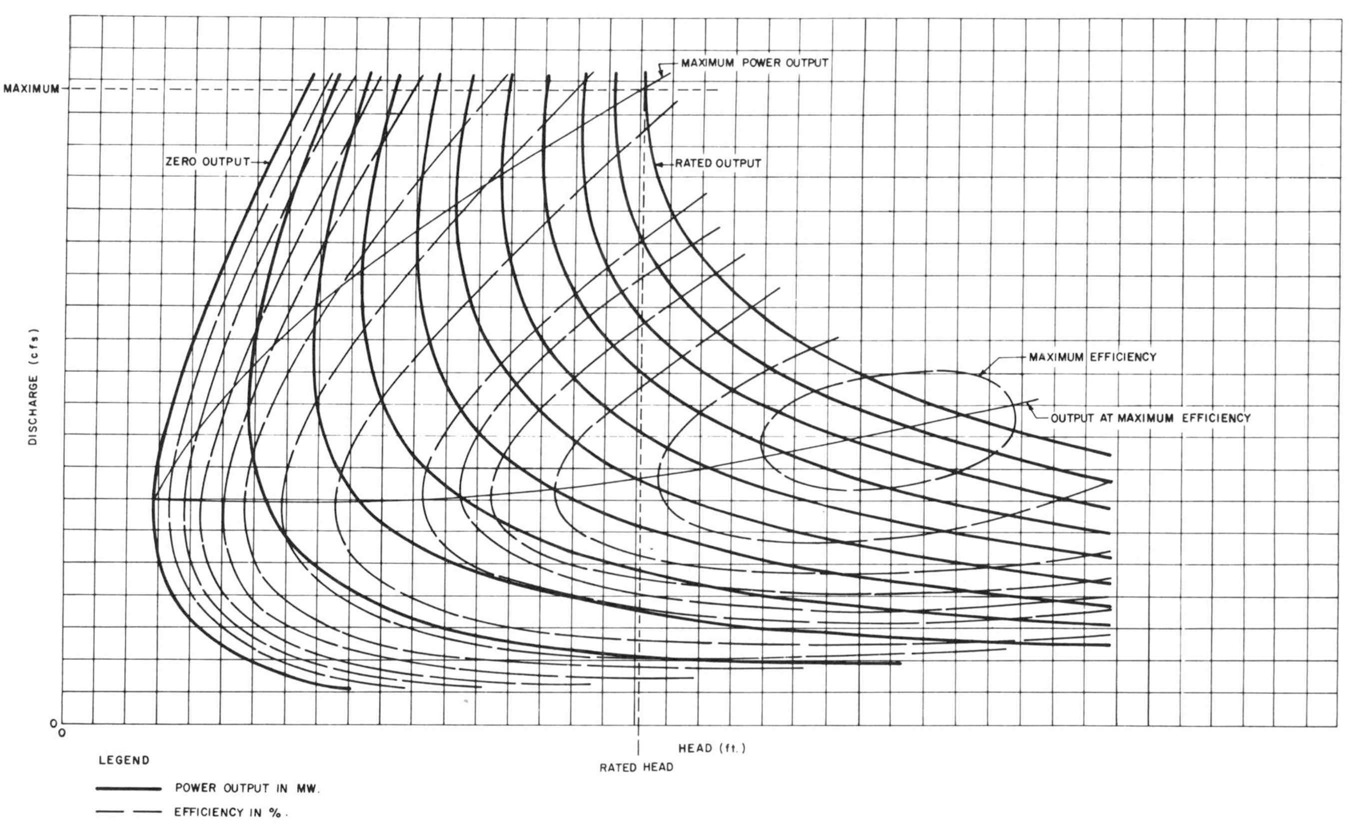

Fig. 17. Typical hill chart.

An important key in the development of tidal power is the use of axial-flow horizontal units able to operate as turbines, pumps or sluices in one or both directions of flow from basin to sea or vice versa.

The bulb unit, though not the only type of axial-flow machine having these characteristics, is the most widely used. Many bulb-type units such as would be required for the Bay of Fundy tidal power plants are now in service. They have been used successfully in the Rance tidal power station for several years and in many run-of-river plants. The characteristics and hill curves for bulb-type units, analogous to those in the Rance development, were utilized in Bay of Fundy studies.

The hydraulic characteristics and efficiencies of other types of axial-flow machines (such as tube units and units with the rotor of the generator surrounding and secured to the turbine runner under consideration or development by several manufacturers) are similar to those of the bulb unit; the results reported herein can be considered equally valid for these other types.

In addition to the foregoing, the elevation of the runner must be established sufficiently low to avoid cavitation. The choice of the elevation at which the generating units must be set is based on cavitation requirements. Two critical values for the cavitation sigma were considered:-

1. Sigma for the minimum (i.e. rated) head ensuring the nominal capacity of the generator.
2. Sigma for the maximum head that can occur at the tidal power plant.

The definition of σ is

$$\frac{H_a - H_s - T_v}{H}$$

where H_a = atmospheric pressure

H_s = runner setting with reference to the minimum downstream water level (as a first approximation, the chart datum)

T_v = water vapour pressure

H = head on turbine.

For bulb units such as considered critical values of σ are in the vicinity of three in the first case, 1.5 to 1.6 in the second. Calculations show that usually the first case is the controlling one, and give the setting C to be adopted as a first approximation for the various sites. C is the elevation of the highest point of the turbine runner with reference to GSCD. Where such a reference was not available, chart datum was taken.

Sluiceway Capacity

Simplifying assumptions in connection with the selection of the sluiceway characteristics had to be made. It was known, from previous studies, that because of the continually changing water levels, a sluiceway comprising crest gates would usually not provide an economical solution and that probably the structure would utilize submerged gates.

Normal operation would be of the submerged type even if in some phases of operation the jet from the sluice was not submerged. As a first approximation, sufficiently precise for the purpose, it could be assumed that the discharge varies as the square root of the head available across the plant:

$$A = Ks\sqrt{2gh}$$

where K is a coefficient depending on the shape of the structure, most probably having a value between 0.8 and 1.2, s is the total cross-section of the sluices, g is the acceleration due to gravity and h is the available head. The capacity of the sluiceway can be adequately specified by the number of cubic feet per second (cfs) it can discharge under a head of one foot.

By analogy with results obtained from earlier tidal development studies, the mean capacity of the sluiceway was determined for the mean degree of equipment

$$\frac{KD^2}{S} = 50 \times 10^{-6}$$

where S is the area of the basin. Larger and smaller values were added to ensure optimization. For larger and smaller degrees of equipment, the capacity of the sluiceway was adjusted for each site taking into account the variation of the capacity of the units when operated as orifices.

Production of Maximum Energy

A family of curves can be plotted giving the variation in energy production, E, one of the three parameters KD^2, nD and N/D^2 being allowed to vary, while the other two remain constant.

This systematic approach to the productivity problem permits the results to be substantially generalized. Optimum conditions can be selected after the whole range of possibilities has been analyzed. In this way, the production of any plant design corresponding to any desired combination of parameters can be determined.

In initial studies, the optimization procedure assumed tidal variations of a prescribed amplitude, the tidal cycle being repetitive in each case. In the first step for each site, the distribution of frequency of the 705 tidal ranges occuring over one year was computed. The yearly variation of energy production per tide could be computed as a function of the tidal range at each site.

Experience indicated that yearly energy production would correspond closely to the production obtained during one tidal cycle with an amplitude equal to the root-mean-square (rms) average of all the tides in a year multiplied by 705. The value of the rms tidal range was computed for each site. Given the tidal range, the problem was to define the shape of the tidal curve at each site. The tidal period was assumed to be 12 hr. 24 min. in all cases.

In later studies, the optimization procedure as well as the dependable power studies were carried out for specified sequences for real tides. Not only was the shape of the tidal curve involved, but also the succession of high and low-water levels and the slight variation in the period of the tide.

The shape of the tidal curves used was based wherever possible on recordings made in situ. It was desirable to select the same recorded tide for this purpose at all locations. Actually the tide for the morning of August 15, 1965 was chosen.

Productivity studies carried out in the initial work were based on a computer programme capable of determining the maximum energy generation from potential tidal power developments. This programme was operated to determine the maximum production of selected tidal power development sites being carried out for various installed capacities in such a way as to determine the most suitable numbers of generating units and sluiceway capacities for each site studied.

The productivity study used during initial work was designed to compute the maximum yearly energy output with either single or double-effect generation with pumping if advantageous. In view of the complexity and length of the calculations required, the study was carried out using an optimization programme run on an IBM 360-50 computer. The programme was designed to select the cycle of operation that would give the maximum energy production over one tidal cycle. The programme operated in the following manner.

For single-effect operation (generating from the basin to the sea), it has been demonstrated that the starting point of operation must be such that the unit must generate with the discharge corresponding to maximum efficiency under the head. The starting point is defined if the starting level is known. It has also been proven that the generation must be continued to the point at which the available head has decreased to the minimum head under which it is possible to insure the no-load synchronous speed of the unit. Once the level of the tidal basin at the beginning of the generating period is fixed, generating operation is determined. For repetitive operation, the generating phase must be followed by a sluicing and sometimes a pumping operation to refill the basin to the starting level.

The programme automatically computed the total output for a set of possible operations for seven arbitrarily chosen starting levels in the tidal basin, and determined the best operating cycle for each of the seven starting levels. An interpolation procedure was then used on the seven results to determine the correct starting level. The ultimate mode of operation was found by repeating the procedure used for each of the original seven levels.

The programme, much more complex than a brief description may suggest, determines the optimum cycle by taking into account the following input data:

1. Amplitude and shape of the tidal cycle.
2. Area of the basin and its variation with water level elevation.
3. Losses of head in the basin (backwater effect).
4. Hill-chart curves for the turbines with requisite extensions to the standard chart to allow for pumping and reversed flow.
5. Laws of discharge for the units when used as orifices in both directions.
6. Runner diameter, rotational speed and generator rating.
7. Number of units.
8. Time needed to start one unit.
9. Discharge capacity of the sluiceway in each direction, time needed to open the gates, maximum head under which the gates can be opened.
10. Possible restrictions on operating water levels (allowable maxima and minima).
11. Maximum increase of the rate of discharge (to avoid the production of a hydraulic bore in the basin for instance).

The complexity of the programme is increased by the wide variety of operating conditions necessarily envisaged, e.g. operation as a turbine (or a pump) at the maximum permissible power, operation of the units as orifices with or without sluicegates, operation of the sluiceway alone, operation in the starting-up or shutting-down phases, straight-out turbine operation under constantly varying head, discharge and power, etc. In all there are more than 15 possible modes of operation to be programmed.

To cover adequately the range through which the optimum installation would be selected, 81 different combinations of equipment would have to be analyzed for each site (three rotational speeds, three generator ratings, three numbers of units, each one with three sluiceway capacities). For each installation, calculations would have to be done for single-effect operation (turbining from the basin to the sea and pumping, if justified, from the sea to the basin) and double-effect operation (turbining and pumping both ways) for various tidal amplitudes to allow the integration of yearly energy production. A large number of cases would have to be run through the computer, a long and costly process. Obviously, some simplification had to be found to minimize the computer work.

It was known that the yearly energy production practically corresponds to the production obtained during one tidal cycle with a tidal amplitude equal to the rms average of all the tides in one year multiplied by 705 (the number of tides during the year). A check made for Sites 7.1 and 8.1 established that the yearly energy derived from the full spectrum of the tidal range is equal to 705 times the energy from the rms tide within plus or minus three percent.

The natural timing of the tides does not correspond exactly to the solar cycle and, hence, human demands for energy and power. In the foregoing studies of energy production, it is worth noting that the concept of single-basin single-effect generation provides generation of energy only, strictly in accordance with the natural timing of the tides. However, a double-effect concept of development using essentially the same civil works but somewhat more sophisticated generating equipment enables not only securing dependable peak output at-site, but also permits generation to be concentrated in the interval from 0600 hr. to midnight, with less than one percent of the annual energy production occurring in the period from midnight to 0600 hr.

Table 3 provides a comparison of single-effect generation versus double-effect generation for Sites 7.1, 7.2 and 8.1.

Table 3
Comparison between Single-Effect and Double-Effect Tidal Plants

Site		7.1	7.2	8.1
Single Effect				
Number of Generating Units		60	36	64
Installed Capacity	MW	1,620	972	2,176
Number of Sluiceways		48	29	51
Sluiceway Capacity Under One Ft. Head	Million CFS	0.970	0.580	1.103
Annual Energy	Million kWh	4,200	2,690	6,500
Double Effect				
Number of Generating Units		108	88	104
Installed Capacity	MW	2,916	2,376	3,536
Number of Sluiceways		95	72	88
Sluiceway Capacity Under One Ft. Head	Million CFS	1.931	1.480	1.790
Dependable Peak Power	MW	1,260	1,022	1.526
Annual Energy	Million kWh	5,402	3,885	7,560

Production of Dependable Peak

An important element in the studies of economic attractiveness of tidal power was the determination of the values of dependable at-site peak power and energy available for potential power markets. To proceed with studies of the problem, the dependable peak and energy capability of a tidal power plant had to be ascertained. Thus, the objective of these studies was the analysis of the peaking possibilities and the optimum operation of the three sites of interest, Sites 7.1, 7.2 and 8.1, taking into account the fact that the value of energy varies according to the time of the day at which it is produced.

Two different types of operation were studied:

1. For at-site dependable-peak power from 1600 to 1800 hr. five days of the week for 60 days during the winter months of December, January and February.
2. For maximum return using a variable value for the at-site energy according to the time of day it is available.

In addition, as Shepody Bay and Cumberland Basin lend themselves to possible development as linked basins, it was decided to study an example of such a possibility. The dependable at-site peak and yearly energy production possibilities of the above and of Sites 7.1 and 7.2 as paired basins, respectively, could thus be compared with their operation as single-basin schemes.

Initial studies were concerned with optimization of energy output for each of the power plants examined. No consideration was paid to the changing value of the energy produced throughout the day. These studies were carried out to compare the respective merits of the various sites analyzed so as to select the most attractive ones for closer scrutiny in subsequent work. For this purpose, neglect of variations in energy value throughout the day was considered of no consequence.

In the later work, the study of energy production had to take into account variations in the value of energy produced by the units operating as turbines or consumed during pumping as a function of the time of day.

A problem that arose at the outset in the later studies was the selection of the tides or tidal sequences. Several possible approaches were examined. These studies resulted in taking sequences of real tides computed in the way outlined making no allowance at this stage for possible reductions in tidal range due to the presence of the power plant.

It was not feasible on account of the length of the computations to analyze the 705 tides in a complete year. Another problem then arose: the period or periods of time selected for analysis had to be representative of the average tidal conditions as a whole, year in and year out.

For this purpose, an analysis of average tidal ranges versus the time of high tide for Site 8.1, mentioned previously, was used as a basis. It was carried out for the year 1967 and the beginning of 1968, i.e. terminating with the period used for the dependable-peak studies. The importance of the time of occurrence of high tide is emphasized by taking the lunar month as a basis for subdivision of the year, the start of any such month being timed to coincide with the occurrence of high tide at 1500 hr.

A study showed that complete lunar months rather than lunar fortnights should be analyzed and that the most representative lunar months in 1967 were the fourth and 10th. These two lunar months were, consequently, selected for the detailed computations of operation for maximum return. The analysis in question was based on a month-to-month comparison of tidal ranges as a function of the time of high water.

Knowing the total production for these two months, a rule of simple proportion gives the annual figures. Although it is true that the resulting figures relate to the calendar year 1967, there is no doubt that they define average conditions for any year very accurately. The situation in respect of year-round energy output is quite different from that in respect of dependable-peak power. In the latter case, the occasionally very small tides that do occur (or very high tides occurring at a critical time of day) are of prime importance in fixing the dependable-peak output. As far as year-round production goes, any one year is as good as another; the occasional extremes of tidal ranges that may occur in one year and not in another are considered insignificant.

The computer analysis was carried out for Site 8.1 as a single-basin scheme with 110 units and a sluiceway discharge capacity of two million cfs under a head of one foot. Simple proportion was then applied (using the results obtained with the maximum energy studies adjusted to the actual conditions of basin area, etc.) to determine the expected annual output for other single-basin sites or for other generating unit capacities at Site 8.1.

Numerical data used in the later computer studies were:

Size of generating units and sluices for Site 8.1

Number of generating units	110
Runner diameter	24.6 ft.
Rated alternator power	34 MW
Speed of rotation	81.8 rpm
Sluiceway discharge capacity	2 million cfs under one foot head.

Turbines and Gates: Operating Features

Time taken to start or stop	7 sec./unit or 770 sec. in all
Total time taken to operate sluiceway gates	600 sec.
Total time required to set the units running as orifices	1500 sec.
Discharge capacity of units as orifices	
(a) Direct flow from basin to sea	6890 cfs per unit under one foot head
(b) Inverse flow from sea to basin	5360 cfs per unit under one foot head
Maximum head for pumping	19.8 ft.
Maximum head under which the gate servomotors can be operated	21.3 ft.
Maximum allowable rate of change of discharge through power station and sluiceway (so as to avoid excessive disturbance in basin water level)	2100 cfs/sec.

Basin

Maximum allowable level + 30.0 ft. GSCD
Minimum allowable level - 29.0 ft. GSCD

The head loss (backwater effect) in the basin is assumed to vary as the square of the discharge and to have the same value, though reversed in sign, for outflows and inflows of equal value. Typical values for a discharge of two million cfs are given below as a function of water level:

Mean water level in basin (in ft. above GSCD)	Mean basin level minus level at power plant on basin side (absolute value in ft.)
-25	2.55
-20	2.02
-10	1.14
0	0.51
+10	0.13
+20	Negligible

The object of the dependable-peak studies was to determine the maximum power generation which could be guaranteed between 1600 and 1800 hr. on weekdays, the plant being operated to achieve this. More specifically, a mode of operation of the tidal plant was sought which, firstly, would give the highest guaranteed output and, secondly, would result in a peak-power duration curve with the largest possible area.

To ascertain the dependable-peak production, it was necessary to calculate this for numerous combinations of tidal amplitudes and times of occurrence of high tide. Fig. 18 shows the time of high tide has far more influence on the peak-power production than the tidal amplitude. Fig. 18 shows schematically the minimum power produced during the peaking period as a function of the time of occurrence of the high tide preceding or following the peaking period.

Two basic approaches were possible in establishing a complete set of combinations of tidal amplitudes and times of high tide. The first approach would have consisted in studying all possible combinations of tidal ranges and hours of high tide. Thus, a total of about 60 cycles would have had to be examined: 23 to 27 cycles when generating from the basin to the sea (direct operation) and 29 to 33 cycles when generating from the sea to the basin (indirect operation). From these results and the statistical distribution of tidal ranges and the time of high tide, energy and peak-power production could be deduced for the winter season.

This method, because of its statistical basis, does not take into account the chronological succession of tides and therefore, gives no insight into the difficult process of filling or emptying the basin in order to shift from direct to inverse operation or vice versa. It is then that the production of peak

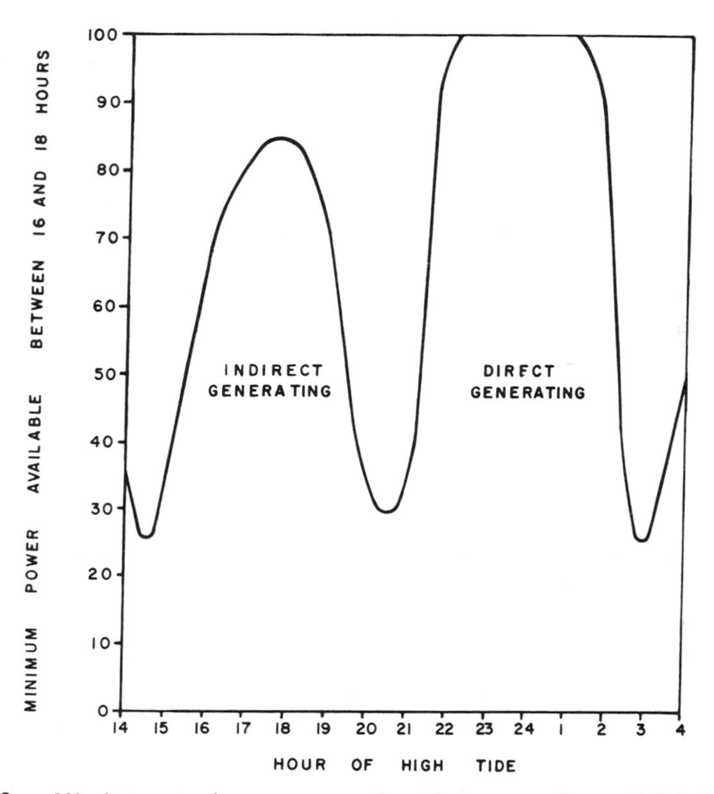

Fig. 18. Minimum peak power production vs. time of high tide.

power drops to the minimum value. This is one of the main reasons why this method was rejected.

The second or calendar approach which was finally utilized consisted in studying the peak-power production for the worst winter season occurring over a period of 18 yr.; this corresponding to the so-called Chaldean period after which the tidal pattern is repeated.

Statistical analysis showed the winter season of 1967-68 to have been the "worst winter", i.e. the one with the smallest tidal amplitude in the 18 yr. period from 1951 to 1968. To better cover the worst conditions possibly occurring in the future, the week ending Jan. 21, 1961, was added to the studies. It was thought that this week, as it shows the largest tides recorded over 18 yr. and occurs at the most difficult time for power production from 1600 to 1800 hr., made it particularly difficult to switch from direct to indirect operation or vice versa, a fact certainly worth investigating.

Peak production was thus simulated for three consecutive months of actual tides. This procedure enabled numerous combinations of tidal amplitudes and hours of high tide to be covered. It also took account of the natural succession of tidal amplitudes. During the calculations, this last factor was seen to have a significant effect on the final stage of emptying or filling of the reservoir during change over from direct to inverse operation or vice versa.

The choice of such a period represents very severe operating conditions and results in a minimum peak-power production having a probability of occurrence of the order of four to five times per century.

The principal restrictions imposed by the interconnected system into which the tidal-power plants would be linked on the operation of a power plant as a peaking station were assumed to be as follows:

1. The energy production of the plant could be absorbed from 0600 to 2300 hr. only;
2. For pumping purposes, power can be drawn from the network at half the installed capacity of the plant from 0700 to 0900 hr. and from 2100 to 2300 hr. Pumping at the full installed capacity is permissible from 2300 to 0700 hr.

For dependable-peak production, the operation of the units follows on the hill chart a path along the curve of maximum output as the head increases until the limiting capacity of the generators is reached and subsequently follows the curve of constant output. When the head decreases, the same path is followed in the reverse direction.

The units are operated as pumps according to a curve on the hill chart which gives the maximum discharge for a given head if sufficient power is available.

The limiting head of 21.3 ft. under which the gates can be opened or closed is the only limitation taken into account in the operation of the sluiceway. This limitation affected the production of peak power in a few cases of

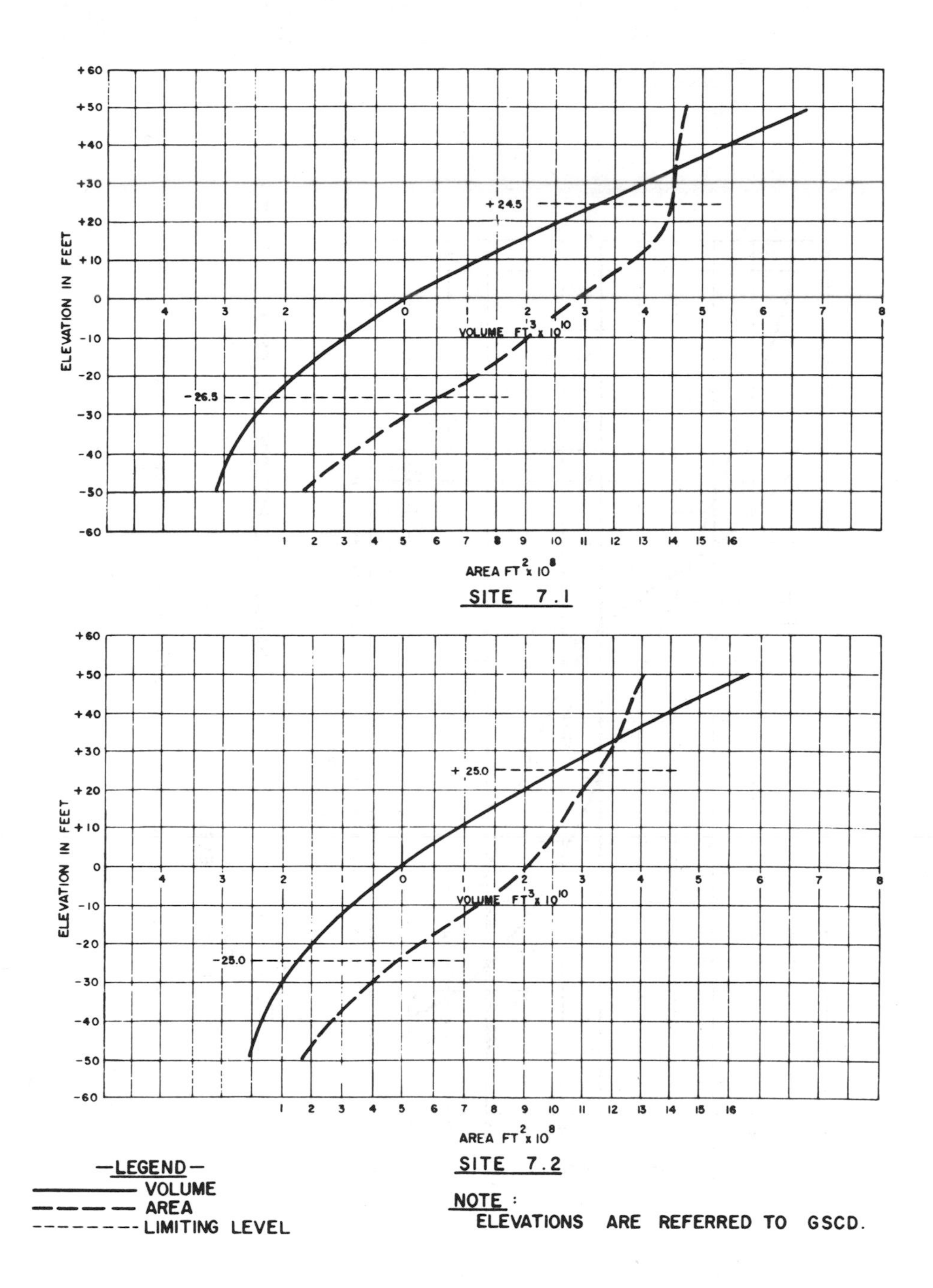

Fig. 19. Basin capacity and area for sites 7.1 and 7.2.

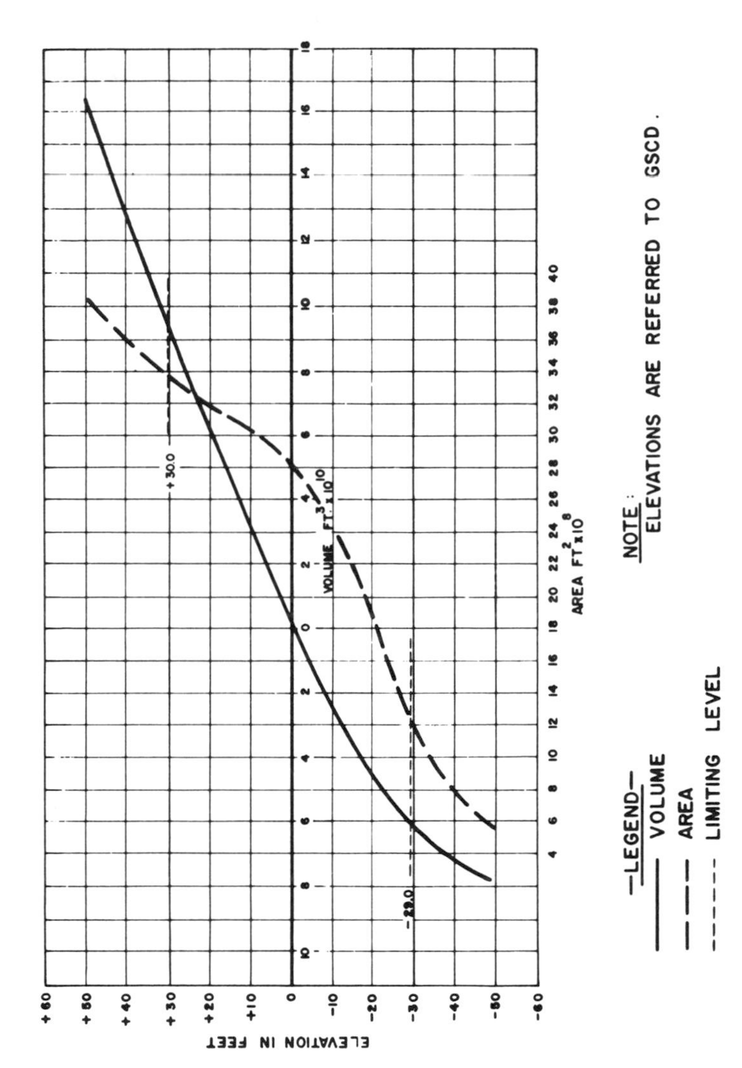

Fig. 20. Basin capacity and area for site 8.1.

complete filling or emptying of the basin in a relatively insignificant manner. In these calculations rather extreme modes of operation were simulated; these might prove objectionable because of the waves and rates of water level variations caused in the basin.

The basin area curves used in these calculations are shown by Figs. 19 and 20.

The sites, the power and sluiceway capacities, and the periods of time for which the power outputs produced between 1600 and 1800 hr. on weekdays were calculated, are tabulated in Table 4.

After the first calculations for Site 8.1 were analyzed, it was found that in order to establish a good basis of comparison between the installations at different sites it was sufficient to calculate the peak power for the synthetic lunar month of January, 1968 (1961)*, only; this gives an average figure for the three months previously considered.

Since the principal objective in the computation of peak power was the production of the highest possible power between 1600 and 1800 hr. on weekdays, this goal was achieved at the cost of a very low overall energy production if not of a deficit in the energy output owing to the consumption of the pumping energy needed to reach suitable water levels. The emptying or filling of the reservoir during the changeover from direct to inverse turbine operation (or vice versa) is always an energy-consuming operation which may take more than a day in cases of an installation with low sluiceway capacity.

Results of calculations are presented on Fig. 21, which shows the guaranteed peak power and gives the number of days as a percentage during which a given output is maintained or exceeded between 1600 and 1800 hr. for Sites 7.1, 7.2 and 8.1 respectively.

Effect of Assigned Energy Values on Dependable-Peak Production

Results determined for a plant equipped with 110 units of 34 MW each at Site 8.1 corresponding to the maximum return for the specified energy-value curve obtained by optimizing over a period of two lunar months in 1967 indicated that the peak-period power output between 1600 and 1800 hr. is frequently zero. In fact this occurs about 37% of the time. An output of 1000 MW corresponding to about 27% of the installed capacity of 3740 MW was reached only 50% of the time.

Consequently, an effort was made to increase the power output between 1600 and 1800 hr. by attributing a much higher relative value to this output in the optimization procedure. This procedure does improve the unfavourable situation mentioned above, at the expense of the overall output considered on a 24 hr. per day basis.

---

*See Table 4 .

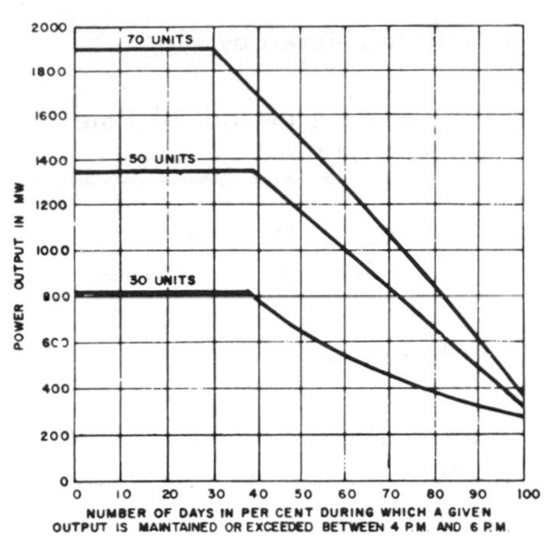

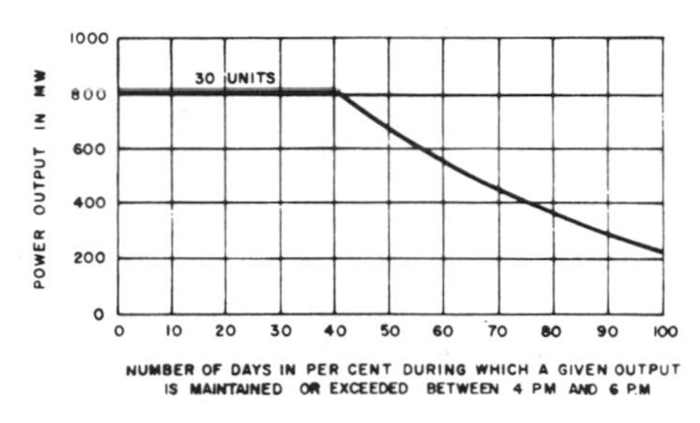

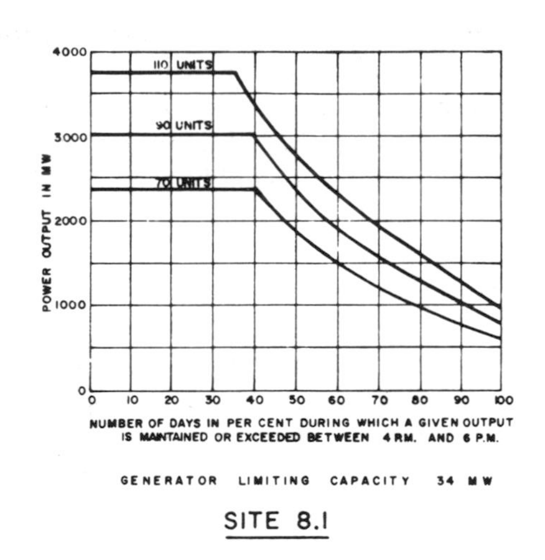

NOTE:
DERIVED FROM CALCULATION FOR THE LUNAR MONTH OF JANUARY 1968-61.

Fig. 21. Duration curves of dependable-peak power per-cent-of-days basis for sites 7.1, 7.2, and 8.1.

Table 4

Sites, Power and Sluiceway Capacities and Periods Studies for Dependable Peak

Site	No. Units, K	Generator Ratings, MW	Sluiceway Capacity V Million CFS Under One Foot Head	Peak Capacity, (95%) MW	Dec. 1967	Jan. 1968	Feb 1968	Jan. 1968–1961
7.1	110	27	1.97	1,307				X
	90		1.56	1,069				X
	70		1.17	832				X
	50		0.80	590				X
	30		0.50	356				X
7.2	102	27	1.98	1,212				X
	90		1.54	1,069				X
	72		1.19	855				X
	30		0.50	356				X
8.1	110	34	2.0	1,646	X	X	X	X
	90		1.46	1,346				X
	70		1.12	1,047	X	X	X	X

Note: – The lunar month of January 1968–61 is composed of 3 weeks of January 1968 (from January 7 to 27, 1968) and one week of January 1961 (from January 15 to 21, 1961).

The analysis was carried out for the 15 day period commencing at 0000 hr. March 29, and ending at 2400 hr. on April 12, 1967, the first lunar fortnight referred to in the second paragraph above. The exact relative values attributed to power produced or consumed are shown by Table 5.

Table 5
Site 8.1, Single-Basin Double-Effect Scheme
Relative Values of Energy as Used for Scheduling of Generation and Pumping to Achieve Maximum Revenue for Energy Production

Hours of Day		0-3	3-6	6-8	8-10	10-11	11-13	13-16	16-18	18-20	20-24
Gener.- ating	VC 1	0	0	31	100	100	140	100	140	100	31
	VC 2	0	0	31	100	100	140	100	400	100	31
	VC 3	0	0	31	92	92	140	92	800	92	31
Pump- ing	VC 1	70	70	140	460					460	140
	VC 2	70	70	140	460		PROHIBITED			460	140
	VC 3	62	78	140	460					460	140

The overall totals attained for three energy-value cases were as follows:
VC1: 345.188 million kWh
VC2: 294.085 million kWh
VC3: 179.669 million kWh

A feature of much interest is the distribution of this production throughout the day. This has been evaluated for the entire fortnight for each of the three cases studied, the cut-off points being chosen to coincide with changes in the value for power produced as shown by Table 6. The negative values in the second column indicate that, on balance, more pumping than generating was carried out for the two energy value curves concerned.

Table 6
Percentage of Total Output for Single-Basin Double-Effect Scheme at Site 8.1 for Specific Hours, Maximum - Revenue Operation

Hour	0-6	6-8	8-11	11-13	13-16	16-18	18-20	20-24
VC 1	0.1	12.6	21.2	11.0	13.4	13.0	13.7	15.0
VC 2	-0.8	18.7	18.3	6.8	11.6	20.0	14.9	10.5
VC 3	-9.4	9.7	8.7	0.4	12.0	42.8	25.8	10.0

The results for the peak period confirmed what was to be expected: namely, the production within the two hours concerned increased steadily as the relative values for energy produced increased from 140 to 400 (for VC2), subsequently to 800 (for VC3), the precise figures being:

VC1: 44.180 million kWh between 1600 and 1800 hr.
VC2: 58.807 million kWh between 1600 and 1800 hr.
VC3: 77.002 million kWh between 1600 and 1800 hr.

The degree of firming up in the peak power due to the revised energy-value curves is of even greater interest. The results computed for the 30 hr. in the peak periods over the 15 days studied are shown by Fig. 22. The same illustration shows the duration curve for dependable power for 110 units of 32 MW each at Site 8.1 for the synthetic month of January 1968/61. Salient features of this illustration are noted in Table 7.

Table 7
Effect of Assigned Value Curve on Production of Dependable Peak at Site 8.1 Operated as a Single-Basin Double-Effect Scheme

Value Curve		Power Generation, in Percentage of Time Between 1600 and 1800 Hr.. During which a Given Output is Reached or Exceeded			
	3,740 MW	3,000 MW	2,000 MW	1,000 MW	NIL
VC 1	9	24	38	50	36
VC 2	15	31	53	69	17
VC 3	22.5	42	72	91	0

For the VC3 energy-value curve, some power is always produced between 1600 and 1800 hr. Naturally this result has been achieved at the expense of power production at other times. As already noted, the overall output for VC3 is about 52% only of that for VC1 (179.669 million kWh as against 345.188 million kWh).

It may be asked whether the above results obtained for one lunar fortnight only would be modified very much should a longer period be taken. The very considerable amount of computer time involved precluded a more extensive study of the two modified energy-value curves, VC2 and VC3, from being undertaken. However, the results relating to VC1 in Table 8 may be of interest; they relate to the foregoing percentages of total output occuring between stated hours of the day.

Dependable-Peak and Energy Production from a Paired-Basin Scheme

If two or more tidal plants are interconnected to the same electrical network, the best utilization of these plants may differ from the operation of a single basin. Such electrically interconnected plants are defined as paired basins.

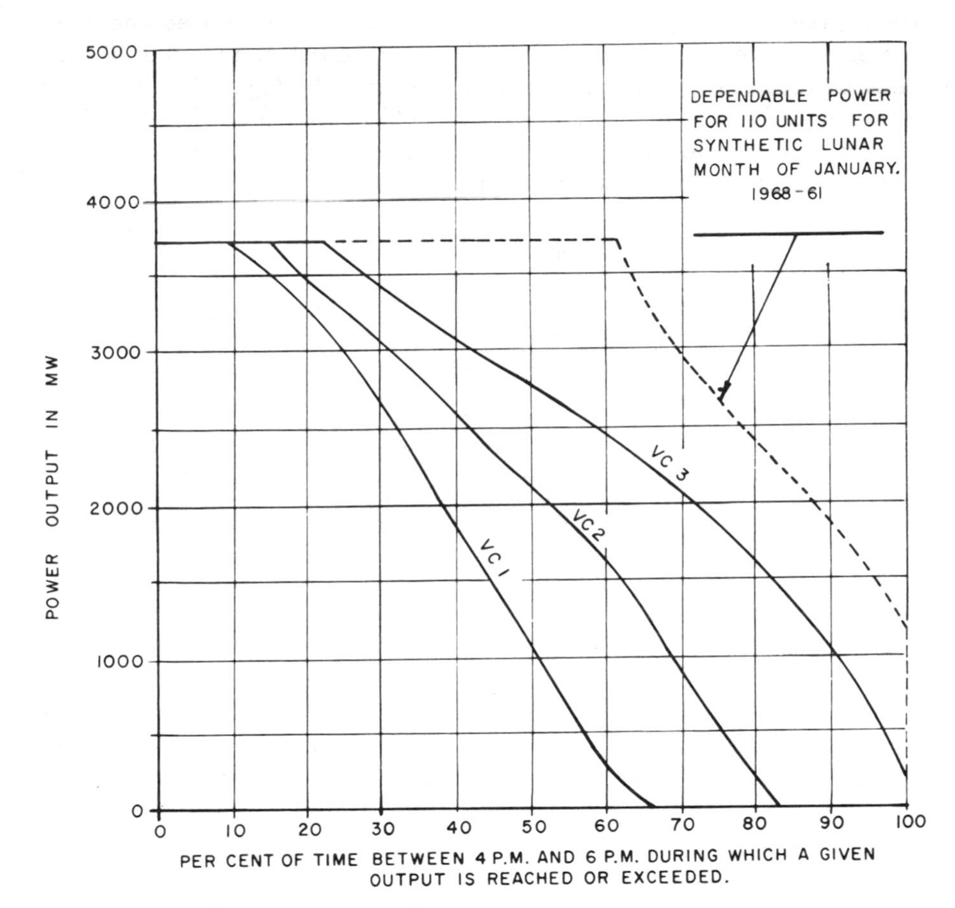

FOR FORTNIGHT CONCERNED (15 x 24 HRS.) TOTAL OUTPUT FOR
"VC1" = 345.188 MILLION KWH.
"VC2" = 294.085 " "
"VC3" = 179.669 " "
INSTALLED CAPACITY: 110 GENERATING UNITS
GENERATOR LIMITING CAPACITY 34 MW.

Fig. 22. Duration curves of dependable-peak power for four different operating modes, site 8.1.

Table 8
Effect of Duration of Production Period
on Percentage of Total Output
for Single-Basin Double-Effect Scheme
at Site 8.1 for Specific Hours

Hours	0-6	6-8	8-11	11-13	13-16	16-18	18-20	20-24
First lunar fortnight only	0.1	12.6	21.2	11.0	13.4	13.0	13.7	15.0
Two lunar months	0.8	10.9	21.3	11.8	14.8	11.6	12.5	16.3

Paired basins can allow a continuous production of energy during the day by the judicious choice of water levels in the basins. Such an operation achieved at the cost of a loss of return on energy is unlikely to be desirable in a large network.

Operation of Sites 7.1 and 7.2, for instance, as paired basins gives an operation for maximum energy production in no way different than if the sites were operated separately, other conditions of load demand being the same. The maximum energy production is simply the sum of the productions of both plants.

Actually, the main advantage of paired basins over an equivalent single-basin is in dependable-peak power operation.

Results of peak-power operation of single-basin schemes have shown that the minimum power production always occurs at either the beginning or the end but never in the middle of the peak period. On the most difficult days, direct or inverse operation produced a roughly symmetrical diagram. Therefore, the operation of two plants in a paired-basin scheme, with one plant generating from the basin to the sea, the other generating from the sea to the basin, produces a more regular diagram of peak power.

Calculations were made for the lunar month of January 1968-61 and for an installed capacity of thirty 27 MW units at each of Sites 7.1 and 7.2. For each site the sluiceway capacity was 0.5 million cfs under one foot of head. These show that such a scheme of development provides great flexibility in producing dependable peak.

Figs. 23 and 24 show the duration curves for the selected operation.

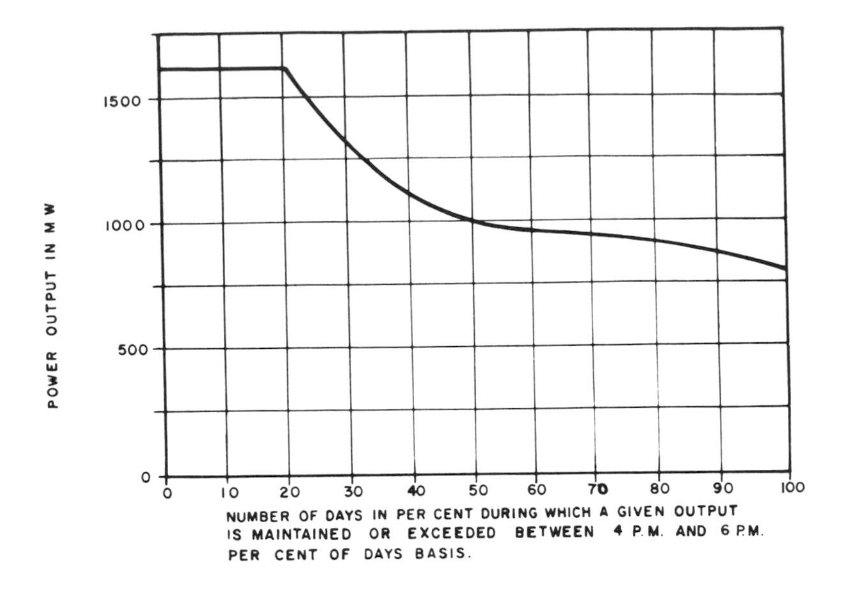

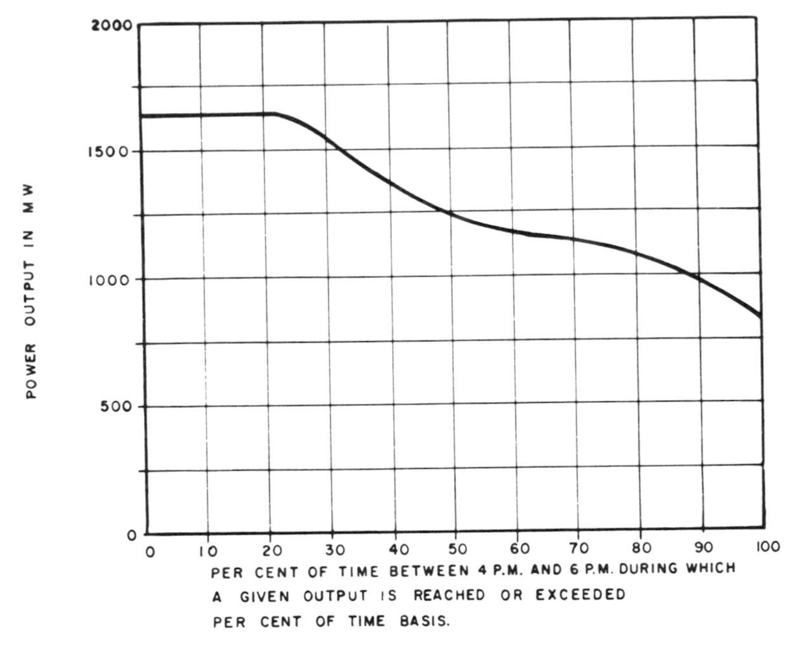

DERIVED FROM CALCULATION FOR THE LUNAR MONTH OF JANUARY 1968-61
INSTALLED CAPACITIES: (NOT AN OPTIMUM DEVELOPMENT)
SITE 7.1; 30 GENERATING UNITS.
SITE 7.2; 30 GENERATING UNITS.
GENERATOR LIMITING CAPACITY 27 MW.

Fig. 23. Duration curves of dependable-peak power for sites 7.1 and 7.2 operated as a paired-basin scheme.

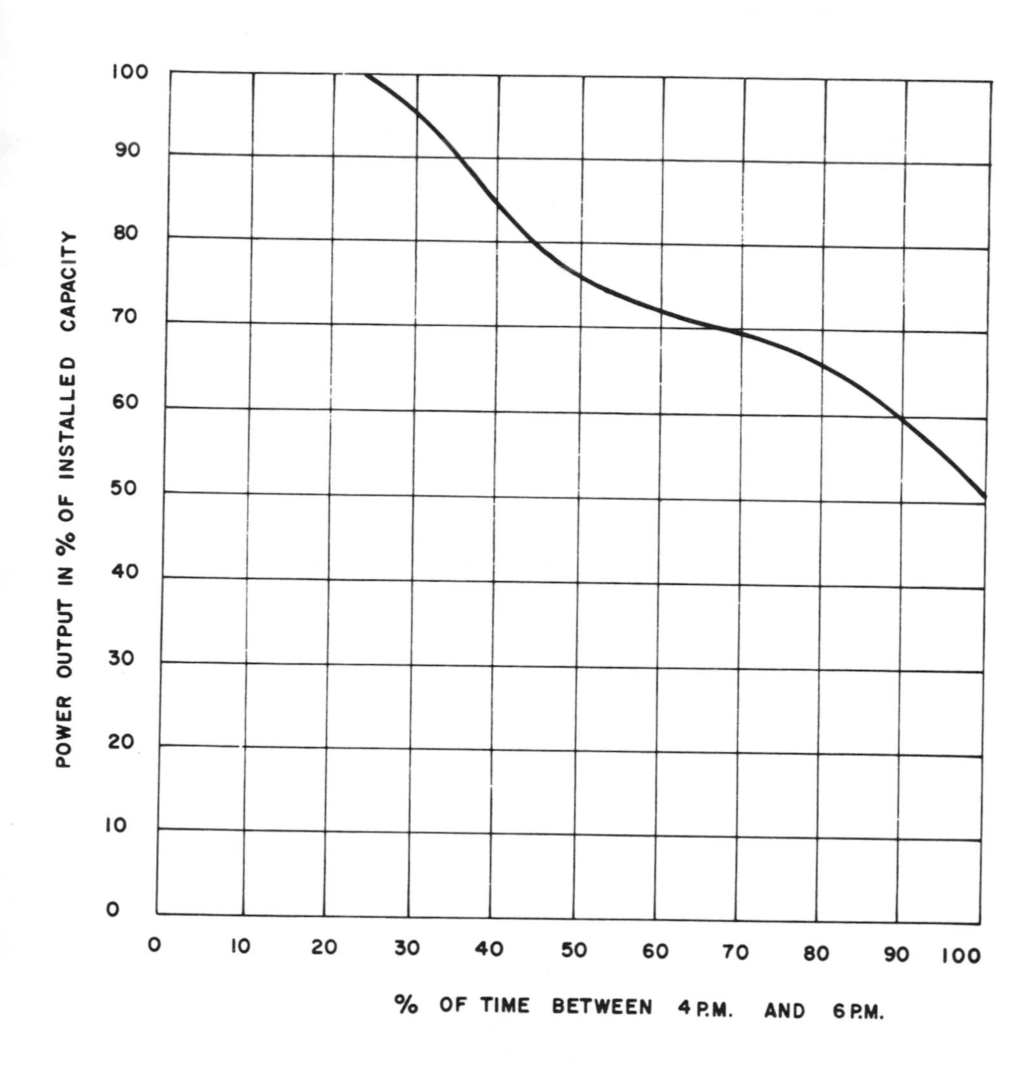

DERIVED FROM CALCULATION FOR THE LUNAR MONTH OF JANUARY 1968-61.
INSTALLED CAPACITIES:
SITE 7.1 30 GENERATING UNITS.
SITE 7.2 30 GENERATING UNITS.
GENERATOR LIMITING CAPACITY 27 MW.

Fig. 24. Duration curve of dependable-peak power in per cent of installed capacity for sites 7.1 and 7.2 operated as a paired-basin scheme.

Dependable-Peak and Energy Production from a Linked-Basin Scheme

For the development of Shepody Bay and Cumberland Basin as linked basins, the power house would be located on the sea side south of Cape Maringouin between appropriate rockfill dams. On one side, a rockfill dyke with a sluiceway along a line reaching Ragged Pt. isolates Cumberland Basin as the high basin and on the other side a rockfill dyke with a sluiceway along a line reaching Mary's Point cuts off Shepody Bay as the low basin. (See Fig. 25)

Initial studies were based on the selection of Cumberland Basin as the high basin in a linked-basin development because:

1. There would be a better match between basin volumes if Cumberland Basin is utilized as the high basin and Shepody Bay as the low basin.
2. Cumberland Basin appears to contain a very large volume of erodible sediments in the lower levels and operation as the high basin would avoid disturbing these levels.
3. A number of populated areas, the largest being Moncton, on or immediately adjacent to the Petitcodiac and Memramcook Rivers would encounter difficulties with sewage disposal if Shepody Bay were used as the high basin.

Contrary to other 1968 analyses and because of delays in the surveys required for making the necessary corrections, surface areas of the two basins were used as during the 1967 work with a correction for the actual layout of the development.

The power plant could be operated for either maximum-energy production over lengthy periods of time or for maximum dependable-power production for predetermined periods of time, at peak hours for instance. A theory based, as with the single-basin schemes, on the theory of the calculus of variations was used to determine the optimum operation, that is, the changes of basin levels, the operation of sluiceways, and the plant that lead to the maximum production.

Computations of the maximum dependable power for a daily two-hour period were carried out using a trial-and-error process involving selection on the unit hill chart of the highest curve of constant power output compatible with the head available for a particular tidal cycle.

Calculations were made to obtain:

1. Maximum annual energy at constant value;
2. Maximum dependable-peak power during a two-hour period.

An analysis of the typical variation in levels vs time for maximum energy production is shown on Fig. 26.

The module of the generating equipment is a bulb unit of 24.6 ft. diameter, running at 75 rpm, with an alternator of 27 MW limiting power. Wicket gates and adjustable blades allow for a high specific flow under the continuously varying operating conditions leading thus to a higher power

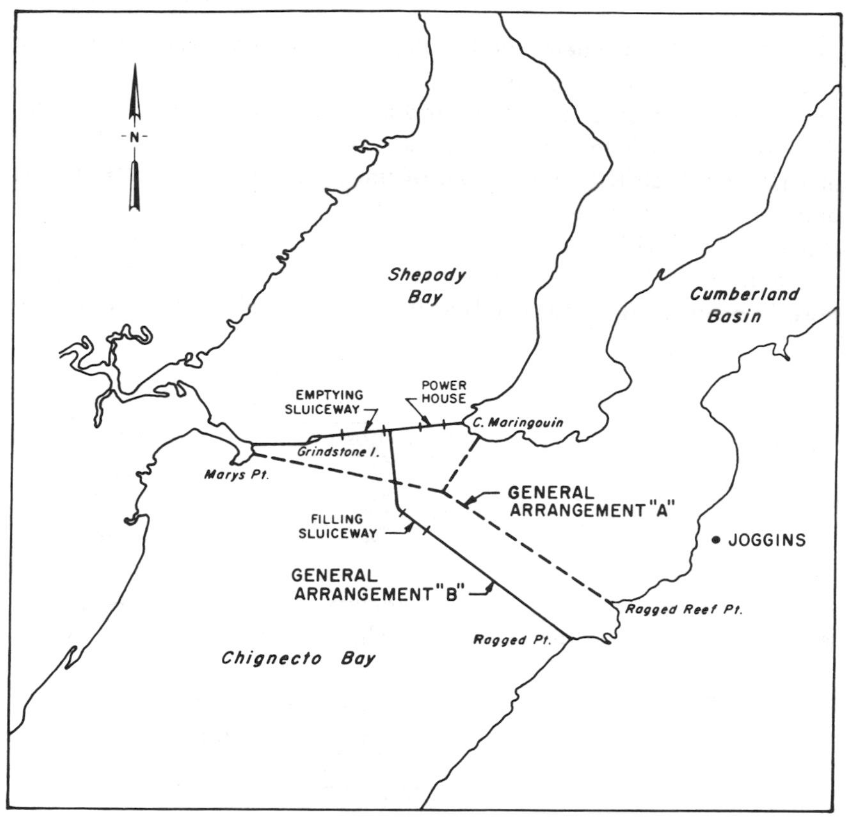

Fig. 25. Linked-basin scheme utilizing Shepody Bay and Cumberland Basin.

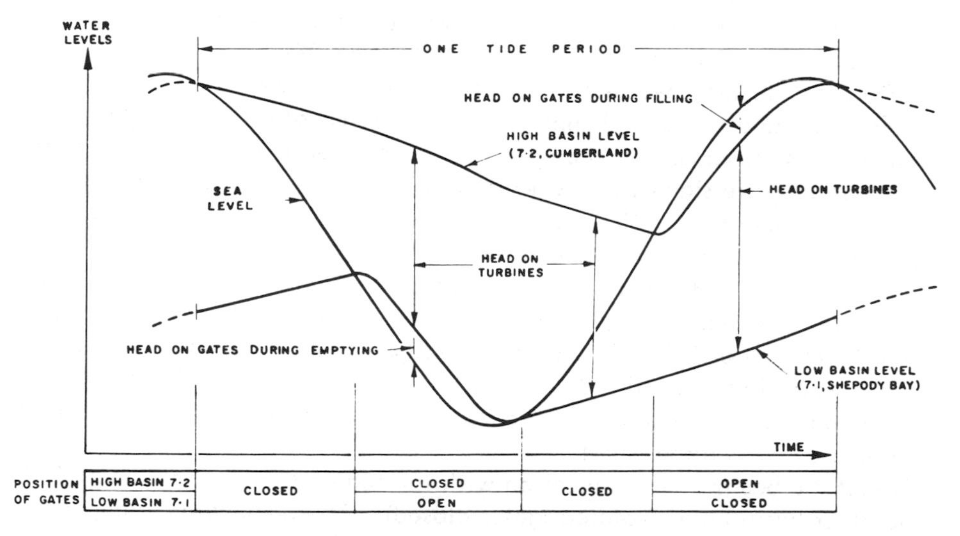

Fig. 26. Details of production technique for linked-basin scheme using Shepody Bay and Cumberland Basin for energy production.

production. Production of energy and of dependable-peak power is expressed as a function of the number of units.

Comparisons with other studies should be made on the basis of an equivalent turbine area (that is to say, on the basis of a constant value of the parameter KD^2 in which K is the number of units and D the diameter of the runner) and not of the installed power capacity.

After having considered the economic influence of the increase in sluiceway capacity on the total cost of the project and, on the other hand, the increase in productivity that resulted from it, the following values were adopted:

Number of Units	Sluiceway Capacity (million cfs under one foot head)	
	High Basin	Low Basin
28	1.09	1.26
34	1.30	1.52
45	1.74	2.02

To determine the influence of the sluiceway, the annual energy production has been computed with the following values:

Number of Units	Sluiceway Capacity (million cfs under one foot head)	
	High Basin	Low Basin
34	1.09	1.26

A straightforward comparison can thus be made of the energy production either for a number of units (34) with two sluiceway capacities or a sluiceway capacity with two numbers of units (28 and 34).

It was not deemed necessary to take into account the head losses in both approaches and tailrace channels, i.e. the backwater effect. This approximation is reasonable as the error thus introduced is the balance between two errors of opposite sign as far as heads on the turbines and total energy in one period is concerned. This is so because the operation of the plant can be divided into two fundamental phases:

1. With both high and low-basin gates closed, the plant is operated between the two basins and the backwater effect tends to decrease the head. The high basin will only be affected to a small degree by the backwater effect as the water level is rather high. At this stage, the low basin is in a filling period with increasing levels so the time during which the depth of water is small is rather short.
2. One of the basins is in connection with the sea and the plant is operating, the other basin being isolated (gates closed). The backwater effect in the open basin is the resultant of two operations, filling and emptying, due either to the gates and the turbines, respectively, for the high basin or

to the turbines and the gates, respectively, for the low basin. The resultant leads always to an increase of head as the flow capacity of the gates is much greater than that of the turbines. The backwater effect in the closed basin tending to decrease the head is still small in the high basin and of short duration in the low basin.

Fig. 27(a) shows the maximum annual energy as a function of the flow capacity of the power plant expressed in terms of:

$E = f(KD^2)$, in kWh
$K =$ Number of units
$D =$ Diameter of units (in metres)

A second scale on the abscissa shows the number of units corresponding to a diameter of $D = 24.6$ ft. Fig. 27(b) shows the sluiceway capacity used for each number of units.

Highlights of the analysis are:

1. For a given number of generating units, there is a tidal range above which the power plant is working at full capacity all the time. All the units are operated at a constant power equal to the limiting capacity of the generator. In other words, for these tidal ranges it is possible to find a mode of operation such as to maintain a minimum head equal or superior to the nominal rated head of the units, whatever be the fluctuation of reservoir levels.
   The smaller the number of units, the sooner this limit of tidal range is reached. For the basic installation mentioned, it is approximately:

Number of Units	Limit of Tidal Range ft.
28	42
34	47
45	55

   At the site under investigation, the extreme tidal ranges were, it must be remembered:
   Minimum tidal range (1967): 20.4 ft.
   Maximum tidal range (1967): 42.4 ft.
   Rms tidal range: 32.4 ft.
2. For a given installation there is, on the other hand, a tidal range below which energy production is increased by reducing the number of units in operation during part of the cycle. The smaller the tidal ranges, the larger will be the number of units to be switched off.
   This situation can be emphasized by stating the maximum number of units for optimum operation during both phases of plant operation with one sluiceway open or with both sluiceways closed. For a few tidal ranges,

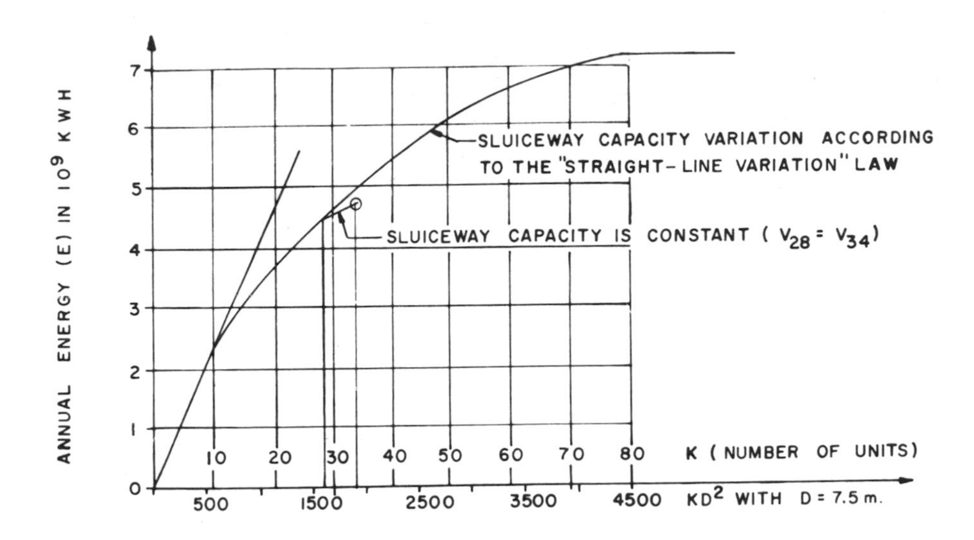

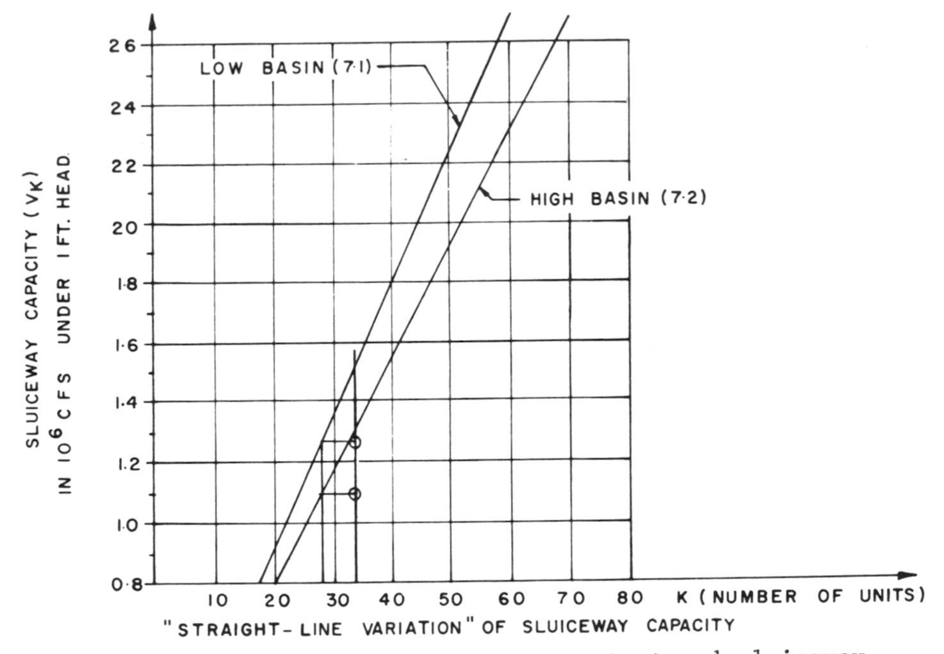

Fig. 27. Relationship between energy output and sluiceway capacity for linked-basin scheme using Shepody Bay and Cumberland Basin.

whatever is the installed capacity, the results are approximately as follows:

Tidal Range, ft.	Number of operating units for optimum operation	
	Both Sluiceways Closed	One Sluiceway Open
24.0	30	Installed capacity*
28.0	38	Installed capacity*
32.4	55	Installed capacity*
40.3	80	Installed capacity*

The conclusion then is that at a given site there is an absolute maximum above which any increase in the installed capacity of the plant does not increase the energy production. This limit, in the present case, has been estimated to be approximately 80 units.

3. The influence of the sluiceway capacity is shown by Fig. 27(b). As an example, computations have been made for 34 units:
   (a) with the sluiceway capacity as shown by the straight-line law, and
   (b) with the sluiceway capacity used for 28 units according to the same law.

   Representative points are circled. It appears that, for a reduction in sluiceway capacity of slightly more than 20%, there is a decrease in energy production of about four percent. The relation between annual energy, E, and the number of units shows that the marginal increment of energy has an appreciably different slope whether the sluiceway capacity is maintained constant or follows the law of proportionality represented by the straight lines.
4. In the foregoing study, the optimum operation of the plant has been computed on the basis of a repetitive process as if the tidal range and period were constant as was done for single-basin schemes during the initial studies on maximum energy output.
5. As an illustration of the variation of power output during this mode of operation, Fig. 28 gives the dependability of power output with 34 units operated for maximum energy production.

With a linked-basin tidal scheme, it is possible when only peak power for a two-hour period is required to isolate the two basins from the sea, the high basin being filled up to the level of the highest high tide during the 22 hr. preceding the peaking time, the low basin emptied down to the level of the lowest low tide during the same period. Thus, the largest available

---

*Depending upon the value of installed capacity, production of energy may be discontinuous during this phase. This occurs for large numbers of units and large tides.

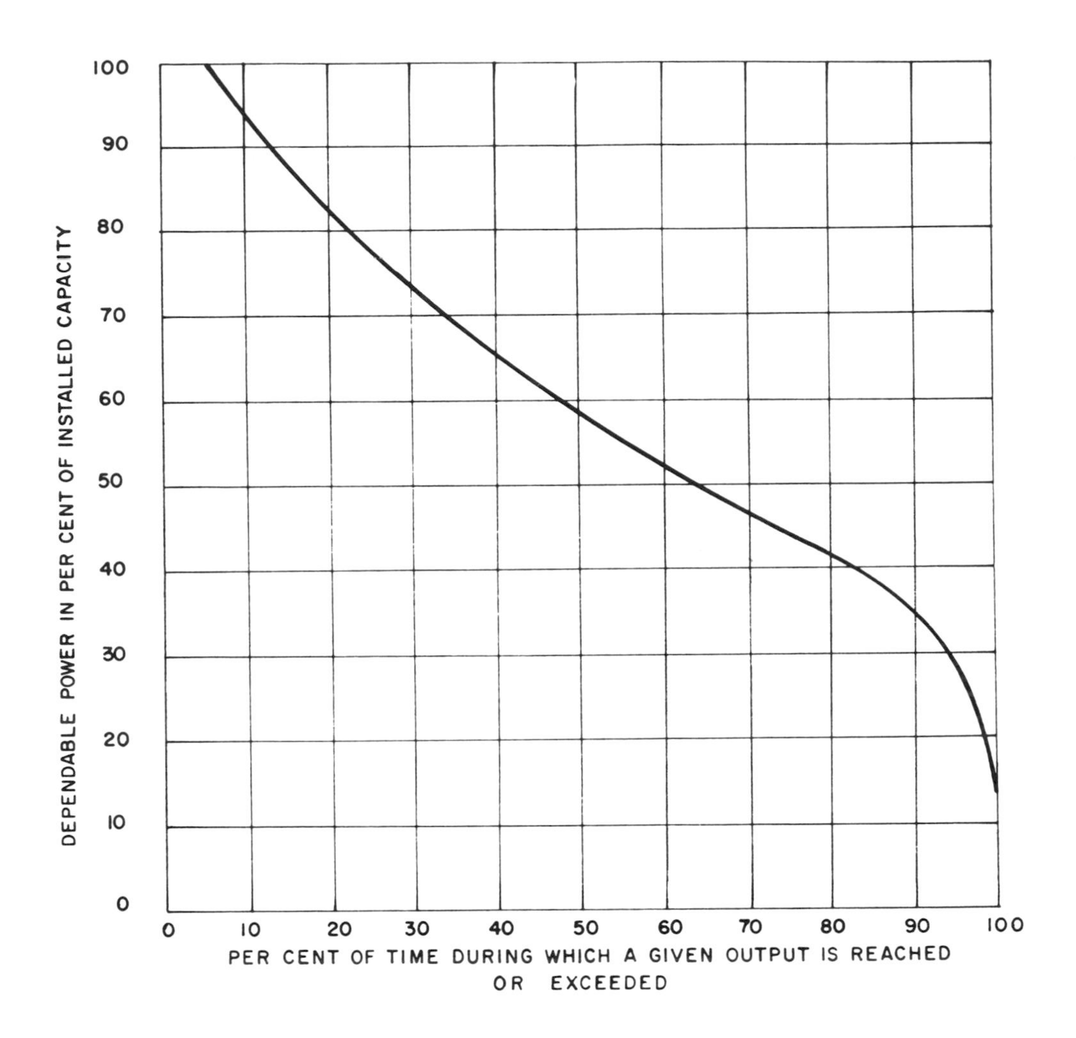

NOTE: INSTALLED CAPACITY 34 GENERATING UNITS
GENERATOR LIMITING CAPACITY 27 MW

Fig. 28. Duration curve of power output for linked-basin scheme using Shepody Bay and Cumberland Basin operated for maximum energy.

head is maintained to the time at which peaking operation will start. If, at this time, the tide is low or high, proper operation of one sluiceway will allow the slowing down of the variation of water level inside the corresponding basin, maintaining a higher head on the tidal plant.

The most adverse conditions of operation occur when both sluiceways are closed. Thus, dependable-peak power has been determined for this phase of operation when there is no communication with the sea.

Since the dependable power is defined as the minimum instantaneous power during the peak hours considered, there is an advantage in keeping the power output constant for the period considered. It is evident that the power production depends on the difference of water levels in the two reservoirs which has been reached at the beginning of the peak period.

There are cases where, for the filling of the high basin and the emptying of the low basin, use is made of both tide cycles included between the peak periods in order to further increase the starting head. Such an operation is possible only if energy production is disregarded.

Fig. 29 shows the frequency of occurrence of heads available for peak-power production as deduced from the 1967 Tide Tables. Working out the power available for various heads and several numbers of units, it is then possible to determine the available peak power as a function of the number of units. The degree of dependability of a given power output is, by definition, the percentage of peak periods during which this power is available.

Fig. 30 represents the whole of the results of peak-power studies for the linked-basin arrangement as a function of the number of units. A complementary picture of the available peak power is shown on Fig. 31 where the degree of the dependability is set in evidence for certain numbers of units. Fig. 31, using relative values, shows the degree of use of the same range of installed capacities as a function of the dependability.

The several illustrations represent the production of the power plant when either maximum at-site annual energy or at-site peak power is aimed at separately. A tentative estimate of the combined production of both was made for the following:

1. Numbers of units installed in the power plant:
   K = 28, 34, 45, 70 and 110.
2. Peak-power production from 1600 to 1800 hr. on working days with a degree of dependability of 95% during December, January, February.
3. Energy production during:
   (a) the nine other months - 24 hr. a day, seven days a week.
   (b) the off-peak hours of the winter months when the production of energy is not detrimental to the peak-power production.

On this basis, the combined production estimate leads to the orders of magnitude shown by Table 9.

Recognition of the advantages of an alternative concept of development gave rise to an evaluation of the significance of increasing the area of the

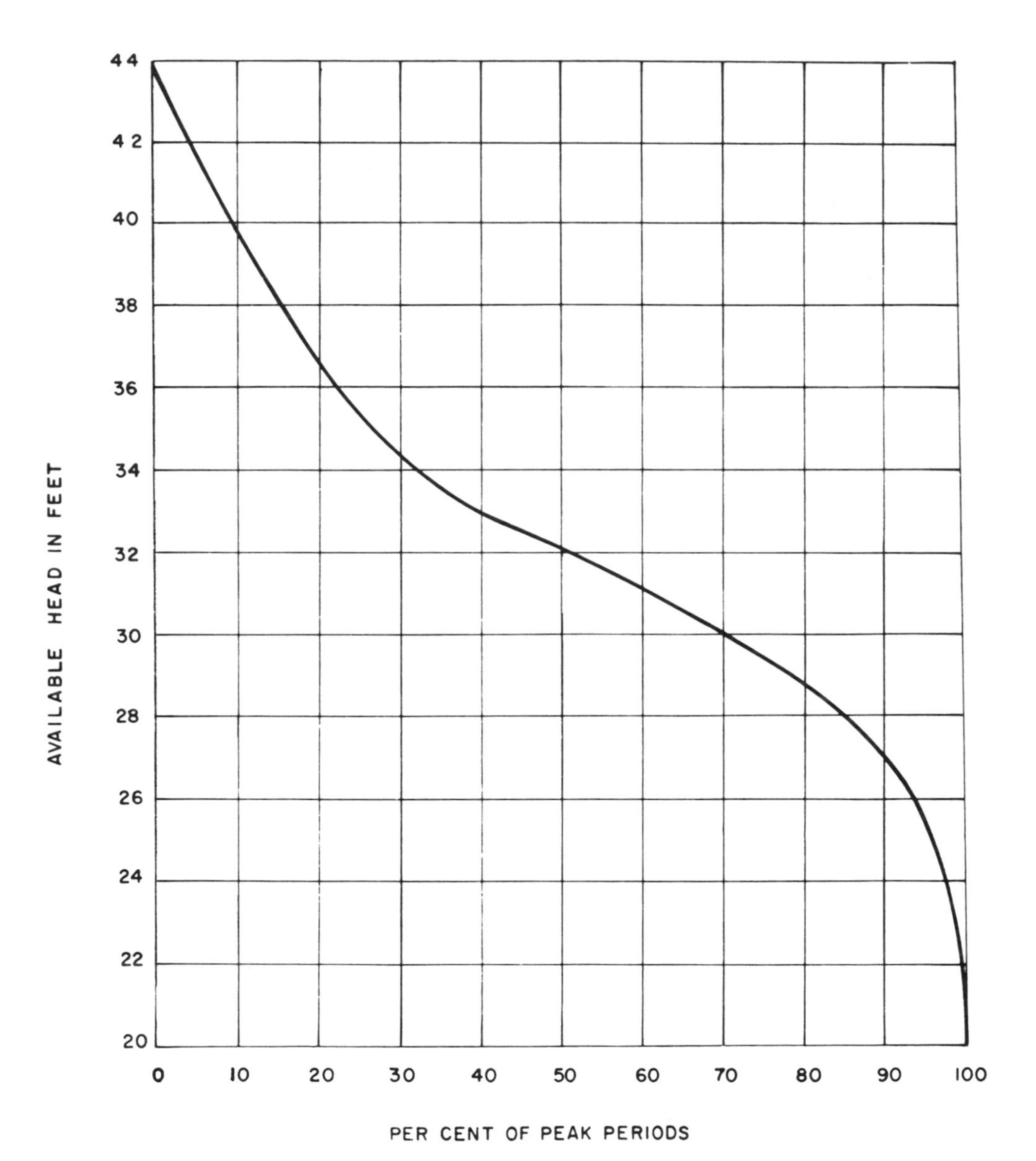

Fig. 29. Duration curve of available head for linked-basin scheme using Shepody Bay and Cumberland Basin operated for dependable-peak output.

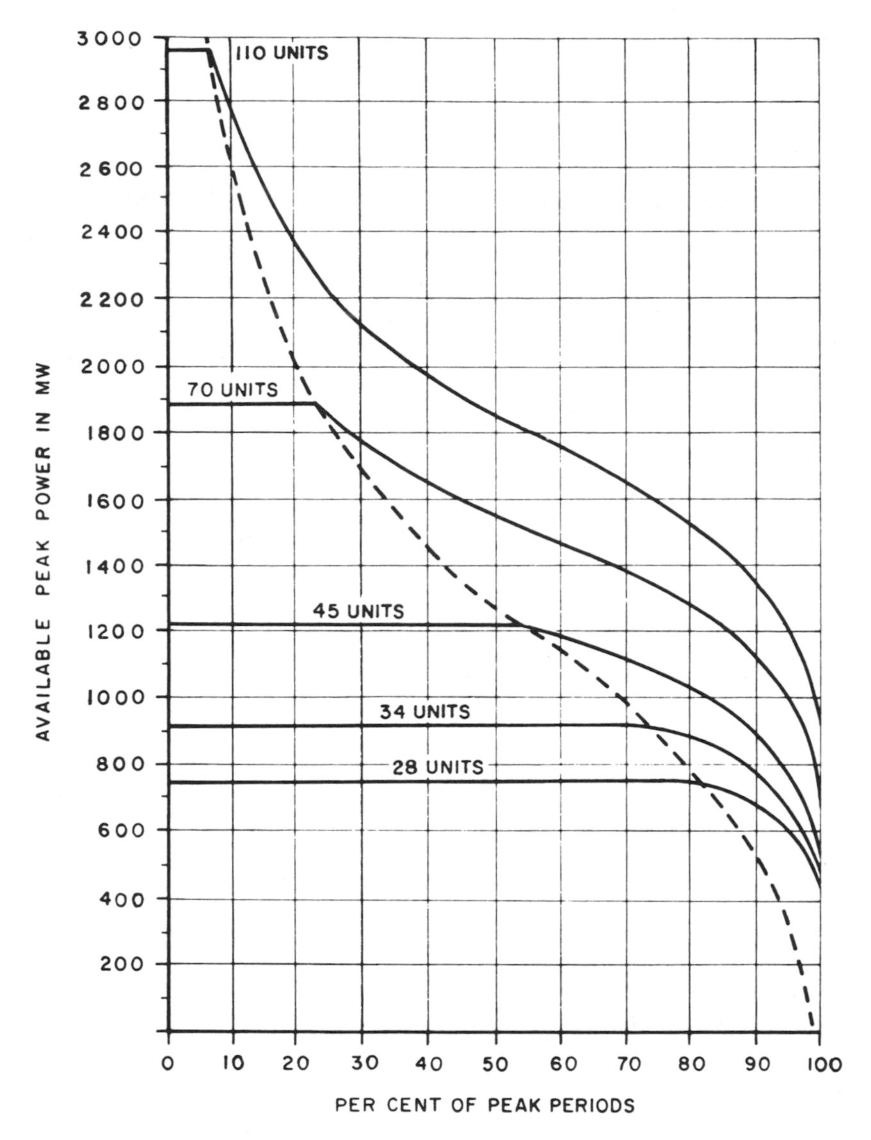

NOTES :- ON THE LEFT OF THE DOTTED LINE, THE AVAILABLE PEAK POWER EQUALS 100 % OF THE INSTALLED CAPACITY.

POWER IS CONSTANT DURING EACH PERIOD.

EACH PEAK PERIOD IS FROM 1600 TO 1800 HRS ON WORKING DAYS (261 DAYS).

GENERATOR LIMITING CAPACITY 27 MW.

SPEED OF ROTATION 75 R.P.M.

TURBINE RUNNER DIAMETER 24·6 FT.

Fig. 30. Duration curves of dependable-peak power for linked-basin scheme using Shepody Bay and Cumerland Basin.

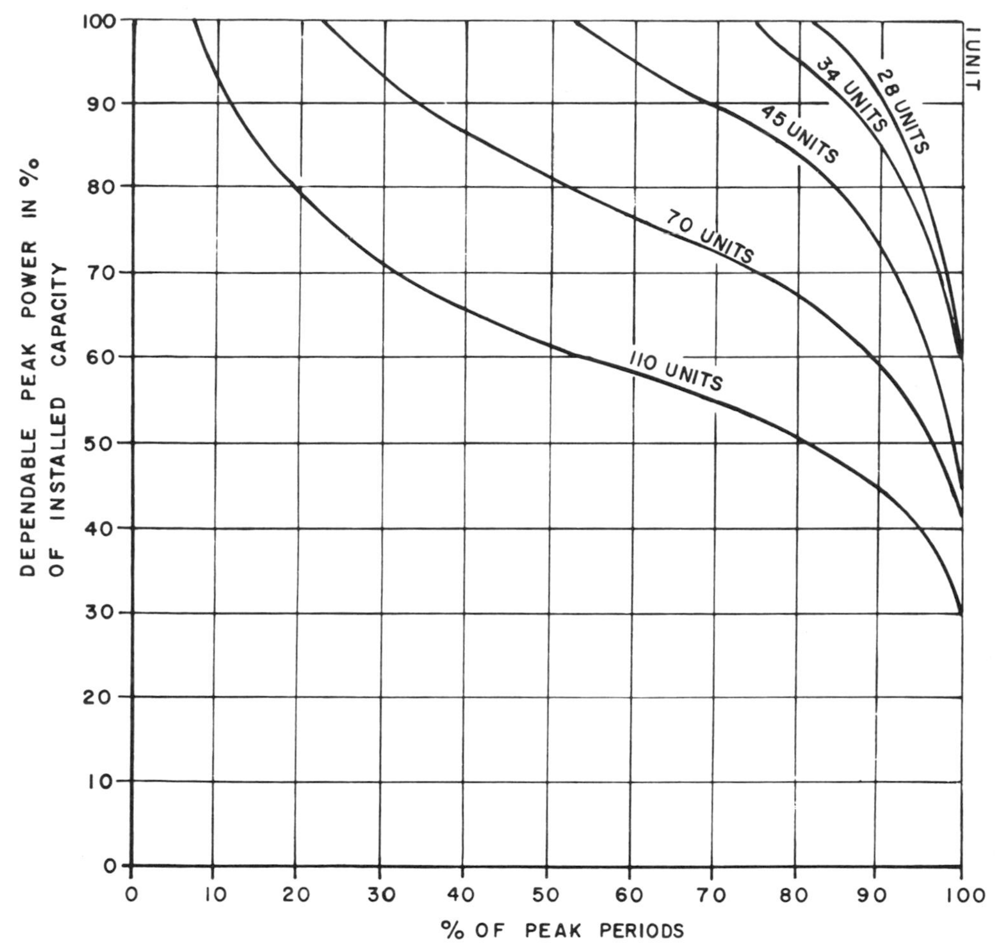

POWER IS CONSTANT DURING EACH PEAK PERIOD.
EACH PEAK PERIOD IS FROM 16 HOURS TO 18 HOURS.
ON WORKING DAYS (261 DAYS).

GENERATOR LIMITING CAPACITY	27 M W.
SPEED OF ROTATION	75 R.P.M.
TURBINE RUNNER DIAMETER	24.6 F.T.

Fig. 31. Duration curve of dependable-peak power in percent of installed capacity for a linked-basin scheme using Shepody Bay and Cumberland Basin.

high basin. Consideration was given to the effect of increasing the area of the basin without limit, since intuition suggested that if the output can be shown to increase in this extreme case, then a partial increase in basin area would tend to achieve the same result.

Table 9
Production for Various Generating Capacities in a Linked-Basin Development at Shepody Bay and Cumberland Basin*

Number of Units	Peak-Power Production with a Dependability of 95%, MW	Production During Peaking Hours, 2 Hr. per Day, 60 Days a Year, Millions kWh	Yearly Energy, Production Outside of Peaking Hours, Millions kWh	Total Yearly Energy Production, Millions kWh
28	600	100	3940	4040
34	680	120	4400	4520
45	735	150	5110	5260
70	910	200	6240	6440
110	1180	250	6420	6670

A typical diagram of linked-basin operation is that shown on Fig. 26. This actually represents a linked-basin scheme with 34 units of 27 MW each, the upper basin sluiceway capacity being 1.30×10^6 cfs under one foot of head and the lower one 1.52×10^6 cfs. The tidal range is an average one: 32.4 ft., the high-tide level reached being +16.5 GSCD.

For the optimized operation shown on Fig. 26, high-basin level varies between +1.8 and -16.0 GSCD. Were the upper basin area to be increased indefinitely, the level in this basin would of course settle down to a constant value. Assuming the same discharge in and out of the upper basin per cycle, this level would be +13.2 GSCD at least.

It could now be assumed that the level variation in the lower basin (i.e. the discharge in and out of this basin) remains unchanged. It is easy to compute the new power output per cycle: it is increased by 20% over the original value (4,900 million kWh annually, according to Fig. 27). As this result is not an optimized one, the increase in output in actual fact could be still higher. It is, therefore, reasonable to assume that any increase in high-basin area, even without an increase in the sluiceway capacity between that basin and the estuary, would lead to more output.

Table 10 provides a comparison of the basin areas originally used for the linked-basin scheme according to the initial concept of development utilizing

---

*Prior to adjustment for 5 to 5.5% reduction due to modification of tidal regime.

Table 10

Basin Areas for Linked-Basin Studies

		Cumberland Basin		Shepody Bay	
Water Level, ft.	Surveyed Basin Areas, sq. mi.	Original Data for Linked-Basin Scheme sq. mi.	1968 Studies (Single Basin) sq. mi.	Original Data for Linked-Basin Scheme sq. mi.	1968 Studies (Single Basin) sq. mi.
+25	60.7	43.2	-	-	-
+20	59.0	42.1	39.5	-	49.5
+15	57.7	40.9	-	-	-
+10	56.2	39.6	36.8	-	45.2
+ 5	54.7	38.0	-	52.2	-
0	52.6	35.8	33.3	49.7	38.5
- 5	49.9	33.3	-	46.8	-
-10	45.0	-	26.8	42.8	32.5
-15	34.6	-	-	38.2	-
-20	31.4	-	19.9	33.4	26.7
-25	30.1	-	-	27.8	-

It will be noted that MWL is + 0.27 GSCD (Site 7.1) and + 0.32 GSCD (Site 7.2)

Shepody Bay as the low basin and Cumberland Basin as the high basin.

It is clear from the above that in an alternative arrangement for the linked-basin scheme using Shepody Bay as the high basin and Cumberland Basin as the low basin, basin area values for Cumberland Basin would greatly exceed the previous areas over the range of tailwater levels (from about -5 to -15 GSCD) for which most of the production is likely to be done. As the area for Shepody Bay is smaller than the area originally used from the point of view of power production there is no doubt that Cumberland should be the low basin and Shepody the high one. Making this change, there is a very fair presumption that the net power output will not change very much from the original linked-basin values which should be reduced by about 5% to 5.5% to give the net output. In effect:

(a) The high-basin areas (Shepody Bay) would be higher than those assumed for Cumberland Basin.
(b) The new low-basin areas would also be in excess of those initially assumed for Shepody Bay down to a little below -11 GSCD, the greatest adverse difference in the area occuring at -15 and amounting to about 10%.
(c) The new Cumberland Basin areas would cause a rather higher loss of tidal range.

Working on the assumption that an area similar to that revised one for Shepody Bay could be considered, it would at first sight seem that power output would be less than the figures previously derived. For one thing, the considerable loss in the low-basin area would hardly be offset by the increase in the high-basin area. It would be rather difficult to make an estimate of the loss.

Detailed consideration of the operation of the alternative arrangement for a linked-basin scheme as shown by Arrangement B, Fig. 25, comprising Shepody Bay as the high basin and Cumberland Basin as the low basin, led to the following appraisal.

The correct area relationship for Shepody shows the low-level area is much less than that assumed for the linked-basin study based on the initial arrangement; this partly because of the general reduction in Shepody area and partly on account of the shifting of the dam on the Shepody side from the alignment assumed in the initial arrangement to that in the alternative arrangement.

On account of the similarity in the revised basin areas, especially the low one with those originally assumed, it is not difficult to estimate the new power output per rms tide.

The result is an increase of 1.5% in maximum energy output. It is anticipated that this increase would be very closely offset by the adverse effect on tidal regime. With the revised basin areas and the change between high and low basins, the power output should remain unchanged at least for the rms tide and the case of 34 units which seems typical.

Energy Production with Single-Regulated vs Double-Regulated Units

The object of this phase of the investigation was the determination of the difference in energy production between:

1. a double-regulated bulb-type generating unit (the turbine having adjustable runner blades and wicket gates), and
2. a single-regulated bulb-type generating unit with fixed runner blades (comparable to a propellor-type unit).

A generating unit with fixed runner blades cannot be operated as a turbine, as a pump, and as an orifice in either direction of flow from basin to sea or vice versa. Such a unit can be operated as a turbine in one direction only. It is impossible to use such units to pump from the sea to the basin because with fixed blades the torque when starting the unit as a pump is too great to allow the operation. Thus the capabilities of units with fixed runner blades are much more limited than those of double-regulated bulb units. With a single-basin scheme, they can be used for single-effect operation only. For this reason, single-effect operation for maximum energy production had to be used.

An analysis was carried out for Site 8.1 for a tidal range of 37.3 ft. with the basic assumptions previously used. The characteristics for the double-regulated bulb units were also the same as in previous work. By comparison, the single-regulated bulb turbine was identical except for the fixed runner blades. Analysis led to the selection of $i = 30^\circ$ as the optimum setting for the runner blades; the corresponding hill chart was used; the speed of rotation was 81.1 rpm. However, it must be noted that, due to more severe cavitation conditions with fixed-blade operation, the setting of the unit must be approximately 17 ft. lower so the elevation of the highest point of the turbine becomes -67 ft. (GSCD) instead of the previous -50 ft.

Computations were carried out for the following numbers of units and sluiceway capacities:

Number of Units	Sluiceway Capacity
70	1.12 million cfs under one foot head
90	1.46 million cfs under one foot head

To avoid the necessity of introducing into the programme the hill-chart corresponding to the fixed-blade operation, calculations for optimization of energy production were done by hand. Otherwise the same method and steps as previously were followed except that single-effect operation only was pursued.

Table 11 gives the maximum energy production as well as the pertinent characteristics of the generating and sluicing facilities.

Table 11

Maximum Energy Production and Pertinent Characteristics of Generating and Sluiceway Facilities

Number of Units	Sluiceway Capacity Under One Foot Head in Million cfs		Sluicing Capacity of Units Under One Foot Head, in Million cfs		Energy Production for 37.3 Ft. Tide, Million kWh		Difference in Energy Production Percent
	Single-Regulated	Double-Regulated	Single-Regulated	Double-Regulated	Single-Regulated	Double-Regulated	
70	1.12	1.12	0.342	0.376	10.535	10.920	-3.5
90	1.46	1.46	0.451	0.484	13.125	13.785	-4.8

With fixed runner-blade operation, the larger turbine discharges used under a low operating head at the beginning and the end of the production cycle lead to a slightly larger decrease in basin level. However, power production remains very similar for the two types of units.

For single-effect operation for maximum energy production, whatever the time of production and the extent of the installation assumed, single-regulated units with fixed runner blades are nearly as efficient as double-regulated ones. However, they require a deeper setting; the flexibility of operation required to produce dependable peak power or to concentrate energy production during the time of day when its value is higher is lost.

INTEGRATION OF TIDAL POWER PLANTS WITH EXISTING POWER SYSTEMS

The effective integration of tidal power plants into existing power systems is a rather complex one due to the pervasive nature of the problem in general and the particular characteristics of tidal power.

The single-effect single-basin tidal power plant is an intermittent producer of energy, the periods and nature of production depending on the tidal cycle. However, this energy can be processed through an associated pumped-storage facility to make the complex provide the same type and qualities of output as a double-effect single-basin tidal power plant without the complications of the latter due to the influence of the varying times of high tide. In other words, the pumped-storage tidal power complex based upon a single-effect single-basin tidal power plant as a supplier of energy can be an extremely flexible producer of dependable peak on a variable time base and of energy of optimum value in the market place.

Much the same result can be achieved where the tidal power plant can be integrated with a power network containing hydro-electric plants having large storage reservoirs. In such a case, the output of the tidal power plant can be substituted for draw-down of storage in the reservoirs by displacement of the normal output from the hydro-electric plant with tidal energy.

Another mode of integration of a tidal power plant into an existing power system is, of course, by the displacement of the output from existing sources.

Space does not permit going into this subject to any great length; it is being treated in a companion paper by the author at this Conference.

EFFECTS OF TIDAL POWER DEVELOPMENT ON OTHER INTERESTS

As required under the Intergovernmental Agreement, an examination was made of the effects of a tidal power development on transportation, fisheries, marshlands, and existing marine installations. In addition, a preliminary economic survey was carried out in order to assess the impact of a large tidal power development on the region. This work was done by, or carried out for, subcommittees of the Committee.

Tidal power developments would have little bearing on navigation requirements, except in the case of that at Site 7.1 where it would be necessary to include minor facilities costing about $6 million to maintain the present standard of navigation, mostly serving Moncton.

In the case of Site 7.2, commercial navigation is practically non-existent and, although there are a number of fishing boats operating from Joggins, it would be feasible to change the location of this activity.

In the case of a development at Site 8.1, there is no navigation in Cobequid Bay which would be cut off by the development. However, changes in the tidal regime would have some bearing on existing ports at Kingsport, Port Williams, Hantsport, Windsor, Walton, and Parrsboro. Necessary dredging and corrective measures have been estimated $310 thousand for those ports requiring remedial work, i.e. Hantsport, Walton, and Port Williams.

A preliminary assessment of the effects of a tidal power scheme on ground transportation in the region indicates that Site 8.1 would permit shortening the distance between Halifax and southern Nova Scotia to New Brunswick and other parts of Canada by about 15 mi. With a development of Site 7.1, the distance would be shortened by about five miles. It is estimated that annual savings in costs to road users at 10 cents per vehicle mile would amount to about $369 thousand for a crossing at Site 8.1, $91 thousand for a crossing at Site 7.1, and $181 thousand for crossings at Sites 7.1 and 7.2. However, the provision of necessary highways to take advantage of these crossings, built to Trans-Canada Highway standards, would involve expenditures of $1,100,000 at Site 7.1, $900 thousand at Site 7.2, and $1 million at Site 8.1 exclusive of the substantial additional cost of structures to accommodate such roadways on the developments. New roads necessary to utilize the crossings would run to three miles at Site 8.1 and about six miles at Site 7.1 or Sites 7.1 and 7.2 at some $225 thousand per mile.

A careful enquiry into the possible changes on the environment arising from the construction of tidal power developments and the effect on the fisheries of the Bay of Fundy have indicated that there would be virtually no change if any.

An examination of the potential impact of a development on the region indicated the following as the principal findings:

1. A new investment of $100 in the two-province region would produce $154 of regional income, i.e. the regional income multiplier is 1.54 based on an analysis of annual data for the years since 1949. A similar estimate indicates that an increment of $1,000 in gross national product (GNP) would lead to an increase of $40 in the gross regional product (GRP) of the Nova Scotia/New Brunswick region.
2. The combined impact on the GRP of the two-province region over the

construction period of those portions of the project cost that represent the purchase of regional resources and those involving expenditures made outside the region but within Canada could be expected to approach $465 million. The year-by-year income impacts amount to approximately 3.75% of the present level of GRP.

3. Primary employment increases resulting directly from project requirements and derived mainly from engineering estimates would amount to 11,000 man-years of local (Nova Scotia and New Brunswick) construction labour over the nine-year construction period and 485 man-years of local engineering and supervisory services, that is 11,485 man-years in all. Direct project requirements would create a further 1280 man-years of employment elsewhere in Canada.
4. Rising levels of GRP are generally associated with rising levels of employment and vice versa. Analysis of historical time series revealed that for every $8200 increase (1969 dollars) in GRP employment increased by one man-year. But the regression equation also reflects the fact that each year technological progress increases the investment required to create an additional job in the economy. Thus, each year a given increment in GRP is associated with 2.5% fewer jobs than the same investment in the previous year. Thus, it was determined that the indirect employment effect of the project would be the creation over the nine-year construction period of 34,000 man-years of employment in New Brunswick and Nova Scotia, and approximately 3900 more in the rest of the country. The total direct and indirect regional employment impact of project expenditures was estimated, therefore, at almost 47,000 man-years; the total impact on the Canadian economy as a whole at 52,000 man-years.

Post-construction project impacts were considered in two parts: the impact of project-related industries; and the impact of the project on other related activities including agriculture, tourism, forestry, fisheries, ground transportation and navigation. Consideration was also given to the long-run regional benefits of three construction facilities of permanent value as well as of the project operation and maintenance budget.

At the present time and with its estimated cost, a tidal power development would not have a significant post-construction impact on the economy of the region.

The effects of a tidal power project on agriculture, forestry, fisheries, ground transportation and navigation were briefly investigated and, in general, were found to be minimal. Low cost power could have a favourable impact on the region's agricultural development by reducing the cost of frozen fruit and vegetable processing. Forestry is expected to benefit from lower transportation costs through better project-related access roads including the roads over the structures themselves. Construction of Site 7.1 could reduce the highway distance from Saint John to Halifax by five miles and the use of Site 8.1 by 15 mi.

A tidal power development should produce a significant increase in total tourist visits and thus to tourist expenditures in Nova Scotia and New Brunswick by virtue of its unique engineering features. However, to produce maximum results from the tourist aspect, imaginative planning of auxiliary tourist-serving facilities would be required.

There are three facilities which, if built during the course of project construction, would be of permanent value to the regional economy: a deep-water wharf, an extensively equipped plant to assemble and finish the generator units, and a hydraulic model test laboratory. A decision to engage in extensive assembly and finishing of the turbine components at site could increase the local impact by as much as 1170 man-years and shift $31.6 million in expenditures on equipment from the central Canadian to the local market.

Expenditures on operations and maintenance for a project of this type typically amount to less than one percent per annum of total project costs. The continuing income impact of the operating and maintenance expenditures on the region would therefore be relatively small, about $3.0 million. The possibility of some 115 related permanent jobs is probably of greater significance.

PRINCIPAL FINDINGS

Of the 23 potential sites examined, three sites showing the best possibilities for economic development were more extensively investigated and preliminary development plans were prepared.

These sites comprised Shepody Bay, termed Site 7.1; Cumberland Basin, Site 7.2; and Minas Basin at the mouth of Cobequid Bay, Site 8.1.

All three sites were evaluated as single-effect single-basin and as double-effect single-basin schemes. Sites 7.1 and 7.2 were also investigated as a paired-basin scheme. In addition, Shepody Bay and Cumberland Basin were studied as a linked-basin scheme. The economic significance of using external pumped-storage in combination with the tidal power development was examined. All these schemes were found to be feasible from the engineering and construction points of view.

Two basically different methods of construction were involved in respect of the power house and sluiceway sections. The most economic methods were found to be construction in the dry behind cellular steel-pipe cofferdams for Sites 7.1 and 8.1 but, because of the foundation conditions, prefabricated caissons floated to the site and sunk onto prepared foundations proved to be the economic choice for Site 7.2.

At an interest rate of seven percent none of the tidal power schemes investigated was found to be competitive with alternative sources of generation as indicated by Table 12 and 13.

Table 12
Cost of Dependable Peak and Energy At-Site from
Double-Effect Single-Basin Schemes at Sites 7.1, 7.2 and 8.1

Item	Site 7.1	Site 7.2	Site 8.1
Dependable Peak*, at Site, $/kW	32.65	40.29	25.20
Energy Credit, mills./kW	2.31**	2.31**	2.31**

A single-effect tidal power plant at Site 8.1 would provide energy at the lowest at-site unit cost of the schemes examined. Careful comparative analyses indicated that a single-effect single-basin tidal power plant operating in conjunction with a pumped-storage plant would offer more advantages than a double-effect plant. Firm peaking capability could be provided more economically, however, by a pumped-storage plant using energy obtained by operating existing steam-electric plants at slightly higher capacity factors.

It was found that a rise in interest rates of two percentage points from five to seven percent would increase the estimated annual costs for tidal power developments by about 43%. It further became evident that the interest rate would have to drop to about half its current value to make tidal power economically attractive.

Despite the foregoing, there are a number of other influences affecting the economic viability of a tidal-power scheme, all tending to operate in favour of tidal power. These are changes in fuel costs for alternative power sources, increased costs of controlling environmental pollution from alternative power sources and technological changes influencing capital cost comparisons.

Studies of the costs of transmission using high-capacity, high-voltage DC links for the transmission of power excess to Maritime needs indicated that annual charges for transmission would be in the order of $9.00/kW/yr. of capacity.

---

*Available 95% of time (total hour basis), two hr./day, 60 day/yr.

**Derived by a weighted averaging of displacement energy values in the Maritime and northeastern United States markets.

Table 13

At-Site Costs of Dependable Peak from a Double-Effect Single-Basin Tidal Power Plant at Site 8.1 and from Alternative Sources (1)(2)

Source of Generation	Installed Capacity, MW	Dependable Peak At Site MW	Annual Energy At Site, Millions kWh	Dependable Peak $/kW/yr.
1. Double-Effect Tidal Power Plant at Site 8.1	3,536	1,526	7,560	25.20 (3)
2. Single-Effect Tidal Power Plant at Site 8.1	2,176			
with 1,526 MW Pumped-Storage, Two Hours per		1,526	6,271	21.17
Day, 60 Days per Year	1,526			
3. Double-Effect Tidal Power Plant at Site 7.1	2,916	1,260	5,402	32.65 (6)
4. Single-Effect Tidal Power Plant at Site 7.1	1,620			
with 1,526 MW Pumped-Storage, Two Hours per		1,526	3,971	21.43
Day, 60 Days per Year (4)	1,526			
5. Gas-Turbine Power only Providing 1,526 MW,				
Two-Hours per Day, 60 Days per Year	1,526	1,526	-	9.86
6. Pumped Storage Using Incremental Oil-Fired	1,526			
Steam-Electric Energy Providing 1,526 MW Dependable Peak, Two-Hours per Day, 60 Days per Year (5)	-	1,526	-	7.32
7. Pumped Storage using Incremental Candu-Type	1,526			
Nuclear Energy, Providing 1,526 MW Dependable Peak, Two-Hours per Day, 60 Days per Year	-	1,526	-	6.95

(1) Cost of money is assumed to be 7%. (2) Annual costs of alternatives include all direct and indirect charges. (3) The cost of dependable peak has been derived by deducting from total annual costs a credit for energy at 2.31 mills/kWh. (4) Pumped-storage capital costs taken at $85.00/kW and annual costs at $6.80/kW. (5) Cost of energy from existing steam-electric plants operated at higher capacity factors or incremental energy from oil-fired, steam-electric sources taken at 3.2 mills/kWh. (6) Cost of incremental energy from candu-type nuclear reactors taken at 0.58 mills/kWh.

ACKNOWLEDGMENTS

The investigations and studies on which this paper is based were carried out under the general direction of the five-member Atlantic Tidal Power Programming Board constituted under the Intergovernmental Agreement of August, 1966, between the governments of Canada and of the Provinces of New Brunswick and Nova Scotia. Responsibility for the supervision of the engineering and management aspects was vested in the five-member Atlantic Tidal Power Engineering and Management Committee. Development of investigations and studies and supervision of work by consultants and staff was the responsibility of the author in his capacity as Study Director.

Grateful acknowledgment is made of the many contributions to the successful accomplishment of the author's task by those noted. The author's presentation is, in fact, a synthesis of many varied contributions.

BIBLIOGRAPHY

Anon. Energy Supply and Demand in Canada and Export Demand for Canadian Energy - 1966 to 1990. National Energy Board, Ottawa, Canada (1969).

Bernstein, L. B., Tidal Energy for Electrical Power Plants. Translated by the Israel Program for Scientific Translations, U. S. Department of the Interior and the National Science Foundation, Washington, D. C. (1965).

Bonnefille, R., and Chabert d'Hières, G., "The Design of a Rotating Model of a Coastal Sea Area - Its Application to the Chausey Islands Tidal Power Plant", Committee of Soc. Hydrotechnique de France (1967).

Bretschneider, C. L., Engineering Aspects of Hurricane Surge, Proc. Tech. Conference on Hurricanes, Am. Meteorological Soc. Miami Beach, Florida (1968).

Bretschneider, C. L., Hurricane Design - Wave Practices, Trans. Am. Society of Civil Engineers, 124, 2965 (1959).

Bretschneider, C. L., Hurricane Surge Predictions for Chesapeake Bay, U.S. Army Corps of Engineers, Beach Erosion Board, Paper 3-59 (1959).

Hudson, R. Y., Laboratory Investigations of Rubble-mound Breakwaters, Proc. Am. Soc. Civil Engineers, Waterways and Harbours Division, 85, WW3, No. 2171 (1959).

Vantroys, L., "Interference with Tidal Regimes Caused by the Operation of a Tidal Power Plant", Les Usines Maremotrices Francaises (1967).

ECONOMICS OF TIDAL POWER

F. L. Lawton*

INTRODUCTION

Tidal power plants are characterized by the influence of the tidal cycle dictated by the astronomical forces controlling the tides. The output, entirely predictable many years in advance, follows the lunar cycle and hence gradually moves out of and then back into phase with the solar cycle. The solar cycle shapes the energy requirements of the community.

A tidal power development is marked by a continuously varying usable head. This variation in head is increased by the fluctuation in basin level resulting from the operation of the plant. The power and energy which can be developed depend on this varying head, the area of the basin controlled, the capacity of the sluiceways used to fill or empty the basin, the capacity of the generating units and the particular modes of operation utilized. The installed generating capacity is fixed by economic considerations rather than by the available flow. The sluiceway capacity is selected so as to allow proper filling and emptying of the basin for optimization of energy or dependable-peak power production rather than, as in the conventional water-power plant, for a specific maximum discharge. The energy available varies continuously from nil to a maximum which is achieved only a small part of the time. Consequently, a selection of the appropriate size of the generating unit requires a rather complex economic analysis.

GENERAL ECONOMIC CONSIDERATIONS

The studies carried out by the Atlantic Tidal Power Programming Board, which form the basis for this presentation on the economics of tidal power, were feasibility studies involving not only a choice between alternative tidal power schemes, utilizing different sites, but also comparisons with a number

---

*formerly Study Director, Atlantic Tidal Programming Board, Halifax, Nova Scotia.

of alternative possibilities for provision of the same service. In brief, tidal power can only be considered attractive in today's economic environment if it is competitive with alternative sources.

Investment costs involved in economic comparisons of alternative facilities are necessarily based on estimates. Relatively little reliance can be placed on historical costs or on unit costs achieved in the construction of similar facilities in the past. Costs of labour and material are changing rapidly; technological advances are rapid. In view of the importance of reaching a correct decision before proceeding with an economic alternative arising from the investigation, sound estimates are the only reliable method. At a later stage hydraulic-model tests of various types, network-analyzer investigations, micro-sytem studies, and other aids to conversion of engineering concepts into dollar estimates would be required.

Economic studies such as those involved in the investigation of the feasibility of tidal power, and its economic viability, are basically estimates of the revenues required to cover anticipated costs. These annual revenue requirements, called for by alternative sources of power generation, may differ both in amount and in their timing. The comparison of alternative plans requires determining the present worth of all future revenue requirements, theoretically to eternity but practically to much lesser periods of time. The studies involve the determination of the cost of capital. This may be defined as the rental or return, plus the tax, in the form of income tax, paid to retain the rental plus the cost of recovery of the capital in the form of depreciation charges. Often these factors are approximated roughly on the basis of the rationalization that if there is an error all alternatives are similarly affected. However, this is a dubious procedure because the effect of the cost of capital is often the factor to be evaluated.

Management choices are not the only factors necessary in an economic study. Neither is the cost of capital the only factor. However, these are vital to a sound decision.

The impact of the foregoing general economic considerations will be apparent in the treatment of the subject of this paper under capital costs, annual costs, optimization and verification, and related aspects. They will again be apparent in the discussion on the cost of transmission of tidal power in considering the costs of alternate sources of power. These factors will also appear with the discussion of the comparative position of tidal power and alternatives as sources of pollution.

CAPITAL COSTS

The capital cost estimates were based on prices in effect during the year 1968. The economic comparison for the cost of power and energy, whether from tidal power development or from alternative sources, was taken at 1968 price levels.

Although the significance of inflation was fully considered, it was decided that the feasibility investigations would not take into account future inflation in the cost of capital.

A common basis was used for the preparation of direct construction costs for all sites and all schemes of development. Due regard was paid to the differences in transportation costs and other conditions peculiar to particular sites, such as foundations, access to sources of armourstone and other materials, etc. Common design criteria were utilized.

Estimates were based on the quantities of work required and the application of realistic unit prices such as might be bid by competent and experienced contractors. These unit prices included full provision for all camp and construction facilities, all items of expense including contingencies, insurance and a reasonable profit, which competent contractors would include in realistic tenders. These unit costs were derived from many sources, such as consultants, construction organizations, manufacturers of equipment and others.

Contingencies were carefully assessed in view of the uncertainties arising from the preliminary nature of feasibility investigations. An allowance of 10% was provided for civil engineering work and five percent for manufactured items.

As customary, engineering and administration costs were taken at rates reflecting all costs associated with engineering and the administration required in the preparation of definitive construction plans and specifications as well as in the related purchasing, inspection and expediting of all purchases. Provision was made for general administration in connection with the development. The cost estimates reflect an allowance of five percent for engineering and administration applied to the estimated direct construction costs plus the contingency allowance.

It will be realized that, if investigations were carried further, to the design stage, a number of model investigations, sediment surveys and other studies would be required over and above the provision in the engineering and administration costs.

Since the cost estimates were based on 1968 price levels, it was logical that the rate of interest chosen for the determination of interest during construction should be derived from the average interest rate then in force for federal government loans to crown corporations.

ANNUAL COSTS

As usual, annual costs were taken as including interest on investment, amortization of investment, allowances for major renewals and operation and maintenance costs. The interest rate was taken at seven percent, the same rate as utilized for interest during construction. The same rate was applied in working out amortization charges.

Although there is little direct experience from which to assess the economic life of various elements of tidal power developments, there is much experience from the closely related hydro-electric power plants which indicates that well designed and constructed plants built some 40 to 50 years ago have an economic life of 75 years or more. Since the periodical replacement of certain elements in the generating units, other major equipment and the regular maintenance of the civil works offset physical depreciation, tidal power developments in the Bay of Fundy were assumed to have an economic life of 75 years.

Rigorous assumptions were made as to the frequency of renewal of such items as surface concrete on structures subject to erosion, treatment by sandblasting and grouting of structures not subject to erosion, armourstone protection, turbine runners, generator stator and rotor windings, generating unit guide and thrust bearings, transformer windings and bushings, miscellaneous mechanical and electrical equipment. The basis utilized for estimating the cost of replacement was the use of 1968 costs and their reduction to a present worth at the time of completion of plant construction, the present worth cost so determined being converted to a fixed yearly cost for renewels.

Operation and maintenance costs were derived as estimates of the actual expenditures involved, these reflecting personnel costs, administration and other miscellaneous charges. Maintenance, of course, was taken as covering running maintenance charges, independent of allowances for major renewals.

OPTIMIZATION AND VERIFICATION

An optimizing procedure was utilized to determine the best scale of installation for each site to produce the optimum revenue. This procedure utilized the results of the power production studies along with estimates of annual costs for the various levels of installation at each site to determine the installation that would produce power at the lowest unit cost.

In optimizing for maximum revenue, operation of the tidal power plants for dependable peak part of the year and energy during the remainder, a more complex method was used. For these cases, the revenues from at-site energy production were first deducted from the annual costs leaving the remainder of the costs to be applied against the production of a dependable-peak capability. This procedure optimized the cost of at-site dependable-peak capability against the size of installation. All optimization procedures were carried out on the basis of at-site production. In the optimization, no consideration was given to transmission requirements or to limitations imposed by interconnected systems.

The initial step in the procedure consisted in the determination of the optimum installation of generating and sluiceway capacity for several development concepts at the different sites. The development concepts studied included single-basin single-effect and double-effect generation at Sites 7.1, 7.2 and 8.1. For the foregoing, computations were normally made using three different numbers of generating units, K_1, K_2, and K_3, each with its associated sluiceway capacity.

From the estimates covering the several numbers of generating units and sluiceway capacities, the optimum number could be derived. For a limited number of cases four rates of interest were used, i.e. three, five, seven, and nine percent, to establish a complete analysis of the development concept for one of the sites after the several optimization processes. However, the rate of seven percent was normally used.

Subsequently, the optimization procedures and preliminary designs were reviewed using modified designs introduced, where desirable, for the purpose of improving and verifying costs for the optimum installation.

The bases underlying the evaluation of cost-revenue relationships for the several concepts of development, modes of operation and methods of construction examined for tidal power Sites 7.1, 7.2, and 8.1 in order of sequence were:

1. Statement of the basis of site design;
2. Preparation of an estimate of direct construction costs for use in the optimization procedure, for K_2 generating units;
3. Preparation of cost estimates similar to (2) above, for K_1 and K_3 generating units, in simplified form;
4. Preparation of estimates of total capital cost, including interest during construction for K_1, K_2, and K_3 generating units;
5. Development of annual costs for K_1, K_2, and K_3 generating units;
6. Derivation of K_{opt} and preliminary cost of dependable at-site peak power;
7. Preparation of complete capital cost estimate for development corresponding to K_{opt}, taking into account any refinements or adjustments in design, at a seven percent interest rate;
8. Preparation of annual costs associated with (7) above;
9. Derivation of cost of dependable at-site peak power.

COMPARATIVE CAPITAL AND ANNUAL COSTS

Comparative capital and annual costs are presented in Table 1 for three single-basin single-effect and for three single-basin double-effect schemes based on developments at Sites 7.1, 7.2, and 8.1, as shown by Fig. 1.

The costs of developments at Sites 7.1 and 8.1 were based on the construction of the sluiceways and power plants in the dry behind construction cofferdams while at Site 7.2 prefabricated caissons built in on-shore facilities, floated, towed to site and sunk onto prepared foundations were assumed.

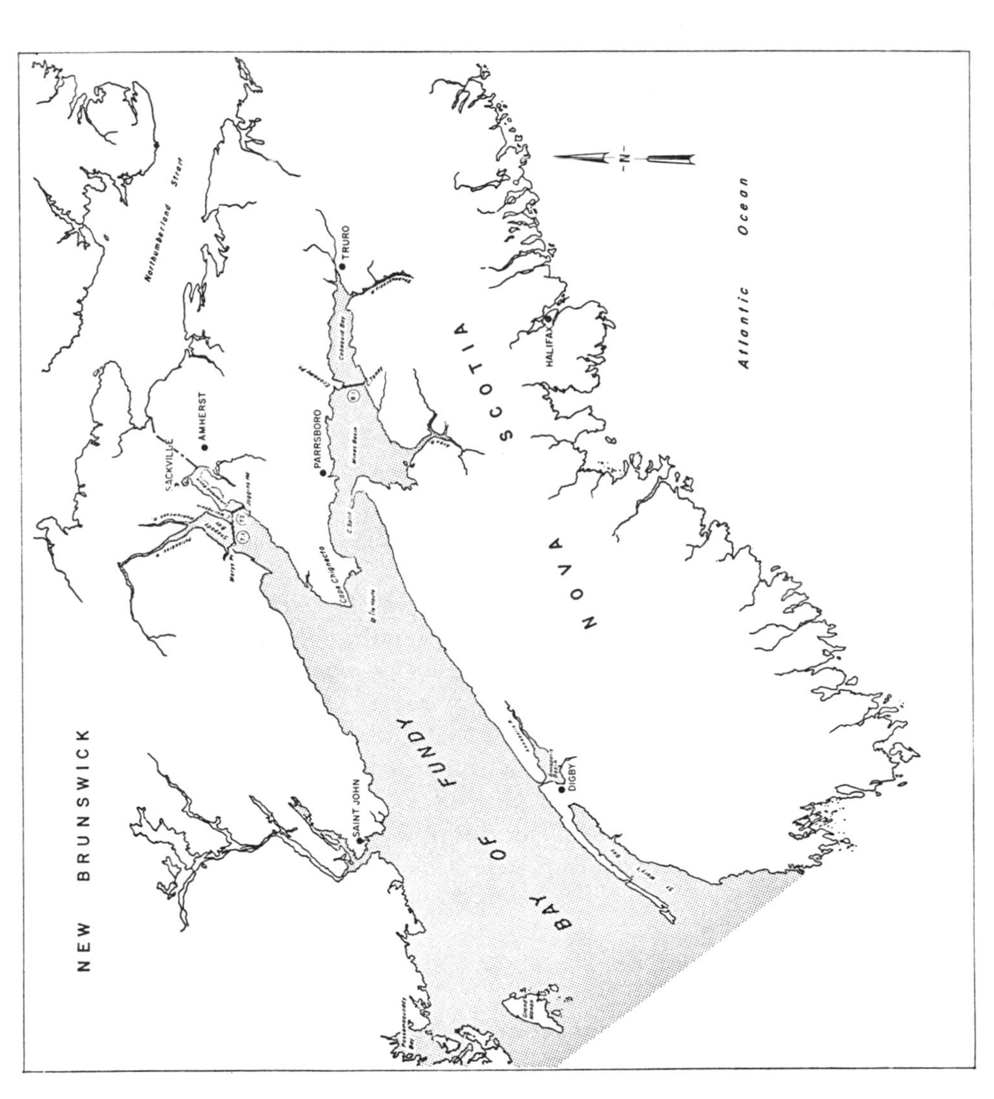

Fig. 1. Locations of tidal-power schemes.

Comparison of the direct capital costs, before application of charges for contingencies, engineering and administration, and interest during construction at seven percent for the two types of developments at the three sites noted is provided by Table 2.

Table 1
Comparative Capital and Annual Costs for Three Single-Effect and Three Double-Effect Single-Basin Tidal Power Schemes
(Millions of Dollars)

	Site 7.1		Site 7.2	
Item	Single-Effect	Double-Effect	Single-Effect	Double-Effect
Capital Costs	414.0	699.7	311.1	660.1
Annual Costs	31.5	53.6	23.4	50.1

	Site 8.1	
Item	Single-Effect	Double-Effect
Capital Costs	473.8	723.6
Annual Costs	36.4	55.9

Table 2
Comparison of Direct Capital Costs for Single-Effect and Double-Effect Single-Basin Schemes Built in the Dry at Sites 7.1 and 8.1 and in the Wet at Site 7.2
(Millions of Dollars)

Site	Single-Effect Schemes Method of Construction	Direct Capital Cost
7.1	Cofferdam	301.0
7.2	Caisson	223.7
8.1	Cofferdam	344.8

Site	Double-Effect Schemes Method of Construction	Direct Capital Cost
7.1	Cofferdam	511
7.2	Caisson	474.4
8.1	Cofferdam	528.5

SIGNIFICANCE OF INTEREST RATES

An analysis of the impact of interest during construction on capital costs shows that for single-basin single-effect schemes at Sites 7.1, 7.2, and 8.1 the interest amounted to 21.7, 22.8 and 21.5 percent of the capital cost prior to application of interest. For the same sites assuming single-basin double-effect schemes, interest during construction was respectively 21.4, 23.3 and 21.4 percent.

An examination of the influence of interest charges on a double-effect scheme at Site 8.1 showed these to amount to 71.2% of the total annual costs with an interest rate of three percent, 84.8% at five percent, 90.6% at an interest rate of seven percent, and 93.5% at an interest rate of nine percent. In brief, at the seven percent level, a reduction of two percentage points in the interest rate would result in a saving of about 28% on the annual cost.

COST OF ENERGY AND DEPENDABLE PEAK

The costs of energy on an at-site basis for the three single-effect single-basin schemes are shown by Table 3. The costs of dependable-peak output of the three double-effect single-basin schemes are shown by Table 4. In both cases, the power-capability figures reflect the reduction in the potential output of a plant due to changes in the tidal regime arising from the construction of the plant.

Table 3
Cost of Energy At-Site from Single-Effect Single-Basin Schemes at Sites 7.1, 7.2 and 8.1

Item	Site 7.1	Site 7.2	Site 8.1
Cost of Energy At-Site Mills/kWh	7.5	8.7	5.6

* No firm capacity is produced.

Table 4
Cost of Dependable Peak and Energy At-Site from Double-Effect Single-Basin Schemes at Sites 7.1, 7.2 and 8.1

Item	Site 7.1	Site 7.2	Site 8.1
Dependable Peak,* at Site, $/kW	32.65	40.29	25.20
Energy Credit, Mills/kWh	2.31**	2.31**	2.31**

It should be noted that the optimized developments were determined on the assumption that all of the output from any of the schemes could be marketed. The unit cost of power and energy were based on the total output, determined from power production studies.

The unit costs of energy, for each of the schemes providing a dependable peaking capability, were computed by deducting from the annual costs a credit of $9.50/kW of dependable capacity. That is, the computation of the at-site unit cost of energy assumed the dependable capacity would have an at-site value of $9.50/kW/yr. This assumed capacity revenue credit represents the estimated cost of peaking capacity from other new sources in the Maritime Power Pool, based on current financing, operation and maintenance costs of the utilities in the Pool. On the other hand, the at-site unit costs of dependable capacity for each of the schemes were computed by assuming an energy value of 2.31 mills/kWh. The value of 2.31 mills/kWh for displacement energy was derived from a weighted averaging of displacement energy values in the Maritime Power Pool and in the northeastern United States market contiguous to the Pool, with due allowance for monetary exchange rates and the proportions of displacement energy which might be sold in the Maritimes and in the United States.

The dependable peak capacities given in Table 4 are only available for two hours daily during 60 days in the three winter months of December to February. During the remainder of the year, the plants would be operated for maximum energy production. The use of the two modes of operation provides the optimum return for single-basin double-effect schemes of development.

---

*Available 95% of time (total hour basis), two hr/day, 60 day/yr.
** Derived by a weighted averaging of displacement energy values in the Maritime and northeastern United States markets.

COST OF MARKETING AND TRANSMISSION

Because of the very large resources involved in tidal power developments in the Bay of Fundy, to achieve optimum economy, much of the power produced, whether in the form of energy or dependable-peak capability, must be sold in markets outside the service area of the Maritime Power Pool. The contiguous markets of interest are those of Quebec and of the northeastern United States, this latter area including the New England States and New York.

Studies were made of the existing loads and their development over the period to 2005.

For the provinces of New Brunswick, Nova Scotia and Prince Edward Island, the 1968 Pool load of 1161 MW and 6445 million kWh is estimated to grow to almost 15,000 MW and 83,000 million kWh by 2005, this approximating an average growth rate of about seven percent per year. It is expected that the average annual load factor will change with industrial, commercial and residential development. It was taken at about 66% for New Brunswick, 63% for Nova Scotia and 47% for Prince Edward Island.

In the Hydro-Quebec service area, forecast loads will increase from approximately 8800 MW and 52,000 million kWh in 1970 to 39,100 MW and 231,000 million kWh by 1990, an average anticipated growth equal to about 7.7% per year. The average annual load factor is estimated to be about 67.5%.

The 1970 system requirements for power and energy in the northeastern United States are estimated to amount to 28,860 MW and 153,140 million kWh, increasing to 89,400 MW and 479,200 million kWh by 1990, the terminal annual rates of growth for the period being 7.7% in 1970 and about 5.8% in 1990. The current average annual load factor of about 62% is expected to increase gradually in the next 20 years.

In the marketing studies, an evaluation of the load and generation characteristics of the northeast United States utilities indicated that, by 1980, for about 12 weeks (five days per week and two hours per day), peak loads would occur which may be about 1000 MW in excess of the two-hour peak loads outside the 12 week winter period. Capacity installed to serve the peak load, which normally occurs during December to February, may be of limited use to neighbouring systems during other periods of the year. For the purpose of this study, it was assumed that this peaking requirement might be supplied by tidal power if the price was competitive with all forms of capacity such as might be derived from pumped-storage generation, gas turbines, and other modes of generation. It was also assumed, for the purposes of the study, that Quebec might absorb some 500 MW of peaking power from a tidal development and accept displacement energy if found desirable from an economic and operating standpoint.

Tidal generation is dependent upon the 24 hr. 50 min. tidal cycle.

While production can be related to the hourly load on the interconnected system, the determination of the amount of power and energy from a tidal plant that could be absorbed by the Maritime Power Pool would depend upon the type of development selected and the operating mode chosen.

In a single-effect single-basin scheme of development, the output occurs in the form of isolated slugs or irregularly shaped blocks of energy as illustrated by Fig. 2. These energy slugs have peak values corresponding to the installed generating capacity. This fixes the transmission requirement for single-effect schemes at 1620 MW, 972 MW and 2176 MW for Sites 7.1, 7.2, and 8.1, respectively. On the other hand, in the case of double-effect single-basin schemes, the transmission facilities would serve two purposes: firstly, the transmission of the tidal power plant output, and secondly, the supply of energy from the system for the operation of the tidal power plant in the pumping mode. Thus transmission requirements are governed by the installed capacity of the plant for both single-effect and double-effect single-basin schemes.

It will be apparent that a tidal plant can be operated in such a manner that its energy production occurring at times when it would have a relatively low value to the system could be used as the input to a pumped-storage facility. The transmission requirements to interconnect the tidal plant, the pumped-storage plant and the contiguous power systems, would depend on the capacities installed and upon the manner in which the operation of these facilities is integrated with that of other system facilities. There are numerous possibilities in this respect. In fact, the transmission requirements cannot accurately be defined without detailed system studies.

Comparisons were made of the possible power-supply arrangements in the Maritime Power Pool with and without the incorporation of tidal power developments of the several types and capabilities. It was evident that the final establishment of such transmission requirements would require comprehensive load-flow and transient-stability analysis of each alternative for the critical years under review. However, for the feasibility study required by the tidal power investigation, it was not considered necessary to make detailed system analyses of the types mentioned. Instead, transmission-line requirements were assessed by using broad guidelines outlining the probable transmission paths and the required transfer capabilities. For the 25-year period under review generation-expansion programs were developed for the Maritime Power Pool and for the individual member systems, taking into account tidal installation of various types and capabilities, comparison being made with the alternative of system development without tidal additions. This study took into account generation schedules determined by the use of the Maritime Power Pool annual load-duration curve. Nuclear units were added only when the incremental growth of load and energy needs required a new unit. Thermal units using fossil fuels were assumed to supply the intermediate load portion of the curve. Hydraulic generation, where known potential developments

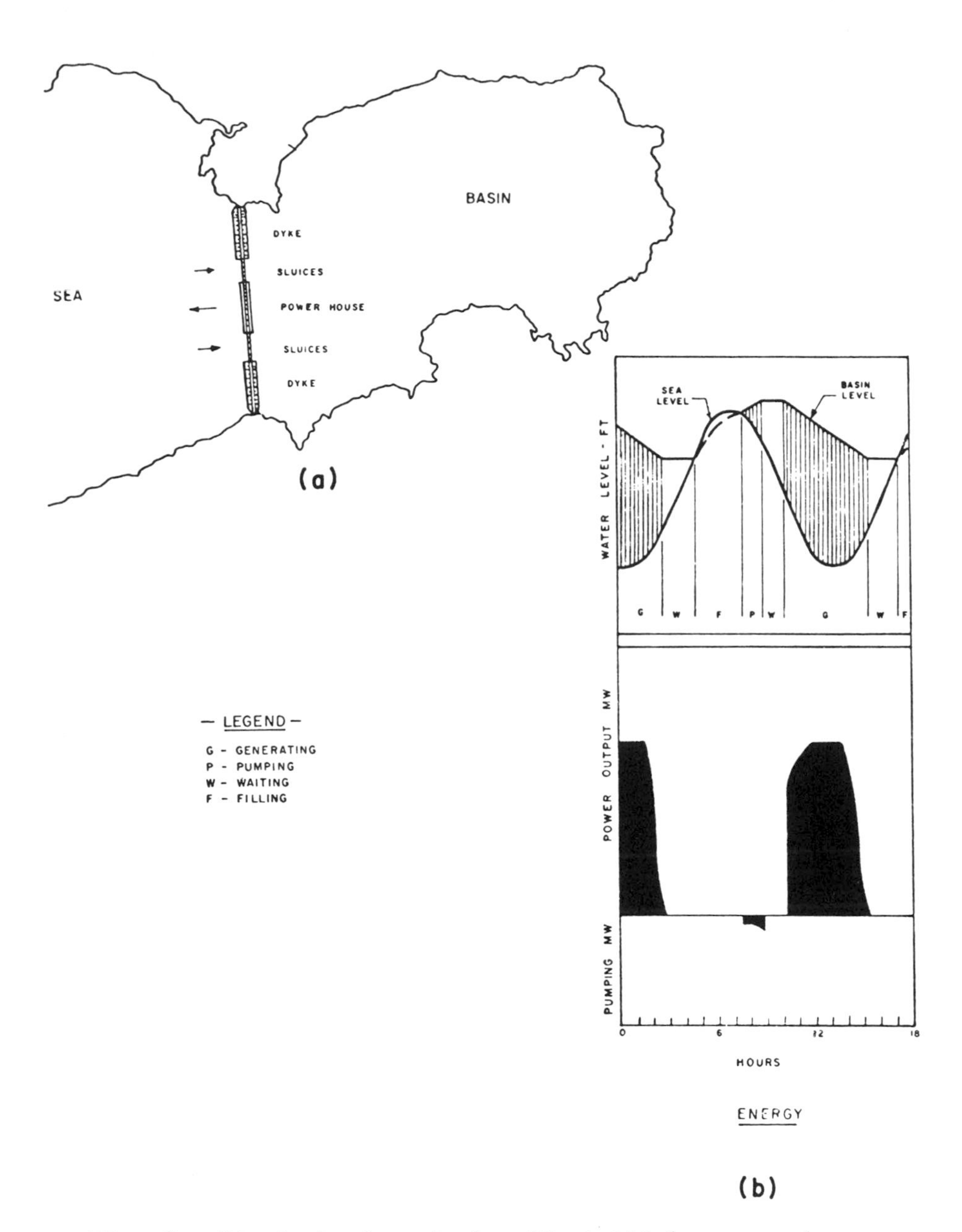

Fig. 2. Single-basin, single-effect tidal-power scheme.

were available, was added for peaking. Gas turbines were allocated to provide for generation reserves and to supply some power and energy at peak load periods.

Once the load and generation patterns were established, for each of the selected years in the 25 yr. period under review, the loads were allocated to the main geographic load points within each province and the generation additions were made either by expanding existing plants or by building new plants at suitable locations. The probable transmission paths and the transfer requirements needed to supply these loads from the generation sources were next determined by area demand and generation-flow diagrams. Finally the transmission-facility additions needed to permit these transfers were determined.

Capital-cost studies, covering the transmission expenditures developed from the foregoing investigation, showed that, including interest during construction at seven percent, the facilities to integrate a maximum of 800MW of tidal output with the Maritime Power Pool system plus facilities for the transmission of 500 MW to the Hydro-Quebec system would cost an estimated $31 million in addition to which an expenditure of $32 million would be needed for additions to a facility providing a direct-current tie between the Maritime Power Pool and Quebec systems. Finally, a 450 mi., ± 450 kV, HVDC, 1080 MW capacity line to a load center serving the northeastern United States would cost $124 million with additional lines costing $124 million per each additional line.

In brief, the difference in the cost of the transmission facilities with or without tidal power is estimated to be about $187 million or an annual cost, taking into account operating and maintenance charges, of about $8.80/kW.

If a double-effect plant at Site 8.1 were considered with the total output being transmitted beyond the service area of the Maritime Power Pool the transmission capacity from the market to the plant would have to be at least equivalent to the installed capacity of the plant, i.e. 3536 MW, in order that the pumping energy required for double-effect operation could be transmitted to the tidal power plant and its dependable-peak energy output could be marketed. Two additional lines, at least, would have to be added. This would bring the total cost of transmission facilities for this site to about 60% of the capital cost of the tidal plant. Summarizing, annual charges of the order of $9.00/kW must be added to the production cost at-site of the tidal-power development schemes to determine the overall cost of capacity or energy in those cases of development involving the transmission of a major part of the output to Quebec and the northeastern United States.

The foregoing cost does not take into account many other benefits which would accrue from strong transmission interconnections with the market areas contiguous to the Maritime Power Pool. If effective coordination of generation planning and operation could be developed, these strong ties could provide additional benefits from the exchange of dependable peak and energy, the sharing of system reserve capacities, and for unit installation participation with systems outside the Maritime Power Pool in order to achieve the economies of scale. However, no monetary values were assigned to these benefits because they are not fundamentally related to the economic evaluation of the tidal power projects.

Reference has been made to HVDC transmission in both the ties from the New Brunswick system of the Maritime Power Pool to the Hydro-Quebec system and the northeastern United States utilities. It may be technically feasible to use HVAC lines to the northeastern United States but preliminary studies indicate the cost will be of the same magnitude as those associated with HVDC transmission.

TIDAL POWER COMBINED WITH PUMPED STORAGE

This section points out some of the interesting aspects of a pumped-storage facility in association with a single-basin single-effect tidal power development although the high cost of energy from tidal power presently precludes the economic feasibility of a pumped-storage plant based on the use of tidal energy as pumping energy.

Although pumped storage requires additional capital investment, its ability to store low-value, off-peak energy for subsequent re-use as high quality peak energy of maximum value makes it an attractive adjunct to a tidal power development. A pumped-storage facility provides flexibility to assign energy to best advantage in the system in meeting load requirements.

There is a number of potential pumped-storage sites in the Bay of Fundy region for which the average capital cost is estimated to be \$85/kW of installed capacity. Assuming adequate pumping energy available from the system, the net annual cost of a pumped-storage plant, neglecting the cost of pumping energy, would be about \$6.80/kW.

The capacity and method of operation selected for the pumped-storage plant is determined by a number of factors such as the volume of the storage reservoir, the characteristics of the output required by the system, and the installed capacity of the tidal power plant. As an example, Fig. 3 displays the net annual cost of dependable peak from a tidal power pumped-storage complex involving single-effect schemes at Sites 7.1, 7.2, and 8.1, for various capacities of a pumped-storage plant up to the value at which the entire output of the tidal plant is consumed in pumping. For the cases shown, the dependable-peak capability would be available for four hours per day, five days per week. The curves demonstrate that since the unit cost of installed capacity of a pumped-storage plant is substantially less than that of a tidal power plant, the unit cost of dependable-peak capacity produced by a tidal power pumped-storage complex would decrease as the proportion of the pumped-storage capacity to tidal plant capacity increases. The same

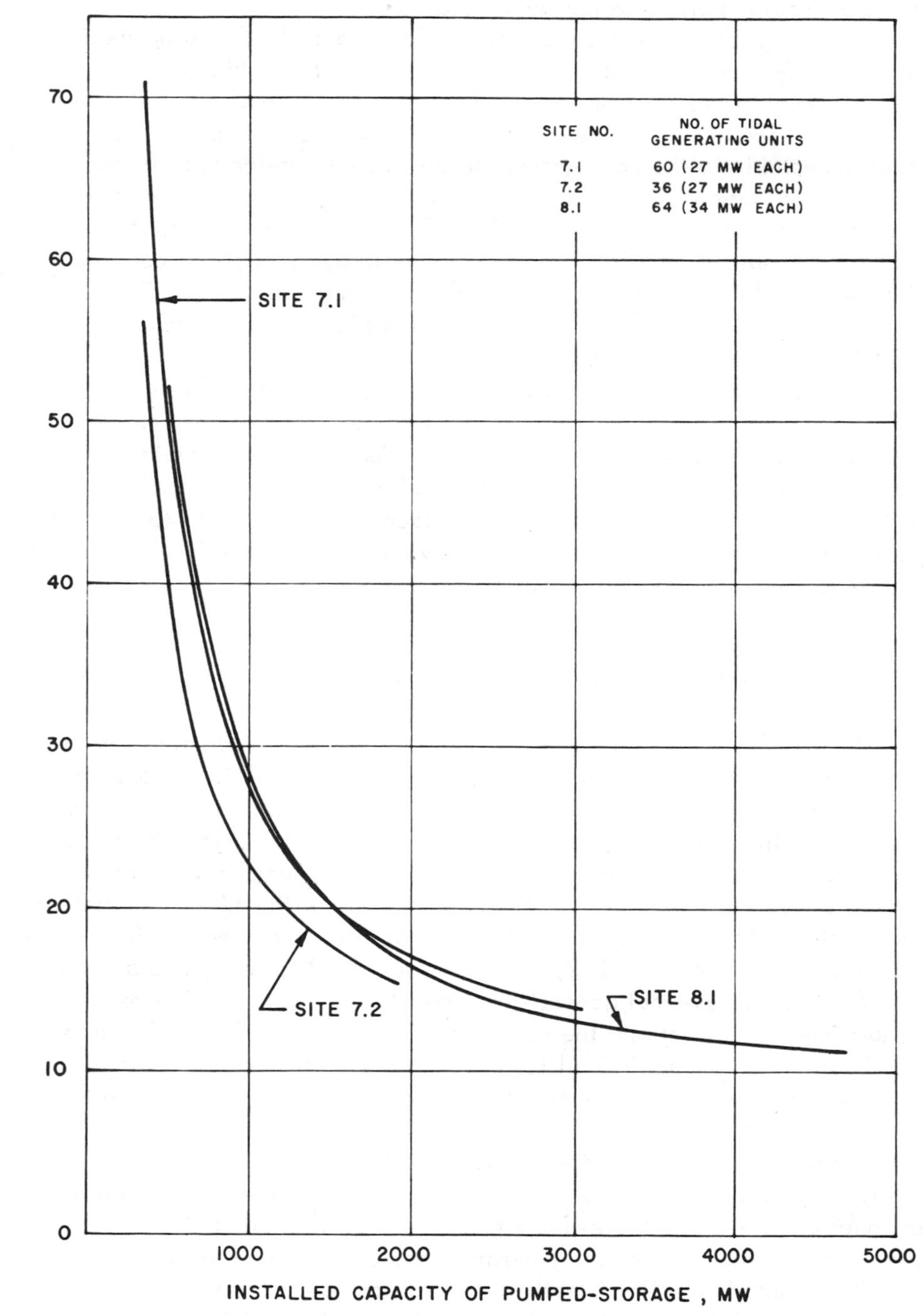

Fig. 3. Net cost of dependable peak from a tidal-power pumped-storage complex based on single-effect single-basin schemes at sites 7.1, 7.2, and 8.1.

conclusion would apply for other peak durations.

Processing the output from a single-effect tidal plant by a pumped-storage facility produces a much better quality of output which can be more readily utilized in the Maritime Power Pool than the output from a double-effect plant. The dependable output of such a complex could vary from an amount acceptable to the Pool system to an amount considerably in excess of this in which case a significant portion of the output would have to be marketed in the contiguous systems. There is a wide range of output possibilities of such complexes consisting of a pumped-storage facility in conjunction with a single-effect tidal power development at any of the sites. These possibilities would have to be investigated in conjunction with detailed market studies to select the optimum complex.

For any given pumped-storage installation, the cost of the peak capability is related to the cost of the input energy since pumped storage uses off-peak energy to fill the storage reservoir. Pumped storage could be used to improve the quality of the output of a tidal power plant. It could also be used to process off-peak energy from other sources of generation to produce dependable capacity. Actual application would be governed by the relative economy to the system in which applied.

COST OF TIDAL POWER AND ALTERNATIVES

The lunar cycle which governs the tides is constantly drifting into and out of phase with the solar cycle and hence with industrial power demands which are geared to the latter cycle. The output of a single-effect, single-basin tidal plant is in the form of large "slugs" of energy during ebb tides with no firm energy or dependable-peak capability. Such an output would command a relatively low price in the power system with which the plant would be connected. As indicated in Table 3, the at-site cost of energy from single-effect schemes at Sites 7.1, 7.2, and 8.1 would be 7.5, 8.7, and 5.6 mills/kWh. Study of the thermal generating capacity that will probably exist in the Maritime Power Pool and the contiguous power systems by 1980 indicates there should be no undue difficulties in increasing, by three percentage points, the annual capacity factor of such generation to provide 6,500 million kWh, i.e. the energy equivalent to that from Site 8.1.

A further illustration of the economic position of a tidal power scheme developing energy only is shown by Fig. 4. This shows the rate of return on additional capital expenditures for a tidal plant as a function of the cost of energy from alternative thermal generating sources. The curve was developed by the "discounted cash flow" method which is suited to the economic analyses of alternative projects for the supply of power. As is usual in this approach, all future costs are taken at current price levels. If the rate of return on the additional investment capital is less than the cost of financing, such invest-

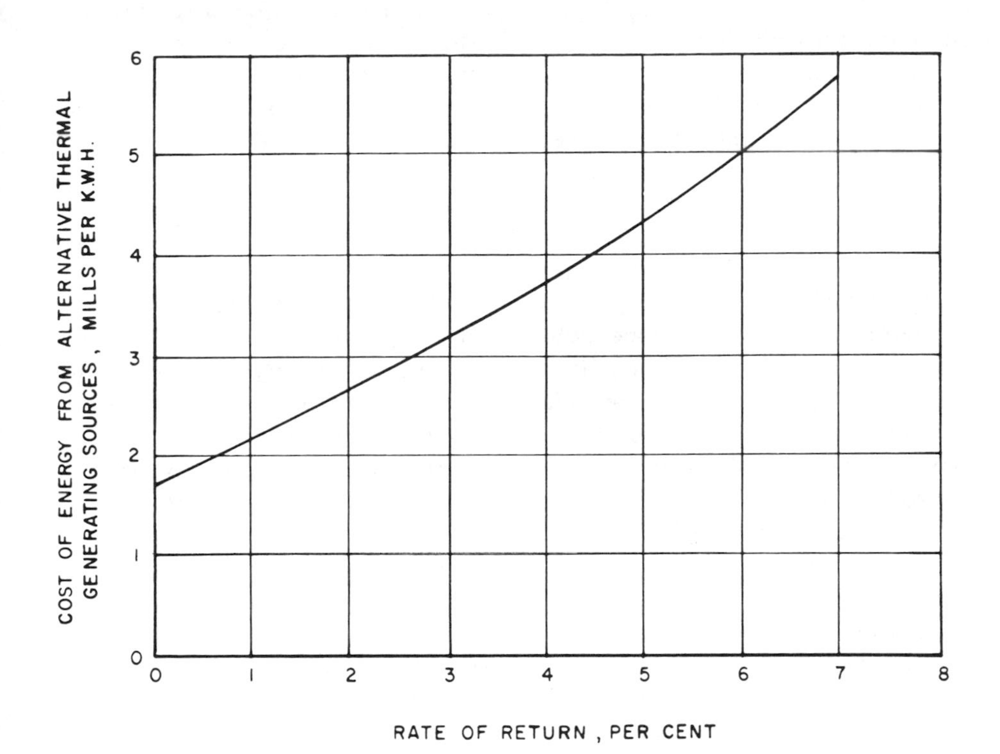

NOTE SINGLE-EFFECT TIDAL PLANT AT SITE 8.1
VS.
ENERGY GENERATED INCREMENTALLY BY THERMAL PLANTS.

Fig. 4. Rate of return on additional capital investment for a single-effect tidal power plant at site 8.1 vs. cost of energy from alternative sources.

ment is not justified.

It can be seen that the rate of return on additional capital required by the tidal plant would be about three percent for a cost of alternative energy of about 3.2 mills/kWh and would become nil if energy could be obtained at 1.7 mills/kWh. As indicated by Fig. 4, the cost of energy from alternative sources must approach six mills if a rate of return of seven percent on additional capital expenditures for a tidal plant is to be achieved. Naturally, it is inherent in this type of analysis that a change in the cost of one alternative relative to the other would change these figures.

Single-basin schemes, with generating units designed for double-effect operation and either single or double-effect pumping as well as paired and linked-basin schemes, were investigated as a means of controlling the timing of the output. Using double-effect generating units, it would be possible, by proper control of basin levels, to secure a dependable peaking capacity. In our studies, production of dependable capacity was scheduled for two hours per day from 1600 hr. to 1800 hr. five days per week. A longer period would have substantially reduced the dependable-peak output while a shorter period would have made the product considerably less attractive to the interconnected system. The variability of the output of a plant when operated for peak-power production between 1600 and 1800 hr. for the worst lunar month likely to be experienced is displayed by the duration curve on Fig. 5.

It is important to note that the pumping energy required by the plant from the system to bring the basin levels to the appropriate values at the proper times in order to achieve a peaking capability would be about equivalent to the energy produced by the plant during the period it is operating in a peaking mode. Thus, the only revenue attributable to a double-effect plant during peaking operations would be derived from its peaking capability. Therefore, with a view to maximizing the potential revenue, the selected operating schedule provided for a two-hour dependable output for 60 days (five days per week) of the high load months, December through February, and maximum energy production during the remainder of the year.

Assuming a capacity credit of $9.50/kW/yr., the at-site unit cost from double-effect single-basin schemes would be 7.7 mills/kWh for Site 7.1, 10.4 mills/kWh for Site 7.2 and 5.5 mills/kWh for Site 8.1. If the energy output were credited at 2.31 mills/kWh, the at-site unit cost of peak capacity from these schemes would be $32.65/kW/yr. for Site 7.1, $40.29/kW/yr. for Site 7.2 and $25.20/kW/yr. for Site 8.1.

The operation of a double-effect plant in the "peaking mode" requires a transmission capability equivalent to the installed capacity of the plant to bring energy from the system to the plant for pumping, and to deliver the energy production from the plant to the system. This transmission requirement would be substantially greater than that needed for delivery of the dependable-peak only and represents a severe penalty against a double-effect plant.

Our studies indicated that, in the circumstances as they are expected

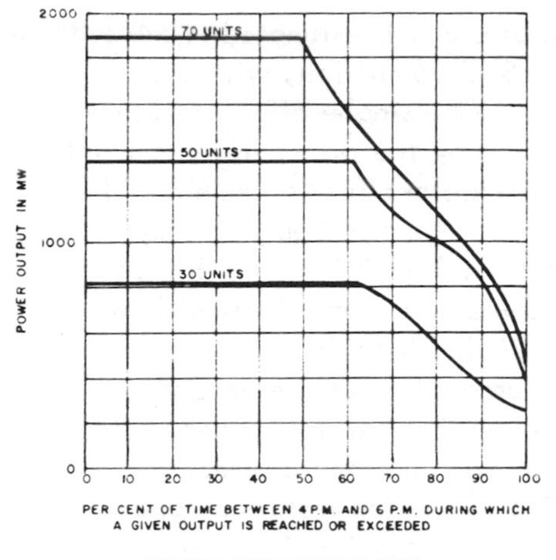

SITE 7.1

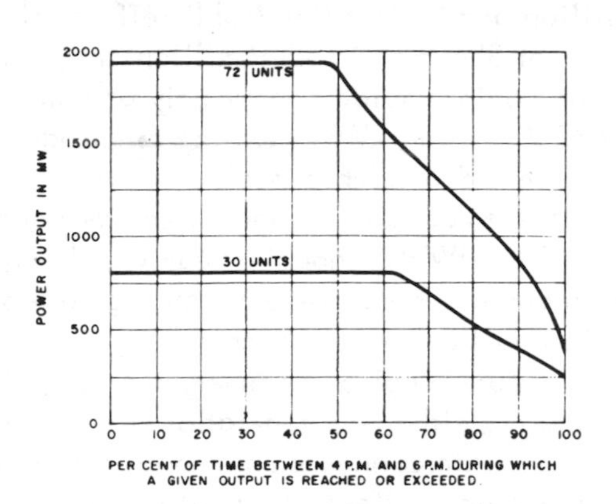

SITE 7.2

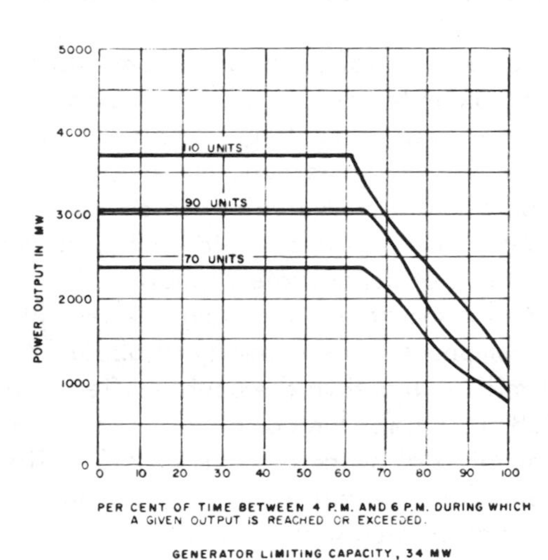

SITE 8.1

NOTE:
DERIVED FROM CALCULATION FOR THE LUNAR MONTH OF JANUARY 1968-61.

Fig. 5. Variability of dependable-peak output of a double-effect single-basin tidal power plant.

to exist in the market areas considered, a single-effect tidal power plant operated in conjunction with a pumped-storage plant is much more attractive than a tidal plant designed and operated in a double-effect mode. This is illustrated in the following discussion.

The cost of the single-effect scheme at Site 8.1 is estimated at $473.8 million and that of the double-effect scheme $723.6 million, with annual costs of $36.4 and $55.9 million, respectively. The single-effect development would provide energy only at times dictated by the tides, in the amount of 6500 million kWh. On the other hand, the double-effect scheme could provide a dependable-peak capability of 1526 MW for two hours per day, 60 days a year and an annual energy generation of 7560 million kWh. Thus the 1526 MW dependable-peak capability and the additional 1060 million kWh annual energy requires an increase of $249.8 million in capital cost and $19.5 million in annual cost. Moreover, the production of the dependable peak would require, at times, the full utilization of the installed generating capacity, either for generation or for pumping, with power for pumping drawn from the interconnected system at a rate equivalent to the installed capacity of the double-effect plant with payment made at the applicable values for energy. Thus to produce a dependable peak of 1526 MW, a transmission capacity of 3536 MW would be required, which is 2.3 times the dependable-peak capability.

By contrast, it is assumed that 1526 MW of pumped-storage capacity could be installed at an approximate cost of $85/kW and operated in conjunction with a single-effect plant so as to meet equivalent dependable-peak requirements, i.e. 1526 MW for two hours/day, 60 days/yr. The capability of such a complex would be essentially equivalent in energy production to that of a single-effect plant at Site 8.1 and would be equivalent to a double-effect plant at that site in terms of dependable capacity. The capital cost of the complex would be that associated with a single-effect plant ($473.8 million) plus 1526 MW of pumped-storage capacity at about $85/kW, about $603.5 million or $120.1 million less than the double-effect plant. Annual costs would be about $46.8 million for the complex compared to $55.9 million for the double-effect plant (after allowing a credit at 2.31 mills/kWh for the additional 1060 million kWh of output from the latter plant). Moreover, the transmission requirement associated with a double-effect plant would be at least 1400 MW greater than that for a single-effect scheme at Site 8.1. The energy component of a double-effect plant could, at times, be of marginal value since alternate generation must be available to pick up and back off load in relation to generation from the tidal cycle. By contrast, a single-effect tidal power and pumped-storage plant complex provides firm capacity and energy.

A comparison of the at-site costs of a single-basin double-effect development at Site 8.1 with alternative power sources producing the same amount of peaking capability and energy output is shown by Table 5.

Table 5

At Site Costs of Dependable Peak from a Double-Effect Single-Basin Tidal Power Plant at Site 8.1 and from Alternative Sources (1)(2)

Source of Generation	Installed Capacity MW	Dependable Peak At Site MW	Annual Energy At Site Millions kWh	Dependable Peak $/kW/yr.
1. Double-Effect Tidal Power Plant at Site 8.1	3,536	1,526	7,560	25.20[3]
2. Single-Effect Tidal Power Plant at Site 8.1 with 1,526 MW Pumped-Storage, Two Hours per Day, 60 Days per Year	2,176 1,526	1,526	6,271	21.17
3. Double-Effect Tidal Power Plant at Site 7.1	2,916	1,260	5,402	32.65[6]
4. Single-Effect Tidal Power Plant at Site 7.1 with 1,526 MW Pumped Storage, Two Hours per Day, 60 Days per Year[4]	1,620 1,526	1,526	3,971	21.43
5. Gas-Turbine Power only Providing 1,526 MW, Two-Hours per Day, 60 Days per Year	1,526	1,526	-	9.86
6. Pumped Storage Using Incremental Oil-Fired Steam-Electric Energy Providing 1,526 MW Dependable Peak, Two Hours per Day, 60 Days per Year[5]	1,526 -	1,526	-	7.32
7. Pumped Storage using Incremental Candu-Type Nuclear Energy, Providing 1,526 MW Dependable Peak, Two-Hours per Day, 60 Days per Year	1,526 -	1,526	-	6.95

(1) Cost of money is assumed to be 7%. (2) Annual costs of alternatives include all direct and indirect charges. (3) The cost of dependable peak has been derived by deducting from total annual costs a credit for energy at 2.31 mills/kWh. (4) Pumped-storage capital costs taken at $85.00/kW and annual costs at $6.80/kW.
(5) Cost of energy from existing steam-electric plants operated at higher capacity factors or incremental energy from oil-fired, steam-electric sources taken at 3.2 mills/kWh. (6) Cost of incremental energy from candu-type nuclear reactors taken at 0.58 mills/kWh.

A paired-basin scheme was investigated to determine to what extent dependable-peak capability could be enhanced by developing two similar single-basin double-effect schemes for coordinated operation. For this study and to secure a measure of the improvement that might be obtained with such operation, Sites 7.1 and 7.2 were selected, each with an installation of thirty-two 27 MW double-regulated generating units, making a total installed capacity of 1728 MW. This installation was selected as an example to give a dependable-peak output close to that which would be marketable in the northeastern United States. With coordinated operation, these two double-effect plants could provide a dependable-peak of 941 MW for 2 hours per day. When operated as a peaking plant, five days/week for 60 days/year, the total energy produced during the remainder of the year, when each plant could be operated independently, would be 2940 million kWh.

There can be considerable flexibility in the coordinated operation of such plants. The operation chosen for this example gave the highest guaranteed production from 90 to 100% of the peak demand periods. The dependable peaking capability of this example of a paired-basin scheme would be about 25% greater than that of the combined capability of the schemes operated independently. The unit cost of dependable peak was found to be $38.90/kW/yr. after allowing an energy credit of 2.31 mills/kWh for the energy production.

It is concluded that two or more basins with about the same installed capacity can be coordinated to provide a greater dependable-peak capability than would be provided by the sum of their peaking capabilities when operated independently. Therefore, the unit cost of dependable-peak capability would be less for a paired-basin scheme than the average of such unit costs from the same developments operated as single-basin schemes. If double-effect developments were to be given further consideration, the pairing of two single-basin developments would merit further investigation.

The advantage of a linked-basin scheme to a system is that it can be operated to provide continuous power and energy. Although Shepody Bay and Cumberland Basin would appear to be suited to such a scheme, the high construction cost relative to its output was shown to make it a considerably less attractive development than the double-effect single-basin schemes or the single-effect single-basin schemes combined with a pumped-storage plant.

Tables 3 and 4, showing at-site costs only, indicate that either a single-effect or double-effect scheme at Site 8.1 could produce power and energy more economically than any of the other schemes, provided all of its output could be marketed. The unit costs of energy from single-effect developments at Sites 7.1 and 8.1 would tend to become equal for scales of development less than the optimum in each case.

The determination of the economic feasibility of tidal-power development has been based on comparisons and evaluations of at-site costs. The transmission of the output to markets would require major expenditures as noted earlier. If tidal power were to become competitive on an at-site basis with alternative sources of generation, additional, more detailed analyses involving power-system studies would be justified.

Tidal power developments, while capital-intensive, have a long economic life. This has been placed at 75 yr. in this investigation. On the other hand, alternative generation sources such as fossil-fuel-fired and nuclear thermal power plants are not likely to have economic lives in excess of 30 and 25 yr. respectively.

When a fossil-fuel-fired or a nuclear thermal power plant is considered as an alternative to a tidal power development the fossil-fuel-fired plant would have to be completely replaced 1.5 times and the nuclear thermal power plant twice in order to provide the same length of service as the tidal power plant. Consequently, the capital costs of the replacement plants are subject to increased cost if inflation occurs during the time from the date of initial construction to the time at which the replacement plants are built. This is a particularly significant aspect for the tidal power plant since its capital cost, although tending to be high, is incurred only once during its economic life of 75 yr. In connection with the foregoing, it is considered that neither the economic life of a nuclear power plant nor its true cost to the community are known with certainty.

If inflation were to occur at the rate of 2.5% per annum, the first replacement for the fossil-fuel-fired power plant would cost 2.1 times that of the initial plant and for the nuclear power plant the first replacement would cost 1.85 times that of the initial plant. Technological advances would tend to offset the effect of inflation.

Inflation in fuel costs might become of considerable significance during the course of 75 yr. The tidal power plant has no fuel costs, while thermal alternatives do. Such fuel costs sometimes constitute as much as 80% or more of the production costs. Inflation in fuel costs might range to as much as one percent per annum. The effect of this, using 0.5% as an example, would be to increase fuel costs over 75 yr. by 1.45 times. Any such inflation in fuel costs would tend to improve the economic position of tidal power.

COMPARATIVE RELIABILITY OF TIDAL POWER AND ALTERNATIVES

A tidal power plant would have a somewhat greater reliability than most other forms of generation. This results mainly from two factors: the many small generating units comprising the plant; and the predictability of the tides.

Experience has shown that several small generating units need less reserve equipment than one larger unit of equivalent total capacity. In a tidal power development, the low head and great amount of water available require a large number of relatively small units. As a result, there would be a certain reliability credit due to this characteristic. Moreover, the tides are predictable so that it is possible to determine the energy and dependable-peak available from a tidal power plant for several years in advance.

The greater reliability of hydraulic generating units was demonstrated in striking fashion during the great blackout of November 9, 1965, in Ontario and the northeastern United States.

COMPARATIVE POSITION OF TIDAL POWER AND ALTERNATIVES AS SOURCES OF POLLUTION

During the last few years the pollution caused by various types of thermal power plants has been a subject of increasing concern. It is receiving closer scrutiny from many elements of the community. Tidal power, however, has the attractiveness of being free from any aspect of pollution, whether water-borne or atmospheric.

Fossil-fuel-fired power plants can cause environmental pollution in three ways: by the emission of gaseous effluents; by the exhaust of particulate matter to the atmosphere; and by increasing water temperatures of rivers and lakes into which the cooling water may be discharged. A fourth minor problem is the disposal of slag and fly ash from coal-fired plants. Nuclear power plants give rise to greater problems of water temperature increases and, in addition, pose a further ecological problem of the safe disposal of radioactive waste from the burn-up of the fuel elements.

In fact all thermal power plants are potential sources of air, land and water pollution in one form or another. As such their potential effect on the health of the human community, and on the ecology in general, warrant close attention.

INCOME OR SAVINGS

Our modern civilization is based on energy and to a greater extent each year on electric power because of its convenience and cleanliness, its unique ease of application to the lightening of mankind's physical burdens, and augmentation of his comfort and ease.

It is, however, all too oftenforgotten that most sources of energy and of electric power rely on fossil-fuels--coal, oil, natural gas and uranium -- that is, on fuels which are depleted with use. When we consume them, we are digging into our savings account provided by a beneficient Creator to serve the world for as long as it may exist. Evidence of approaching exhaustion of these depletable fuels is accumulating. Costs are increasing. Fortunately there are a few sources of energy which are essentially income-type fuels: interest, as it were, on our inheritance. These are water power, geothermal power, ocean thermal power, solar power, wind power and tidal power. Of these, only water power, geothermal power and tidal power can be readily tapped for man's use. Interestingly enough these income-type energy sources involve no air, water or land pollution.

TIDAL POWER PRODUCTION: SOME EQUIVALENTS

Furnace oil is delivered in Halifax in 1,000 gal. lots at about $0.16/ imperial gal. Assuming heat content of 168,000 Btu per gal., furnace efficiency of 50%, a gallon is equivalent to 24.6 kWh or, in other terms, a kilowatt-hour of energy in the form of heat costs 4.05 mills, neglecting maintenance and equipment costs.

Taking 35 imperal gal. per barrel or six million Btu per barrel and a heat rate of 9,200 Btu/kWh, which corresponds to the rate obtainable from modern 500 MW fossil-fuel-fired thermal units, the yearly output of a single-effect single-basin scheme of development at Site 8.1 would correspond to 10.1 million barrels of oil yearly.

ACKNOWLEDGMENTS

The investigation and studies which form the subject of this paper were carried out under the general direction of the five-member Atlantic Tidal Power Programming Board constituted under the Inter-governmental Agreement of August 1966, between the governments of Canada and of the Provinces of New Brunswick and Nova Scotia. Responsibility for the supervision of the engineering and management aspects was vested in the five-member Atlantic Tidal Power Engineering and Management Committee. Development of investigations and studies, supervision of work by consultants and staff, was the responsibility of the author in his capacity as Study Director.

Grateful acknowledgment is made of the many contributions to the successful accomplishment of the author's task by those noted. The author's presentation is, in fact, a synthesis of many varied contributions.

MATHEMATICAL MODEL OF TIDAL REGIMES IN THE BAY OF FUNDY

F. E. Parkinson*

INTRODUCTION

The mathematical model which is the subject of this paper was one of several studies carried out recently by our company for the Atlantic Tidal Power Programming Board. In turn, our total contribution made up only a part of the input to the A.T.P.P.B. overall study to evaluate the economics of tidal power development on the Bay of Fundy.

Steps preceding this model included selection of the tidal parameters, project sites and equipment characteristics, as well as a preliminary comparison and finally optimization of the absolute maximum energy production, irrespective of the time of production, for the sites retained (see Fig. 1). The tidal regime model was introduced at this stage to establish the modifications of tidal amplitudes and phases caused by construction and operation of the various schemes. Reductions in production could then be evaluated and applied to the results from the earlier optimization studies.

These corrected values then allowed a further selection of the most promising sites, which were later subjected to more detailed computer comparisons. This advanced work was based on optimum yearly return, taking into account the variation of energy values with respect to time of day, as well as the dependable late afternoon peak power. It also included consideration of the tidal regime modifications.

The work described in this paper was done with technical assistance by Sogréah, of Grenoble, France. Their previous work on the Rance Estuary served as the basis for the programs applied to the Bay of Fundy study.

---

*Vice-President and Development Manager, LaSalle Hydraulic Laboratory, 0250 St. Patrick Street, LaSalle, Quebec, Canada.

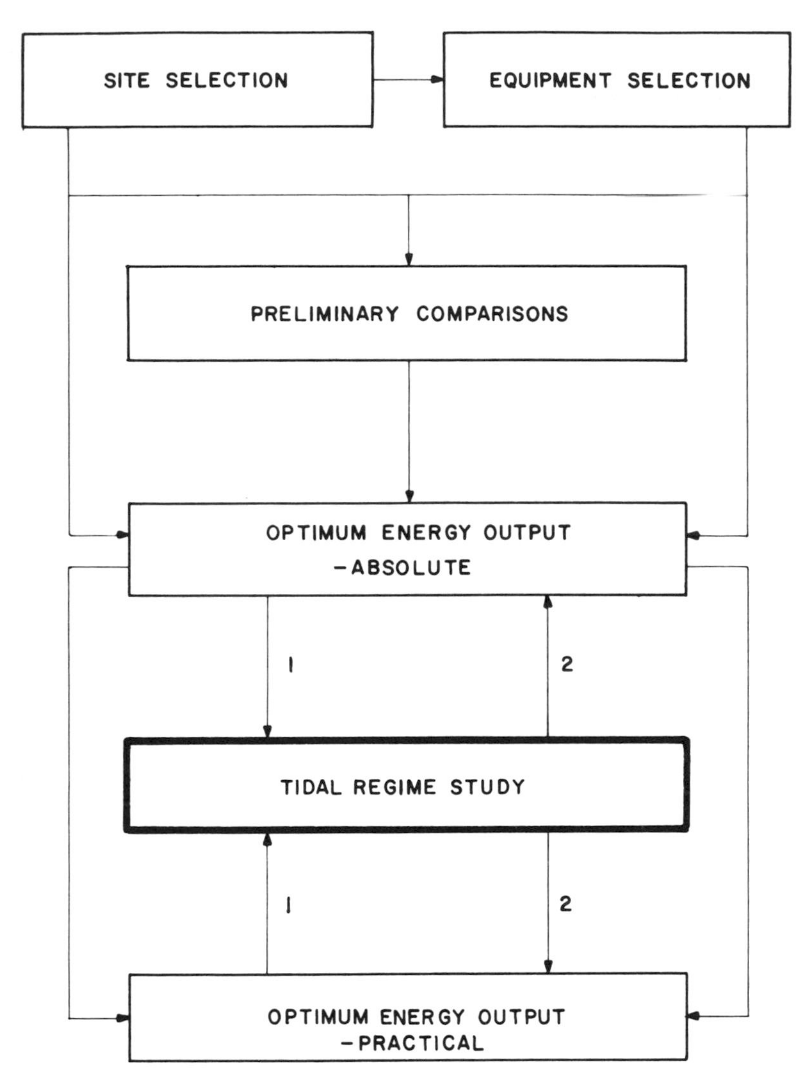

Fig. 1. Bay of Fundy. Block diagram of study series, by Lasalle Hydraulic Laboratory.

PRINCIPLE OF THE MODEL

The amplification of tidal ranges that occurs going from the mouth to the head of the Bay of Fundy is characteristic of the bay in its natural state. It may be attributed to:

1. the gradual shelving and narrowing of the bay (Fig. 2 and 3);
2. near-resonant conditions for the semi-diurnal tidal fluctuation (the main semi-diurnal component has a period of 12 hr. 25 minutes). In fact, the bay is a little on the short side as far as resonance is concerned.

Blocking off part of the bay near its head may understandably lower tide ranges somewhat, partly on account of a further shift, due to shortening, away from full resonance, and partly (when the tidal plant is running) on account of the artificial ebb and flow imposed on the natural currents and more or less at variance with these.

The basic principles involved in a mathematical representation of tidal propagation are well known and have been described in numerous publications. In our case, the two basic equations (those of continuity and of motion, including of course the Coriolis acceleration) were written in finite difference form, and the resulting system of algebraic equations was solved on a step-by-step basis.

The term in the equation of motion corresponding to variations in velocity head could have been, but was not, introduced into the computations; this was done in order to economize computer time. While resulting in possible phase changes, it is not believed that the neglect of this velocity head term would affect tidal ranges appreciably. The effect would be much smaller still on the tidal differentials resulting from the construction of a power plant.

The assumption of a uniform velocity distribution at any vertical section is basic to the mathematical representation used. Consequently local discrepancies may arise when there is a sudden large change in width in the prototype, capable of modifying the velocity pattern in plan (the flow near Cape Split is an example). The mathematical model does not represent possible vortex formation resulting in this way. The effect on tidal range is likely however to be much smaller than that on local current patterns.

The foregoing representation results from many years of experience on the part of Sogréah. The accuracy of this type of mathematical model has been fully confirmed by comparison with physical models (specifically, the rotating model of the English Channel built in France).

The initial state of the bay must be chosen arbitrarily in the very first computer run; in actual fact, a state of rest was chosen. The tidal wave resulting from the constantly varying level differences applied at the outer boundary of the model then propagated into the bay, and the resulting disturbance settled down to a constantly repeating pattern after a certain time (it was assumed that the imposed root mean square tidal range remained constant with time (see Fig. 5). This equilibrium condition was usually reached after

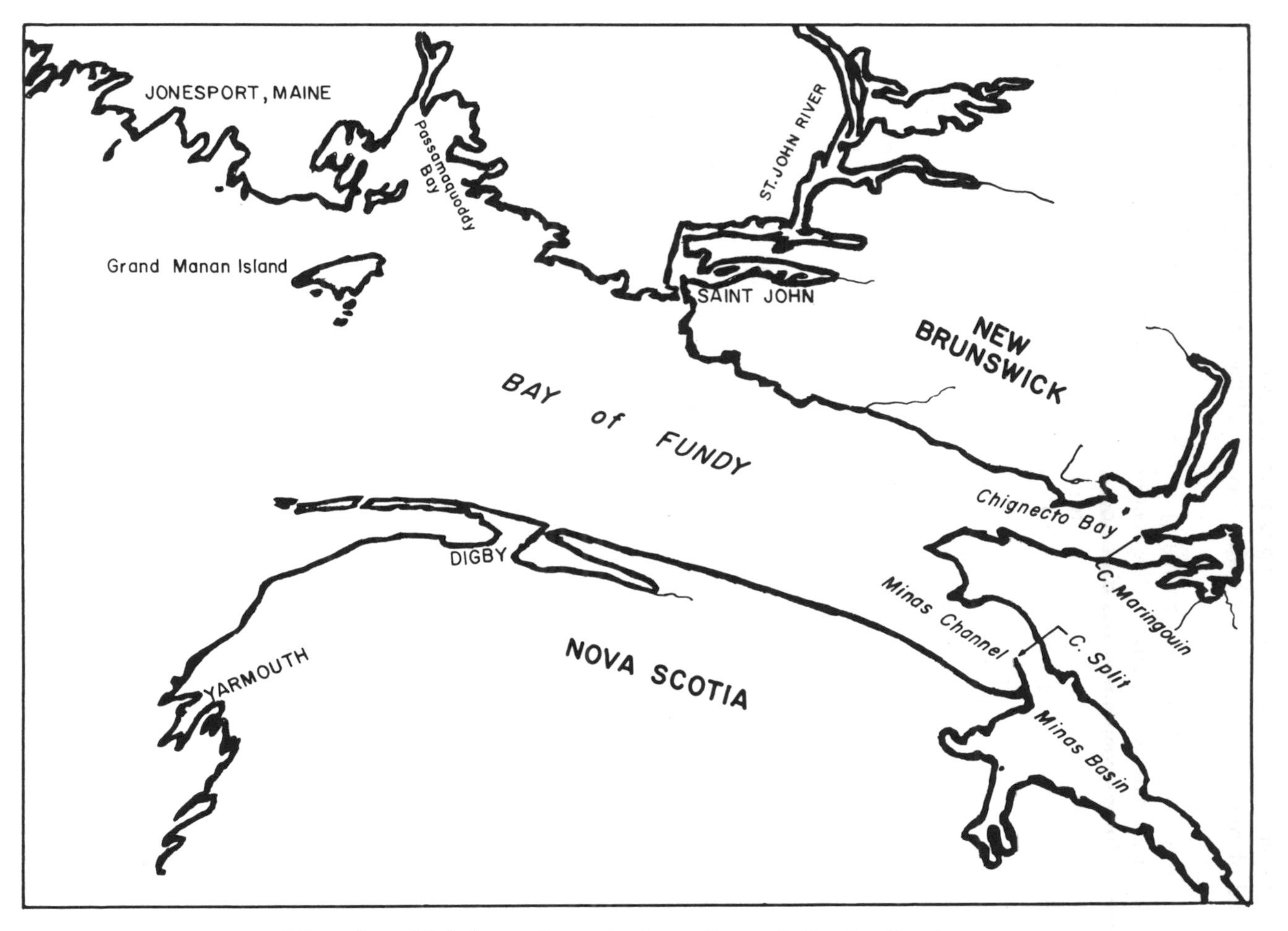

Fig. 2. Tidal regime study. Bay of Fundy Region.

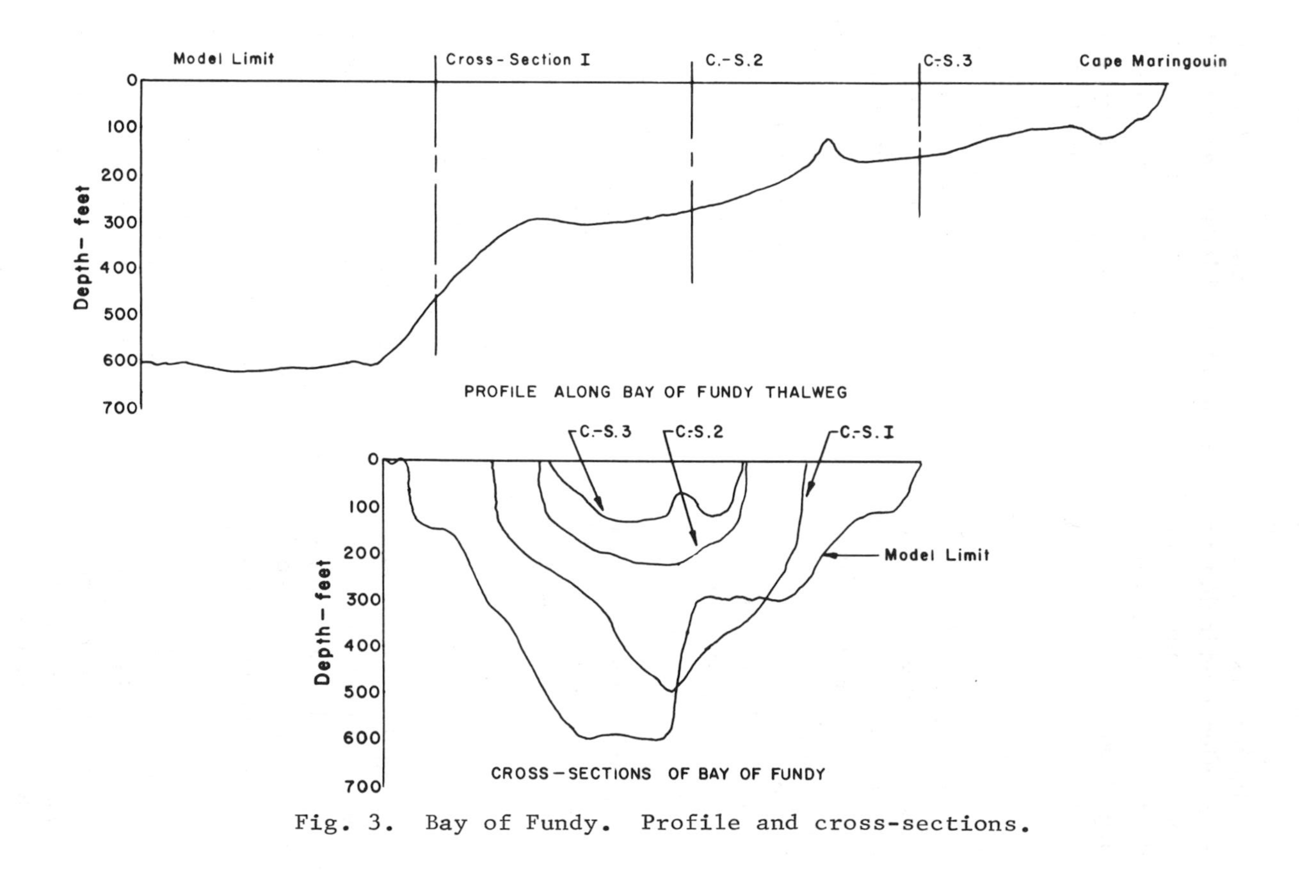

Fig. 3. Bay of Fundy. Profile and cross-sections.

12 to 15 tide cycles. Subsequent computer runs, e.g. with a tidal power plant in place, could be started from a later state in the tidal propagation process by storing intermediate data from the first run, with a consequent saving in computer time at no loss in accuracy.

Particular features in the present mathematical model are as follows:

1. The grid system used includes points at which tidal levels were computed interspersed with those at which currents were computed. Each current point features two vector components computed simultaneously (with a given grid spacing, the currents were thus more accurately represented than if one component were to be evaluated at each alternate current point in the grid); (see Fig. 4).
2. The grid was curvilinear orthogonal and could hence be adapted better to an irregularly shaped estuary than a square grid of unchanging size and angular inclination. This approach meant that the mesh length between neighbouring "current" and "level" points was not constant, but ranged from a maximum of 26,000 ft. to a minimum of 8,000 ft. approximately.
3. The shape of the estuary could be approximated closely by using the allowance in the model for truncating the grid system at any point either along or at right angles to the grid or, under certain restrictions, for cutting across the grid at 45° to the curvilinear coordinate system.
4. At points on the periphery of the mathematical model, i.e., either on the ocean side, or at the inlet to bays (such as Passamaquoddy Bay) not fully represented, it was possible to select at will values of levels, currents or discharges or alternatively, to incorporate so-called absorbers (simulating a transmission and loss of energy outwards from the model).
5. The construction of a tidal power plant for example could be readily simulated by truncating the "natural" model of the bay at the proper points, as under (3) above, and then by specifying the appropriate discharges through the newly created barrier using the facilities described in (4).
6. The seaward limit of the model was fixed as a straight line running from Yarmouth (more precisely, from a point 43° 45' N 66° 07' W) to Jonesport, Maine. This boundary actually cut across the main reference grid of the model at 45°. It should be noted that the selection of a proper seaward limit for the model, the imposed tidal conditions at which are supposed to be unaffected by any of the changes studied within the model itself, is of much importance. If placed too far out to sea, the seaward limit would lead to an inordinate increase in the duration of the computer runs; if too close in, the model would yield unduly optimistic results. Any reduction in tidal amplitude that may be felt in fact at the seaward limit would then be eliminated, thus amplifying modified tidal ranges within the model above their true new values.
7. On the inner side, the area of water covered by the model reached past

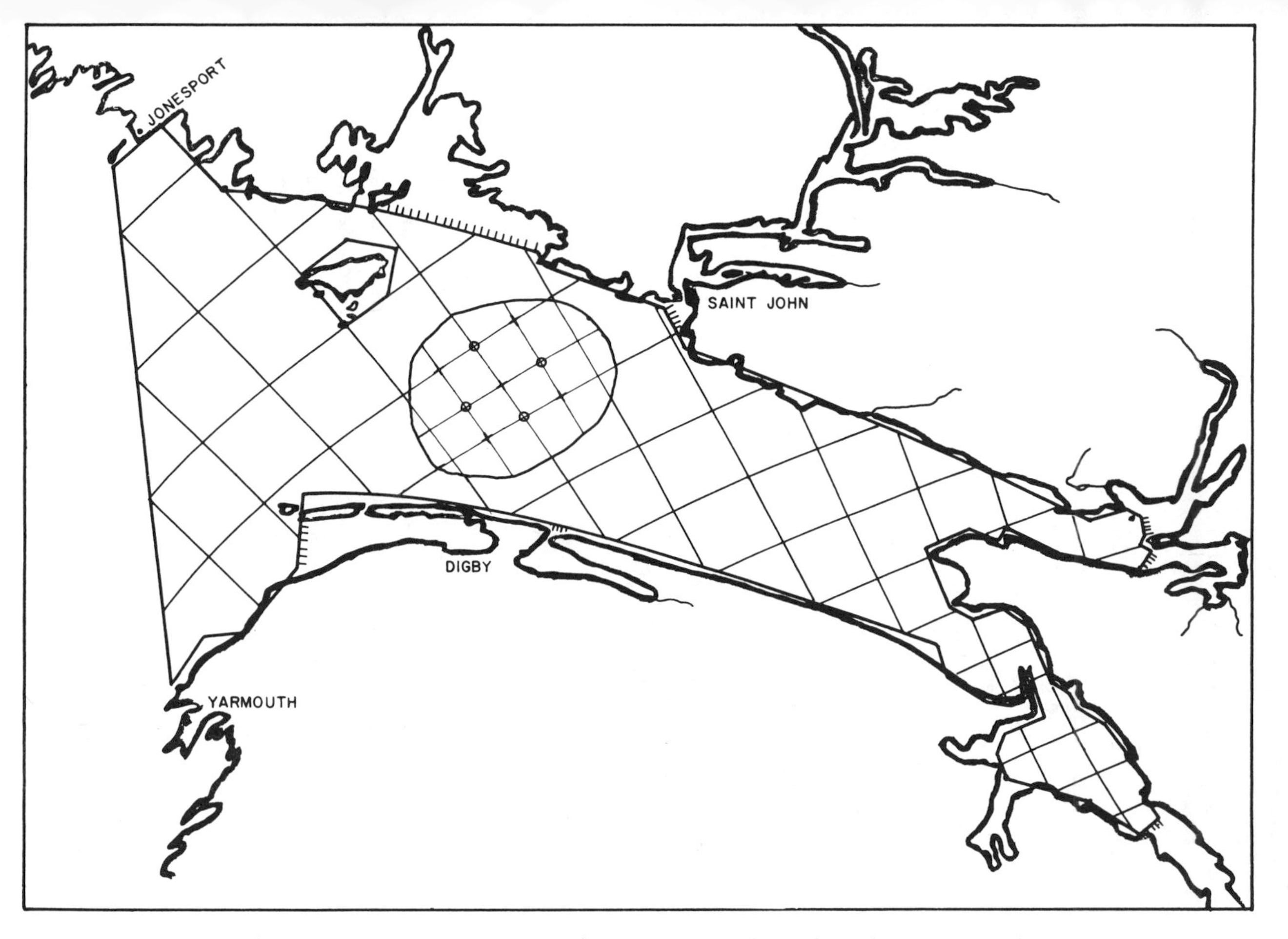

Fig. 4. Bay of Fundy. Tidal regime study, sketch of grid layout.

Grindstone Is. on the Chignecto Bay side and past Economy Point on the Minas side. Grand Manan Island was represented for example by a pointed five-sided overlay blocking out a few meshes in the grid system. Another typical feature, Cape Split, was represented by a promontory with a tip embracing an angle of 45°.

TIDAL, GEOGRAPHICAL AND HYDROGRAPHIC DATA

Descriptions of the geographic shape of the bay and the depths of the sea bed as presented on the standard hydrographic charts prepared by the Canadian Hydrographic Service, were adequate for setting up the model. Given such information and knowing the hydraulic friction everywhere within the Bay, a complete mathematical model could in theory be prepared giving, for any tidal variations imposed at the outer limit of the model (i.e., on the ocean side), the tidal level and tidal stream variations within the model as a function of time.

However, the frictional resistance was not known within the area, and this called for an adjustment procedure based on other information, and particularly, on level data (current data are a little more difficult to interpret, seeing that the model was concerned with average velocities over the whole depth. Measurements of currents, on the other hand, relate to definite depths, often near the surface at which the velocity may markedly exceed the mean).

One particular requirement which was hard to fulfill relates to the outer boundary of the model, on the ocean side. It was necessary to assume either a tidal stream or a tidal range distribution there. Finally, an assumed distribution of tidal ranges and phase lags was used. This information could not be based on measurements, which in themselves would have been very difficult to carry out (small tidal ranges would have to be recorded in water of considerable depth at a considerable distance from the coast in most cases) even had the limits to be chosen for the model been known a long time ahead.

Despite these difficulties, it is believed that a satisfactory mathematical representation of the bay was achieved. The data actually used came from a number of sources. Most important were:

1. tidal differences contained in the Canadian Tide and Current Tables;
2. a large number of hourly tidal readings (and in some cases copies of actual tide gauge recordings) for a number of recent years, the most complete coverage being for 1965, date of an extensive Tidal and Current Survey by the Bedford Institute of Oceanography.

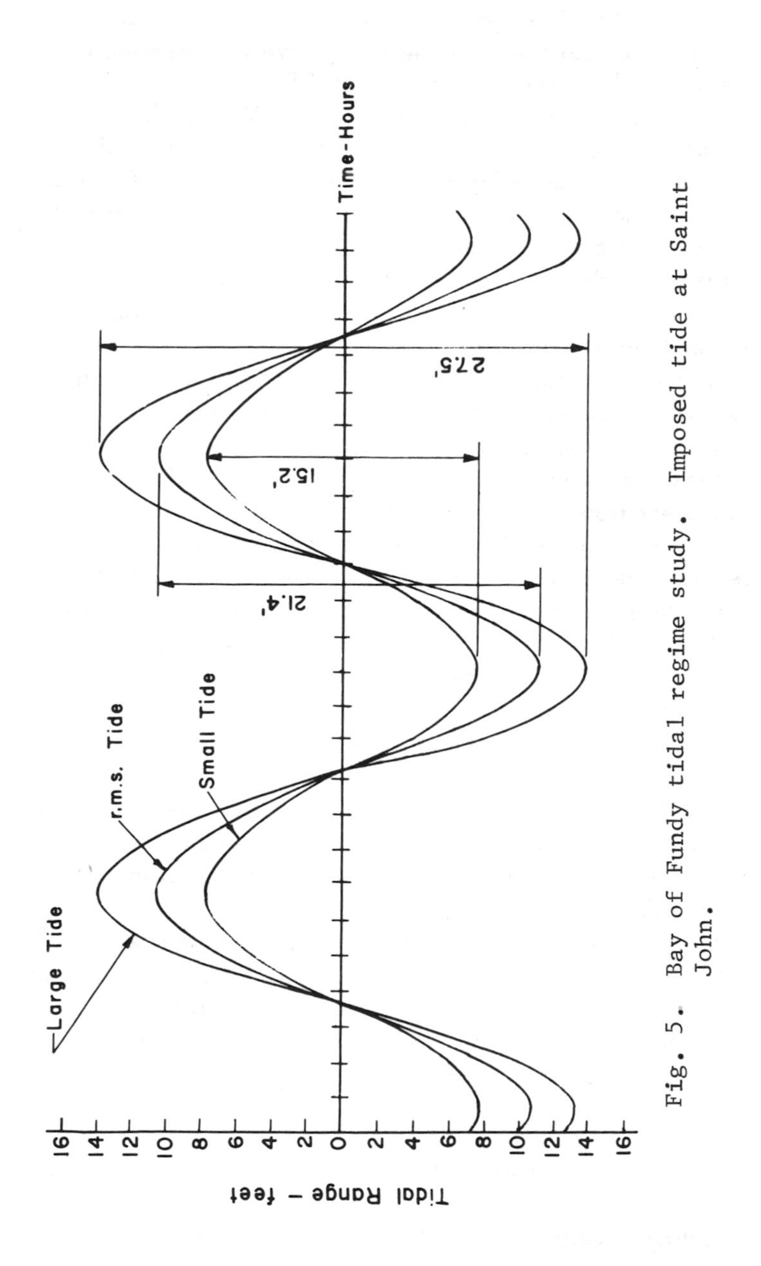

Fig. 5. Bay of Fundy tidal regime study. Imposed tide at Saint John.

ADJUSTMENT OF THE MATHEMATICAL MODEL

Tidal recordings and tide tables were used to deduce tidal ranges at various points around the model for the root mean square tide of 21.4 ft. at Saint John, N.B. The real shape of the tidal wave was imposed on the model rather than a sinusoidal approximation (see Fig. 5).

On the seaward boundary, running from near Yarmouth, N.S. to Jonesport, Maine, as previously mentioned, no information at all on tidal ranges was available at intermediate locations across the Gulf of Maine. Taking adequate estimates of the tidal range and phase at either end of the line, it was supposed that the phase differences were evenly distributed across the Gulf, and that the tidal range increased steadily from either side towards the centre (a parabolic distribution was used, with the amplitude at mid-point about 16% higher than at Yarmouth).

Appropriate tidal conditions had to be assumed at the entrance to those parts of the Bay which for various reasons (drying up at low tide is one of them) could not be represented on the mathematical model. The assumptions made are not completely arbitrary; the required law for filling the basins is known approximately from continuity considerations.

Once the foregoing elements had been fixed, the only item remaining to be specified was the bottom roughness. The roughness used was calculated on the basis of prior experience on models of the Rance Estuary in France, and in the English Channel. It is worth pointing out that no adjustment had to be made to this value, which was selected prior to the adoption of the basic data, hence prior to knowing how the model would actually work out; the excellent agreement between model and the prototype tides is thus a justification of the roughness chosen.

Fig. 6 shows the 11 points on the model at which the tide ranges and times of high and low water were computed and checked. The four points shown with the actual levels and times given serve as indicators of the degree of precision of the model reproduction.

It is quite probable that an equally satisfactory agreement could be reached with a somewhat different value of roughness coupled with somewhat different boundary conditions on the Gulf of Maine in particular. However, there was little point in pursuing this matter further, unless substantial additions to the field data had been available.

STUDIES CARRIED OUT

The preceding developments had provided the working model of the bay's tidal movements under the existing natural boundaries. Tidal ranges under those conditions had been used for the energy optimization studies, so that the next step, and the object of the present models, was to compute what differences

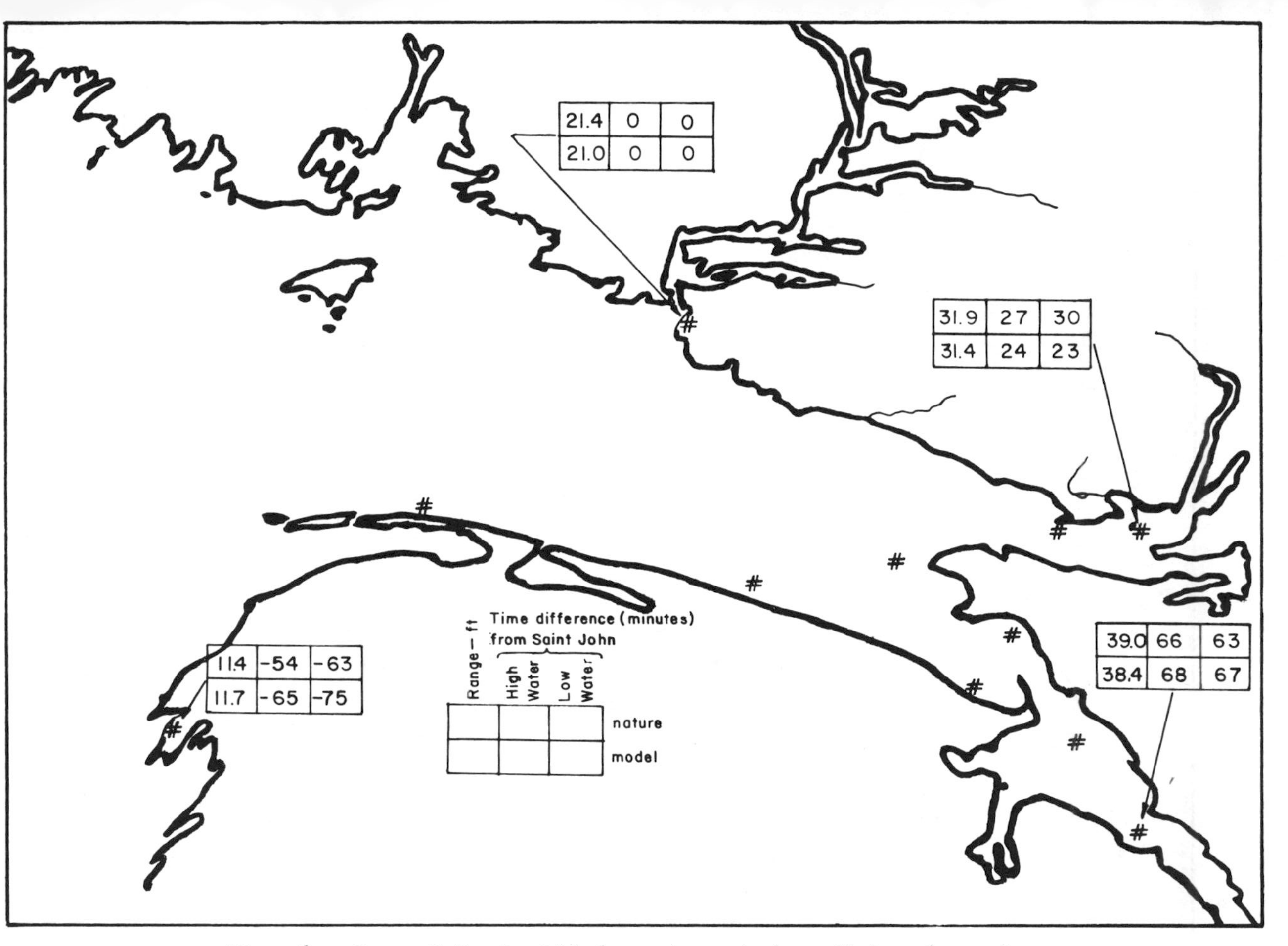

Fig. 6. Bay of Fundy tidal regime study. Natural regime adjustment results.

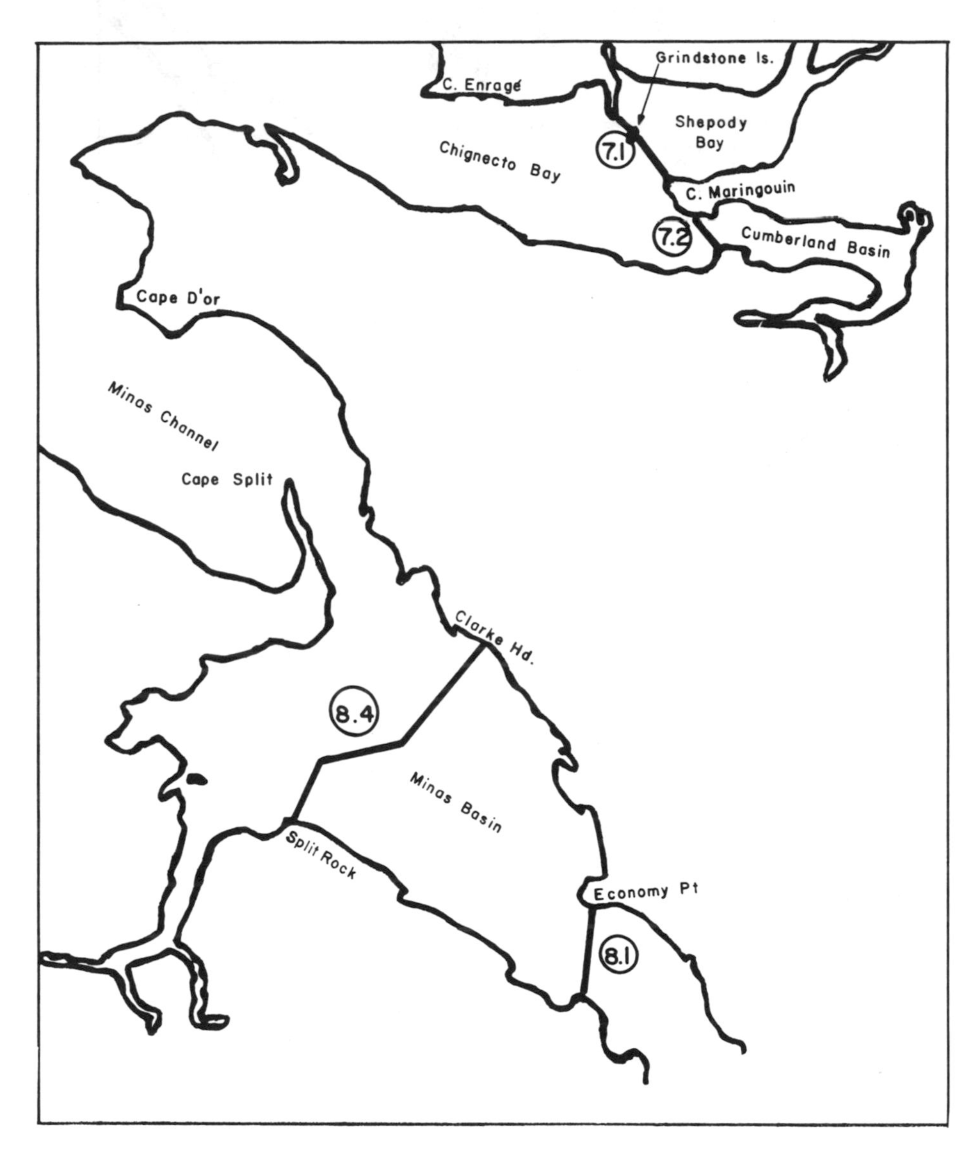

Fig. 7. Bay of Fundy. Tidal regime study, site locations.

would be introduced to these ranges by construction of the power schemes.

The following plant sites (see Fig. 7) and operating conditions were therefore put into the model and subjected to the r.m.s. tide:

Site 7.1 - Shepody Bay

Plant stopped; 50, 100 and 150 units in operation.

Site 8.1 - Economy Point

Plant stopped; 45 and 93* units in operation.

Site 8.4 - Minas Basin

170 and 358 units in operation.

Sites 7.1 and 7.2 - Shepody Bay and Cumberland Basin

(150/102)* units, respectively.

Sites 7.1, 7.2 and 8.1 - Shepody Bay, Cumberland Basin, Economy Point

(150/102/93)* units, respectively.

All the above cases were analyzed for a 21.4 ft. tide under natural conditions at Saint John, those with an asterisk being treated for larger and smaller tidal ranges (15.2 ft. and 27.5 ft. at Saint John). The large and small tidal ranges actually resulted from adjustments of ±30% (see Fig. 5) to the initial mean ranges set on the outer boundary of the model running, as will be recalled, from Yarmouth to Jonesport.

Four other cases were treated to check suggestions on the possibility of improving the tidal range in some parts of the Bay. The following cases were studie d(see Fig. 8):

1. Effect of the removal of Cape Split.
2. Effect of placing dykes between Cape Chignecto and Ile Haute and between Ile Haute and Cape d'Or.
3. Effect of a spur dyke situated on the North side of Minas Basin and penetrating about 25,000 ft. into the basin.
4. Effect of a spur dyke located off Cape Enragé and penetrating about 17,000 ft. into Chignecto Bay.

The first modification led to a fairly marked reduction in tidal range in Minas Basin and an appreciable one elsewhere. The probable explanation is that excavation of Cape Split reduces by about 15 to 20 minutes the time lag of the tides within Minas Basin and so accentuates the departure from resonance.

The second modification led to a small reduction of the tidal amplitude in Chignecto Bay and a slight increase in Minas Basin. The increase corresponds most likely to an increase in the tidal lag which brings Minas Basin slightly closer to full resonance.

Neither of the two dykes described under 3 and 4 altered appreciably the tidal regime. The only effect is a very slight reduction; a few tenths of a foot in the tidal amplitude.

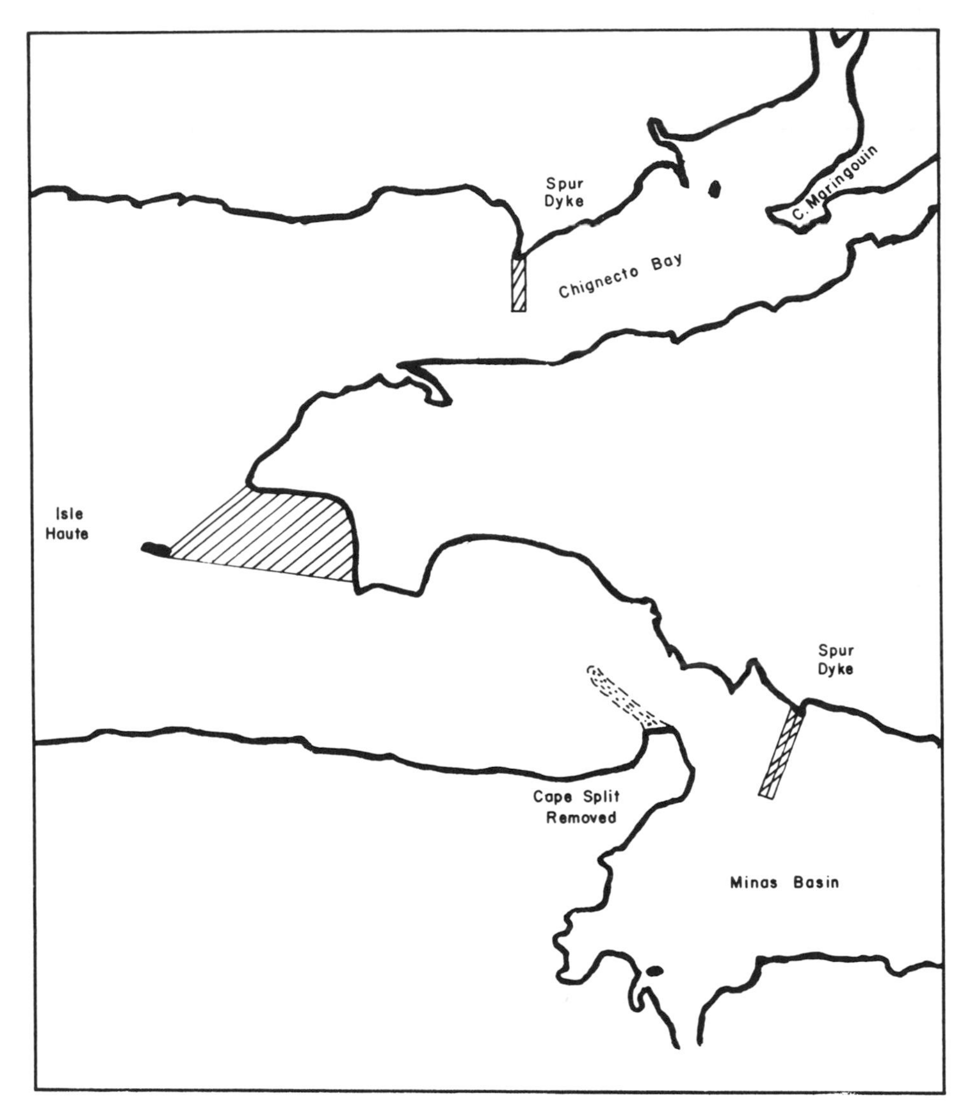

Fig. 8. Bay of Fundy tidal regime study. Artificial boundary modifications.

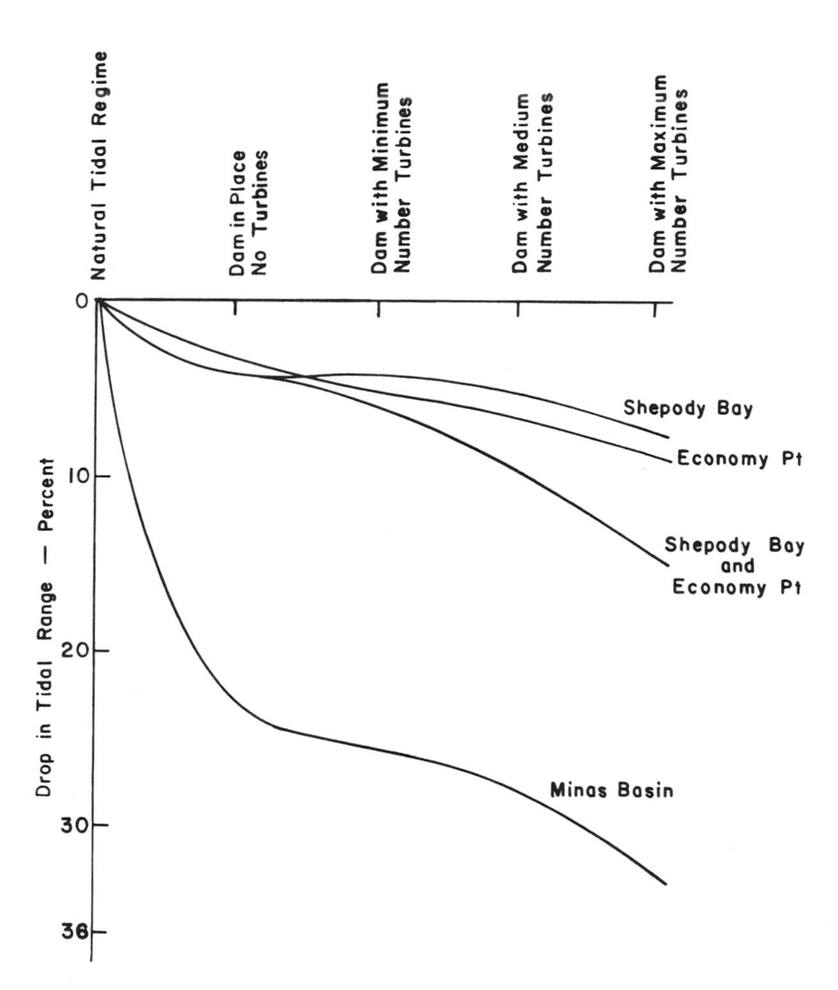

Fig. 9. Drop in R. M. S. tidal range at the sites caused by projects.

RESULTS OF MODEL COMPUTATIONS

The complete results from the various computation runs took the form of a large volume of graphs, which it is obviously not possible to reproduce for the present discussion. Therefore, five graphs (Fig. 9 to 13) have been prepared which show in either concentrated or partial form the steps taken in this study.

Effects at the Individual Sites

This was the most important part of the study as concerns direct and immediate applicability of the results to the optimization computations. It was done considering each individual site alone, and the effects the project construction and operation would have at that position. As was explained earlier, the basic comparisons between sites were done using the r.m.s. tide, and this same approach retained for the present discussion.

Fig. 9 shows the percentage drop in the r.m.s. tidal range at the various sites for single phase power generating operations with various degrees of equipment, as well as for the dams simply being in place.

The curves for the two first projects at Shepody Bay and Economy Point, along with their combined operation, show the typical effects. With the dams alone in place, the only mechanism operating to reduce the tidal range was the change of length of the bay, hence further removal from resonant conditions. Then, with the increasing number of turbines in operation, progressively more water was transited through the dam out of phase with the tides, and the resulting tidal range dropped accordingly.

The larger project in Minas Basin, however, was far more seriously penalized by its geographic position. Being further down the bay, the length of the upper basin was much greater. This automatically infers a greater departure from the resonant length (time). On Fig. 9, this appears by showing a 24% range drop with the dam alone in place. The turbine operations then increase the drop in about the same manner as the other schemes.

Effects of Different Imposed Tidal Ranges

Although all the comparative energy production computations were done using the r.m.s. tide, it was necessary to know what effects there would be with larger and smaller tides (see Fig. 5), in case some marked irregularity should occur. Fig. 10 shows one of the examples computed, using the Economy Point project, with maximum turbine operations. The tidal drops are shown along the Nova Scotia shore over the whole length of the Bay of Fundy, -- the "head of bay" values being taken at Economy Point.

It is interesting to note that about two thirds the way along the bay, the small tide shows a much greater drop in its range. This no doubt results

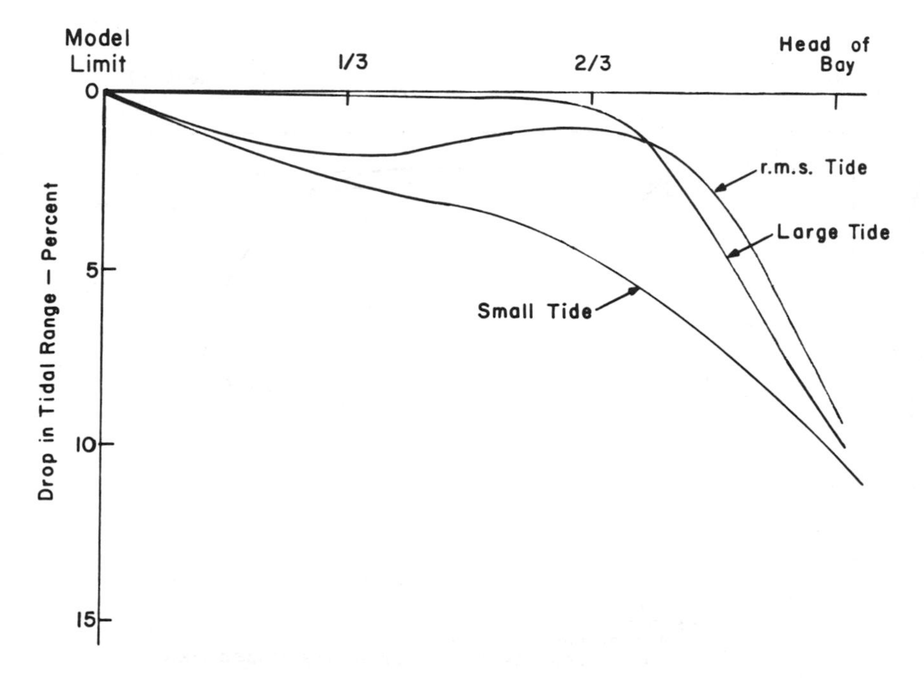

Fig. 10. Drop in tidal range along Nova Scotia shore caused by operation of project at Economy Pt. - maximum number of turbines.

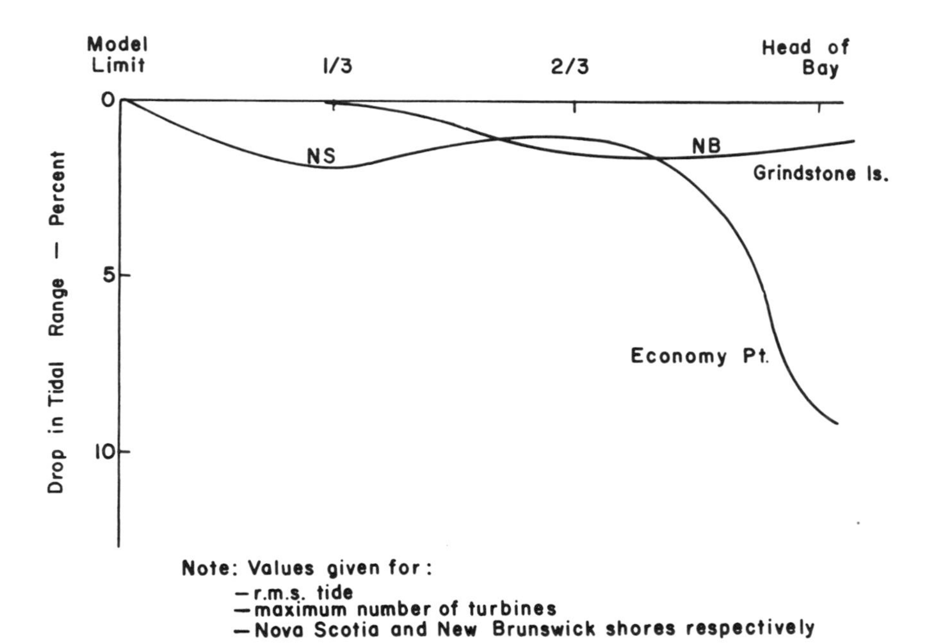

Fig. 11. Bay of Fundy tidal regime study. Drop in R. M. S. tidal range caused by operation of Economy Pt.

from the fact that the turbine discharges represent a relatively greater portion of the tidal discharges, which combined with some resonant phenomenon at this point in the bay.

On the other hand, at the head of the bay, which was of prime importance for our studies, the three tide forms had very nearly the same percentage reduction in their ranges. In fact, for the whole series of computations in this sub-program for the various power schemes, the "head of bay" values for the three tides fell within a spread of little more than one percent. This was sufficient justification at this stage for the retention of the r.m.s. tide for the optimization and regime change studies.

Effects Along the Bay of Operation of Economy Point Project

Our concern for the power production capacity of the Bay of Fundy automatically directed all our interest toward the head end where the projects were being considered. However, secondary considerations had to evaluate the effects along the bay, and Fig. 11 sets out the tidal range modifications along the two shores of the bay for the Economy Point scheme operation.

The curve for the New Brunswick shore shows minimal effect, which seems logical, since it is the furthest away from the project. Conversely, the Nova Scotia shore registers a greater drop over most of its length, increasing sharply toward the project itself above Minas Basin.

Effects Along the Bay of Combined Operation of the Shepody Bay and Cumberland Basin Projects

Similar to the object outlined in the preceding paragraph, this sub-study defined the tidal range changes along the bay shores for operation of the two projects on Chignecto Bay.

For the first time, a reduction of the range at the model limit appears. This was an undesirable discovery, but the value was small enough to fall within acceptable accuracy limits (see Fig. 12).

Although the projects were on the New Brunswick shore, nearer the mouth of the bay, there appears to be less effect than on the Nova Scotia shore. It seems likely that there is some directional influence of the discharge out of Chignecto Bay, favoring flow toward the opposite shore.

Conversely, in the upper reaches, the curves take on the logical relative positions, with less effect at Economy Point, (furthest away from the schemes) than at Grindstone Island, (in the centre of the Shepody Bay project).

Effects Along the Bay of Combined Operation of the Shepody Bay, Cumberland Basin and Economy Point Projects

Following the same reasoning set out in the two preceding paragraphs,

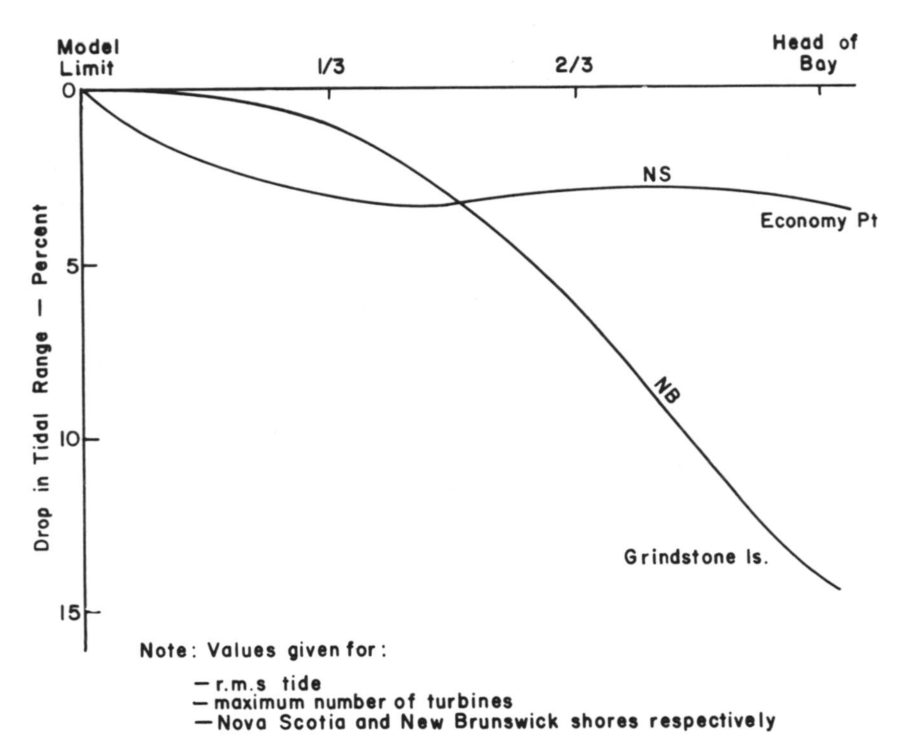

Fig. 12. Drop in R. M. S. tidal range caused by combined operation of projects at Shepody Bay and Cumberland Basin.

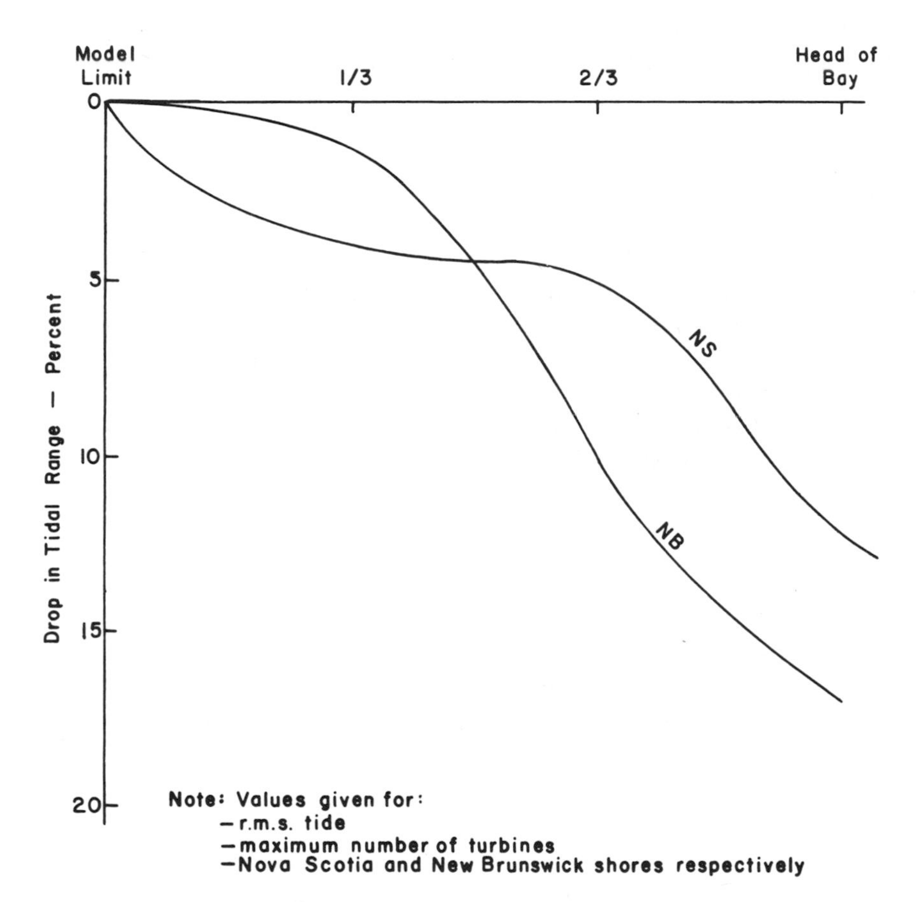

Fig. 13. Drop in R. M. S. tidal range caused by combined operation of projects at Economy Pt., Shepody Bay, and Cumberland Basin.

this sub-study considered the ultimate tidal power equipment evolved in our overall study; that with both of the head end arms of the Bay of Fundy fully developed (see Fig. 13).

Once again, the New Brunswick shore nearer the mouth of the bay shows less effect than the Nova Scotia shore, and reason would seem to imply some directional flow effects from Chignecto Bay. Further toward the head end, however, the New Brunswick shore shows more tidal range drop, due to a combination of two parameters: first, the greater total capacity of the Shepody Bay and Cumberland Basin schemes, and second the shorter distance of the schemes from the bay entrance.

CONCLUSIONS

1. The tidal regime model was built on the basis of rather sparse data on certain parameters, which no doubt introduces a certain margin of error. However, the adjustment tests and successive calculations with differing boundary conditions gave very consistent and rational results, so it is felt a satisfactory reproduction of the tidal phenomena was obtained.
2. The same set of assumed conditions were used for the energy production calculations, so once working within this framework, comparisons to a finer degree of accuracy were possible.
3. The tidal regime model was a valuable tool for determining the necessary corrections to the energy production calculations, thereby facilitating the absolute economic comparisons.
4. It also provided descriptions of the tidal range reductions to be expected along the length of the bay, which were valuable for assessing any changes these might introduce in these areas.

SEDIMENTATION PATTERNS IN THE BAY OF FUNDY AND MINAS BASIN

B. R. Pelletier* and R. M. McMullen**

INTRODUCTION

This study summarizes some earlier investigations in the Bay of Fundy and Minas Basin, and presents new sediment data particularly from the Minas Basin. The earlier investigations are those by the following: Forgeron (1962) on bottom sediments, Swift, Cok and Lyall (1966) on subtidal sandbodies in Minas Channel; Miller (1966) on suspended sediments; McMullen and Swift (1966) on large-scale rhomboid ripples in Minas Basin; McMullen, and Lyall (1967) on tidal deltas in the Minas Basin; Swift and Lyall (1968a, b) on bedrock studies with the use of the sub-bottom seismic reflection profiler; Swift, Pelletier, Lyall and Miller (1969) on sediments of the Bay of Fundy; a manuscript under review by Swift et al on Quaternary sediments in the Bay of Fundy; and a manuscript under review by Pelletier on sediment sampling in the Bay of Fundy with a diver lock-out submersible.

The purpose of this paper is to present various ideas on sedimentary trends in the Bay of Fundy and Minas Basin system, particularly those views dealing with sedimentation in a high energy, hydrodynamic regime which have not been covered earlier, and to compare these two subsystems.

THE BAY OF FUNDY

The Bay of Fundy is a funnel-shaped body of water which lies between Nova Scotia and New Brunswick. Approximately 70 percent of the fresh water entering the Bay of Fundy is contributed by the Saint John River. The Bay is 144 km. long and 100 km. wide at the base. The northeast end bifurcates into Chignecto Bay and Minas Basin (Fig. 1). The Bay occupies a

---

*Chief, Marine Geology, Atlantic Oceanographic Laboratory, Bedford Institute, Dartmouth, Nova Scotia, Canada.
**Head, Scientific Information Services and Library, Atlantic Oceanographic Laboratory, Bedford Institute, Dartmouth, Nova Scotia, Canada.

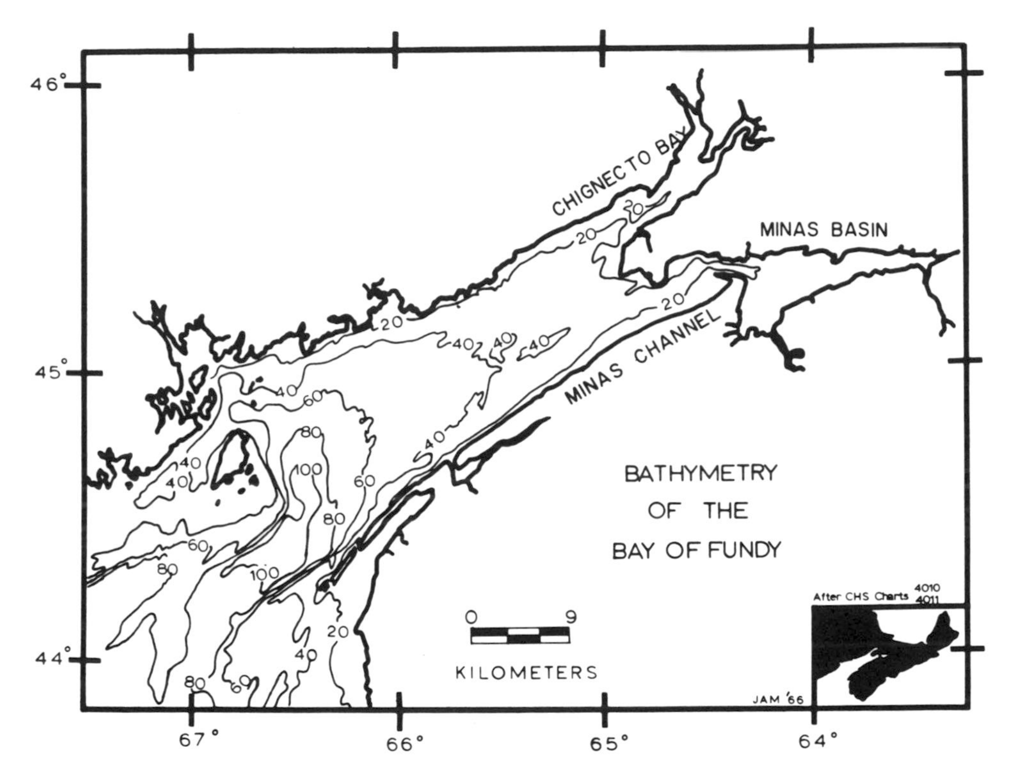

Fig. 1. Bathymetry and location of the Bay of Fundy and Minas Basin. Bathymetric contours in fathoms after Canadian Hydrographic Service Charts 4010 and 4011.

Triassic half-graben structure composed of red continental mudstones and sandstones, and basalts. Most of the sediment cover on the floor of the Bay was emplaced under subaerial conditions during the repeated glacial episodes of the Pleistocene. Now these relict materials exposed on the Bay floor are evolving into sediments adjusted to the modern hydraulic regime. Where this process has run to completion, the resulting deposits bear the distinctive impress of a high-energy, tidal regime.

Physiographic Setting

A pocket of Lower Cretaceous sediment, which presently occurs as fill in an ancient tributary valley, indicates that the physiography of the Fundy region has been inherited from the Mesozoic but the original erosional surface, as interpreted from seismic records, has been modified beyond recognition by Pleistocene glaciation. This pavement slopes 2 m./km. or less, and has been incised by an ancient river system and later modified by glaciation as are the embranchments at the head of the Bay. This sequence of events has yielded the present-day physiography shown in Fig. 1.

The shoreline consists primarily of an intertidal wave-cut terrace of bedrock up to 300 metres wide, which is veneered with less than one metre of coarse, variable, poorly sorted detritus of mainly local origin, and is fronted by a sea cliff. These cliffs are receding up to 2 m./yr. (Churchill, 1924), although locally this rate may be only several centimetres and decimetres per year. Bedrock lows are occupied by glacial drift and outwash gravel, either of which may produce shingle spits and bars. In protective coves and estuaries, intertidal sedimentation is forming extensive tidal mud flats backed by marshes, and where hydraulic conditions are suitable, intertidal sand bodies have developed.

Tides and Currents

Tides in the Bay of Fundy are 6.4 metres at the mouth and 17 metres at the head, and are a consequence of the axial dimension of the Bay, which is almost equal to the critical length required for resonance (Defant, 1961). Its critical length to the very head is 296 km. which closely agrees with the measured length of 300 km.; also, the natural period of the Bay is 6.29 hours which is close to one-half the semi-diurnal period, or 6.21 hours. Thus the conditions for resonance are approximated and the entering tidal wave is amplified to the point where constructive interference is balanced by tidal friction. The tidal wave in the Bay of Fundy is a standing wave rather than a progressive one and consequently high water is attained nearly simultaneously throughout the Bay.

The tidal currents (Fig. 2 data from Forrester, 1958, and Langford, 1966) are parallel to the axis of the Bay and at half-flood and mid-depth they aver-

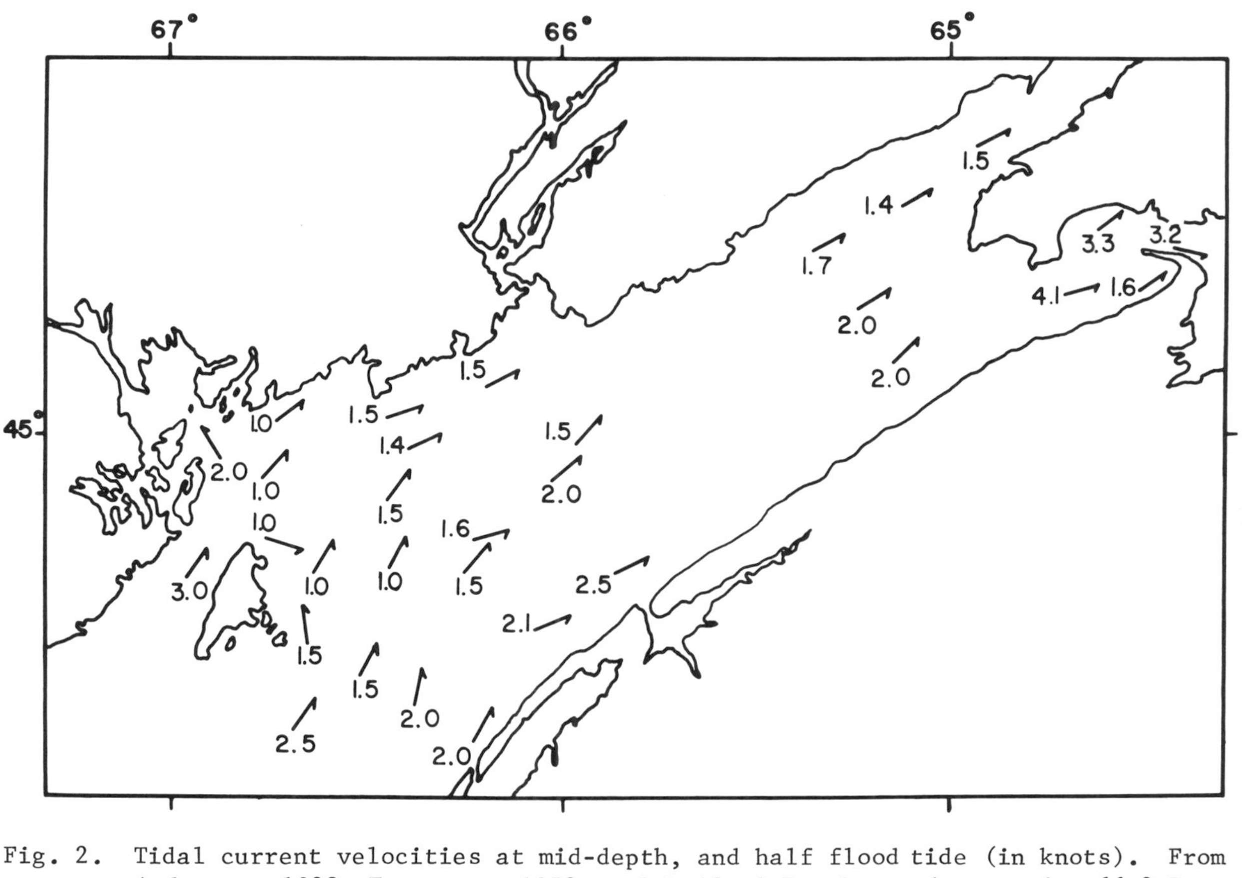

Fig. 2. Tidal current velocities at mid-depth, and half flood tide (in knots). From Anderson, 1928, Forrester, 1958, and Bedford Institute data series 66-2-D, Dartmouth, N. S., (1966). Compiled by J. A. Miller (1966), p. 44-46.

age 103 cm./sec. (2 knots) on the south shore, and 77 cm./sec. (1.5 knots) on the north shore. Towards the head of the Bay, shoaling and narrowing retard the tidal wave, and current velocities increase to 206 cm./sec. (4.0 knots) in Minas Channel and up to 556 cm./sec. (11 knots) in Minas Passage. A counterclockwise pattern of residual currents (Fig. 3) was determined from drift bottle studies (Mavor, 1922; Hachey and Bailey, 1952; Bumpus, 1959; Chevrier, 1959; Bumpus and Lauzier, 1965).

Waves and Ice

Summer winds in the Bay of Fundy blow mainly from the south and southwest; during the rest of the year they blow mainly from the west and northwest, hence Fundy's coasts are generally exposed coasts. Winter waves are over .3 metres high 90 percent of the time, over 1 metre high 50 percent of the time, and over 4 metres high 10 percent of the time. Waves of all heights are 5 to 10 percent less frequent in summer.

Drifting ice plays an active role in the transportation of sediments in the Bay of Fundy. During the winter months ice forms in the upper portions of the bay and remains there until the latter part of March. The ice moves back and forth along a path which is the resultant of the effects of wind stress and the tidal excursion. Wind shifts are responsible for the departure of the ice from the upper bays into the main bay in the spring. The drifting ice collects its load by direct incorporation of muddy water, and by grounding on tidal flats where it acquires sand and gravel as well as mud.

Hind (1875) studied the ice that forms in the mouths of the Avon, Cornwallis and Shubenacadie Rivers of the Minas Basin. After collecting and melting ice samples in Windsor Bay in order to determine their sediment load, he concluded that a cubic metre of ice with a load acquired by means of grounding carried an average of 4.54 grams of sediment, and a similar amount of ice with a load acquired by direct accretion carried 2.38 grams of sediment.

Bottom Sediment

Bottom sampling of the Bay of Fundy was carried out by the Atlantic Oceanographic Laboratory and the Dalhousie Institute of Oceanography between the years 1961 and 1966. About 500 samples were collected on a 2-km. sample grid using a .2 m^3 Van Veen grab sampler (Fig. 4). The sand and silt-clay fractions were subjected to sieving and pipette analyses, respectively (method by Krumbein and Pettijohn, 1939), and the present study is based partly on these results.

Textural facies or aspects of bottom sediments in the Bay of Fundy are shown in Fig. 5 and 6, the latter showing the percentage probability of finding the namesake sediment type within each textural province. The probability is lowest for the coarsest sediments and highest for the finest. This was confirmed

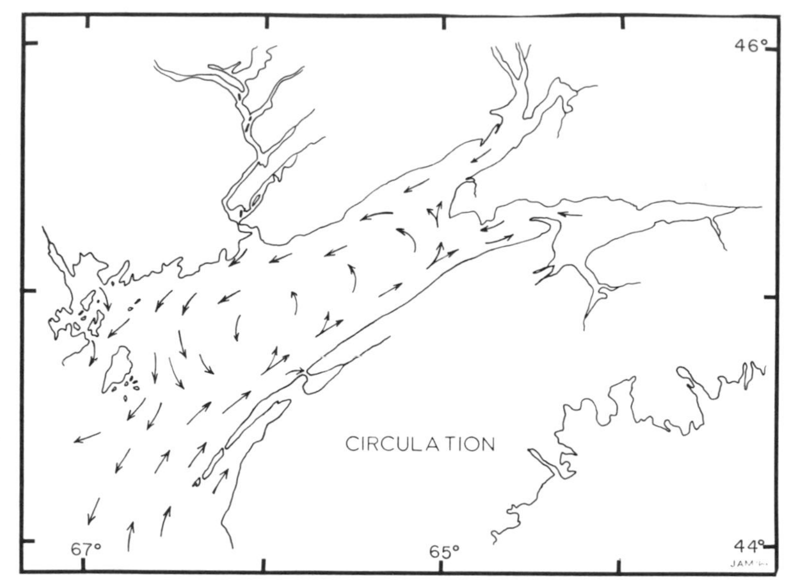

Fig. 3. Residual circulation pattern, Bay of Fundy, June 15-18, 1932. Compiled by J. A. Miller (1966) p. 47-48.

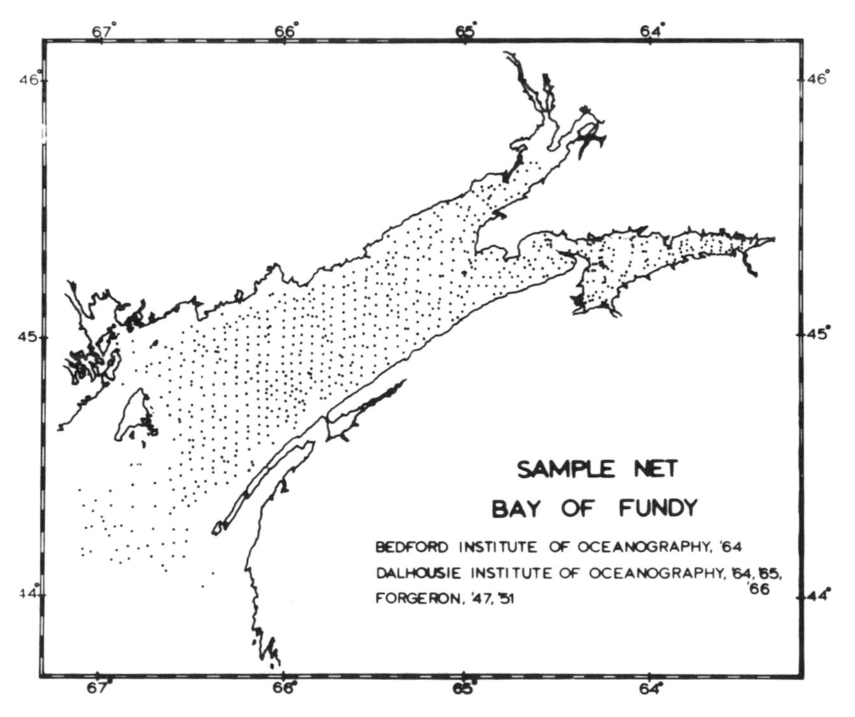

Fig. 4. Grab sample locations for Bay of Fundy and Minas Basin.

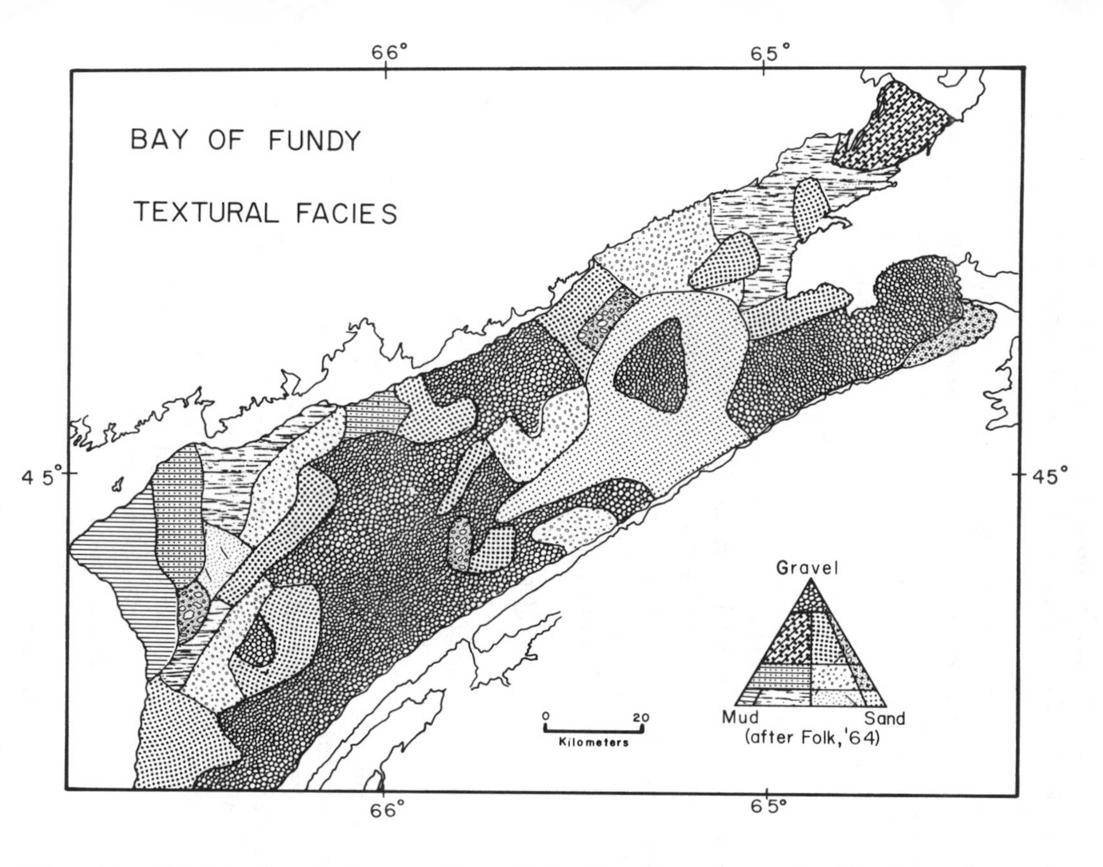

Fig. 5. Textural provinces, Bay of Fundy. Key shown in the insert sketch, lower right, with apices of triangle representing 100 percent of gravel, mud, and sand respectively. See Fig. 6 for explanation of legend.

by sampling from a diver-lockout hatch in the research submersible SHELF DIVER in 1969 (Pelletier, In Press). For example, in silty sand, 14 out of 16 samples were similar; in muds, 15 out of 16 samples were similar.

Textural provinces in the Bay of Fundy fall into three main groups as follows: gravel, sand, and mud. Generally gravels occur over most of the floor of the Bay, whereas the sand occurs in the central part of the bay, and forms transitional zones between the mud and gravel provinces. Mud provinces mostly extend along the northwestern margin of the bay. Thicknesses of the sediments in excess of 30 metres as determined by the seismic reflection profiler, are widespread only beneath the mud provinces. The thinner section which exists beneath the sand and gravel provinces, exhibit patterns on the sub-bottom profiles which are interpreted as till (massive, "sharkskin" pattern, due to numerous intersecting parabolas generated by point sources), or as glaciofluvial outwash (stratified pattern). These interpretations have locally been verified by tracing them into the various reflectors, representing the various sedimentary bodies, occurring in the intertidal zone where they can be examined in outcrop. Thus, the Holocene sand and gravel provinces appear to be veneers of reworked Pleistocene material, generally thinner than can be resolved by the sub-bottom profiler. Only in the mud provinces do significant buildups of Holocene materials occur.

Gravel Provinces

Medium to coarse pebble gravels, sandy gravels, and muddy sandy gravels occupy 58 percent of the Fundy sea floor. They occur mainly beyond the 40-metre contour where they are immune to wave action. The one- to two-knot tidal currents are capable of moving very fine to fine pebbles, but the bulk of the bay-floor gravels have median diameters coarser than this, and are presumably relict from Pleistocene low stands of the sea.

The major petrographic components of Fundy gravels in order of abundance are granitic and gniessic rocks, red sandstone and shale, drab brown sandstone and shale, and basalt. The first group is of pre-Carboniferous origin, the second group is primarily of Carboniferous origin, and the third and fourth groups are primarily of Triassic origin. The composition of gravel samples generally reflects the composition of the underlying bedrock, as well as its resistance to abrasion. Gravel samples from pre-Carboniferous and Carboniferous terranes contain 20 percent or less of far-travelled material. But for gravel samples taken from friable red Triassic mudstones, the percentages are commonly reversed, unless there is abundant outcrop on the sea floor.

Sand Provinces

Sands, gravelly sands, and muddy sands occupy 22 percent of the bay floor. Median grain size is medium to very coarse, except in the vicinity of

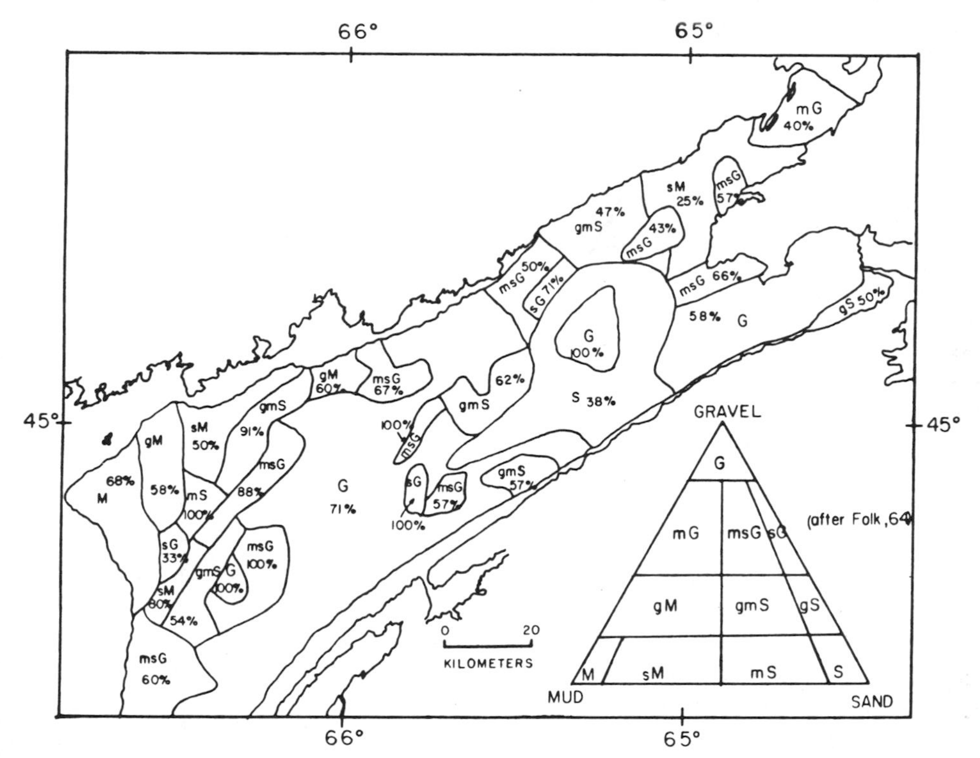

Fig. 6. Percentage probability of finding namesake texture for textural provinces, Bay of Fundy. Note: G = gravel; mG = muddy gravel; msG = muddy sandy gravel; sG = sandy gravel; gM = gravelly mud; gmS = gravelly muddy sand; gS = gravelly sand; M = mud; sM = sandy mud; mS = muddy sand; and S = sand.

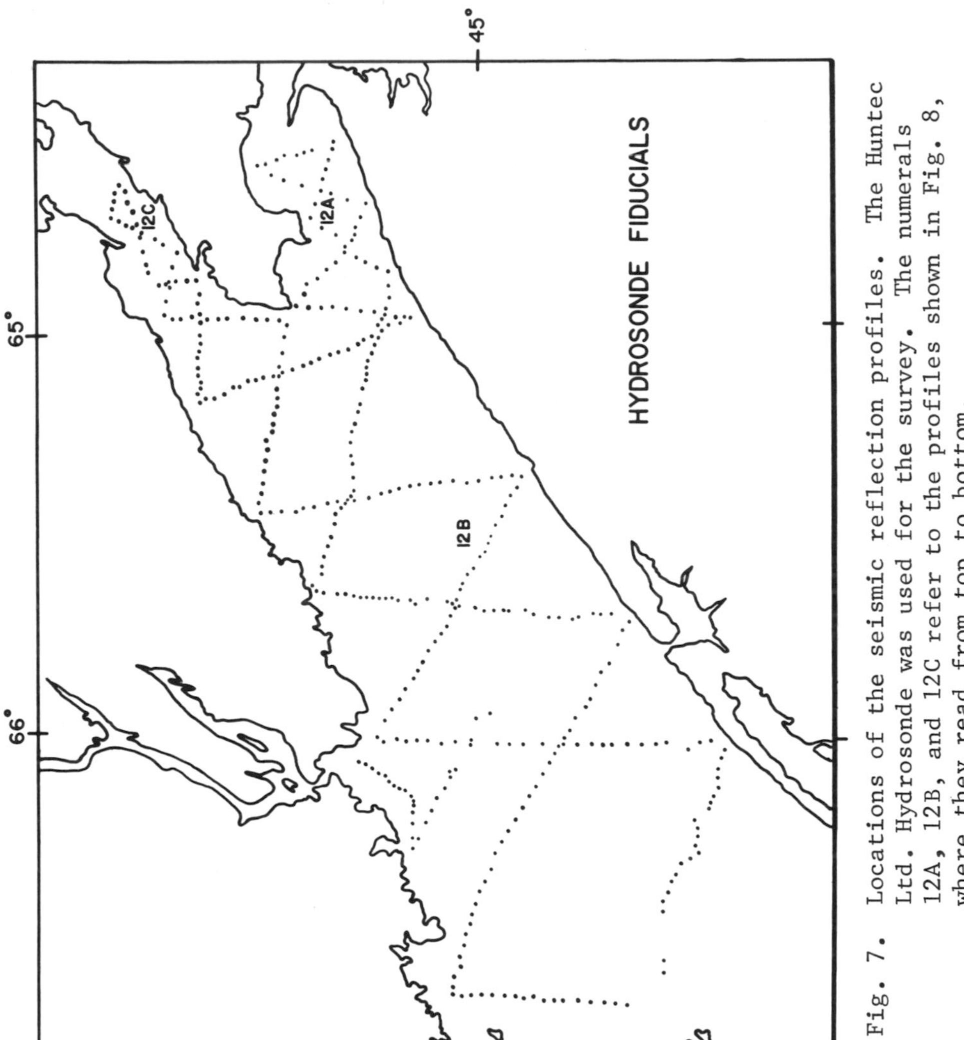

Fig. 7. Locations of the seismic reflection profiles. The Huntec Ltd. Hydrosonde was used for the survey. The numerals 12A, 12B, and 12C refer to the profiles shown in Fig. 8, where they read from top to bottom.

mud deposits where it is fine to very fine. Two major sand deposits are present; one occurs in a transverse north-south band towards the head of the bay, and the second also occurs as transverse band which extends across the eastern corner of the bay. Where they border mud deposits, these sand provinces may be in part hydraulically maintained transitional zones.

Forgeron (1962) collected 28 gravity cores in the Bay of Fundy, up to two metres in length. His data indicate that the contact between the mud provinces and the fine sand provinces dips gently southeast; thus the fine sands appear to be transgressing the muds, the latter of which appears to be compacted. Forgeron suggests that the transitional fine sands are in fact lag deposits that were generated during the retreat of the zone of mud accumulation through the late Holocene as Fundy's tidal currents intensified (see Swift and Borns, 1967, for a discussion of the ontogeny of the tidal regime). Forgeron reports sand-filled borings within the buried mud, and interfingering of thin beds of mud and sand (Forgeron, 1962, p. 109).

The main mass of sand toward the head of Fundy may be an outwash delta generated by a late Pleistocene periglacial river, whose channels have been traced by sub-bottom profiles (Swift and Lyall, 1968B), or generated by meltwater from a local late Pleistocene ice cap centered on southern Nova Scotia (Hickox, 1964), or by both. Sub-bottom profiles (Fig. 7 and 8) from this area reveal irregular solitary sand waves with amplitudes up to two metres, localized by bedrock or till highs. They are mainly transverse forms with flood asymmetry.

In the Minas Channel and Minas Basin, the friable Wolfville sandstone (basal Triassic) forms the bedrock, and the overlying tills are very sandy. Sand released from the till and exposed bedrock has been swept by the side into plano-convex sand bodies with upper surfaces that may reach into the intertidal zone. These surfaces are deformed into longitudinal sand bars that extent for several kilometres and bear extensive fields of sand waves (see Swift, Cok, and Lyall, 1966, and Swift and McMullen, 1968).

Mud Provinces

Muds and muddy sediments occupy 20 percent of the floor of the Bay of Fundy. They occur along the northeast side of the Bay of Fundy from Chignecto Bay to Grand Manaan Island. On sub-bottom profiles they are well stratified and are acoustically transparent and up to 100 metres thick.

Lithofacies Ratios

From the gross petrologic plot of the samples shown earlier it was decided to carry out a series of surface trend analyses based on 300 samples obtained by the Atlantic Oceanographic Laboratory. To carry out the lithofacies analyses, McMullen devised the following classification: gravel (more than 80 percent of the sediment has a diameter greater than 2 mm.); sandy gravel (50-80 percent

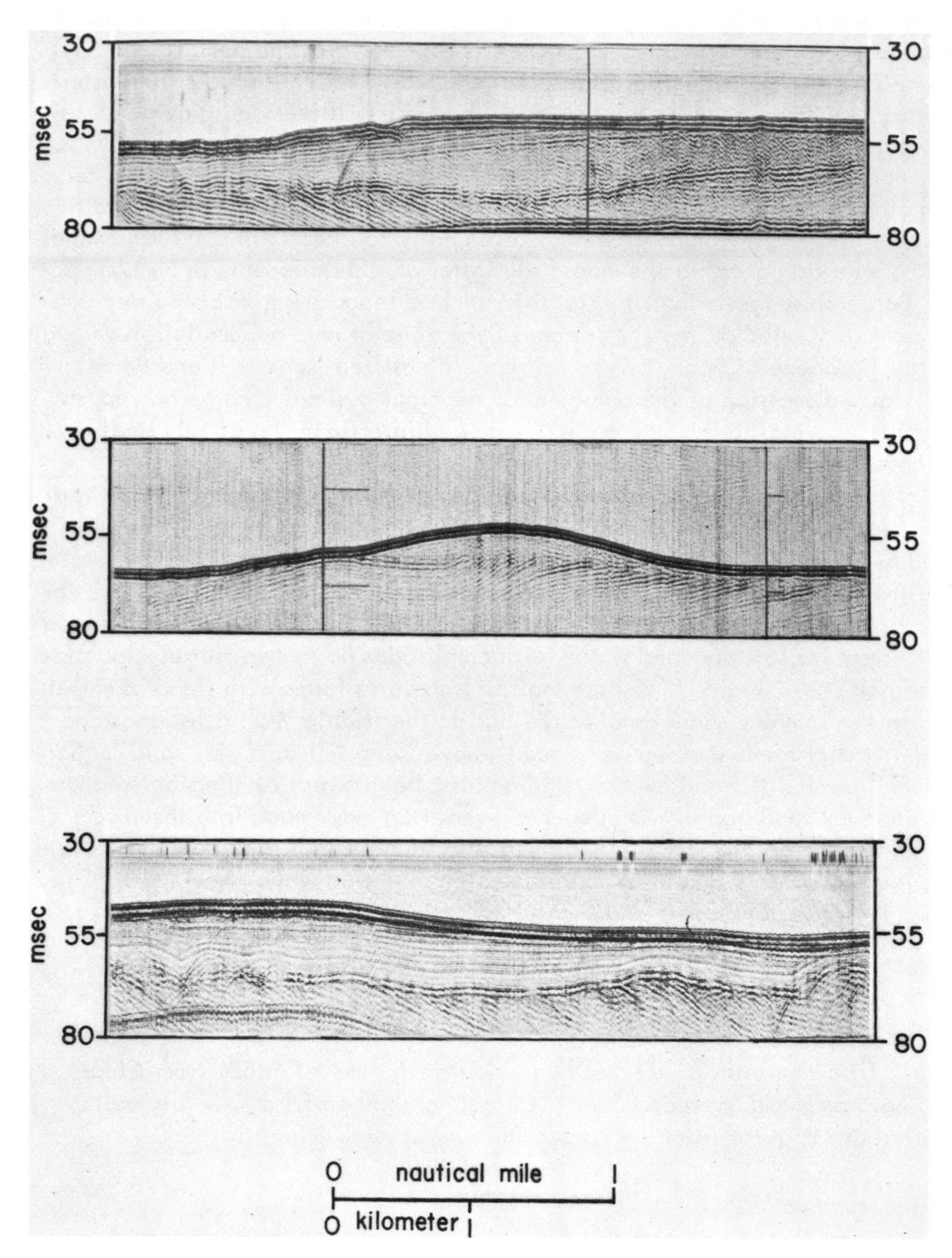

Fig. 8. Sub-bottom profiles from Bay of Fundy: (A) stratified drift overlying till, central Fundy; (B) solitary sand-wave localized by Triassic bedrock high, central Fundy; (C) stratified muds, lower Chignecto Bay. Bedrock is Triassic, and second echo is visible at lower left.

of the sediment has a diameter greater than 2 mm.); gravelly sand (10–50 percent of the sediment has a diameter greater than 2 mm.); sand (more than 90 percent of the sediment has a diameter of 0.063–2.0 mm.); muddy sand (10–50 percent of the sediment has a diameter less than 0.063 mm.); sandy mud (50–75 percent of the sediment has a diameter less than 0.063 mm.); mud (more than 75 percent of the sediment has a diameter less than 0.063 mm.).

First the lithofacies ratio of sand to gravel was plotted (Fig. 9) and it became apparent immediately that the sediments occurred in more-or-less alternating areas occupied by respectively coarse and fine material. These trends occur oblique to the south shore in a north-south trend and perpendicular to the north shore,and the distances between their axes are more-or-less equidistant along the length of the Bay. The sand/mud ratio was plotted (Fig. 10), and this reflected a similar trend, although generally the coarser material is in the upper part of the bay.

Sediment sizes were graded according to a logarithmic scale to the base 2, and these were coded into so-called phi units in which 1 mm. is 0 phi, 2 mm. is -1 phi, 4 mm. is -2 phi, etc., and .5 mm. is +1 phi, .25 mm. is +2 phi, etc. Moment measures were then calculated and a plot of the second moment (standard deviation) (Fig. 11) representing the relative degree of sorting showed a good correlation, as in the case of the trends of the lithological ratios, with the better sorted sediments in the upper part of the bay. However, the pronounced, almost equally spaced, trends suggest that a combination of factors is responsible for these trends.

The median diameter (Fig. 12) was also plotted in phi units and again the strong sedimentary trends were apparent. This parameter is a sensitive indicator of trends and supports the findings of the other studies above. We think this may be due to erosion of the relict Pleistocene under conditions of a standing wave, one which may have migrated in time, as the spacing between the trends is almost one-quarter of the length of the Bay of Fundy.

Suspended Sediment

Water samples were collected for suspended material from 43 stations at half flood, half ebb, at the bottom, 1 metre from bottom, 10 metres from bottom and at the surface (after Miller, 1966) (Fig. 13). Concentrations varied from 0.2 to 30.4 mg./l.with an average value of 6.6 mg./l. for 263 samples. The suspended sediment concentration for the entire water column throughout the tidal cycle (Fig. 14) reveals that turbidities in excess of 8 mg./l. occur in a strip along the New Brunswick coast north of Saint John.

Fig. 14b shows net concentration of turbidity, and indicates which phase of the tide (flood or ebb) is more important in the suspension of particulate matter. The distribution is obtained by comparing the weight-per-volume values of the flood- and ebb-tide samples, and plotting the higher concentration in terms of flood or ebb tide. As previously indicated, the half-flood tide

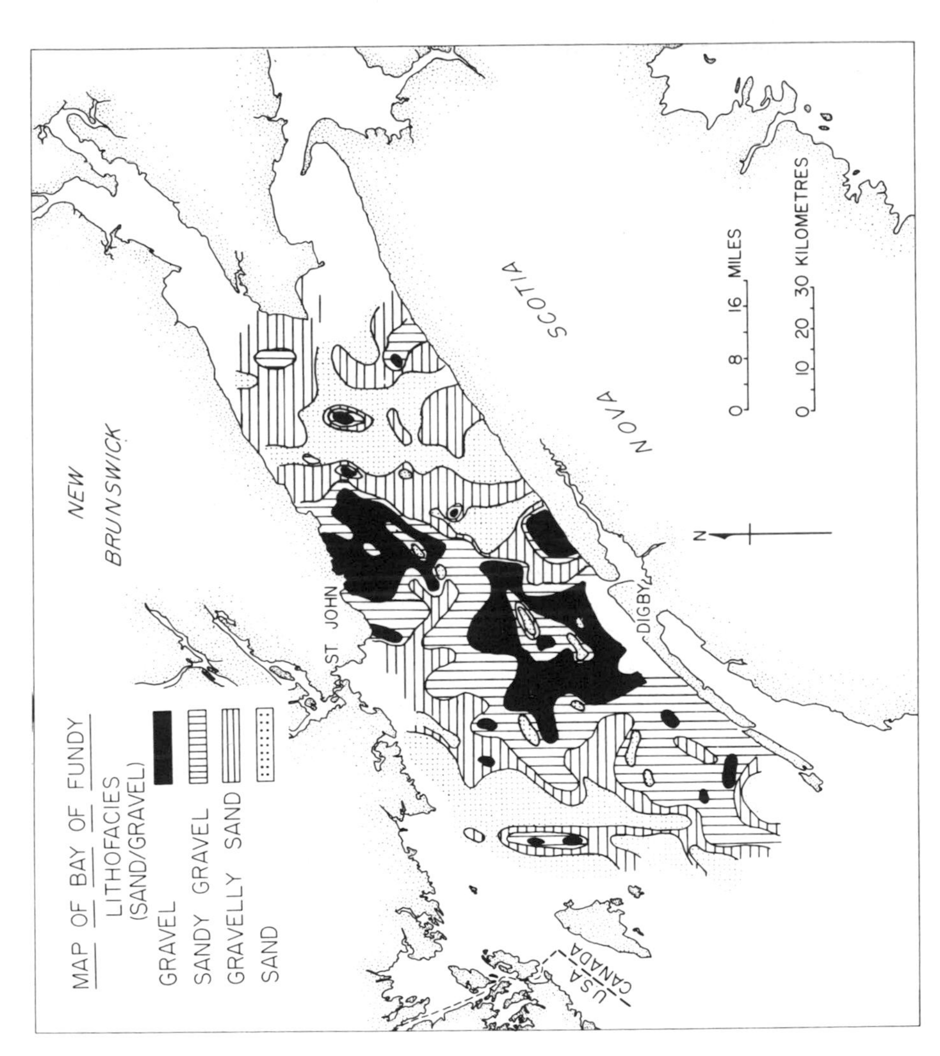

Fig. 9. Sand and gravel occurrences in Bay of Fundy.

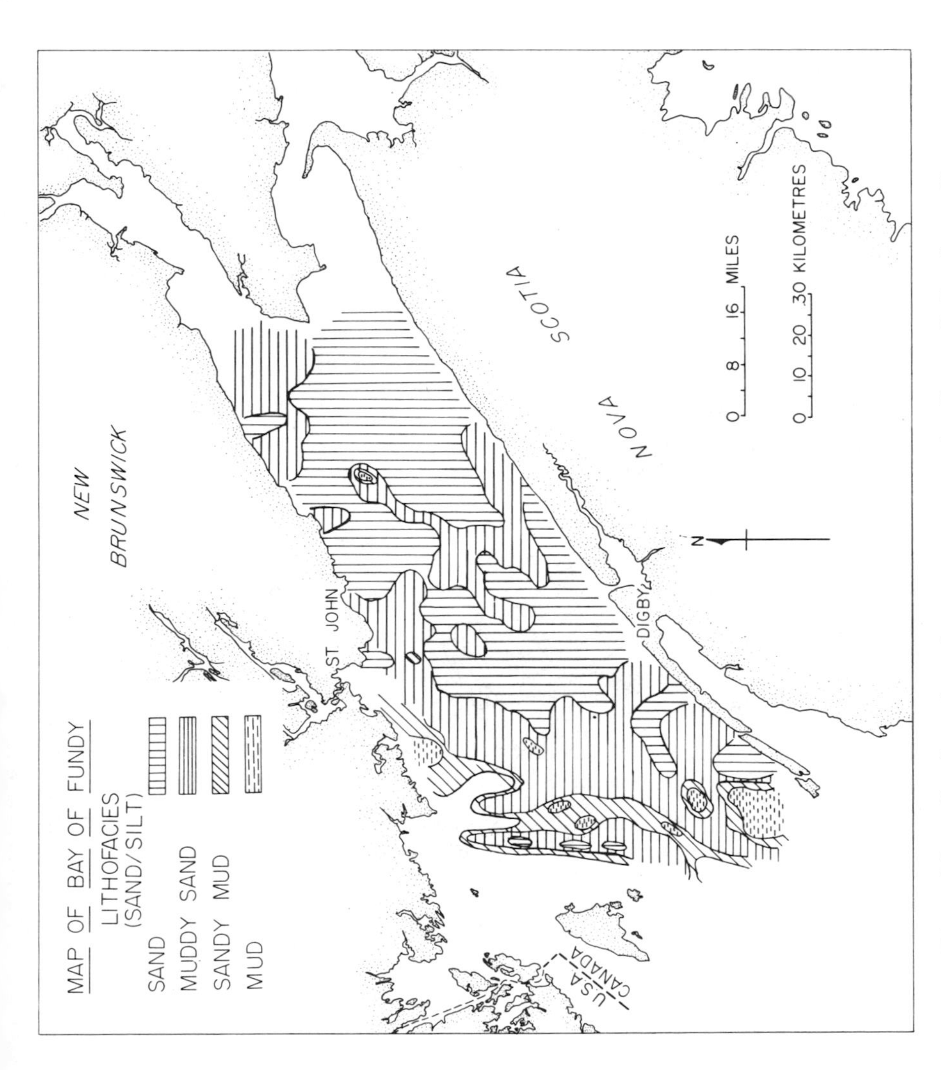

Fig. 10. Sand and mud occurrences in Bay of Fundy. Note similarity of pattern to those in Figs. 9-12.

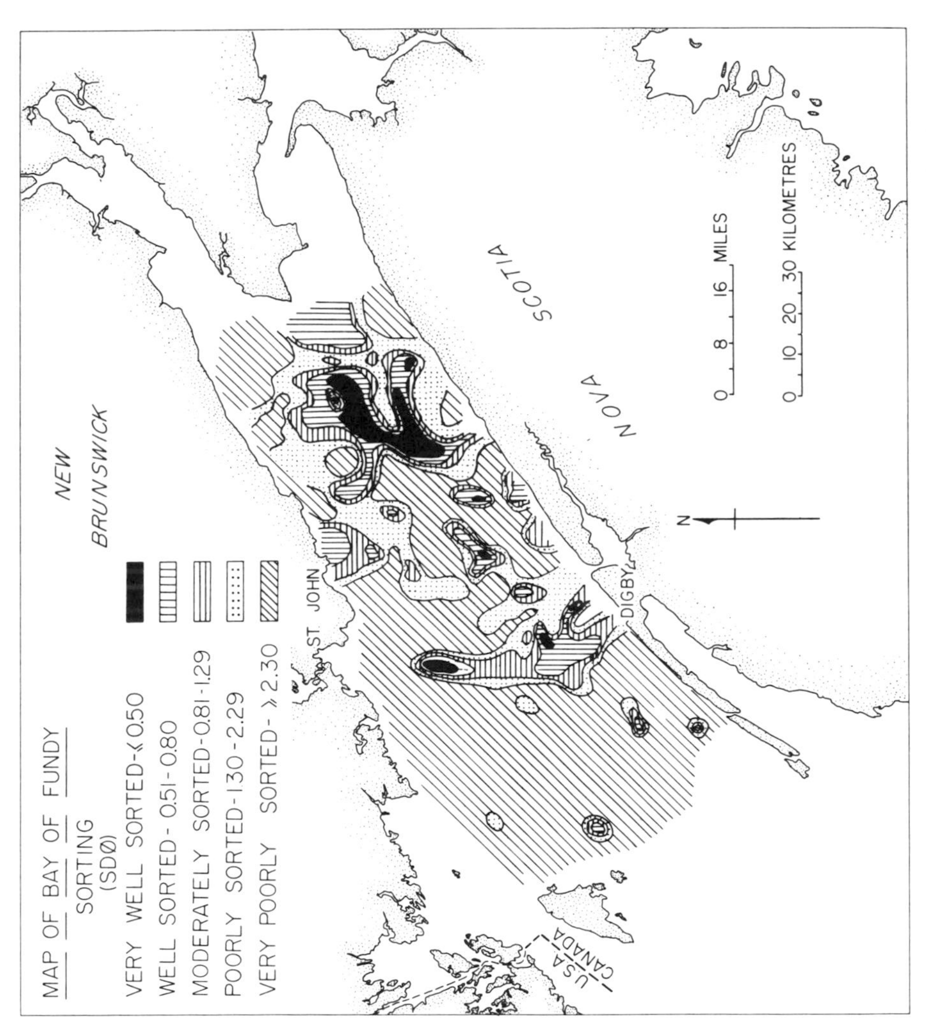

Fig. 11. Sediment sorting based on standard deviation (SDϕ) in Bay of Fundy. Note similarity of pattern to those in Figs. 9-12.

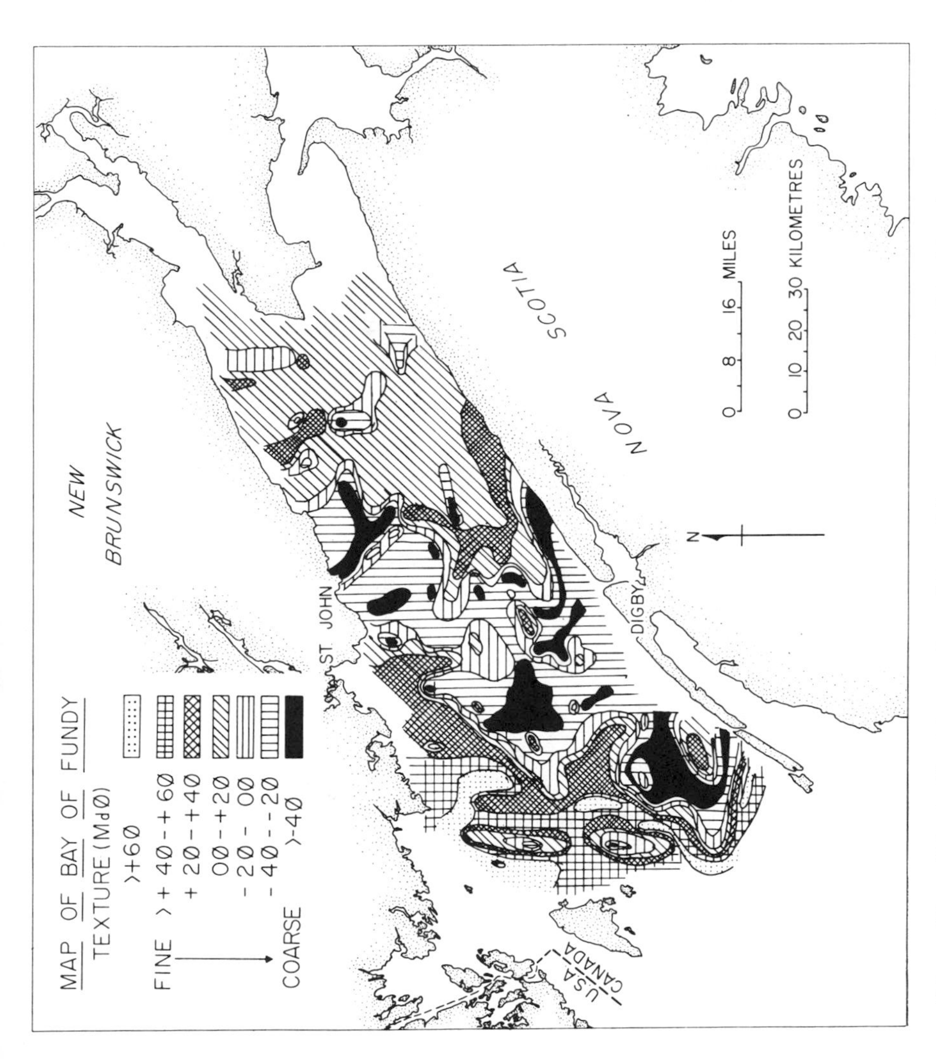

Fig. 12. Distribution of median diameters of individual samples. Note similarity of pattern to those in Figs. 9-12.

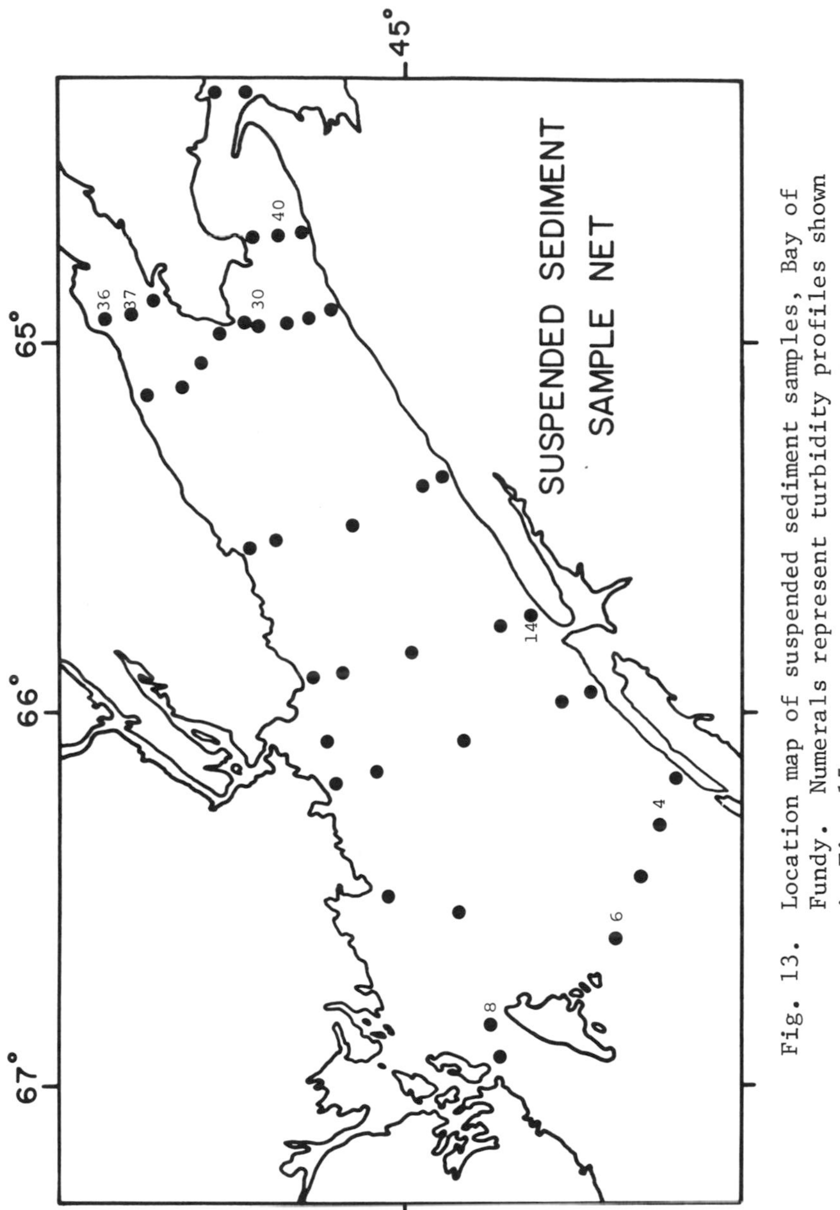

Fig. 13. Location map of suspended sediment samples, Bay of Fundy. Numerals represent turbidity profiles shown in Fig. 15.

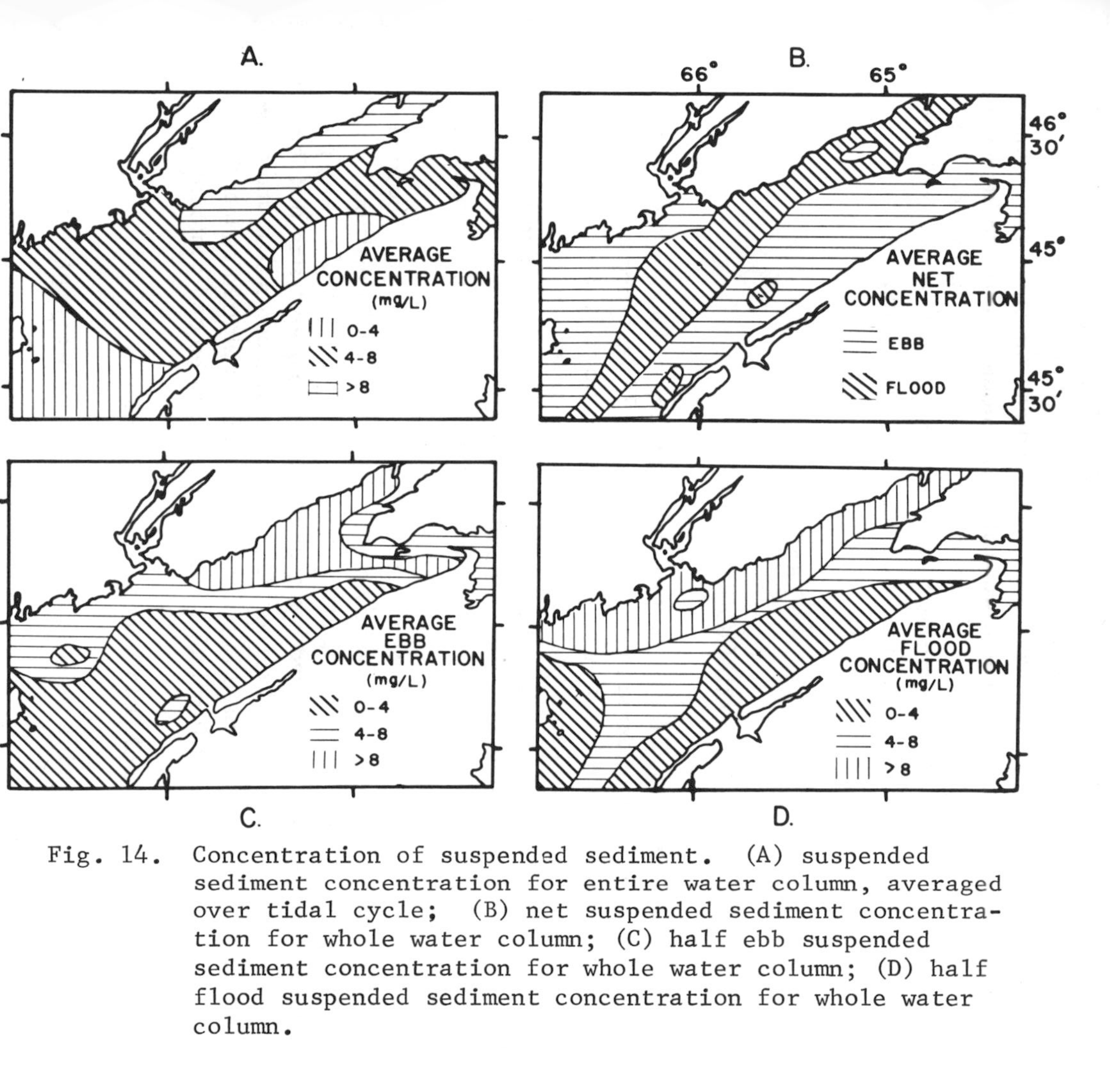

Fig. 14. Concentration of suspended sediment. (A) suspended sediment concentration for entire water column, averaged over tidal cycle; (B) net suspended sediment concentration for whole water column; (C) half ebb suspended sediment concentration for whole water column; (D) half flood suspended sediment concentration for whole water column.

stations are the most turbid on the northwest side of the Bay, except for the sheltered sector south of Saint John. Southwest of Saint John Harbour and on the southeast side of the Bay the ebb tide is the most turbid, probably because during this period each station receives from the north and the adjacent shoreline more turbid water than can be generated locally.

Vertical turbidity profiles are shown in Fig. 15. Those profiles from the upper bay show marked gradients, but are negligible in lower bay stations, suggesting turbidity is well travelled and homogenized throughout the water column. Some suspended sediment was noted by Forgeron (1962) to be moving out of the bay and into the Gulf of Maine. A great transfer or exchange of sediments with that of the watermass takes place in the area of the tidal flats where mud is picked up by the flood tide and deposited during ebb.

MINAS BASIN

INTRODUCTION

Minas Basin (Fig. 16) is a part of the larger Bay of Fundy system and, therefore, must be considered in this context. It is 77 km. long and 31 km. wide (at its widest) and has a very high tidal range, up to 17 metres. This tidal range plays a dominant role in the sedimentary processes within the Basin; it, together with the Coriolis Force, frictional forces and bathymetric configuration establishes the water circulation pattern. The tides also set up currents of considerable competence and velocity within this circulation pattern, which are able to move sediment particles.

About 90 samples from various locations in the Basin (Fig. 4) have been collected and analyzed mechanically according to grain-size distribution. The samples were subjected to standard laboratory sieving and pipetting procedures and the data so obtained has been processed to give a size distribution curve and standard statistical moments. Two of these, median size and standard deviation (both expressed in phi units), have been plotted on maps and contoured. A scheme of lithofacies has also been plotted and contoured.

Plots of Data from Mechanical Analysis

This is a comparable study to that carried out on sediment from the Bay of Fundy.

Median Diameter:

The most striking features of this map (Fig. 17) is the large area of relatively coarse sediment on the west side of Economy Point, extending out into the Basin, and the large area of relatively fine sediment on the east side of Cape Blomidon. Much of the rest of the Basin is floored by sand-size material with dispersed patches of coarser and finer sediment.

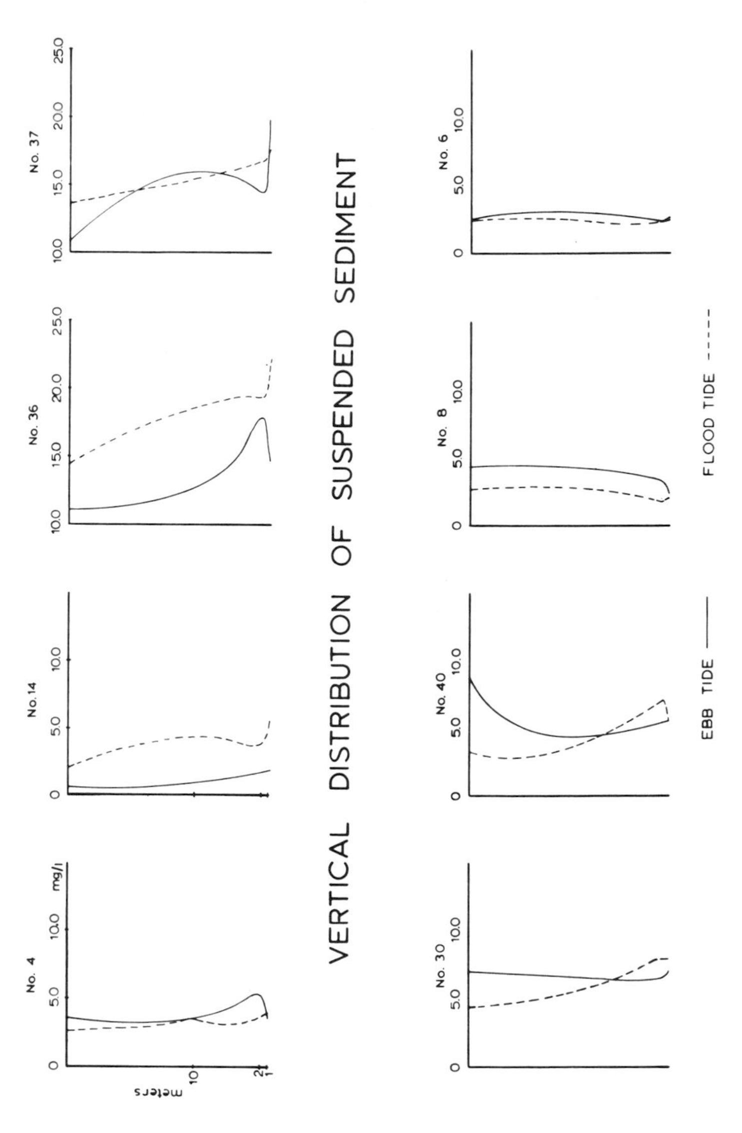

Fig. 15. Turbidity profiles from selected stations in Bay of Fundy. Locations refer to Fig. 13.

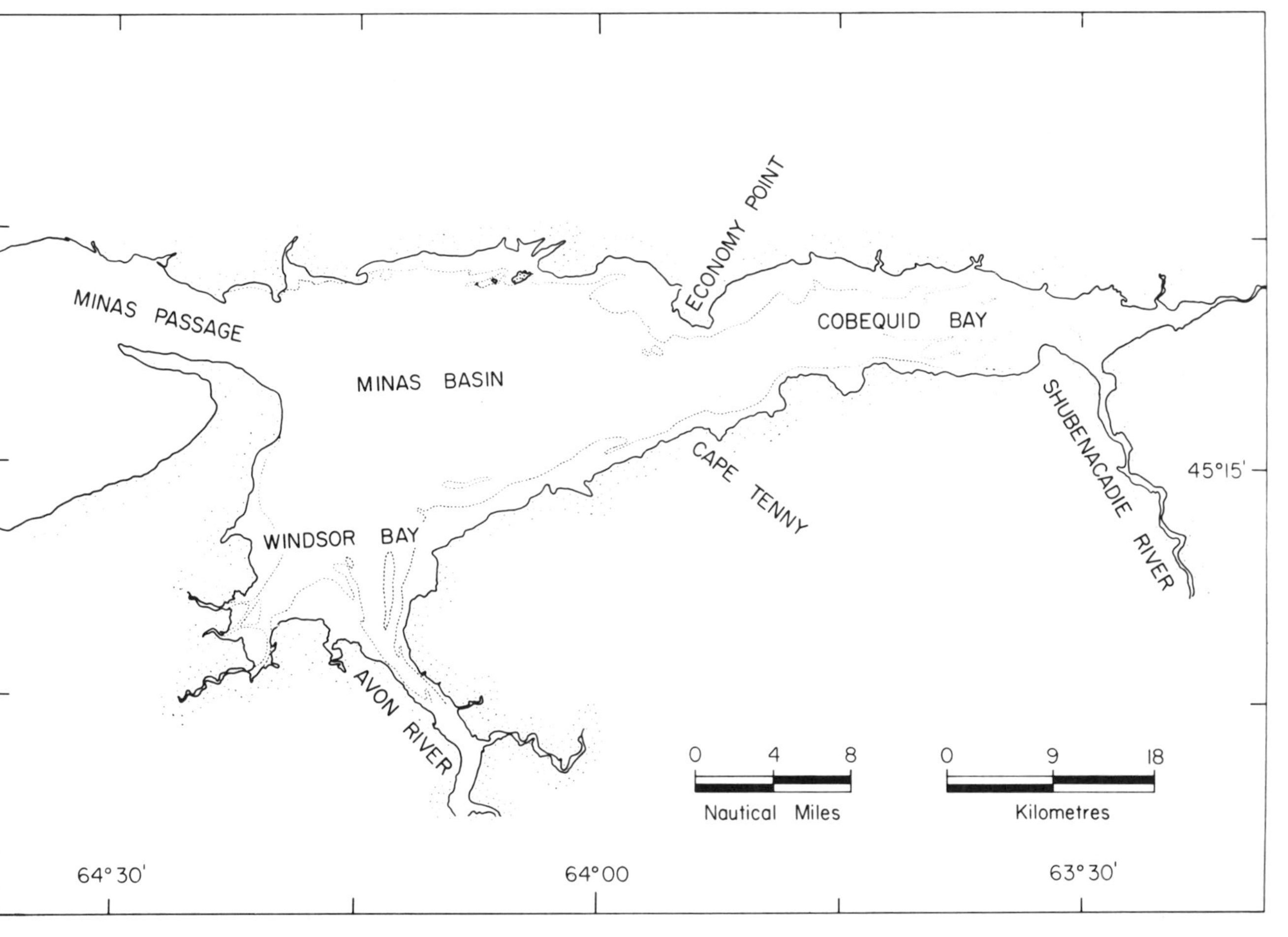

Fig. 16. Geographical references in Minas Basin study.

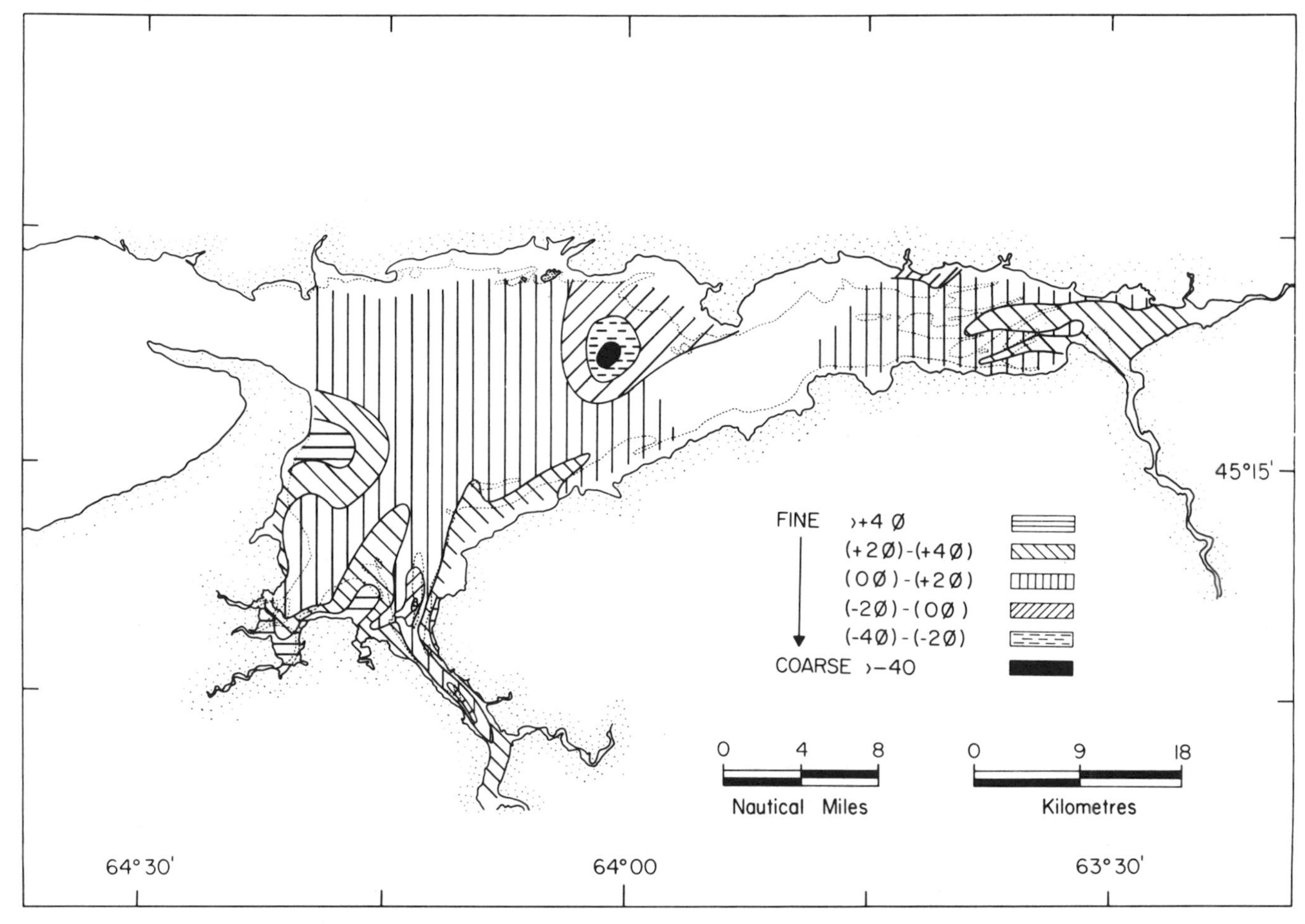

Fig. 17. Distribution of median diameters for individual samples. Note similarity of pattern to those in Figs. 18 and 19.

Lithofacies:

The map of the lithofacies distribution (Fig. 18) shows much the same features as that of the median diameter, except that it is slightly more sensitive. Again, the principal features are the large area of gravel and sandy gravel to the west of Economy Point and the large area of muddy sand and sandy mud to the east of Cape Blomidon. However, almost the whole of the remainder of the Basin is floored with sand, except for finer sediment in the estuaries and near the shoreline, and occasional patches of coarser sediment in the form of bars. These bars are generally oval in shape and are probably elongated in the direction of the dominant, local current flow.

The classification used presupposes a two-component system, sand plus gravel or sand plus mud, with the third component (either gravel or mud) not being present at all or in not significant amounts (less than 10 percent). However, two of the samples had all three components present and these are indicated separately. The precise mechanism for the deposition of this sediment is not known, but it could be either a patch of glacial till or an area that had undergone two successive stages of deposition.

Standard Deviation or Sorting:

When this data is plotted and contoured (Fig. 19), a significant pattern emerges. Much of the central part of the Basin, and other local areas, have sediment that is well or very well sorted. The nearshore areas, the estuaries, and those parts of the Basin west of Economy Point and east of Cape Blomidon have sediment that is moderately to very poorly sorted. In addition, in Cobequid Bay, there is a central core of moderately to poorly sorted sediment, which divides areas of sediment with better sorting.

Suspended Sediment:

Sixty water samples were collected from various locations in the Basin and at various times within the tidal cycle. These samples were filtered in order to retrieve most of the particulate matter in suspension, which was then weighed. The concentration values obtained varied from 72 grams per cubic metre of water to 2,680 grams per cubic metre of water. However, all but three samples had more than 90 grams per cubic metre, and more than half were between 100 and 200 grams per cubic metre. Most of the higher values were for samples collected at almost low tide, near the sediment-water interface. Those with the highest values were collected after the tide had turned and was returning over the bared sediment surface, agitating it considerably and throwing much sediment into suspension. Most of the material in suspension is silt and clay, but there is some fine and even medium sand, particularly in the samples with the highest concentrations.

A study of values obtained indicates that water near the surface in the Minas Basin at or near high tide has about 125 ± 25 gm./m^3 of sediment in suspension. This is a considerable amount of sediment (water from the open sea contains about 2 gm./m^3) when one considers the large amounts of water that move in and out of the Basin every day and indicates that this is an impor-

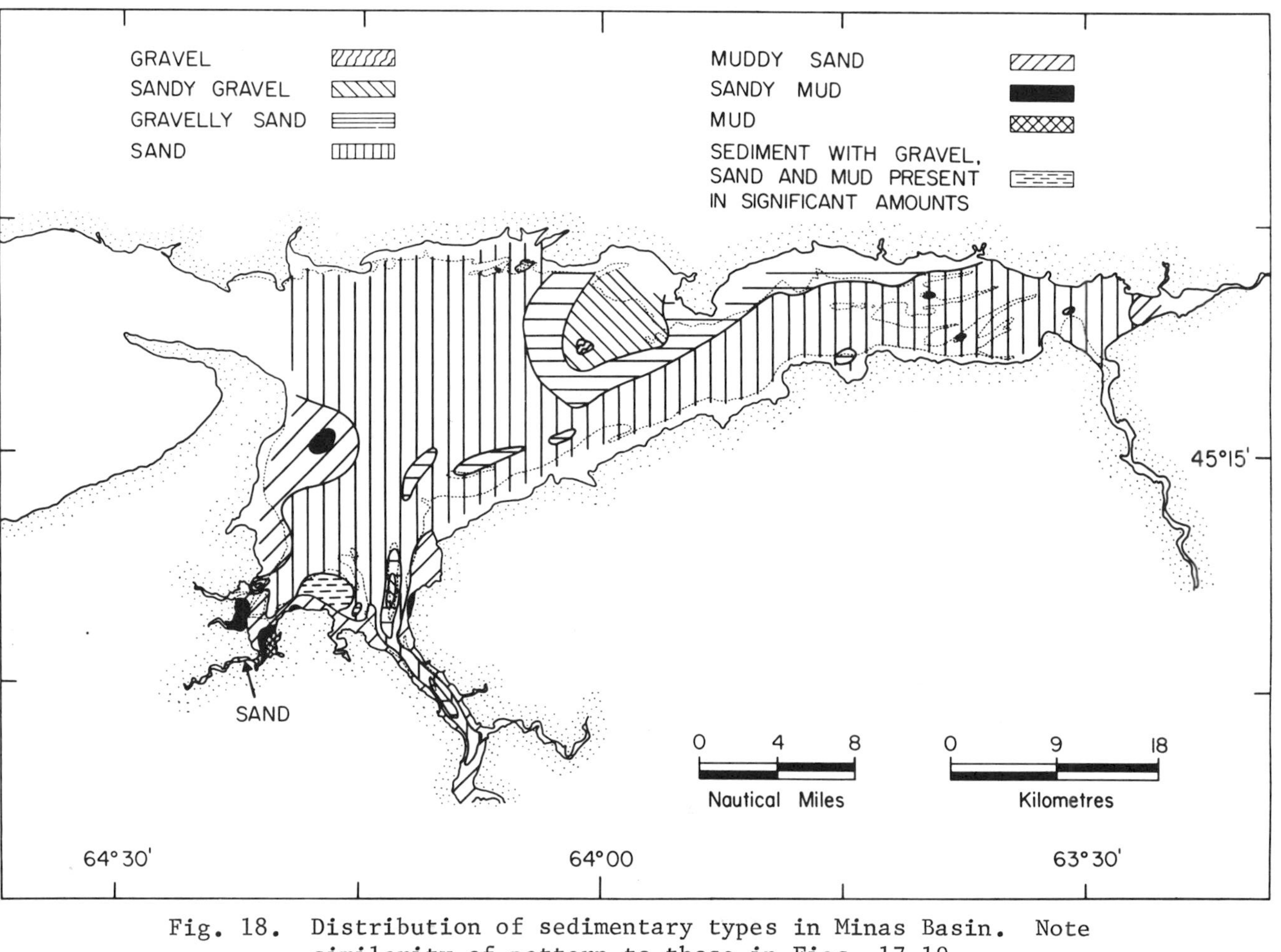

Fig. 18. Distribution of sedimentary types in Minas Basin. Note similarity of pattern to those in Figs. 17-19.

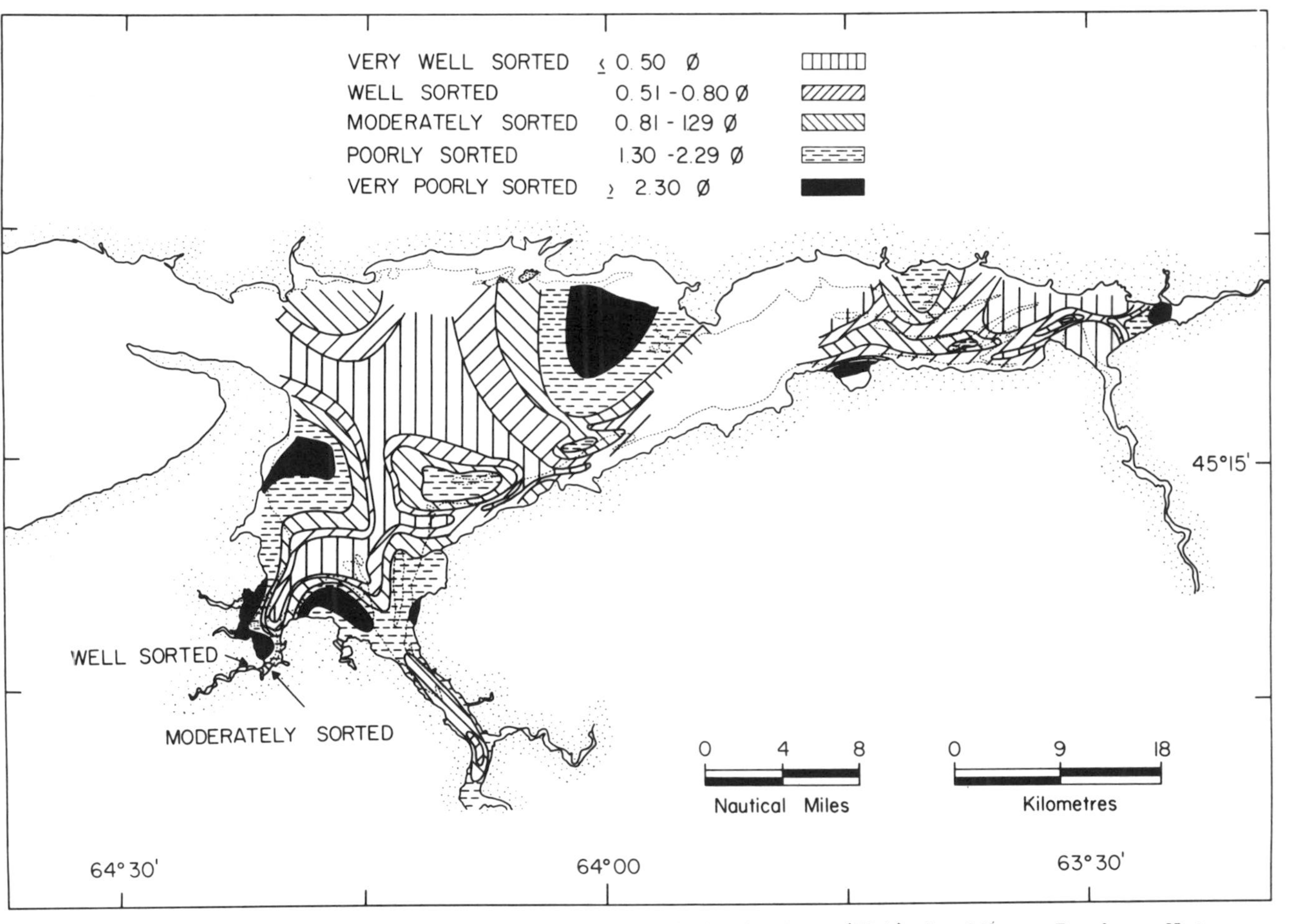

Fig. 19. Sediment sorting based on standard deviation (SDϕ) in Minas Basin. Note similarity of patterns to those in Figs. 17 and 18.

tant factor in sediment movement and deposition.

Interpretation:

Poor sorting in conjunction with fine sediment indicates very little movement or agitation of the water, allowing relatively large amounts of silt and clay to settle out. This then, is the case to the east of Cape Blomidon. The water enters through the Minas Passage (Fig. 20) and goes well out into the Basin before it curves around to the south in response to the Coriolis Force and other factors mentioned earlier. Thus, a shadow effect is produced in the lee of the Cape and the sediment deposited is relatively fine and poorly sorted. Similarly, as the currents approach the shoreline, they generally lose competence allowing the deposition of finer sediment. A small patch of fine sediment with poor sorting near the head of Cobequid Bay is probably the result of shearing between currents going in opposite directions, thus setting up a weak eddy.

Poor sorting in conjunction with coarse sediment is the result of either a "dumping" effect through loss of current competence or of selective removal of sand size and finer material, leaving the gravel component. The large area of poorly sorted, coarse material to the west of Economy Point is thought to be the result of the dumping effect, compounded by what appears to be a standing wave effect. The current from Cobequid Bay to the Minas Basin is restricted between Economy Point and Burntcoat Head, thus increasing in competence. However, the Basin widens beyond Economy Point, causing the current to lose much of its sediment load. The more-or-less isolated patches of coarser material are probably the result of the selective removal of fine material and the building of gravel bars, through local amplification of current competence.

Another interesting aspect of the sorting plot is the repetitive well sorted-poorly sorted pattern along the north side of the Basin, forming a trend which is oriented at a high angle to the shoreline. The distance between each area of poor sorting is approximately the same and there are four such areas on the north side. The pattern appears to represent nodes, at approximately one-quarter wave-length intervals, of a standing wave. This indicates that the current is of a different quality on the north side of the Basin than on the south side.

Present Current Pattern and Sediment Transport:

Through an analysis of the data, particularly the sorting values, a probable current pattern in the Minas Basin has been plotted (Fig. 20). This is in general agreement with the current pattern shown by Swift and McMullen in their study of the Minas Basin in 1968. However, a much more detailed sketch of the current's path is now possible. In the earlier paper, Swift and McMullen (1968) pointed out that the return current along the north side is probably weaker than the current along the south side, going toward the head of the bay. This is borne out by the present analysis which indicates that the current on the north side is of a different character from that on the south side.

Sediment, then, is transported from Windsor Bay, toward the head of

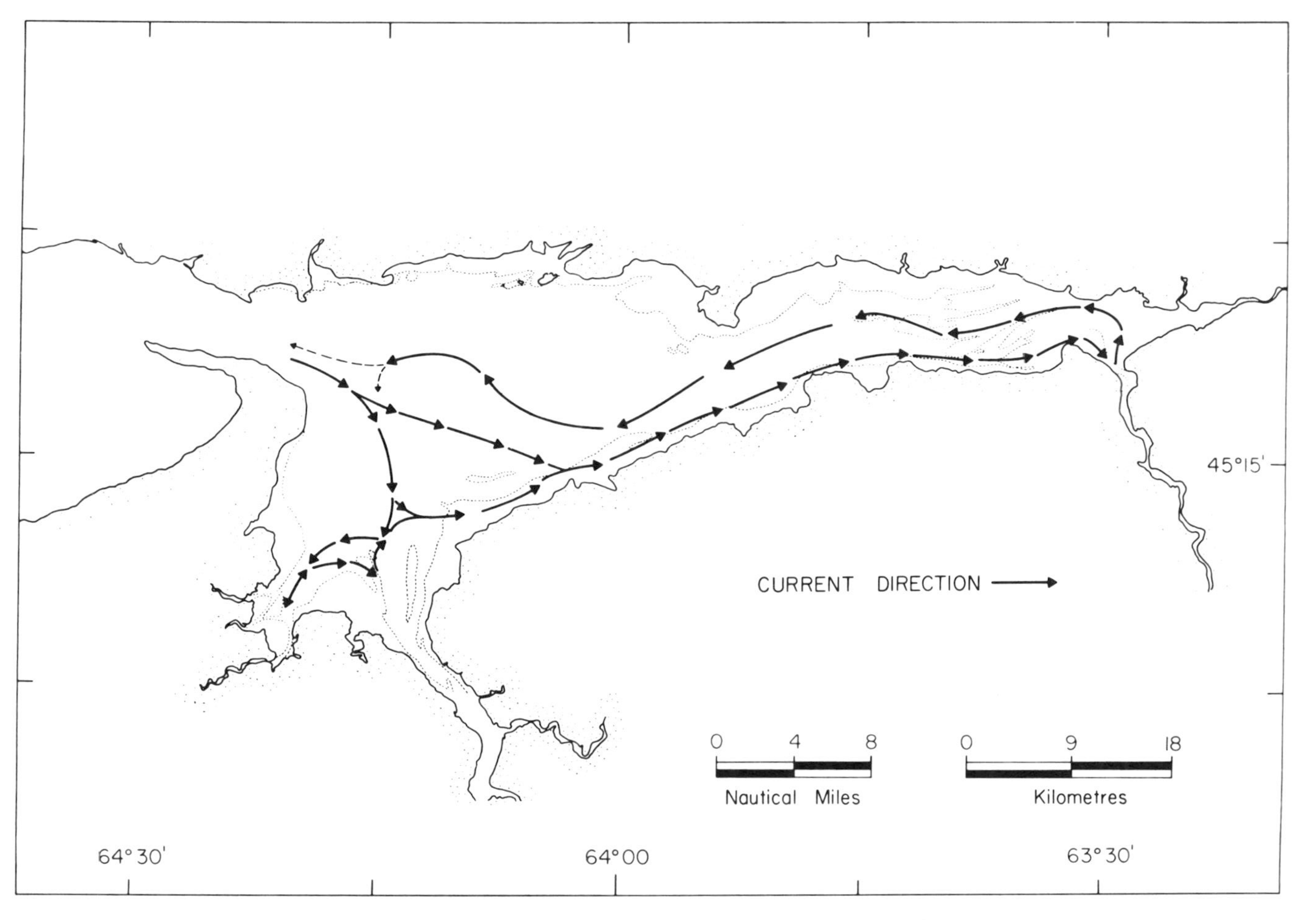

Fig. 20. Current directions in Minas Basin.

Cobequid Bay and then back along the north shore, where it either leaves the Basin or is recirculated. The sediment is transported both in suspension and as traction load, or by saltation near the sediment-water interface. No quantitative estimate is available with regard to the relative importance of the mechanisms, although qualitatively suspension is probably quite important overall and most important for the finer material.

The source of the sediment is three-fold:

1. it is brought in through the Minas Passage from the rest of the system;
2. it is brought down by the river systems emptying into Cobequid and Windsor Bays; and
3. it is eroded from the shoreline.

Although no quantitative data are available, it is known that the shoreline in places (especially where it is composed of glacial deposits or friable Triassic sandstones) has been cut back considerably (up to 2 metres per year). Also, several of the rivers (Avon, Salmon and Shubenacadie) contribute significant amounts of material, particularly in the finer grades. Therefore, it is apparent that material brought into the Basin from the Bay of Fundy proper (sediment concentrations of about 6 gm./m^3) is the least important source of the sediment in the Minas Basin.

This indicates, therefore, that from a sedimentological point of view the Minas Basin is an almost closed system which is gradually filling up with sediment. This is particularly true of Cobequid Bay. In the middle 1800's large sailing vessels used to be able to get to Truro and into the Shubenacadie estuary, but this is not possible now.

Possible Changes in Current-Pattern and Sedimentology which could be caused by the construction of a Tidal Power Dam:

If a tidal power dam were to be built across the Minas Basin between Economy Point and Cape Tenny (Fig. 21), the effect on both the water and sediment circulation patterns would be considerable. The immediate effect would be to break up the water circulation pattern into two units. The area between Minas Passage and the dam would retain a more or less normal current pattern and competence. The circulation pattern in Cobequid Bay might be relatively unaffected; however, the velocity and competence of the currents would be lowered because, although the water would flow into Cobequid Bay normally, it would not flow out normally. This lowered ebb-current competence would allow the deposition of more and finer sediment than is now being deposited there, thus accelerating the filling-in process. Furthermore, sediment loads brought in by the Salmon and Shubenacadie rivers would most likely be deposited in Cobequid Bay, rather than being largely transported through as at present. As the sediment-water interface rises through sediment deposition, the following events would probably occur: a lesser volume of water would be accommodated, and this would further lower the current competence and accelerate the filling-in process. No estimates on the rate of these

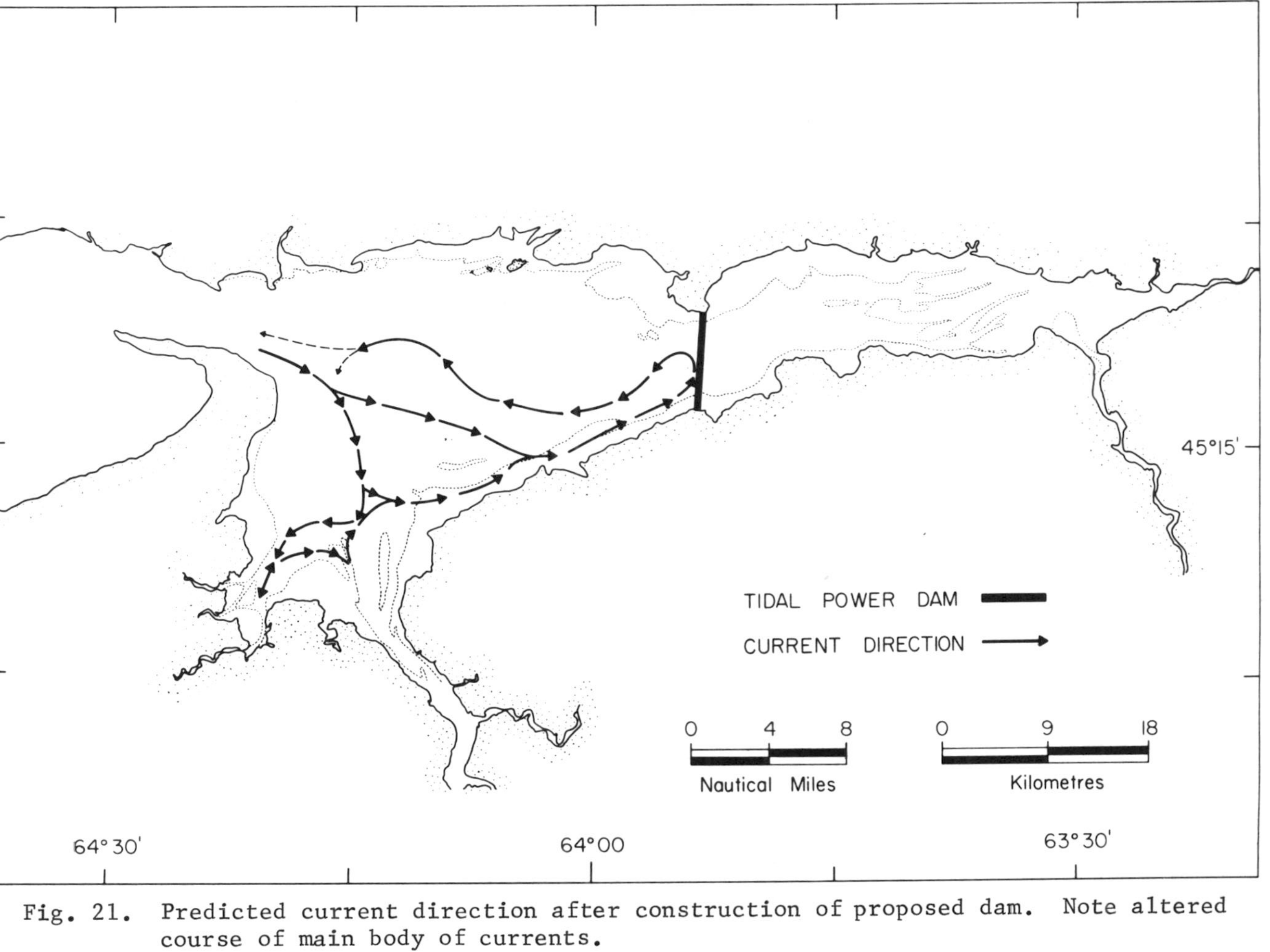

Fig. 21. Predicted current direction after construction of proposed dam. Note altered course of main body of currents.

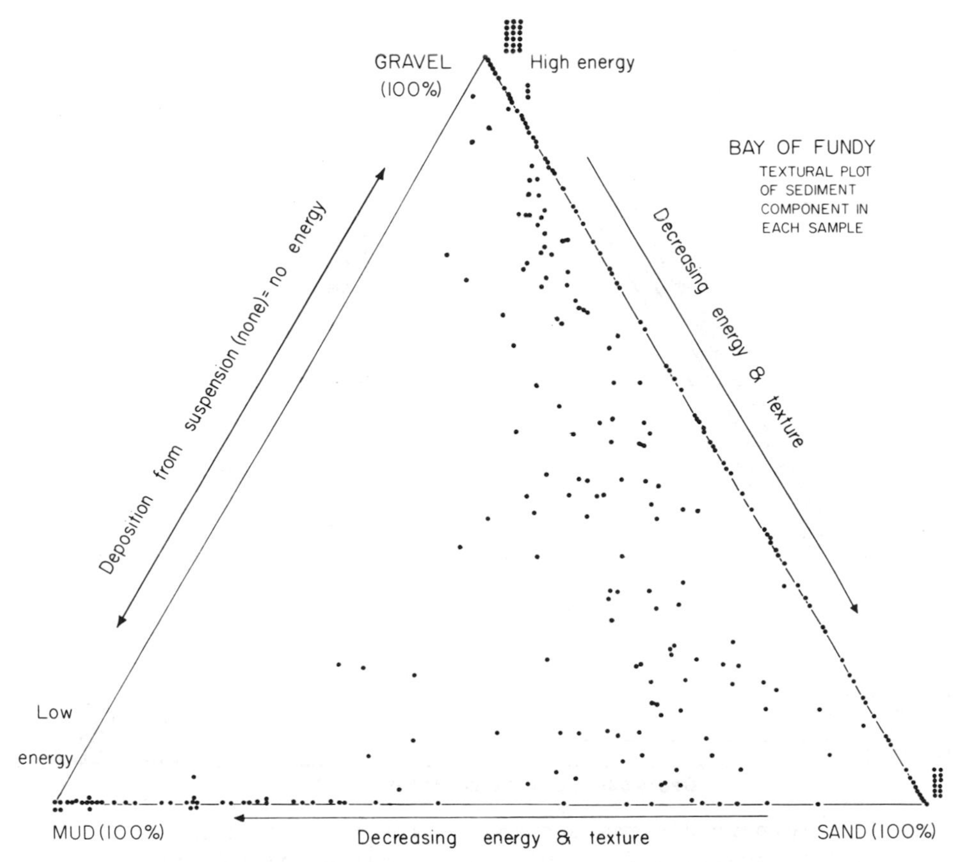

Fig. 22. Ternary diagram showing relationship of textural composition for individual samples and the relative energy available to erode at the sampling site in Bay of Fundy.

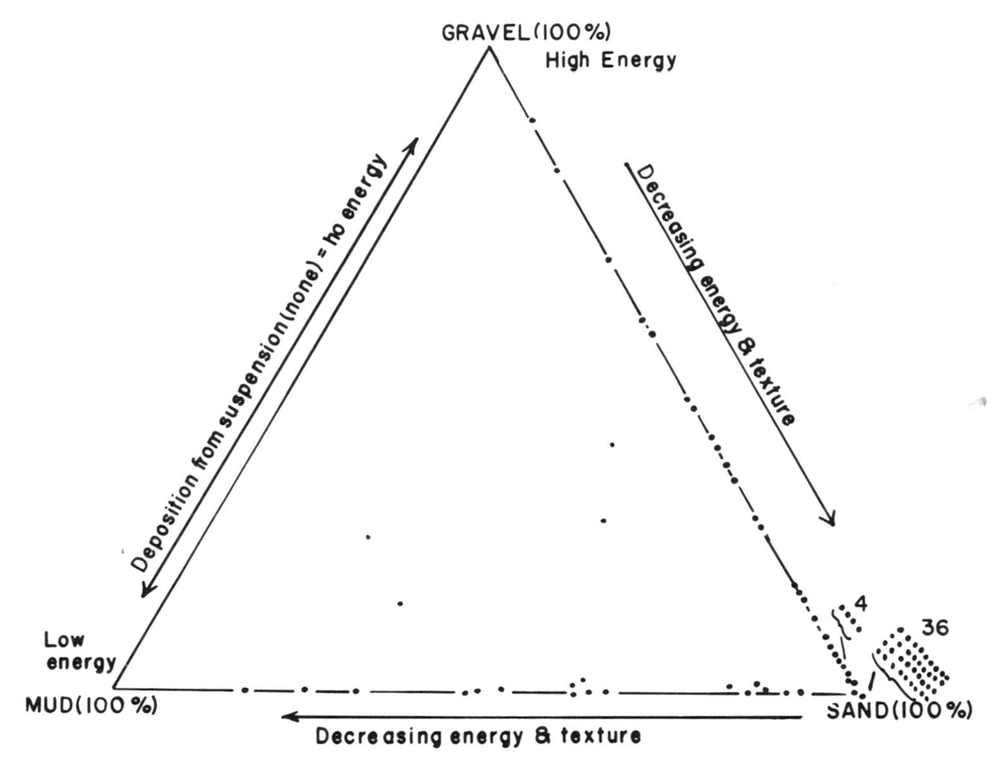

Fig. 23. Ternary diagram showing relationship of textural composition for individual samples and the relative energy available to erode at the sampling site in Minas Basin.

processes are available so that terms must be considered as relative only. However, this basin filling would reduce the hydraulic head as silting on both sides of the dam would probably take place.

COMPARISON OF BAY OF FUNDY AND MINAS BASIN

Although the sedimentological systems are somewhat comparable in these two water bodies, they differ in two ways:

1. Fundy has less suspended matter, and
2. it is not as close to hydrodynamic equilibrium as Minas Basin.

This may be shown by an analysis of the two systems based on texture and relative hydrodynamic vigour. In Fundy (Fig. 22) sediments plot mainly along the major dynamic axes but do include some poorly sorted material that reflects conditions of low hydrodynamic vigour. In Minas Basin (Fig. 23) the sediments appear to be in perfect equilibrium with the hydrodynamic system and any blockage of this system could lead to increased amounts of silt and mud being deposited. This limit of restriction is not known.

SUMMARY

At present the sedimentary regime in the Minas Basin is in a state of near equilibrium. It is an almost closed system, sedimentologically speaking, with most of the sediment coming into it through erosion of its shoreline or brought in by the several relatively large rivers. There is probably net sediment deposition, most of which is taking place in Cobequid Bay. If a tidal power dam were built across the entrance to Cobequid Bay, this delicate balance would, of necessity, be upset, leading to a gain in net deposition. This would accelerate the filling-in of Cobequid Bay which would, in a long but finite time, be completely filled in.

REFERENCES

Bumpus, D. F. (1959) "Sources of Water in the Bay of Fundy Contributed by Surface Circulation", Washington and Ottawa, International Passamaquoddy Fisheries Board, International Joint Commission, App. (1), 6, 11.

Bumpus, D. F., and L. M. Lauzier (1965) "Surface Circulation on the Continental Shelf off Eastern North America between Newfoundland and Florida", Atlas of the Marine Environment, Am. Geog. Soc., N. Y.

Chevrier, J. R. (1959) "Drift Bottle Experiments in the Quoddy Region", Washington and Ottawa, International Joint Commission, App. (1), 2, 13.

Churchill, F. J. (1924) "Recent Changes in the Coastline in the County of Kings", Proc. Trans. Nova Scotian Inst. Sci., 1b, 84-86.

Defant, A. (1961) Physical Oceanography, Vol. 2, 598, London, Pergamon Press.

Forgeron, F. D. (1962) "Bay of Fundy Bottom Sediments", M.Sc. Thesis, Carleton University, Ottawa.

Forrester, W. D. (1958) "Current Measurements in Passamaquoddy Bay and the Bay of Fundy, 1957-1958", Canadian Hydrographic Service, Tidal Pub. 40, 73.

Hachey, H. B., and W. B. Bailey (1952) "The General Hydrography of the Waters of the Bay of Fundy", Fisheries Research Board of Canada, Repts. Biol. Sta., 455 .

Hind, H. Y. (1875) "The Ice Phenomena and the Tides of the Bay of Fundy", Canadian Monthly and National Review, 8, 189-203.

Krumbein, W. C., and F. J. Pettijohn (1938) Manual of Sedimentary Petrography, 549, Appleton - Century - Crofts, New York.

Langford, C. J. (1966) "Bay of Fundy Report on Tidal and Current Survey", Bedford Inst. Oceanography Report, Nova Scotia.

Mavor, J. W. (1922a) "The Circulation of the Water in the Bay of Fundy, Part I. Introduction and Drift Bottle Experiments", Contr. Canadian Biol., Nova Scotia, 1, 106-124.

Mavor, J. W. (1922b) "The Circulation of the Water in the Bay of Fundy, Part II. The Distribution of Temperature, Salinity and Density in 1919, and the Movements of Water Which They Indicate in the Bay of Fundy", Contr. Canadian Biol., Nova Scotia, 1, 353-375.

McMullen, R. M., and D. J. P. Swift (1966) "An Occurrence of Large-Scale Rhomboid Ripples, Minas Basin, Nova Scotia", Jour. Sedimentary Petrology, 27, 705-706.

Miller, J. A. (1966) "Suspended Sediment System in the Bay of Fundy", M.Sc. Thesis, Dalhousie University, Halifax, Nova Scotia.

Swift, D. J. P., and H. W. Borns, Jr. (1967a) "Genesis of the Raised Fluviomarine Outwash Terrace, North Shore of the Minas Basin, Nova Scotia", A Preliminary Report, Maritime Sediments, 3, 17-23.

Swift, D. J. P., and H. W. Borns, Jr. (1967) "A Raised Fluviomarine Outwash Terrace, North Shore of the Minas Basin, Nova Scotia", Jour. Geology, 75, 693-710.

Swift, D. J. P., Cok, A. E., and A. K. Lyall (1966) "A Subtidal Sand Body in the Minas Channel, Eastern Bay of Fundy", Maritime Sediments, 2, 175-180.

Swift, D. J. P., and A. K. Lyall (1968a) "Reconnaissance of Bedrock Geology by Sub-Bottom Profiler, Bay of Fundy", Geol. Soc. America Bull., 79, 639-646.

Swift, D. J. P., and A. K. Lyall (1968b) "Origin of the Bay of Fundy", Marine Geology, 6, 331-343.

Swift, D. J. P., and R. M. McMullen (1968) "Preliminary Report on Intertidal Sand Bodies in the Minas Basin, Bay of Fundy", Can. Jour. of Earth Sciences, 5, 175-183.

Swift, D. J. P., McMullen, R. M., and A. K. Lyall (1967) "A Tidal Delta with an Ebb-Flood Channel System in the Minas Basin", Maritime Sediments, 3, 12-16.

Swift, D. J. P., Pelletier, B. R., Lyall, A. K., and J. A. Miller (1969) "Sediments of the Bay of Fundy - A Preliminary Report", Maritime Sediments, 5, 95-100.

L'USINE MARÉMOTRICE DE LA RANCE

Georges Mauboussin*

PREAMBULE

C'est de Saint-Malo qu'en 1535 le célèbre navigateur Jacques Cartier s'embarquait à bord de la "Grande Hermine" pour découvrir le Canada et remonter le cours du St-Laurent jusqu'au petit village indien : Hochelage dominé par une colline qu'il appela le Mont-Royal (Montréal), en l'honneur du roi de France : François ler.

C'est d'une région où les marées sont parmi les plus fortes du monde qu'il partit pour trouver de l'autre côté de l'Atlantique, sensiblement sur le même parallèle, une autre région où les marées sont encore plus importantes.

Et si, 430 ans plus tard, la première usine marémotrice du monde a été construite à l'embouchure de la Rance, à quelques kilomètres de Saint-Malo, il n'est pas déraisonnable de penser que la seconde le sera au Canada, dans la grande baie de FUNDY.

L'ESTUAIRE DE LA RANCE

La Rance est un petit fleuve breton, de 100 kilomètres de longueur, au débit insignifiant, qui, peu après son passage au pied des remparts de la cité féodale de Dinan, débouche dans une large vallée que la mer visite et inonde deux fois par jour.

Ce vaste estuaire, d'une vingtaine de kilomètres de longueur, constitue un ensemble touristique de toute beauté : tantôt, ce sont de grands bassins, de véritables mers intérieures, au bord desquels se nichent de pittoresques petits villages, comme Mordreuc et St-Suliac, tantôt ce sont d'étroits goulets limités par de hautes falaises couvertes, au printemps, d'ajoncs et de genêts en fleurs, comme St-Hubert et Cancaval.

René de Chateaubriand, l'écrivain romantique - né, lui aussi, à Saint-

---

*ancien Directeur de la Région d'Equipement Marémotrice, d'Electricité de France.

Malo, comme Jacques Cartier - écrivait en 1821, dans ses "Mémoires d'Outre-Tombe":

> "Rien de plus charmant que les environs de Saint-Malo dans un
> "rayon de cinq à six lieues - Les bords de la Rance, en remon-
> "tant cette rivière depuis son embouchure jusqu'à Dinan,
> "mériteraient, seuls, d 'attirer les voyageurs - mélange conti-
> "nuel de rochers et de verdure, de grèves et de forêts, d'antiques
> "manoirs de la Bretagne féodale et d'habitations modernes de
> "la Bretagne commerçante".

Dans la région de Saint-Malo et du Mont St-Michel, les mouvements d'eau ont une amplitude très importante, due au fait que l'onde de marée, issue de l'Atlantique et se déplaçant d'ouest en est, vient buter et se réfléchir sur la presqu'ile du Cotentin. Aussi, les marées se font-elles sentir de façon puissante dans l'estuaire de la Rance.

En grande marée de'équinoxe, le flot amène, dans le réservoir créé à l'amont de l'usine, 180 millions de mètres cubes, qui s'étalent sur 22 kilomètres carrés, avec une différence de niveau de 13.50 m entre pleine mer et basse mer. En vive-eau moyenne, il amène 150 millions de m^3, avec une amplitude de 10.90 m. En marée moyenne, 110 millions de m^3 avec 8.50 m. En morte-eau moyenne 85 millions avec 5 mètres.

LE SITE DU BARRAGE

L'usine-barrage est implantée entre la pointe de la Briantais sur la rive droite et la pointe de la Brebis sur la rive gauche. A cet endroit, la largeur de la Rance est de 750 mètres.

Les fonds se situent à environ 12 à 14 mètres sous le niveau des plus basses mers, c'est dire qu'ils sont recouverts d'environ 26 mètres d'eau, au moment des grandes marées.

Le lit est constitué par un granit, de qualité moyenne, mais suffisante pour y asseoir les ouvrages : il est recouvert par une faible couche d'alluvions : sable et galets - .

A environ 150 mètres de la rive droite, sur le tracé des ouvrages, un rocher appelé rocher de Chalibert ne laisse apparaître que sa pointe au moment des pleines mers -.

L'usine-barrage est construite à environ 4 kilomètres à l'embouchure de la Rance, c'est à dire de St-Malo et de Dinard - Cette distance, parsemée de quelques ilôts, ainsi que la direction Nord-Sud du fleuve, font qu'on est totalement à l'abri des vents dominants et des tempêtes du large, venant de l'Ouest.

Lors de la grande tempête du 5 avril 1962, le chantier n'a subi aucun dommage, et pourtant cette tempête fut très violente : on n'en avait pas vu de semblable dupuis 1904, paraît-il : dans la région des digues se sont écroulées,

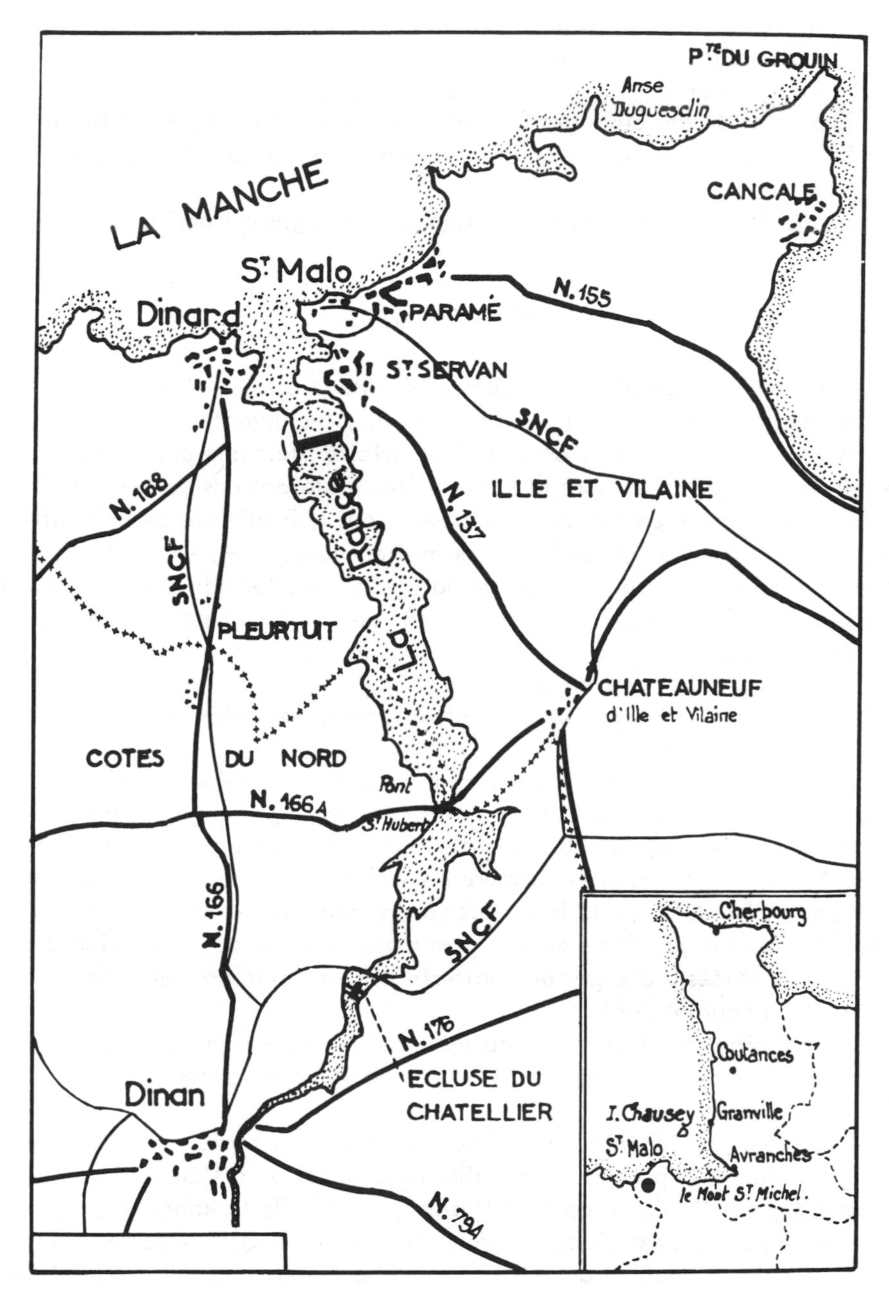

Fig. 1. Plan de situation. (Document E.D.F.)

des parapets ont été emportés, des quais ont été submergés.

Les conditions essentielles, permettant d'envisager la construction d'une usine marémotrice, se trouvent donc bien remplies à la Rance:

- fortes amplitudes des marées,
- vaste réservoir de 180 millions de mètres cubes,
- implantation des ouvrages favorables: longueur relativement faible 750 m; rocher de fondation à un niveau acceptable, ni trop haut, ni trop bas,
- bonne direction du fleuve, mettant les ouvrages à l'abri des tempêtes.

LES ANCIENS PROJETS

Un site aussi favorable à la construction d'une usine utilisant l'énergie des marées ne pouvait que séduire les chercheurs, les inventeurs.

1906 - Et, de fait, en 1906, alors que l'électricité était une chose toute nouvelle, une denrée rare, dont bien des villes n'étaient pas dotées - 1906! la consommation totale en France ne dépassait pas 600 millions de kilowatheures, soit grossièrement 0.5% de la consommation actuelle -, en 1906, donc, un professeur de physique du collège de Saint-Servan, Gaston Boucher, avait dressé un projet sommaire pour l'aménagement de l'embouchure de la Rance, projet très vaste puisqu'il avait en vue trois objectifs:

- la création d'un port en eau profonde,
- la construction d'un pont assurant la liaison St-Malo-Dinard,
- une usine marémotrice.

Suivant Gaston Boucher, l'usine comprenait 192 turbines de 200 CV. L'auteur estimait que les progrès techniques de l'époque permettaient de transporter l'énergie jusqu'à 120 kilomètres du lieu de production, et par conséquent de desservir un grand nombre de villes de Bretagne et de Normandie. Bien entendu, l'irrégularité de la production n'avait pas échappé à l'inventeur, qui résolvait le problème en prévoyant pour chaque agglomération des installations destinées à charger une batterie d'accumulateurs, et la fourniture d'électricité en courant continu.

1918 - Un ingénieur des Ponts et Chaussées - Maynard - reprenait une étude, faite par M. Maire, d'une usine à construire sur la Rance entre St-Malo et Dinard.

1920 - A son tour, un ingénieur de la marine, Mangin, reprenait l'étude de Maynard, et afin d'apporter une certaine régularisation à l'énergie produite par les marées, eut l'idée de coupler l'usine projetée de la Rance avec une usine de même puissance à construire dans la rade le Landevennec (partie de la rade de Brest). Mangin tirait profit du fait qu'il existe un décalage d'environ 2 heures entre la pleine mer à Brest et à St-Malo. Ainsi, moyennant la construction d'une ligne de 200 kilomètres entre ces deux points, on obtenait une énergie déjà un peu régularisée. Compte tenu de l'amplitude des marées à Brest et à St-Malo, il était nécessaire de prévoir un réservoir d'une surface de 72 km² à Landevennec pour 22 km² à la Rance.

Bien évidemment, la régularisation n'était pas complète avec le couplage de ces deux usines.

"Mais, -écrivait Mangin, dans la Revue Générale de l'Electricité "du 8 mai 1920-, le réseau d'Etat projeté s'étendra dans un avenir pas trop "lointain à toute la France et il sera naturel d'y rattacher le couplage "Brest-St-Malo. Ce serait alors les usines thermiques ou hydrauliques de "l'intérieur qui seraient appelées à lui porter secours, et ce secours leur "serait facile en raison de leur énorme capacité globale".-

Dans ces quelques lignes prophétiques, Mangin laissait déjà entendre-, il y a un demi-siècle - qu'avec le développement des lignes électriques et leur inter-connexion il ne serait plus nécessaire de chercher des dispositions onéreuses pour obtenir une énergie continue des marées comme avait cherché à le faire autrefois Bélidor*.

<u>1951</u> - Après des études longues et détaillées, faites par les services d'Etudes et Recherches d'Electricité de France, avec les conseils de M. R. Gibrat -, un projet complet était établi.

L'usine-barrage était implantée entre la Pointe de la Brebis et la pointe de la Briantais. L'équipement comportait 26 groupes Kaplan classiques, à axe vertical, de 8,000 kW de puissance unitaire. Le remplissage du bassin par le flot était assuré par 10 grands clapets automatiques de 15 x 16 m. La production de l'usine était estimée à 500 millions de kWh. De plus, le mode d'exécution des travaux à l'aide de digues successives en enrochements avait fait l'objet d'études assez complètes.

L'exploitation de l'usine était prévue comme celle des anciens moulins à marée** simple effet au vidage. "simple", parce que l'on ne produit de l'énergie qu'une seule fois par marée; "au vidage", parce que c'est en vidant le réservoir que l'on fabrique cette énergie.

Le projet, dit projet 1951 -, nécessitait huit années de travaux et était d'un prix élevé : 26 milliards d'anciens francs (base économique de 1950) soit,

---

*Traité d'architecture hydraulique (1737) par Bernard Forest de Bélidor, général et ingénieur français (1697-1761).

**Les plus anciens moulins à marée, établis sur les côtes de Bretagne, remontent au 12e siècle.
Il existait autrefois en Rance, 14 moulins à marée, construits dans des anses de l'estuaire. Quelques uns fonctionnaient encore il y a une vingtaine d'années: ils mettaient en mouvement des meules à grains.
Dans le département des Côtes du Nord, à Ploumanach, un moulin à marée fabriquait de la glace pour la conservation du poisson - En 1898, il produisait chaque jour 450 kg de glace.

Fig. 2. Moulin à marée de Saint Suliac. (Photo Doucet-Dinard)

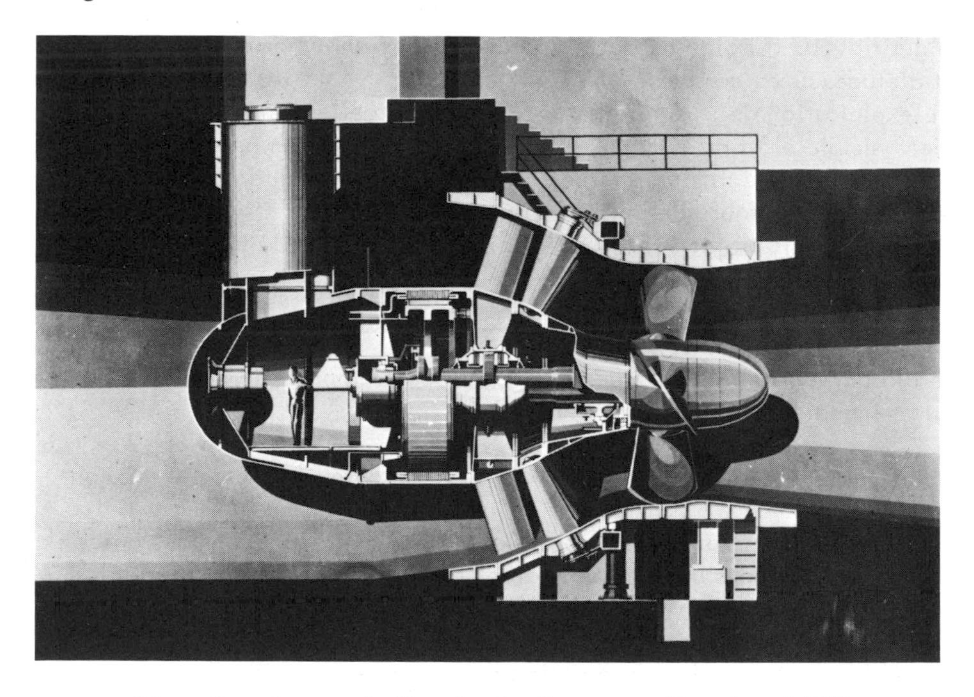

Fig. 3. Groupe bulbe de la Rance. (Janvier 1961)

en gros, 62 milliards d'anciens francs (ou 620 millions de nouveaux francs (base économique de 1960).

Pour diverses raisons, ce projet ne vit pas le jour -, et il fallut attendre 1961 pour entreprendre la construction de l'usine marémotrice de la Rance.

On a mis à profit la période 1952-1961 -, pour poursuivre les études et en entreprendre de nouvelles, afin d'arriver à une solution meilleure et plus économique.

LES ETUDES DEPUIS 1952

1) Etudes sur les Groupes

Il est apparu que l'exploitation d'une usine marémotrice serait grandement améliorée si l'on pouvait produire de l'énergie, non seulement au jusant, à marée descendante, mais aussi au flot, à marée montante. Ne pourrait-on trouver un type de machine qui aurait des rendements acceptables quel que soit le sens d'écoulement de l'eau? Si la chose était possible, on remplirait le réservoir d'abord par les groupes, dès que la mer serait assez haute pour qu'il y ait une chute suffisante entre elle et le réservoir maintenu au niveau le plus bas: on produirait ainsi de l'énergie pendant quelque temps, et l'on compléterait le remplissage en ouvrant toutes les vannes, et également en utilisant les groupes qui, découplés du réseau, fonctionneraient comme des orifices.

Ensuite, on produirait à nouveau de l'énergie lors du vidage du réservoir. On réaliserait de la sorte des cycles de production, dits à "double effet". Il en résulterait une légère augmentation de la production, et, surtout, cela aurait l'avantage de fabriquer de l'énergie plus fréquemment aux heures de forte consommation.

Les turbines classiques "Kaplan", à axe vertical, ne semblaient pas pouvoir répondre correctement au problème posé, il fallait trouver autre chose.

Et c'est ainsi qu'à l'occasion de la construction de l'usine de Cambeyrac, usine de compensation de l'usine de Couesque sur la Truyère, les services de l'Equipement de l'Electricité de France demandaient aux principaux constructeurs français et étrangers d'étudier et de proposer un type de groupe pouvant fonctionner convenablement dans un sens ou dans l'autre*, aussi bien en turbine qu'en pompe.

On devine, de suite, le grand intérêt du pompage. Si, en effet, au moment où le réservoir est plein ou presque plein, on dispose d'énergie électrique sur le réseau général, on peut faire du surremplissage, et cela sans

---

*Des dispositions particulières avaient été prises, dans les ouvrages de Génie Civil de Cambeyrac, pour permettre d'alimenter le groupe par l'amont ou par l'aval.

grande dépense, car la hauteur de refoulement est faible. Quelques heures plus tard, quand la mer aura baissé, le volume supplémentaire accumulé sera turbiné sous une hauteur de chute beaucoup plus forte : par exemple, si l'on pompe sous 1 mètre un certain volume d'eau et qu'on le turbine, quelques heures plus tard, sous 7 m de chute, il en résultera compte tenu du rendement des groups un gain de production égal environ à 50% des kWh dépensés pour le pompage. Mais, ce gain, traduit en argent, sera bien supérieur puisque le pompage aura été fait en heures creuses et le turbinage en heures pleines, peut-être même en heures de pointe.

Et, c'est ainsi que fut mis au point le premier groupe-bulbe dont deux exemplaires, de conception un peu différente, de 5,000 kW sous 10 m de chute furent installés à Cambeyrac.

A la même époque, un groupe bulbe de 14,000 kW sous 16 m de chute fut installé sur la Dordogne, à Argentat, usine de compensation du Chastang; puis un de 8,500 kW sous 11 m à Beaumont-Monteux sur l'Isère.

Enfin, un groupe bulbe de 9,000 kW était installé dans une écluse désaffectée du Port de St-Malo. Mis en service en novembre 1959, il a travaillé pendant quelques années comme une véritable petite usine marémotrice, entre les bassins à flot du port et la mer, en vidant une tranche maximale de 2 m de ces bassins, soit 1,300,000 m^3, et en la reconstituant, évidemment, dans les heures suivantes, soit en fonctionnant en turbine, si le niveau de la mer était suffisamment élevé, soit en pompe ce qui était le cas le plus fréquent. On a pu, également, à St-Malo, vider les bassins en faisant du pompage, en sens inverse, grâce à une vanne réglable, d'un type spécial, placée côté mer, et par dessus laquelle de débit était déversé.

Bref, à St-Malo, tous les modes de fonctionnement ont pu être essayés : en turbine et en pompe, dans les deux sens d'écoulement. Le groupe s'est montré parfaitement stable; son rendement a été trouvé de 3 à 8% supérieur à ce que l'on avait escompté. Et grâce à l'enseignement tiré des mesures mécaniques, hydrauliques et électriques qui ont été faites, les constructeurs ont pu proposer, pour la Rance, un groupe plus puissant quoique de dimensions légèrement plus faibles.

Le groupe bulbe a la forme d'un petit sous-marin, à l'extrémité duquel se trouve une roue Kaplan, à quatre pales. La roue des groupes de l'usine de la Rance a un diamètre de 5.35 m. A l'intérieur de la coque du sous-marin, un alternateur classique de 10,000 kW à 3,500 volts, tournant à 94 tours minute, et fonctionnant dans une atmosphère surpressée : 1 kg/cm^2 au-dessus de la pression atmosphérique*.

Le groupe est placé dans un conduit rectiligne, horizontal, auquel il

---

*Les caractéristiques du groupe expérimental de St-Malo étaient : diamètre de la roue : 5.80 m. - vitesse : 88 tours par minute - alternateur : 9000 kW tournant dans l'air à la pression atmosphérique.

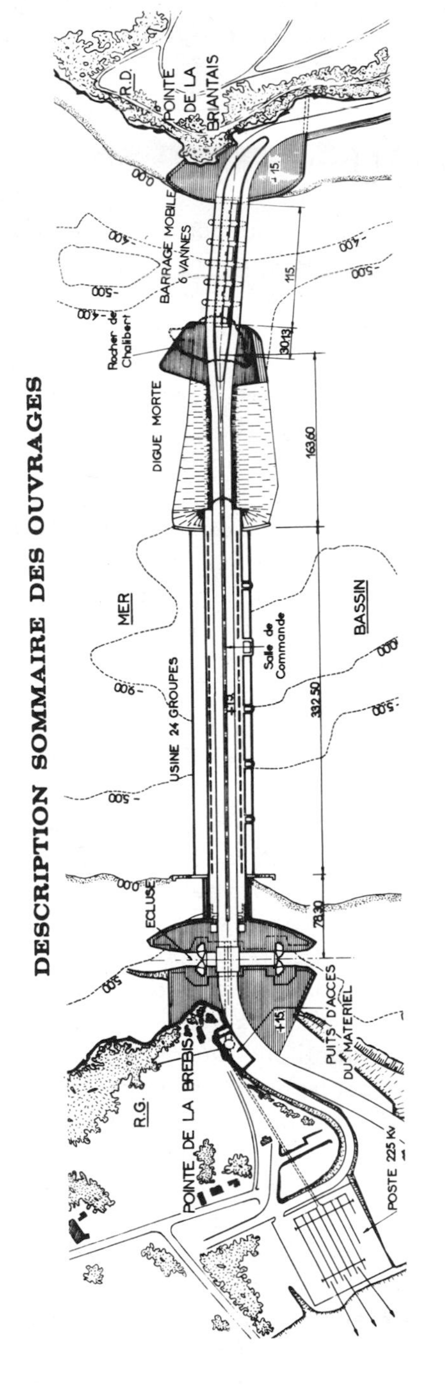

Fig. 4. Plan d'ensemble des ouvrages. (Document E.D.F.)

est fixé par douze directrices fixes. Un puits vertical métallique, débouchant dans la salle des machines permet d'accéder à l'alternateur. C'est par ce puits que sortent les câbles de puissance, la filerie et les tuyauteries qui servent à la commande du groupe.

2) Etudes sur les Ciments et les Bétons

a/Ciments

On sait que la chaux litre dégagée au moment de la prise forme avec l'aluminate tricalcique, contenu dans le ciment, et en présence de certaines eaux contenant des sulfates, un sel expansif, appelé sel de Candlot qui se gonfle démesurément et se transforme rapidement en pâte molle et en bouillie. Le secret de la bonne tenue des ciments romains aux eaux de mer et aux eaux contenant du gypse est dû aux cendres volcaniques, aux pouzzolanes, qui y étaient incorporés et qui fixaient la chaux litre.

Le laitier de haut-fourneau ayant les mêmes propriétés que les pouzzolanes, il a été décidé, dès le départ, que l'on utiliserait à la Rance un ciment du haut-fourneau (C.H.F.) contenant 70% de laitier et 30% de clinker.

Mais, le souci de construire les ouvrages de la Rance de la façon la plus sûre a conduit à rechercher par des essais comparatifs si cette proportion était bien la meilleure, compte tenu des autres conditions requises : résistance mécanique suffisante et retrait minimal. On a voulu s'assurer que, malgré la différence de dureté des deux composants, clinker et laitier, le broyage ensemble de ces deux produits, dans un même appareil, comme cela se fait normalement dans les cimenteries, donnait des résultats aussi bons que le broyage séparé.

Pour contrôler l'indécomposabilité du ciment soumis à l'action chimique de l'eau de mer, qui ne contient, pourtant, que peu de sulfates, on a soumis des éprouvettes à l'essai Anstett, essai réputé tellement sévère, qu'en général les fabricants de ciment ne l'acceptent pas.

b/Bétons

Sous la poussée alternée des eaux, suivant le niveau du réservoir par rapport à celui de la mer, les ouvrages de la Rance travaillent à la flexion, donnant naissance à des efforts de traction risquant de créer des micro-fissures.

On a fait des essais sur des poutres armées, immergées en Rance, au niveau de la mi-marée : elles étaient donc tantôt immergées, tantôt émergées. Elles étaient chargées en leur milieu par un flotteur dont le poids et le volume avaient été calculés pour qu'elles fléchissent vers le haut pendant les périodes d'immersion, vers le bas pendant les heures d'émersion. Des dispositions analogues étaient réalisees en faisant fléchir des poutres dans des bacs que l'on pouvait, grâce à un système de pompage, remplir et vider en quelques heures c'est à dire réaliser artificiellement six marées par jour au lieu de deux et accélérer ainsi la fatigue dûe aux efforts alternés.

On a soumis des éprouvettes à des essais très sévères, en les plaçant dans des bacs remplis d'une eau de mer concentrée, contenant le double de

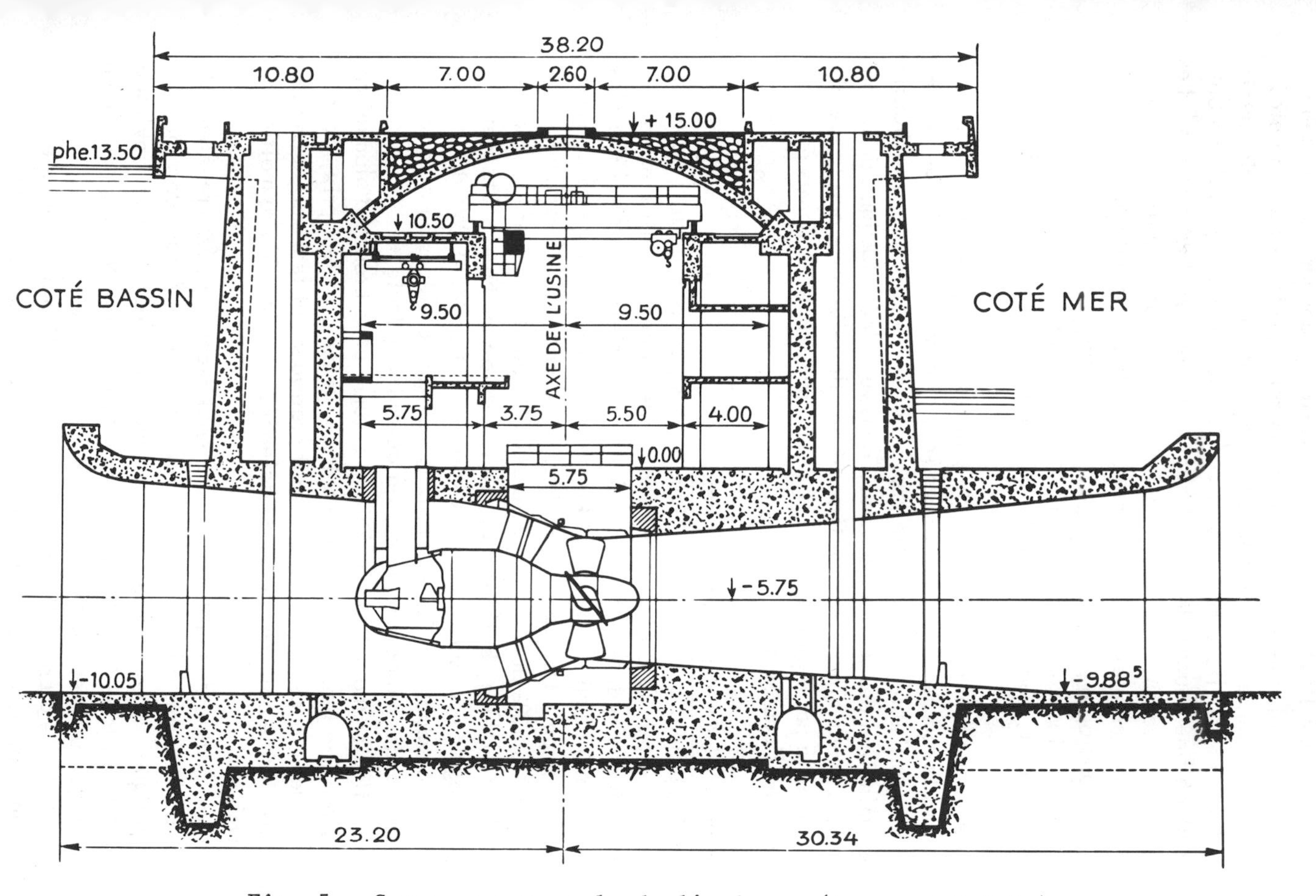

Fig. 5. Coupe transversale de l'usine. (Document E.D.F.)

sels. Tous les huit jours, les éprouvettes étaient sorties des bacs et exposées pendant une semaine à l'action de lampes infra-rouge donnant une température de 60°, puis replongées dans les bacs pendant une autre semaine etc.

En ce qui concerne l'eau de gâchage, la question s'est posée de savoir s'il valait mieux utiliser l'eau douce ou l'eau de mer : des aciers à béton ont été revêtus d'une fine couche de ciment gâché de l'une et l'autre façon. Les différences sont peu sensibles, néanmoins les petites piqûres de rouille ont semblé être plus nombreuses sur les aciers recouverts de ciment gâche à l'eau de mer. On a donc opté pour le gâchage à l'eau douce.

Enfin, toutes les expériences ont confirmé que les conditions que l'on s'était fixé au départ, pour le calcul des ouvrages, étaient raisonnables, à savoir :

- dimensionnement de l'épaisseur de poutres et plaques fléchies, de telle sorte que le béton, considéré seul et comme un matériau homogène, non fissuré, ne travaille pas à plus de 25 kg/cm^2, à la traction.
- détermination des sections d'acier, dans le calcul classique du béton armé, c'est-à-dire en cas de fissuration, de telle sorte que le métal ne travaille pas à plus de 1000 kg. par cm^2, dans les parois en contact avec l'eau de mer.

3) Etudes sur la Tenue des Métaux et Peintures

Le comportement à la mer des métaux et des peintures a fait l'objet de longues études, soit de façon statique en immergeant des échantillons en Rance, soit de façon hydro-dynamique,-si l'on peut s'exprimer ainsi - en installant un petit groupe miniature de 314 m/m de diamètre, alimenté en eau de mer par une pompe. Des pales en métaux différents ont été essayées : aciers ordinaires, aciers à 13% de chrome, aciers à 17% de chrome, 4% de nickel et 1 ou 2% de molybdène, bronzes d'aluminium de divers types. Les métaux ont été essayés egalement sur le groupe expérimental de 9000 kW, de St-Malo : pales en acier inoxydable, pales en bronze, pavage de pièces en acier ordinaire avec des bandes minces, soudées, en acier inoxydable.

En bref, l'acier inoxydable à 17% de chrome et le bronze d'aluminium ont fait preuve d'une tenue remarquable à l'eau de mer, et leur prix, dans le cas du marché des groupes de la Rance étant pratiquement le même, la moitié des groupes de l'usine ont leurs pales en acier inoxydable, l'autre moitié en bronze d'aluminium.

Toute une gamme de peintures ont été également essayées : revêtements bitumineux, revêtements à base de caoutchouc, peintures à base de zinc, peintures vinyliques. De ces essais, il est apparu que les peintures vinyliques donnent entière satisfaction, à la condition qu'elles soient très soigneusement appliquées sur des surfaces parfaitement propres.

Il a été installé à proximité du chantier une station de sablage et de peinture, où étaient maintenues des conditions rigoureuses de température et d'hygrométrie.

Fig. 6. Usine de la Rance. (Juin 1967)

Enfin, on a utilisé, de façon générale, la protection cathodique.

4) Etudes sur Modele Reduit

Il n'était pas suffisant d'étudier et de mettre au point un type de groupe, bien adapté à l'énergie marémotrice, d'étudier la resistance à la corrosion des métaux et peintures, de rechercher à faire des bétons de la meilleure qualité; il fallait encore savoir comment entreprendre et conduire les travaux dans un fleuve qui est en crue deux fois par jour et dans lequel les courants changent de sens quatre fois en 24 heures 50 minutes, il fallait savoir comment se défendre contre l'action violente de débits considérables, pouvant atteindre à certaines époques 18,000 mètres cubes par seconde.

Le meilleur instrument de calcul était le modèle réduit: aussi, dès 1954, un grand modèle, représentant toute la vallée de la Rance à l'échelle de 1/150e était-il construit sur le terre-plein du Naye, à St-Malo. Des appareils très ingénieux furent conçus et construits par le laboratoire national d'hydraulique* pour reproduire fidèlement et sans interruption toutes les marées successives, pendant plusieurs jours ou même une semaine. Grâce au modèle, on a pu connaître avec précision les modifications, apportées par les ouvrages en cours de construction à la propagation de la marée, à la direction et l'intensité des courants et également apprécier les efforts exercés à chaque instant sur les parties partiellement construites, donc encore peu stables.

Le modèle a permis de rechercher et de mettre au point, dans les plus petit détails, les divers moyens envisagés pour la coupure de la Rance, toutes sortes de types de batardeaux ont été étudiés : digues en enrochements, caissons métalliques ou en béton armé, amenés par flottaison et échoués, ouvrages en palplanches etc...

Lorsq'en 1955-1956, les principales entreprises françaises ont été invitées à étudier et proposer pour les batardeaux des variantes au projet d'Electricité de France, elles ont pû, à tour de rôle, utiliser le modèle réduit de St-Malo et vérifier que ce qu'elles imaginaient était, ou non, réalisable.

DESCRIPTION DES OUVRAGES

L'usine-barrage de la Rance se présente essentiellement comme étant composée de deux ouvrages bien distincts.

En partant de la pointe de la Briantais, rive droite, les pertuis de vannage, destinés au remplissage ou au vidage du bassin.

En partant de la pointe de la Brebis, rive gauche, l'usine marémotrice proprement dite.

---

*Le Laboratoire National d'Hydraulique, aménagé à Chatou dans la banlieue parisienne, dépend du Service Etudes et Recherches, d'Electricité de France.

Fig. 7. Usine marémotrice de la Rance. (Novembre 1966)

Ces deux ouvrages ne se rencontrent pas, il s'en faut d'un peu plus de 150 mètres : cette brèche est fermée par une digue en enrochements, appelée digue morte, avec voile intérieur en béton, pour assurer l'étanchéité.

Les pertuis de vannage sont au nombre de six : ils sont fermés par des vannes-wagon de 15 mètres de largeur et de 10 m de hauteur, et l'ensemble avec ses piles est semblable à un barrage mobile classique construit pour un aménagement de basse chute. Les débits qui peuvent transiter par les pertuis sont très importants : ils atteignent 5000 m^3 par seconde, pour une différence de niveau de 1 mètre entre mer et estuaire, au moment de remplissage.

L'usine marémotrice proprement dite qui s'enracine sur la rive gauche se présente comme un long tunnel d'environ 370 mètres de longueur, de 20 m de largeur et près de 15 m de hauteur. Ce grand bâtiment est comme posé sur un socle en béton de 10 mètres de hauteur et 50 mètres de largeur, dans lequel sont découpés tous les 13.30 m les conduits hydrauliques des 24 groupes de l'usine.

A l'intérieur de l'usine, au niveau du sol, c'est-à-dire à 1 m au-dessus du niveau des plus basses mers, on rencontre au rythme de 13.30 m les fosses dans lesquelles se trouvent les parties métalliques des conduits hydrauliques, parties métalliques démontables permettant le montage et le démontage de la roue motrice.

Au-dessus, et sur toute la longueur de l'usine, circulent 4 ponts roulants de 90 tonnes.

De part et d'autre de la ligne des fosses, sont disposés les locaux nécessaires à l'exploitation de la centrale : régulateur, appareillage électrique ... et à la transformation de l'énergie.

Un seul régulateur commande 4 groupes, et deux ensembles de 4 groupes débitent sur un même transformateur. L'équipement électrique de l'usine est de 24 groupes de 10,000 kW et de 3 transformateurs portant la tension à 225,000 volts. La productibilité est de 550 millions de kWh par an.

L'énergie est évacuée par des câbles à pression d'huile vers un poste extérieur de départ, situé sur la rive gauche.

- Les parois côté mer et côté estuaire de l'usine sont des voiles épais en béton armé s'appuyant sur des contreforts espacés de 13.30 m, ces contreforts sont bien encastrés dans les parties pleines du socle.

Le toit de l'usine est réalisé sous la forme d'une voûte continue de 20 mètres de portée : la voûte est calculée pour supporter une route nationale à double voie de circulation. L'effet de cette voûte est particulièrement bénéfique, car les poussées auxquelles elle donne naissance sont précisément en sens inverse de la poussée de l'eau et, de ce fait, réduisent considérablement les efforts de traction, donc les risques de fissures, à la liaison des contreforts avec le socle.

L'extrémité rive gauche de l'usine comprend un certain nombre de travées réservées au décuvage des transformateurs, au montage et au démontage des groupes.

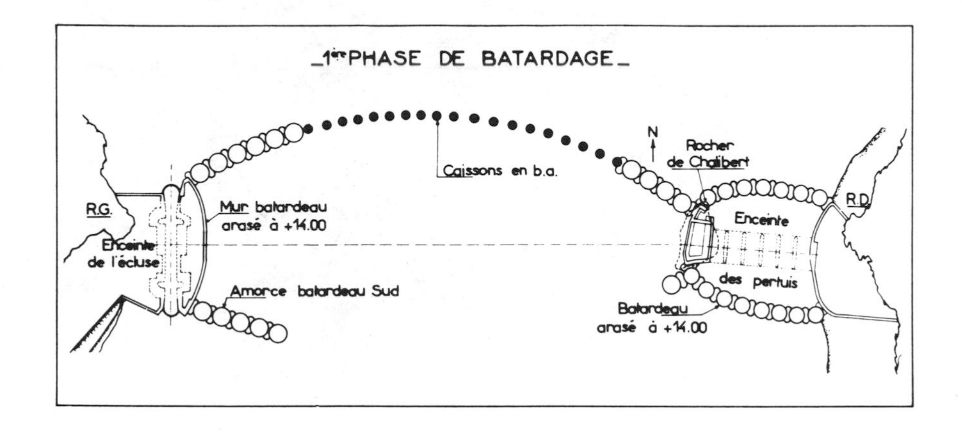

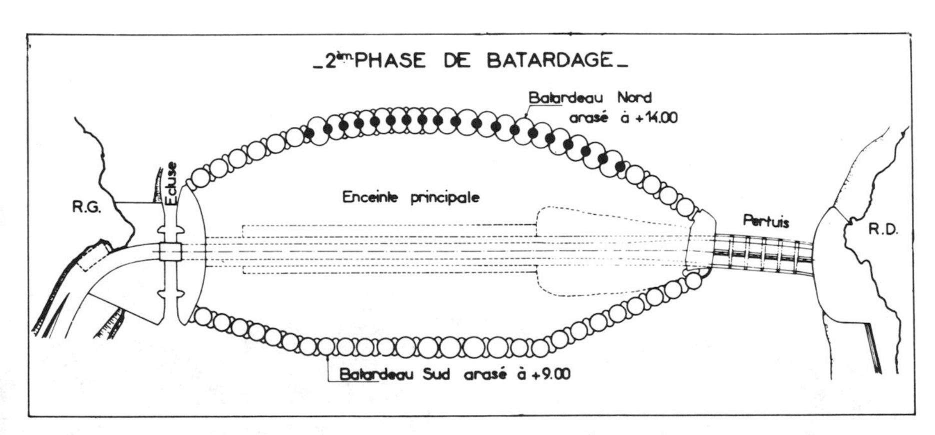

Fig. 8. Phases de batardage. (Document E.D.F.)

Fig. 9. Construction d'un gabion. (Photo Doucet-Dinard)

Fig. 10. Echouage d'un caisson. (Photo Doucet-Dinard)

Enfin, pour assurer la navigation en Rance, une écluse de 65 mètres de longueur, et 13 m de largeur, est implantée dans la pointe de la Brebis sur la rive gauche : les portes de cette écluse sont des secteurs, à axe vertical, s'effaçant dans des évidements des bajoyers. Suivant les niveaux respectifs du réservoir, et de la mer, les bras des portes secteurs travaillent soit à la compression, soit à la traction.

Sous le radier de l'écluse, passe une galerie, au niveau -7 (sous les plus basses mers) : elle aboutit, d'un côté à la salle des machines, de l'autre sous un puits de 22 m de profondeur, abrité par un bâtiment, accolé à la colline qui domine la pointe de la Brebis.

L'accès des grosses pièces des machines et des transformateurs est assuré par le puits et la galerie. Ainsi, il n'y a aucun grand bâtiment en superstructure sur toute la longueur de l'usine-barrage, dont la crête est au niveau + 15, c'est-à-dire à 1.50 m au-dessus de celui des grandes marées d'équinoxe; pour employer une image, le barrage de la Rance apparaît à marée haute comme une "règle posée sur l'eau".

EXPLOITATION DE L'USINE

Il a été dit, plus haut, que la productibilité de l'usine était de 550 millions de kWh : cette production est le maximum que l'on puisse tirer des marées. Mais, il est bien évident que l'énergie n'a pas la même valeur suivant les heures de la journée, et il vaut, peut être, mieux produire moins d'énergie, si, en définitive, la recette est plus grande : cela est possible étant donné les performances du matériel.

Les variables à faire entrer dans les formules pour déterminer le fonctionnement, donnant le maximum de recettes sont nombreuses :

- heures et amplitude de la marée du jour,
- valeurs attribuées au kWh suivant les heures de la journée,
- fonctionnement possible en turbine directe ou inversée,
- fonctionnement possible en pompe directe ou inversée,
- fonctionnement des turbines en orifice,
- manoeuvres des vannes au remplissage ou au vidage,
- nombre de groupes en service,
- sujétions particulières, imposées pour la navigation.

Le recherche de l'optimum, serait une opération inextricable, si l'on ne disposait pas d'un ordinateur.

- Un ordinateur, installé à Nantes, fait chaque jour le calcul de l'optimum, et envoie à l'usine une bande perforée, définissant les manoeuvres à faire tous les quarts d'heure.

Pour simplifier l'exploitation, la bande perforée est directement utilisée pour les manoeuvres : la centrale fonctionne donc en automatique.

Il y a lieu, toutefois, d'apporter quelques corrections aux instructions établies par l'ordinateur de Nantes. En effet, la marée réelle est fréquemment différente de la marée prédite par les tables de l'annuaire des marées, pour diverses raisons : variation de la pression atmosphérique, action des vents ...; de plus, il peut arriver que, par suite d'un incident, soit à l'usine même, soit sur le réseau, le niveau de vidage ou de remplissage du réservoir ne soit pas exactement celui que l'ordinateur avait prévu.

Aussi, un petit calculateur est-il installé à la Rance; il a pour mission d'apporter de légères modifications au programme, pour qu'en fin de compte, le niveau de réservoir, prévu à la fin d'un cycle, soit respecté. Les corrections indiquées par le calculateur se font automatiquement.

On peut dire que sur une longue période, un an par exemple, le fonctionnement de l'usine à recette maximale conduit à une production inférieure d'environ 10% à celle qui donnerait le maximum d'énergie.

LES TRAVAUX

Les ouvrages ont été construits à sec, à l'intérieur de trois enceintes de batardeaux.

- une enceinte sur la rive gauche, pour l'écluse,
- une enceinte sur la rive droite, pour les pertuis de vannage,
- une grande enceinte centrale pour l'usine et la digue morte, cette dernière étant faite avec les déblais rocheux provenant des fouilles de l'usine.

Dans une première phase, ont été faites les deux premières enceintes, ainsi que les ouvrages définitifs correspondants, et en seconde phase, l'enceinte centrale.

Le programme ainsi conçu avait deux avantages :

- ne jamais interrompre la navigation, l'écluse étant mise en service avant que la Rance soit trop encombrée par les batardeaux de l'enceinte centrale, et que les courants deviennent trop violents.
- faciliter la construction des batardeaux de l'enceinte principale, en utilisant le débouché de 900 m^2, offert par les pertuis : d'où une réduction des différences de niveau entre mer et bassin.

1) Enceinte de l'écluse

L'écluse étant implantée dans la pointe de la Brebis, les batardeaux ont été réalisés de façon relativement aisée, en construisant, pendant les heures de basse mer, un mur-poids en béton, accroché sur les flancs de la pointe rocheuse.

2) Enceinte des pertuis

Au départ de la rive et sur l'ilot de Chalibert, des amorces ont été faites en béton, avec les mêmes sujétions de marée que les murs de l'enceinte

Fig. 11. Ecoulement à travers les brèches du batardeau de coupure. (Photo Doucet-Dinard)

Fig. 12. Ecoulement à travers les cellules.

de l'écluse. Entre ces points, les batardeaux ont été formés par des gabions cylindriques de palplanches plates, de 19 mètres de diamètre, remplis de sable, par voie hydraulique : leur hauteur ne dépassait pas 20 mètres. Les palplanches étaient mises en place autour d'un gabarit en charpente métallique, amené par un ponton-mâture; le gabarit était muni de pieds de hauteur réglable, suivant la côte du fond qui avait été préalablement dragué. Lorsque les palplanches étaient toutes en place, elles étaient légèrement battues pour les faire mordre d'un ou deux décimètres dans la couche de surface plus ou moins altérée du rocher. Avant le remplissage en sable, un cordon annulaire en pierres calibrées drainait tout le pourtour intérieur de chaque gabion, et de plus, pour assurer une meilleure étanchéité, un cordon de sacs d'argile était mis en place par des hommes-grenouilles (scaphandriers autonomes) sur le pourtour extérieur, en contact avec la mer.

Chaque gabion était un ouvrage stable par lui-même, et il a suffi, ensuite, de les raccorder entre eux par des arcs de petit rayon.

Commencée au début de 1961, l'enceinte des pertuis, comprenant 17 gabions et leurs arcs a été terminée au mois d'octobre de la même année. Pour une longueur développée de 507 m, le débit de fuite est resté compris entre 400 et 700 mètres cubes par heure, ce qui est très acceptable.

Dès que l'enceinte fut vidée, la construction des pertuis a été entreprise et les vannes montées. Et l'on a pu, comme il était prévu, démolir les batardeaux en temps utile pour mettre en service les 900 m^2 de passage des vannes, le 24 mars 1963, trois jours avant la grande marée d'équinoxe de printemps.

3) Grande enceinte

La grande enceinte a été obtenue par la construction de deux grands batardeaux, se raccordant à l'ouvrage des vannes et à l'écluse. Le premier situé côté mer, est appelé: batardeau de coupure, l'autre est situé côté Rance.

L'ouvrage le plus délicat, le plus difficile, a été incontestablement le premier.

Le modèle réduit avait montré qu'au delà d'une certaine réduction du lit de la Rance, il devenait pratiquement impossible de construire des gabions, comme ceux de l'enceinte des pertuis, car la vitesse des courants devenait de plus en plus forte et les efforts sur les gabions devenaient tels que la stabilité d'un gabion en cours de construction n'était plus assurée. Sur une longueur de 600 mètres, entre l'écluse et les pertuis, on pouvait construire encore 240 mètres suivant le procédé utilisé aux pertuis. Pour les 360 mètres restant il fallait trouver une autre méthode.

C'est à M. Albert CAQUOT, Membre de l'Académie des Sciences, ancien Administrateur d'Electricité de France, que l'on doit le principe de la fermeture de ces 360 derniers mètres. L'idée directrice est la suivante:

1. Disposer d'abord sur le tracé du batardeau des éléments relativement légers, dont la mise en place et l'alourdissement par une masse de sable ne

demandaient que peu de temps afin que l'opération puisse être fait pendant une étale de morte-eau. On obtenait ainsi un certain nombre de points d'appui, espacés de 21 ou 18 m suivant la profondeur des fonds.

Ces points d'appui, très élancés étaient des caissons cylindriques en béton armé de 9 m de diamètre et d'une hauteur variable de 20 à 25 m. Quand ils étaient tous en place, ils ne créaient qu'un obstacle limité au passage de la marée.

2. Transformer ces points d'appui isolés en des points d'appui beaucoup plus stables, en reliant deux caissons voisins par une grande cellule de palplanches remplie de sable. Bien entendu, on ne fermait qu'un intervalle sur deux, en première étape.
3. Fermer, enfin, les intervalles intermédiaires par de grandes cellules de palplanches, comme précédemment.

Ainsi qu'on le voit, le grand intérêt de ce mode de coupure résidait dans le fait que la stabilité des ouvrages, faible au départ lorsque le passage offert à la mer était grand, croissait au fur et à mesure que se réduisait le débouché et que, par suite, augmentaient les dénivellations entre mer et Rance, et vice-versa, donc les poussées.

Les caissons ont été construits dans la forme de radoub de St-Malo et munis d'un tampon métallique pour fermer la partie supérieure afin de pouvoir les faire flotter horizontalement; ils ont été amenés par remorquage jusqu'à leur emplacement d'échouage.

Mais, fallait-il encore préparer sous l'eau une fondation pour les recevoir. A cet effet, un grand caisson, en béton et en acier, de 25 m de hauteur, 22 m de longueur et 15 m de largeur, jaugeant 3000 tonnes, équipé à l'air comprimé, a été construit dans le port de St-Malo, en 1961.

Les fondations, construites à l'aide de cet engin, consistaient en une couronne de béton, avec une margelle formant butée, laissant un jeu total de 0.10 m pour l'emboîtement du caisson.

Il est intéressant de donner quelques explications sur la manoeuvre d'échouage d'un caisson. Deux câbles étaient fixés aux extrémités d'un diamètre du caisson et à sa base. Des hommes-grenouilles venaient faire passer ces câbles dans des chaumards*, scellés dans le rocher lors du travail à l'air comprimé.

Les deux câbles revenaient ensuite à deux treuils, placés sur un ponton bien ancré.

On exerçait alors une traction de 20 tonnes par câble, en même temps que l'on désiquilibrait le caisson en ouvrant une vannette pour le remplir d'eau. Le caisson s'enfonçait progressivement tout en se redressant, puis il se mettait à la verticale et s'emboîtait dans sa fondation. Aussitôt échoué, il était rempli de sable.

---

*chaumard: pièce de guidage pour les câbles et amarres.

Il fallait ensuite fermer par une grande cellule l'intervalle compris entre deux caissons, dans lequel passaient déjà de forts courants. La première opération consistait à couper ces courants, en descendant-dans des rainures, prévues lors de la construction des caissons - des planches en béton, à la base, et des poutrelles métalliques à la partie supérieure.

On pouvait alors construire la cellule à l'aide de gabarits : un pour l'arc côté mer, un pour l'arc côté Rance. La reprise des efforts de traction qui se développaient dans les arcs (et qui pouvaient atteindre 3000 tonnes pour un caisson de 24 m de hauteur) était faite par des tirants de 50 à 60 m/m de diamètre, placés de part et d'autre d'une cloison diamétrale du caisson, et reliant, ainsi, deux épaisses plaques d'acier, dans lesquelles avaient été usinées des encoches ayant le profil d'une serrure de palplanche.

Commencée en octobre 1962, la construction des cellules de première phase, c'est à dire une sur deux, a été achevée au début de mai 1963. La dénivellation constatée entre mer et Rance a été de 1.50 m lors d'une vive-eau de février, mais la mise en service des pertuis de vannage avant la vive-eau de mars a permis de la réduire à 0.75 m pour une même amplitude de marée, conformément, d'ailleurs, aux indications données par le modèle réduit.

Bien que la marée de mai et de juin aient été moins fortes que celles d'équinoxe de printemps, la dénivellation n'avait cessé de croître; les 24 et 25 mai, elle était de 2 m, alors qu'il restait encore à construire 7 cellules; le 24 juin, elle atteignait près de 3 m.

Le 4 juillet, les deux derniers intervalles étaient fermés par les planches de coupure provisoire, et les cellules immédiatement entreprises. Le seul débouché, restant libre, était alors celui des pertuis de vannage : 900 m^2. La dénivellation maximale était de 3.30 m le 8 juillet.

Enfin, les cellules étant terminées, le 20 juillet à 10 heures du matin, les six vannes étaient fermées. La coupure de la Rance était réalisée et l'estuaire transformé en un lac à niveau constant : 8.50 m, au-dessus des plus basses mers.

Ce fut une belle journée, le 20 juillet 1963, on sentait qu'il était temps que la coupure soit faite. Dans les semaines précédentes, alors qu'il restait quelques intervalles à fermer, le spectacle de l'écoulement des eaux était véritablement effrayant : l'eau se ruait dans les passes, la lame plongeait presqu'à la verticale, lisse et rigide comme une glauque coulée de pâte de verre, irrisée de quelques yeux d'écume. Il se produisait comme un effet de succion sur les parois de palplanches, et les grandes cellules, malgré leur masse énorme, vibraient - non, le mot est trop fort -, il vaut mieux dire : frémissaient, comme frémissent les animaux qui ont peur à l'approche de l'orage.

Ces jets puissants s'enfonçaient dans une masse blanche d'écume, qui sur trois à quatre cents metres, brillait comme un immense champ de neige. Parfois, fantaisie de l'écoulement à proximité des vannes, de gros apports d'eau, vert pâle, arrivaient par le fond et venaient crèver à la surface comme des bulles monstrueuses.

Malgré la beauté de ce spectacle grandiose et inoubliable, les ingénieurs étaient impatients d'en voir la fin. On imaginait les dégâts considérables que pourraient faire de tels écoulements furieux; on pensait au rocher dans lequel étaient à peine fichées les palplanches, à peine encastrées les fondations des caissons. On pensait à la faille de diabase constatée en mars, lors d'une plongée sous-marine, le long du 17e caisson et qui avait été affouillée sur 3 m de profondeur. On l'avait alors tapissée avec des sacs de béton, mis en place par les hommes-grenouilles (mille sacs de 30 litres) et recouverte ensuite par du béton amené au fond de l'eau par une benne-preneuse.

On pensait aux érosions de la roche dans les derniers intervalles, et on avait pu mesurer la rapidité de ce rabotage continu du rocher lorsqu'on avait constaté que la première planche de coupure provisoire des dernières brèches, dont le profil avait été dessiné pour épouser sensiblement le fond, se trouvait à 2 mètres au-dessus du sol érodé. On avait pu s'en tirer en déversant des blocs de béton de 5 tonnes.

Quelques mots, maintenant, sur le batardeau situé côté Rance, fait à l'aide de babions de 16, 19 et 21 mètres - Entrepris depuis de longs mois, il restait 350 mètres qui ne pouvaient être exécutés qu'une fois les courants supprimés, donc après la coupure. Le batardeau était achevé le 24 octobre 1963.

Puis l'enceinte, d'une superficie de 100,000 m^2, fut vidée progressivement, en surveillant le bon essorage du sable des gabions et cellules, à l'aide de tubes piézométriques placés au cours de la construction. Le débit de fuite a été au début de 4200 m^3 par heure et après quelques mois, s'est maintenu aux environs de 2500 m^3, ce qui correspond à 0.6 litre par seconde, par mètre linéaire de batardeau.

En javier 1964, l'enceinte était vidée. Fin 1965, soit deux ans plus tard, le gros oeuvre de l'usine était fait à 98%, et une dizaine de groupes étaient presqu'entièrement montés.

Le 14 mars 1966, les ouvrages étaient mis en eau.

Le 19 août, après enlèvement d'une partie des batardeaux, les premiers kilowatheures étaient fabriqués.

Le samedi 26 novembre la première usine marémotrice du monde était officiellement inaugurée par le Président de la République : le Général de Gaulle.

Et en 1967, après quelques petites mises au point du matériel, l'usine fut mise entièrement en service.

Quelques Considérations sur les Délais et les Prix

Les travaux de Génie Civil ont été menés avec une grande célérité par l'Entreprise Tramarance* , et le programme des travaux établi en 1960, a été tenu à quelques mois près.

En ce qui concerne la dépense, le devis de 1960 était de 420 millions de francs (bases économiques de 1960) - Sur les mêmes bases, et malgré les difficultés et aléas dûs à la nouveauté de cet aménagement, le devis d'origine a été tenu à moins de 10% près.

FIN

Un article technique est toujours froid et sévère - Il est dépourvu de la poésie, qui, pourtant, est une nourriture nécessaire à l'homme. Aussi, a-t-on pensé, pour terminer, à faire appel à deux grands auteurs romantiques : René de Chateaubriand (1768-1848) et Victor Hugo (1802-1885).

Le premier, malouin d'origine, a chanté le charme de la Bretagne, la majesté de la mer, et - on l'a vu - les beautés de la Rance, dans ses "Mémoires d'Outre-Tombe".

Le second, que le premier appelait "Enfant Sublime", et qui admirait le premier, en écrivant à 15 ans sur son cahier d'écolier : "Je veux être Chateaubriand ou rien" -, le second a souhaité ardemment que les hommes, au lieu de se quereller et de s'entretuer s'unissent afin d'amérliorer leurs conditions d' existence, et utiliser les richesses naturelles pour le bein-être de tous. Dans son roman "Quatre Vingt Treize", il faisait dire à l'un de ses personnages:

"Utilisez la nature, cette immense auxiliaire dédaignée. Réfléchissez au mouvement des vagues, au flux et au reflux, au va et vient des marées .. ... Qu'est-ce que l'Océan - Une énorme force perdue".

---

*Tramarance - Société constituée par les entreprises suivantes :
- La Sté Générale d'Entreprises - l'Entreprise Fougerolle - les Entreprises Campenon-Bernard - Les Entreprises des Grands travaux Hydrauliques La Sté Françaíse d'Entreprises de dragages et de Travaux Publics.

Les Groupes ont été construits par un groupement comprenant les sociétés suivantes : Alsthom - Alsthom Charmilles - Forges et Ateliers du Creusot - Jeumont - Neyrpic - Matériel Electrique SW.

KISLAYA GUBA EXPERIMENTAL TIDAL POWER PLANT AND PROBLEM OF THE USE OF TIDAL ENERGY

L. B. Bernshtein *

The small Kislaya Guba experimental tidal power plant was started up in the U.S.S.R. during the closing days of 1968. The significance and purpose of this structure are determined by the nature of the unsolved problems associated with the use of tidal energy. These problems will now be examined.

The periodic nature, diurnal and lunar, of tidal phenomena is no longer an obstacle to the use of tidal energy. Thanks to the results of French, Russian and British engineering research it has become possible to match the waves of tidal energy with the waves of energy consumption and to make tidal plants a desired and valuable component of grid systems supplying constant, firm power during all seasons of the year, whether it be wet or dry.

This has been achieved mainly as a result of the works of French specialists headed by the outstanding scientist R. Gibrat, who created the theory of the flexible, controllable utilization of tidal energy, implemented at the presently operating Rance tidal plant. This plant is the result of many years of research and incorporates solutions to design problems by Allard, Vantroys and Gugenheim in tidal power and dynamics; by Voyer and Penel in methods of calculating the cyclic regime of tidal plants; by Gibrat, Auroy, Maubussin and Soulles in the design of the power plant building; by Caquot in the development of a cofferdam from floating columns; by Rath and Surrel in the methods of corrosion control; and, finally, by Kammerlochez, Casacci and Rouville, who developed the most important element in the solution of the entire problem, a reversible flow, bulb housed turbine-generator unit.

This unit, owing to its exceptionally favorable hydraulic contours (small generator enclosed in a bulb) and horizontal position of the axis makes it possible to harness tidal energy to maximum advantage (efficiency up to 91%) during direct turbine operation, which in the case of the Rance tidal

---

*(Chief Design and Construction Engineer of the Kislaya Guba Experimental Tidal Plant), Institute Hydroproject, The Ministry of Power and Electrification of the U.S.S.R., Moscow, U.S.S.R.

plant produces 80% of the total output. The pumping regimes are carried out at a relatively low efficiency, but they enable the tidal plant to operate at peak hours regardless of the phase of the tide, or in a fuel economy regime at jointly operating thermal electric power plants, i.e. at base loads.

Thus the practical possibility of solving the first part of the problem, viz., matching of the tide waves with energy consumption waves, has been proved at the Rance tidal plant, which has already been operating for almost three years.

How does one then explain the recent decision of the French government concerning the priority of construction of atomic power plants over the accomplishment of the Chauze tidal plant project?

The answer to this question is not given either in the address by Minister Marcelin or in subsequent publications in the French technical press. In our opinion, there are two factors delaying the construction of the Chauze tidal plant; first, the need for intramonthly ("intersyzygial") regulation of the tidal plant and second, the high cost of constructing tidal power plants.

French and Soviet investigations show that the tidal power plant is not a direct alternative to the atomic power plant but may, on the contrary, be harmoniously combined with it. This was proven mathematically by Gibrat[2]. The effectiveness of this combination of tidal plant–thermal power plant (atomic power plant) has been proved by examples proposed in our studies for a number of countries whose power resources and natural conditions permit harnessing tidal energy[1].

If tidal plants could meet some substantial part of the peak loads, the existing and future superpower thermal (including atomic) plants could operate smoothly under base-load conditions. This deals with "smooth operation" not in the sense of the technical difficulties associated with the operation of an atomic power plant under peak-load conditions. It is known that these difficulties have already been successfully overcome. What is envisaged is the operation of a superpower thermal power plant (and atomic power plant) under base-load conditions, which is determined by the cost structure of the energy that they produce.

Expensive, rapidly depreciating equipment constitutes the major share of fixed capital outlays at modern thermal electric power plants. Its magnitude does not depend on the number of operating hours of the thermal power plant (atomic power plant). Therefore, it is clear that such plants should operate a maximum number of hours per year so that each kilowatt-hour produced by them involves a minimum value of depreciation allowances. Consequently, in the system consisting of thermal power plant (atomic power plant), tidal power plant and hydroelectric power plant in which the tidal plant and hydro plant components can meet the entire peak-load demands, it is possible to reduce considerably the cost of power from the thermal and atomic power plants owing to an increase in the number of hours of use of the thermal electric power plants. However, to obtain such a harmonious system, it is necessary that the tidal plant can not only synchronize the waves of tidal energy with the periods of consumption (intradaily regulation) but also compensate for the decrease of the tide potential during the weekly period from syzygy to quadrature. This problem can be solved effectively by more extensive use of the reservoirs of run-of-river hydro plants incorporating suitable regulation. Since in France the volume of such reservoirs is extremely small and insufficient for intersyzygial regulation of a tidal plant such as the Chauze and they cannot be created owing to the natural conditions, it seems impossible to include the Chauze tidal plant with weekly power fluctuations from 3 to 12 million kW into the country's power system.

The second factor, cost of tidal plants, is also a considerable obstacle to the accomplishment of the Chauze project. The Rance plant cost 480 million francs, which amounts to 2030 francs per installed kilowatt. Although this is three times the cost per kilowatt of a run-of-river hydro plant, it could be justified if the higher cost of the peak power produced by the Rance tidal power plant is accounted for. However, the entire point is that the amount of this (peak) energy being delivered directly into the system is a relatively small portion (20%) of the entire power produced by the tidal plant. Consequently, for economic substantiation of the possibility of constructing a tidal plant, the problem of achieving a substantial decrease in its cost must be solved. An attempt was made to solve this problem when constructing the Kislaya Guba experimental tidal power plant.

FLOAT-IN-PLACE METHOD OF CONSTRUCTION. KISLAYA GUBA TIDAL PLANT.

The U.S.S.R. has a considerable tidal power potential, 210 billion kWh/year out of 1240 billion kWh in the entire world [4].

The use of the power resources of the White Sea (40 billion kWh/year), which could be incorporated into the integrated power system of the European part of the country, is of real interest in the foreseeable future. The presence of run-of-river hydro plants with seasonal regulation allows using them to compensate for the intersyzygial drop in power output of the tidal plants. In this case tidal plants can be an important component of the system. The damage due to flooding when the basins of tidal plants are created and the possibility of producing appreciable quantities of peak power predetermine the urgency of the problem. However, under conditions in the White Sea, where the sites of the possible location of tidal plants are in uninhabited regions with a rigorous climate, the main obstacle to harnessing tidal energy is the difficulty of constructing tidal plants and their high cost.

Many years of exploring ways to reduce the cost of tidal plants have led to the conclusion that the most workable solution to the problem under these conditions is the float-in-place method, which involves constructing the tidal power plant building at an industrial center and delivering it to the site in a finished form with the equipment already assembled. This solution was based on overseas experience in constructing underwater tunnels (Fraser, Canada; Rendsburg, West Germany) and dams (Delta Project, Holland) by the float-in-place method, and on Russian experience in the construction of floating reinforced concrete docks.

When it was decided, in 1962, to construct the Kislaya Guba tidal power plant, it was appropriate to check this method under actual conditions during design, investigation and construction, in view of the fact that the construction of a tidal plant by the float-in-place method was to be undertaken for the first time in water power construction practice in the U.S.S.R. Since the solution to problems that arose during construction of this experimental plant was of value for hydraulic constructions in general and under conditions of the North in particular, a broad program of comprehensive investigations was worked out and implemented by many organizations and institutes. This program was an integral part of the research plan.

The basic features of the project and the results of investigations obtained during its realization are examined below.

The Kislaya Guba tidal power plant is located in a narrow neck (50 m) connecting the sea with the Ura-Guba Bay (Figure 1). This site is located 600 miles north of Murmansk and was selected for reasons of a relative minimum volume of work and its nearness to the power system, although the height of the tide here varies from 1.3 to 3.9 m, which is less than at other portions of the Murmansk and Mezeń coasts (Lumbovskii Bay 7 m, Mezenskii Bay 9–10 m).

In the natural state the tide filled the bay (1.14 km^2) to full height through the neck within 6 h 12 min, which was accompanied by high speeds (up to 3 m/sec). The discharge reached 300 m^3/sec. In accordance with the project developed by the All-Union Planning, Surveying and Research Institute (Hydroproject) during 1962-66, the float-in-place buildings of the tidal power plant were installed in the neck of Kislaya Guba. In the natural state the depths at the site were 4–5 m. The bottom was composed of unconsolidated marine sediments 5-7 m thick (sand, shells and up to 40% boulders with a diameter as much as 0.5 - 1 m). These deposits were excavated to a depth of 5 m by a floating crane with a lifting capacity of 10 tons, equipped with a set of grab buckets. The total volume of the shaped excavation was 25,000 m^3. Excavation involved preliminary loosening by means of concentrated charges containing 600 kg of explosives each.

Underwater excavation ran into a number of difficulties and its cost proved to be high. Therefore, when selecting sites for future tidal plants it will be necessary to strive for a minimum volume of excavation or its accomplishment in soft ground amenable to excavation by hydromechanical means. For such large excavations, it is now possible to use devices of the Hosar (Holland) and Apsheron type (Baku) which successfully excavate and bore the sea bottom to a depth of 60 m.

Fig. 1. A general panorama of Kislaya Guba and the tidal power station.

The floating open frame of the tidal power plant building, proposed by the author of this article, was erected in a construction dock at Cape Prityka near Murmansk (Figure 2). It is a reinforced concrete box measuring 36 x 18.3 m in plan and 15.35 m high (Figure 3). In this box are located two turbine conduits, in one of which is installed the tidal turbine-generator, delivered by the French firm Neirpik, and the second conduit is used as a bottom discharge conduit until the modified Soviet turbine-generator unit is manufactured (see below). This conduit increases the output of the first unit owing to a decrease in the time needed to equalize the levels when reversing the unit.

The increase of the discharge capacity achieved by using this conduit also increases the dynamic loads on the structure, which makes it possible to test it under the severest conditions and to investigate the propagation of the filling-emptying wave in the river basin.

Above the two conduits are located the span of the surface spillway over the first unit and the closed electrical control room over the second.

The spillway is also of research value for checking the proposed design under conditions of a run-of-river hydroelectric station when the surface spillway is needed for releasing excess flows (for tidal stations at sites on the White Sea coast its use is impossible owing to the danger of clogging with ice).

With this layout, which is governed by technological requirements, the main structural members absorbing the external load are maintained rigid longitudinal with transverse braces (slab, side walls and floor of the building, turbine conduits, longitudinal and transverse diaphragms). On the 20 cm thick bottom (foundation) slab are installed, to full height, two 15 cm thick longitudinal side walls with partitions spaced 1.5 m apart and 9 - 10 m high. In all partitions two holes are cut to accommodate the pipes of the turbine line, which alternately function as delivery pipes and draft tubes. The maximum wall thickness of the pipes is 20 cm. At a height of 10 - 11 m from the bottom the side walls are supplemented with inside longitudinal walls 15 cm thick forming the two extreme and one middle pier.

The total volume of concrete is 1800 m^3, and the weight of the entire structure with assembled equipment is 5200 tons, which gives a draft afloat of 8.3 m.

Calculation of the thin-walled structure, the strength of which is provided by the spatial work of its members, proved to be quite a difficult task. The cross sections were selected by strength-of-materials methods on the assumption that in a longitudinal direction the structure behaves as a beam of complex cross section (equivalent beam) on a perfectly rigid base. It was found in this case that compression (18 kg/cm^2) occurs in the lower part of the structure and tension (6 kg/cm^2) in the upper. The magnitude of normal, tangential and principal stresses reaches 18 kg/cm^2. Calculation of the structure in the transverse direction was carried out as a plane frame with a conditional constant stiffness of the members. A Winkler model of an elastic

Fig. 2. Kislaya Guba Experimental Tidal Power Station. The floating part of tidal power station building erected at a construction dock at Cape Prityka, near Murmansk. The pit is being filled by the incoming tide, July 10, 1968.

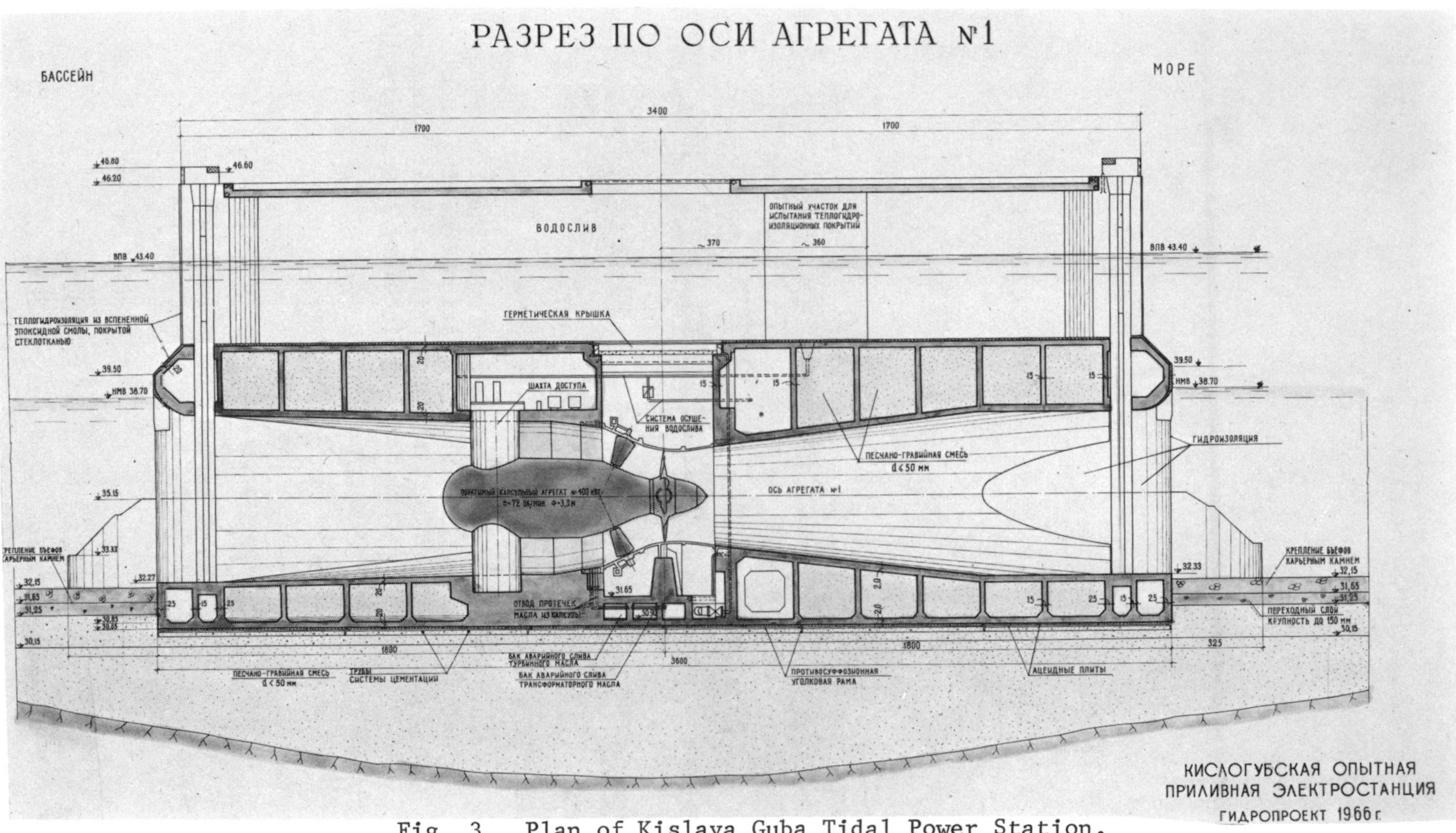

Fig. 3. Plan of Kislaya Guba Tidal Power Station.

base with a variable modulus of subgrade reaction was used. The tensile stresses in the bottom and pier were equal to 9–10 kg/cm^2 and the principal tensile stresses 18 kg/cm^2.

The thermal stresses in the structure, which is subjected to cooling in the upper part and in the zone of variable (tidal) levels and to positive temperatures in the zone of constant submergence, were calculated by theory-of-elasticity methods and showed the presence of tensile stresses up to 20 kg/cm^2 in the middle zone. A check of these calculations by photoelasticity methods (5) gave the same results. Determination of the thermal stresses in the zone of variable levels was also carried out on analog devices (hydrointegrators). In this case it was established that the tensile stresses due to temperature effects on the surface of the members in this zone can reach 26.2 kg/cm^2 (horizontal slab) and 30.2 kg/cm^2 (pier). These same investigations showed that the use of insulation (5 cm-thick boards) reduces stresses to 10.4 kg/cm^2.

Lightweight, thin insulation made of glass-fiber reinforced epoxy resin foam was used to reduce the magnitude of total tensile stresses, which greatly exceeded permissible levels. The bulk weight of the insulation is 200 kg/m^3, the compressive strength is 24 kg/cm^2, tensile strength 18 kg/cm^2, and coefficient of thermal conductivity 0.05 - 0.1 kcal/m h deg. The high efficiency of this insulation is evidenced by the fact that a 5 cm thick layer weighing 10 kg/m^2 is equivalent to a 50 cm thick layer of wood-asphalt insulation weighing 75 kg/m^2, which requires the installation of seats weighing 2 kg/cm^2 (6). It is especially important that the glass-fiber reinforcement of this insulation ensures that it remains intact even when cracks up to 1-1.5 mm open up in the concrete being protected. The insulation withstood 3000 accelerated test cycles under the actual tide conditions in Kola Bay.

Reinforced concrete is the main building material of the hydropower plant building. The use of metal, which would have been more acceptable for the floating structure (lightness, strength, ease of manufacture) had to be rejected on grounds of its service life. Sandy concrete, despite its high strength, impermeability and frost resistance, could not be used either due to the difficulty of its preparation. The concrete developed for the Kislaya Guba tidal plant (7) has the necessary frost resistance (tested at 3000 accelerated cycles under actual conditions), strength (better than 500 kg/cm^2), and workability. It was prepared from high-grade aggregates with the use of sulfate-resistant cement ($Ru = 50$ kg/cm^2) and additives (neutralized air-entrained resins and sulfite-liquor wastes), which improved the workability of the concrete mix with a small slump of the cone (1–3 cm). This is especially important, considering the thickness of the structural members is 15 - 20 cm.

The building for the tidal power plant was constructed by the monolithic method, but the experience of concreting the block showed the expediency and need of using precast reinforced concrete in the case of constructing an industrial tidal power plant with a large number of uniform blocks. The monolithic method required considerable manpower when placing the low-plastic concrete in the densely reinforced (130 kg/m^3), 15 cm thick vertical walls.

Among new materials used in the construction of the tidal plant mention should be made of a hydrophobic mix (sand mixed with fuel oil of grade 80 - 100), which was used for filling the ballast compartments in the zone of variable levels (these compartments were filled with sand in the zone of constant submergence)[9]. The composition of the hydrophobic mix (320 - 350 kg of fuel oil per 1m^3 of sand), which ensures its workability and spreading in horizontal cavities, was determined by filling an experimental cavity.

Protection of the structure of the tidal plant against fouling and corrosion was solved on the basis of the available experience of protecting offshore oil rigs and seagoing vessels and by investigations on stands installed at Kislaya Guba. Special preliminary underwater investigations were carried out by scuba-diving engineers. These studies resulted in cathodic protection of the metal structures of the turbine-generator unit of the tidal plant and of the reinforcement in the concrete in combination with organic and nonfouling coatings. The underwater part of the block, including the turbine conduits, is protected against fouling by marine organisms by means of nonfouling vinyl chloride coatings and specially selected compositions applied on tar-epoxy insulation.

The block, whose lower surface has a metal cutting edge projecting 25 cm, had to be sunk on to a preleveled base of sand-gravel soil. Investigations by the electrohydrodynamic analog method and modeling of the contour showed that the sand and gravel borrow soil which was planned for underwater dumping to form the 0.5 m thick leveling layer, had the necessary anti-piping properties when fractions larger than 50 mm were screened from it. The cutting edge, penetrating into the ground, serves as a sufficient barrier to piping in the event of individual disturbance of the tight contact with the base. Hydraulic investigations of a model of the channel on a scale of 1 : 100 and 1 : 50 made it possible to establish, along with previous investigations, the necessary size of the revetment and intermediate layer in the head race, magnitude of the recovery differentials, velocity conditions, and hydraulic conditions of performing the work.

The experience gained during the process of dumping and underwater leveling the 0.5 m thick foundation layer with exceptional accuracy proved to be most valuable. The soil was delivered in 100-ton lighters and submerged by a grab bucket which opened directly at the bottom. Leveling was done by divers by means of a rake, balanced in the water, which moved along 40 m long guides. The guides were laid with a 2 mm deviation from the horizontal. This was attained by using a specially designed sounding rod.

The operations involved in bringing the floating building of the Kislaya Guba tidal power plant to the site and its submergence on to the base constitute the most important part of the entire experiment and will be described in detail.

The schedule of the entire operation was worked out, studied,

Fig. 4. The dredger completing excavation of the cutting for the removal of the floating caisson; the rigging of the pontoons is completed. (August 17, 1968)

and investigated for a long time at various institutes under the supervision of the author of this report. The main stages of the operation and the basis of the plan consisted of the following. The floating building of the tidal plant was constructed in a pit (construction dock). After completing the construction and insulating works, testing the equipment, and excavating the ground from the cofferdam separating the pit from the Kola Bay, the pit was flooded between July 5-16 by pumping seawater to a depth of 9.7 m.

Ballasting was done as the block began to float and its weight (5200 tons) and draft (8.32 m), which corresponded exactly to the design values, were established. The bottom was also inspected and proved to be completely watertight.

Upon completing the excavation of the breach in the dam, careful measurement, and thorough dredging, the block was moved out into Kola Bay on July 28 at 8 a.m. (two hours before high water of the spring tide). This was done after reducing the draft of the block to 6.3 m by attaching six pontoons, each with a lifting force of 400 tons (Figures 4 and 5). The block was brought out by guiding it into the breach by means of four 4-ton electric winches and a 350 h.p. tugboat moored to the stern pontoon of the block. Despite some unforeseen circumstances, the block was brought out in 1 hr. 36 min., i.e., 6 min. less than calculated.

Twenty minutes before the start of high tide towing began with one tugboat with a 2000 h.p. diesel-electric motor (Figure 6) and then by a second with the same power (Figure 7). The 350 h.p. tugboat participated in towing until the deep water stretch was reached. The total towing time to Kislaya Guba (18 hr. 20 min.) was less than that calculated due to the favourable weather (wind of force 1 - 2); only on the sea stretch of the run (2 hr.) were waves to 1.5 m high and winds to force 4 observed. The transportation of the block from the Kislaya Guba roadstead (Figure 8) to the site began at 6:45 a.m. and was completed by 11:20 p.m. on August 29.

With the start of the outgoing tide the block began to be submerged on to the base without deviation horizontally or vertically, ideally straight and even. After this the water ballast began to be replaced by sand (Figure 9).

The success of the unprecedented operation of moving the Kislaya Guba power plant building and installing it on a prepared underwater base was due to the well-developed plan which was based not only on theoretical calculations and model investigations of towing forces but also on the experience of constructing and moving floating docks. In collaboration with designers, investigators and builders, our sailors were able to perform the operation, which was predetermined by an exact combination (to within several minutes) of its elements with the tide cycle.

The capacity of the tidal plant was not selected on the basis of the complete use of the energy potential, which is determined by the size of the water area of the basin (1.14 km^2) and average amplitude of the tide (2.32 m) and amounts to 2000 kW. This power can be obtained here by installing five units ($L_I = 3.3$ m).

Fig. 5. A floating caisson before removal (August 27, 1968).

Fig. 6. Towing the floating caisson to Kislaya Guba tidal power station, in Kolski Bay (August 28, 1968).

Fig. 7. The floating caisson in deep water.

Fig. 8. The Tidal Power Station building before bringing into alignment by winches (August 29, 1968).

Since our investigations(1) show that the use of the tide cannot be economically substantiated in small plants, such as the Kislaya Guba, the power of this small experimental plant was determined only by the requirements of the experiment, which could have been limited to one unit, especially since the use of the complete energy potential of the tide would have required an extremely large amount of work to increase the discharge capacity of the neck connecting Kislaya Guba with the sea. However, owing to the observed requirement (see below) for further modification of the modern tidal turbine-generator unit at the Kislaya Guba tidal plant, the installation of two units was provided, one Russian and one French. The French unit was supplied by the Neirpik-Alstom Company (Figure 10). It is a bulb housed, reversible-flow turbine-generator with a capacity of 400 kW. The diameter of the runner is 3.3 m. Its speed is 72 rpm and it is connected via a speed booster to a 600 rpm synchronous generator. The planetary booster is manufactured by Krupp (with a connector of the Stekicht system).

Under conditions of the U.S.S.R. coasts, where at possible construction sites for tidal power plants the amplitude is smaller than for the French shores (4 - 10 m in the U.S.S.R., 6 - 13 m in France), the use of a tidal turbine-generator unit with a booster connection is necessary in order to obtain an acceptable generator-shaft speed. This circumstance predetermines the rejection of the use of a more expensive and complex, special small synchronous generator and permits the use of an ordinary synchronous generator, but requires the introduction of a planetary booster, a machine which is also complex but one which has been mastered by industry in the U.S.S.R.

The delivery of an imported machine for the Kislaya Guba tidal plant was substantiated by the need to use the experience gained by French firms during 20 years of work on the development of a tidal turbine-generator unit and also by the need for its modernization applicable to our conditions. The slightly smaller amplitudes of the tide at the sites of possible construction of tidal plants in the U.S.S.R. than in France require the use of a smaller head and an appreciable increase in the range of variation of the heads (the head varies by a factor of 10 - 14). Under these conditions, a substantial increase in efficiency and hence of the use of the potential of the tide can be obtained by conversion to a variable speed machine. Special investigations showed that the problem can be solved by using an asynchronous generator (M.M. Botvinnik's sytem), which is to be installed later at the Kislaya Guba tidal plant.

Research and development of this unit presupposes a study of the power and hydraulic regimes of the tidal plant, for which purpose five tide gauges are installed along the course of the flow, the readings from which are the basis for a check of computer calculations of the optimal regimes of the tidal plant and propagation of the filling and emptying wave of the basin.

The power regimes of the Kislaya Guba tidal plant were calculated on the basis of the performance of the unit which provides reversible turbine,

Fig. 9. Aligning the Tidal Power Station building before sinking (August 1968).

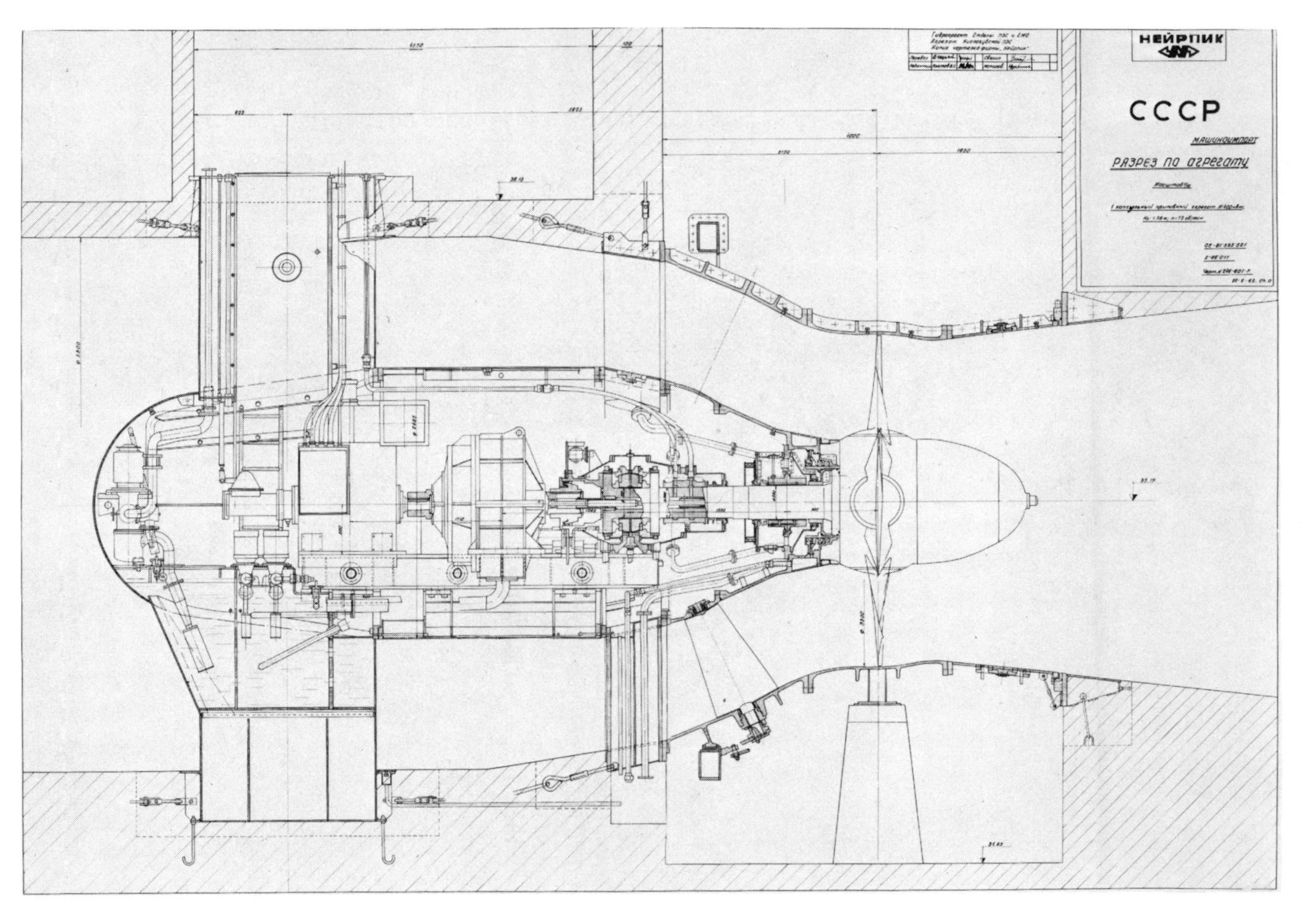

Fig. 10. Bulb turbine at Kislaya Guba tidal power station supplied by Neirpik-Alstom, France 1964.

pump and conduit operation in conformity with the operating characteristics presented by the manufacturer based on model investigations. In this case it was assumed that the unit operates at a head from 1.8 m to 2.5 m, and the pump at a delivery height up to 1.35 m. The operation of the surface and bottom conduits for accelerating the equalization of the levels for maximum performance was also taken into account.

The method of differential dynamic programming was used for optimizing the regimes of the tidal power plant [11]. Calculations by this method enabled us to obtain the operating schedule of the tidal plant in two variants: for maximum performance and for output of peak power. The combination of these two regimes applicable to the conditions of the local power system ensures an output of 1.2 million kWh per year for one unit and 2.3 million kWh for two units.

When the construction of the Kislaya Guba tidal plant was completed, it became a facility of Hydroproject for investigating the problem of the use of tidal energy.

To ensure a comprehensive investigation of the new design of the tidal power plant, 500 monitoring and measuring instruments were placed in the building, which permit studying the forces in the structure, temperature conditions, the energetics of the tidal plant, building materials and means of their protection.

All these investigations are subordinate to one main task - the development of large tidal electric power plant projects and their economic substantiation on the basis of the float-in-place method and our concept of the efficient operation of tidal plants in the power system.

NEW TIDAL POWER PLANT PROJECTS IN ENGLAND, U.S.A., CANADA, AND ARGENTINA

The float-in-place method has given rise to the consideration of several new projects for tidal electric power plants, the basic data of which have been published in the press and at the World Power Conference [12][13]. Among these tidal power projects are:

1. Solway Firth (England) with a production of 3.2 billion kWh/year.
2. Bristol Channel (England) - 50.2 billion kWh/year.
3. Passamaquoddy (U.S.A.) - 1.0 billion kWh/year.
4. Fundy (Canada) - 5 -10.0 billion kWh/year.
5. San José (Argentina) - 12.0 billion kWh/year.

Almost all these plants are once again being reconsidered on the basis of float-in-place structures and there are reasons to hope for their further advance. But it is precisely for this reason that attention should be drawn in these projects to one controversial concept which was put forward by the delegates of England and West Germany at the World Power Conference in Moscow

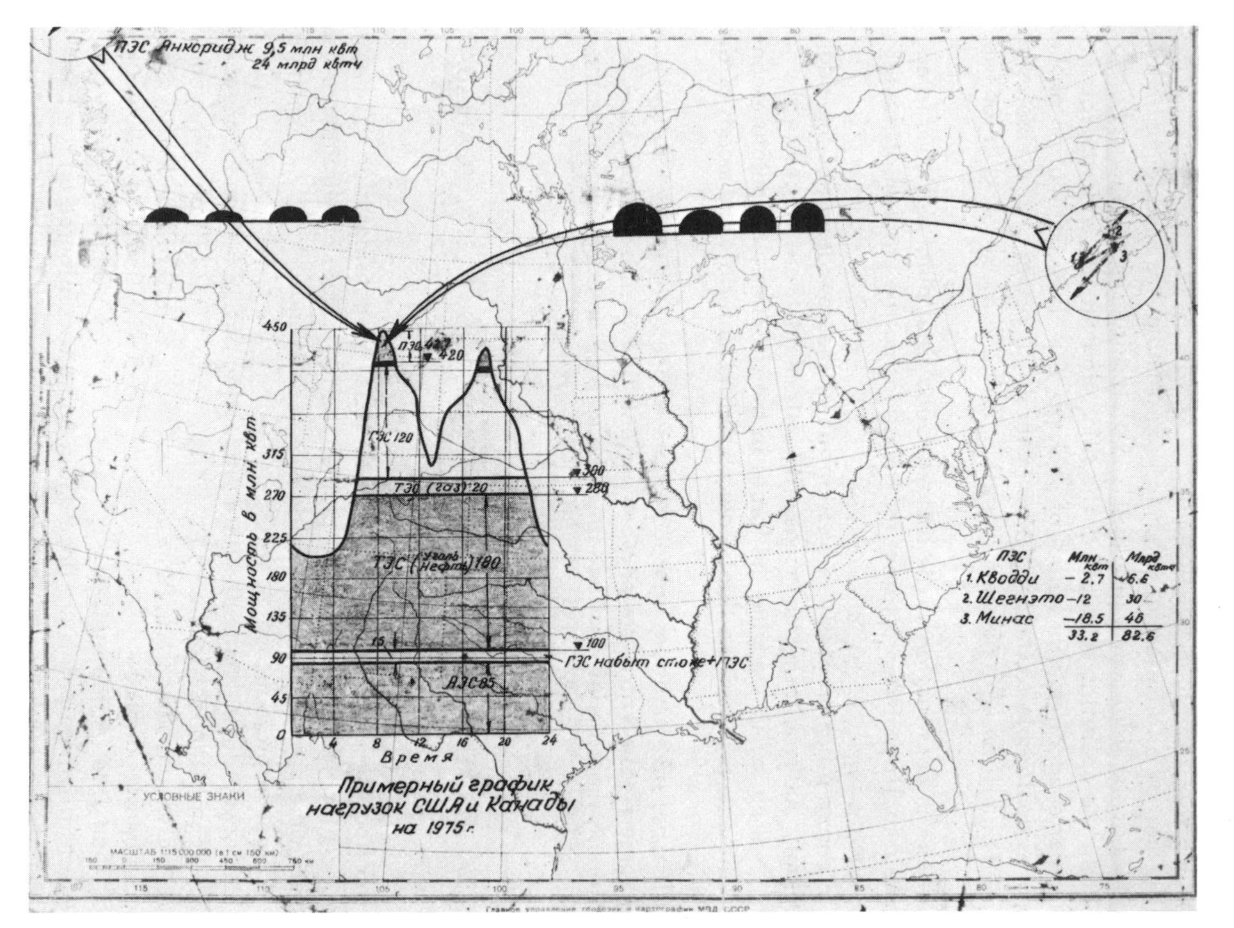

Fig. 11. The formation of tidal energy basins, Fundy and Cooks Inlet.

in 1968. In these reports, the new tidal power projects were invariably related to direct-flow turbine-generator units (generator on the rim of the runner). These units in the papers by H. Fentzloff (West Germany) and Braikevitch (England) were contrasted with the bulb housed units with respect to their simplicity, comparative cheapness and decrease in the overall size of the building for the tidal power plant. However, in the proposed projects (12)(13) the length of the draft tube proves to be equal to 5.5 D, and the distances between axes of the units is 2.5 D, which exactly corresponds to the size of the building in the case of the bulb housed turbine-generator unit. Actually, the direct-flow unit does away with the small, complex and expensive bulb housed generator. However, the generator on the rim of the runner is a more complex machine, even taking into account the recent inventions in England for providing water tightness for the generator. To this must be added the adverse and not the favorable (as H. Fentzloff states) experience using these units during the pre-war and war years in Germany. The experience of manufacturing and operating a large number of these machines (90) installed on the Iller, Lech and other rivers, showed the difficulty of increasing the diameter of the turbine runner.

The rejection of the reversible-flow variant of the tidal turbine-generator unit led the authors of the new tidal power projects in England and West Germany to the need for complete doubling of the power of the tidal plants as pumped storage plants. It can be understood why our British colleagues are endeavoring to compensate for the variations in power output of the grandiose tidal power plants they are developing by constructing pumped storage plants which derives from the absence of run-of-river hydroplants in England and the impossibility of their construction there.

In the case of using the reversible-flow unit, the present storage plants and those under construction could probably be used for solving the problem of intramonthly compensation of the power from the tidal plant. Nevertheless, it seems that the most fundamental solution of the problem lies in realizing the concept of the multipurpose use of tidal and river energy in systems uniting the electric power stations of a number of countries. Our investigations show that the tidal energy generated over the vast expanses of the world ocean cannot be locked up in a small plant serving a coastal village. Its powerful, pulsating but invariably firm flows breach the boundaries of maritime provinces. The flow, transformed from lunar to solar time thanks to the reversible-flow unit, can be directed into superpower grid systems, where it can be integrated with various types of electric power stations, mutually improving their operation and thus creating harmony in the power of countries and continents.

The topography of France does not allow the creation of large reservoirs for intersyzygial regulation of powerful tidal plants which could be built on the shores of Cotentin Peninsula; nor can the problem be solved in the territory of England for regulation of the tidal plants in the Bristol Channel, Solway Firth and other tidal plants on the shores of Great Britain. But in our work

Fig. 12. Kislaya Guba tidal power station as completed with the turbines and surface spillway working (October 1969).

published eight years ago [1], it was shown that this problem can be solved on the basis of international collaboration.

By joining the tidal energy of Cotentin Peninsula and the channels and firths of England with the river energy of Sweden, Norway and perhaps the U.S.S.R., it is possible to realize the solution of the problem of covering peak loads for the power systems of Europe ("English Channel-Mezeń-Ob"), and the integration of the power energy potential of the Bay of Fundy and Cook Inlet with the run-of-river hydroelectric plants of Alaska and western United States would solve this problem for North America (Figure 11).

These projects might possibly be taken as fantasy. But in our time the dream of the engineer very often becomes a reality. The idea of constructing the Kislaya Guba tidal power plant as a ship seemed a fantasy. But today this experiment has been accomplished (Figure 12) and the dream of the engineer has become a reality.

Who can say that the grandiose tidal power projects made of floating blocks will not be accomplished in the U.S.S.R., Canada, England and France in the very near future?!

REFERENCES

1. Bernshtein, L.B., "Tidal Power Plants in Modern Power Engineering", Gosenergoizdat (1961) (Translation National Science Foundation, Washington, 1965).

2. Gibrat, R., "Tidal Power and Tidal Electric Power Plants", Izd. Mir (1964).

3. Revue Francaise de l'Energie ,183, (October 1966).

4. Water Power Resources ,Izd.Nauka (1967).

5. Khesin, G.L., Svast'yanov, V.I., Shvei, K.M. and B.V. Bida, "Investigation of Temperature Stress of the Floating Block of the Kislaya Guba Tidal Power Plant by the Photoelasticity Method", Gidrotekh. Stroitel',No. 12 (1966).

6. Pshenitsin, P.A. and V. I. Sakharov, "Foam Epoxy Insulation and Waterproofing of the Kislaya Guba Tidal Power Plant", Energet. Stroitel', 4, 70 (1967).

7. Ivanov, F.M., "High Frost-Resistant Concrete for the Kislaya Guba Tidal Power Plant", Energet. Stroitel', 4,70(1967).

8. Usachev, I. N., "Concreting Works in the Construction of the Thin-Walled Floating Block of the Kislaya Guba Tidal Power Plant", Énerget. Stroitel', 4, 70 (1967).

9. Kroselev, Yu. V., "Hydrophobic Mixture for Filling the Cavities of the Floating Block of the Kislaya Guba Tidal Power Plant", Énerget. Stroitel', 4,70 (1967).

10. Kulev, I. P., Trifel', M.S., Khanlarova, A.G., Nazirov, R.K., and Yu. B. Ryss, "Electrochemical Protection Against Corrosion of the Structures of the Kislaya Guba Tidal Power Plant", Énerget. Stroitel', 4 , 7 (1967).

11. Silakov, V. N., "Optimization of the Operating Conditions of a Tidal Power Plant in a Grid System ", Izv. AN SSSR, Énergetika i Transport, 5 (1968).

12. Braikevitch, M., Gwynn, J. D., and E. M. Wilson, "Tidal Power with Special Reference of Plant, Construction Techniques, and the Integration of the Energy into Existing Electric Systems", World Power Conference, Moscow, August 20-24.

13. Fentzloff, H. E., "A Fundamental Approach to Tidal Power", Water Power, 8 (1967).

14. Bernshtein, L. B., Direct-Flow and Submerged Turbine-Generator Units, Moscow (1962).

15. Bernshtein, L. B., "Problem of the Use of Tidal Power and the Kislaya Guba Experimental Tidal Power Plant", Gidrotekh. Stroitel', 1, 9-13 (1969).

TIDAL POWER FROM COOK INLET, ALASKA

E. M. Wilson* and M. C. Swales**

INTRODUCTION

Ever since 1798 when Captain James Cook first recorded the extraordinary range of the tides in the Alaskan inlet that now bears his name, there has been speculation about the possibility of extracting energy from the sea there, and Cook Inlet invariably appears in the location lists of writers on tidal energy as a possible site for a tidal power scheme. Although it is reported that several engineers have formulated schemes in the past, the literature contains few references and apart from the passing attention of Bernstein [1], the writers have been unable to find any assessment of the practical possibilities. There are several reasons for this. Alaska is a remote and sparsely populated state. Its electrical load is small and its networking is, as a result, still in an elementary state. The inlets become full of drift ice in the winter, the waters are heavily loaded with silt and the region is active geologically so that earth movements may be expected regularly.

Recently large natural gas fields have been discovered in the area and further exploration for oil and gas continues, so that fuel costs for thermal generating plant are low. Accordingly the costs of tidal energy would have to be correspondingly low before construction could be justified.

One barrage site, however, across Knik Arm in the upper reaches of Cook Inlet, has a multipurpose potential which, the authors believe, might well justify further study. This paper describes a general assessment of the practical possibilities of several sites in the inlet and presents the results of an investigation into the development of the most promising site across Knik Arm.

---

*Dept. of Civil Engineering, University of Salford, Salford 5, Lancs, England.
**Montreal Engineering Co., P. O. Box 777, Montreal 114, Quebec, Canada.

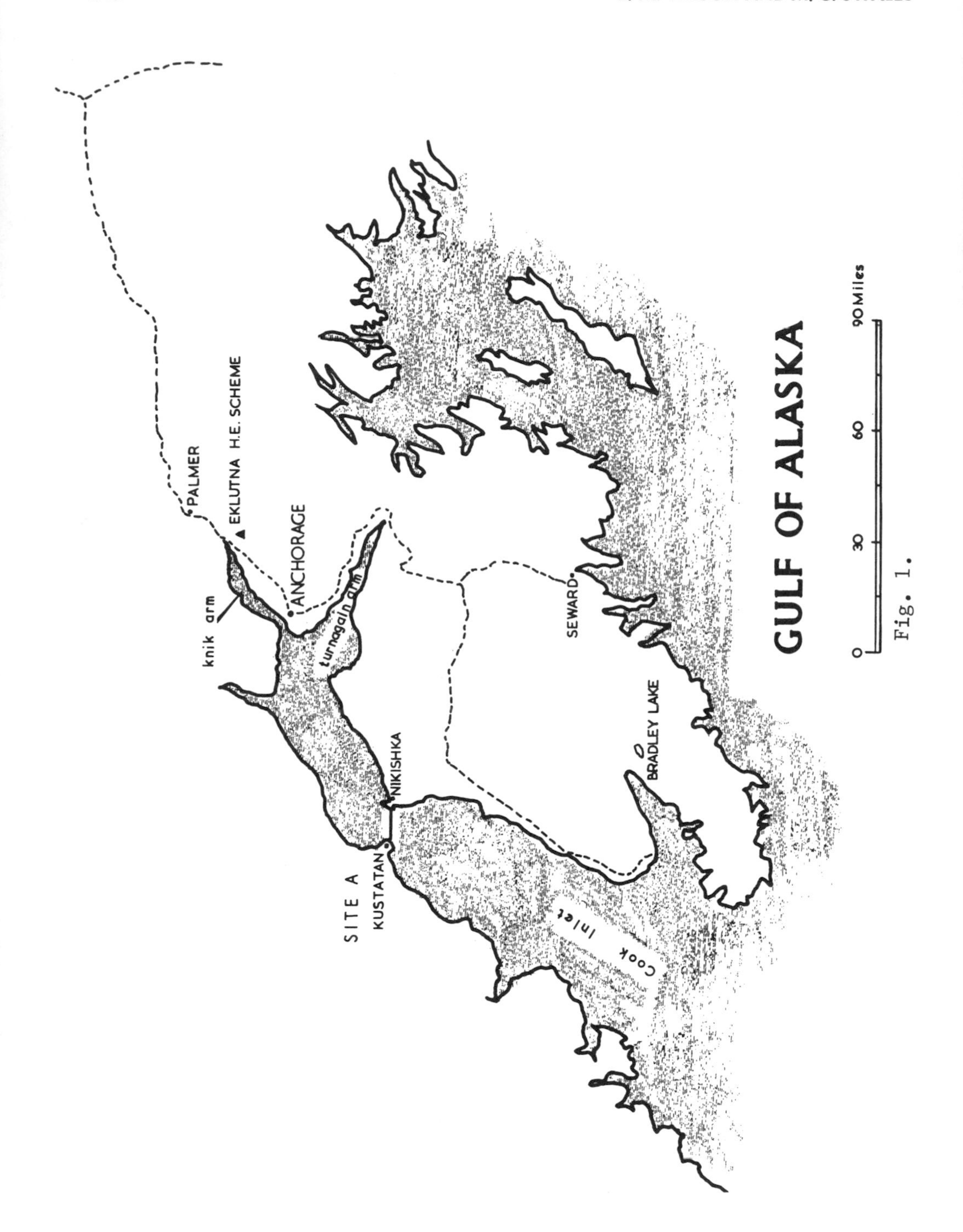
PALMER
EKLUTNA H.E. SCHEME
ANCHORAGE
knik arm
turnagain arm
SEWARD
NIKISHKA
BRADLEY LAKE
SITE A
KUSTATAN
Cook Inlet
GULF OF ALASKA
0
30
60
90 Miles

Fig. 1.

GEOGRAPHY, HYDROGRAPHY AND TIDES OF THE COOK INLET

Cook Inlet is a 230 mile long indention in the southern coast of the Alaskan peninsula with its centerline lying approximately NNE - SSW. The inlet varies in width from about 60 miles at the entrance on the Gulf of Alaska to about 13 miles at the entrance to Turnagain Arm with a local increase immediately landward of the Gulf entrance. The area is shown in Fig. 1. The depth of water at the entrance is generally 450 ft. and this shallows to about 60 ft. at the Turnagain Arm-Knik Arm bifurcation. The mean depth of the whole inlet is approximately 200 ft.

The tides at the entrance have a mean range of approximately 14 ft. (Port Graham 14.4') while at Anchorage this has increased to 25.1 ft. It seems likely that this increase in amplitude is due both to the funnelling effect of decreasing width and also to the condition of near resonance of the tidal wave entering the inlet. The principal constituent of the tide which is the semi-diurnal component recurs every 12.4 hours. Applying the shallow water wave equation to the inlet, the wave celerity C is given by

$$C = \sqrt{gh}$$

where h = average water depth.

If the average depth in the inlet is taken as 200 ft. the celerity is approximately 80 ft. per sec. Hence the wave length L is about 675 miles giving a value for L/4, the theoretical resonant length of 170 miles. Since the theoretical resonant length seems to be less than the actual length, Cook Inlet offers the intriguing possibility of increasing the tidal range by barraging part of the upper end.

The tides of the inlet exhibit marked diurnal inequality which results in differences in level of succeeding high waters and low waters of as much as 6 ft. and, incidentally, makes the task of analyzing the available power and energy from a tidal power scheme more difficult.

SILT AND ICE

The waters of both the Knik and Turnagain Arms are heavily silt laden and generally appear dark grey or black. Large deposits of sediment occur in both Arms though, according to the charts, the bottom is swept clear and is hard or rocky in those regions where barraging may be contemplated. Drift ice is present in the Anchorage area in large blocks (see Figs. 2 and 3) for 5 or 6 months of the year. This ice is apparently formed in laminated sheets on the flat shelving edges of the Arms during successive exposures during low tides, gradually building up to thicknesses of several feet. It drifts in large packs up and down the Arms, particularly in the region of the bifurcation,

Fig. 2. Drift ice, beached at low tide, at the northern extremity of Knik Arm; April 1967.

Fig. 3. Drifting ice floes, seen from 1000' above Knik Arm, off Anchorage; April 1967.

forming a hazard to navigation.

POSSIBLE BARRAGE SITES

In choosing a barrage site, an economic balance must be struck between the cost of barrage construction in which length and depth of water are important, and the energy which may be generated in which the reservoir area and tidal amplitude (including any modification through barraging) are concerned. An inspection of Figs. 1 and 4 will reveal a number of sites which deserve examination:

1. Between the West and East Forelands, at Kustatan - Nikishka, a length of 9.8 miles with a maximum water depth around 60 fathoms (site A on Fig. 1). This would be an immense scheme impounding 1200 square miles of sea and since the remaining unbarraged inlet would be very near to the resonant length, (about 140 miles), the mean tidal amplitude might be of the order of 30 ft. The annual, fully-developed output of such a scheme would be vast, certainly of the order of 75,000 Gwh or about 7% of the entire electrical energy production in the U.S. in 1970. Although such an output is very large for a single scheme, the engineering difficulties do not seem insuperable in the light of advances recently made in hydraulic engineering.
   The fundamental questions to be answered before serious study of this site is made are what might happen to the tidal amplitude at the site and how could such huge blocks of unregulated energy be stored and integrated into the U.S. supply network. Because of the scale of these problems, this site has not been considered further in this paper.
2. A series of potential barrage sites are shown in Fig. 4. The barrage lines B_1, B_2 and B_3 represent alternative sites by which a reservoir created in Turnagain Arm could be exploited (B_1 and B_3). This could also be achieved using combined reservoirs created in Turnagain Arm and Knik Arm (B_1 and B_2). In either alternative the power station would be installed in barrage B_1. The lines B_1 and B_3 are the only feasible way in which Turnagain Arm could be used as the inlet becomes extremely shallow in the upper reaches.
3. Two alternative sites in Knik Arm, C and D, are attractive with respect to the length of barrage required, but site C represents special problems as it is situated at the narrowest point in the Arm. The hydrographic charts indicated depths of 150 ft. or more along the line of the proposed barrage and it is clear that very high velocities exist at mid tide through this channel. Site D at the entrance to the Arm would require a longer barrage but is sited in shallower water approximately 70 ft. deep, the depth required for a turbine caisson unit.

From these possibilites, three sites, B_1 and B_2, B_1 and B_3 and D were

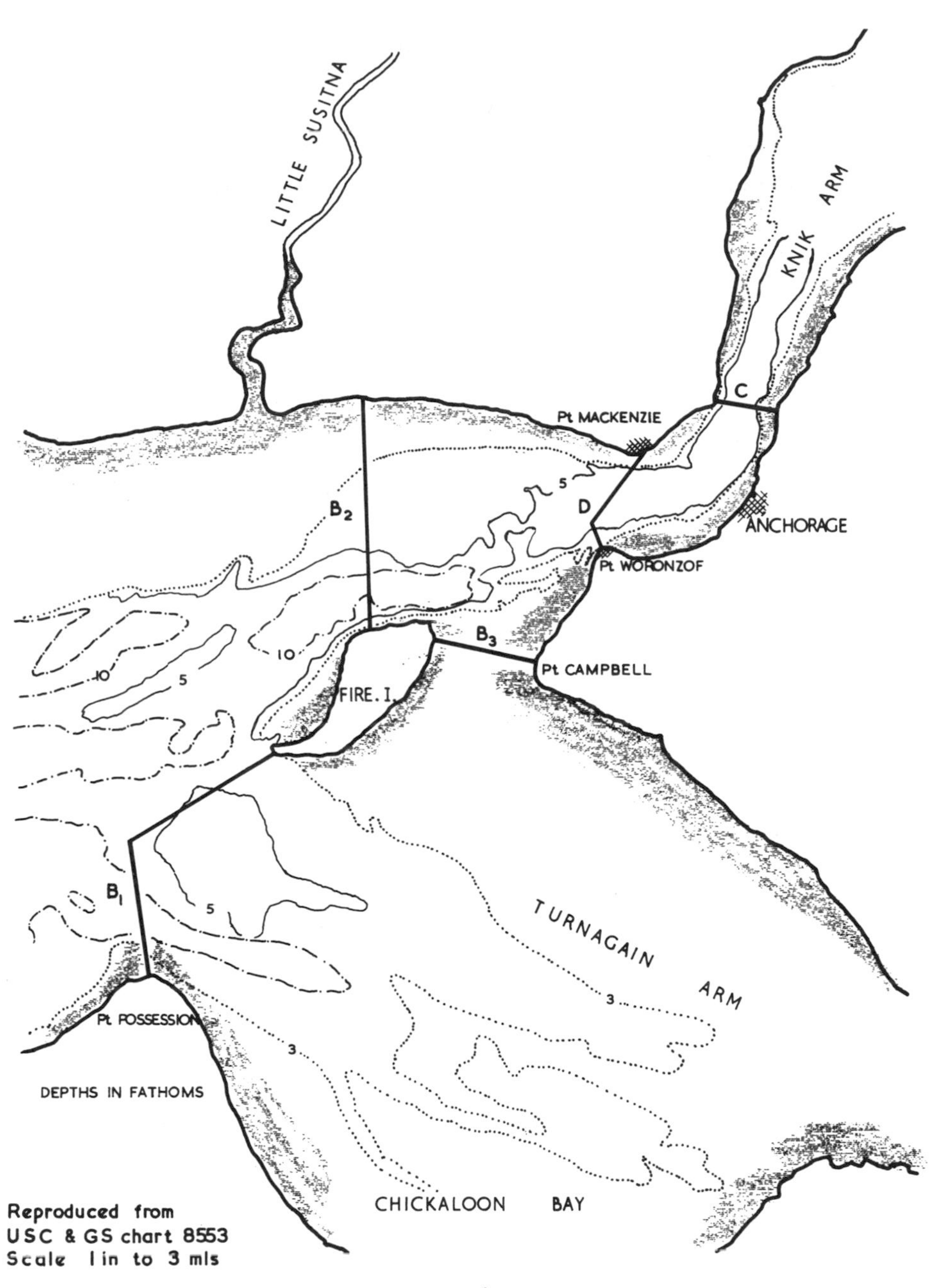
LITTLE SUSITNA
KNIK ARM
C
Pt MACKENZIE
5
D
ANCHORAGE
B2
Pt WORONZOF
B3
10
10
5
Pt CAMPBELL
FIRE. I.
B1
5
TURNAGAIN ARM
3
Pt POSSESSION
3
DEPTHS IN FATHOMS
CHICKALOON BAY
Reproduced from
USC & GS chart 8553
Scale 1 in to 3 mls

Fig. 4.

examined more closely and a preliminary energy analysis including estimates of the probable number of units and the refilling sluice area required were made. The figures presented in Table 1 might vary by up to 10% following a more rigorous economic analysis, but are sufficiently accurate for comparison.

Table 1

Site Reference	High water area square miles	Low water area square miles	Mean Tide range ft.	Energy per annum GWh
B_1 and B_2	460	230	24.5	18,600
B_1 and B_3	330	175	24.5	12,500
D	120	47	24.5	6,000

The mean tidal range assumed in these calculations is approximately equal to the mean range at Anchorage. No reduction of tidal amplitude is assumed as the inlet is longer than the resonant length. Naturally the energy would decrease or increase with a change in mean tidal amplitude.

DETAILED ENERGY COMPUTATIONS

From a survey of the local demand for electricity in Alaska (see Section 8), it was quickly apparent that even the approximate figures presented in Table 1 indicated that considerable development of the electrical load in the area is required before even the smallest scheme could be integrated. Accordingly only site D was considered in detail.

It is not the purpose of this paper to discuss the relative merits of the various modes of operation of a tidal power scheme. Such discussions are presented elsewhere (2)(3). For the purpose of this study, it was assumed that simple, ebb-flow generation coupled with pumped-storage would provide the most economical and practical development. The authors accept that these assumptions are open to argument.

Basic information of the area was gathered from the hydrographic charts of the area prepared by the U.S. Coast and Geodetic Survey. A relationship between the water-surface area and the depth of water in the barrage pool was obtained by planimeter from the charts. Areas were measured at six feet intervals and the water surfaces were assumed as horizontal planes at all elevations. The tidal oscillations were described by a series of cosinusoidal constituents derived at Anchorage by the U.S.C. and G.S. and the phase shifts of these constituents were as derived for July 1964. A histogram of all tidal ranges for a year was obtained by using these constituents.

A computer programme has previously been developed to optimize the operation of an ebb-flow tidal power plant during the research programme completed by Swales [4] and detailed descriptions of the development of this programme have been published [5]. Essentially, the rate of change of the reservoir level at any instant during a tidal cycle is described by a first order differential equation which is solved using numerical integration. The computer programme optimizes the operation of a scheme having a given number of turbogenerators of specific size and characteristics, together with a given area of refilling sluice to augment the refilling of the reservoir during the rising tide period. A number of combinations of turbogenerators and refilling sluice areas may be tried for the same tidal range. By selecting a series of four or five tide ranges, it is possible to compute the annual energy output from a particular combination of turbogenerators and sluice areas by using the histogram of all annual tidal ranges.

The optimization process requires significantly large amounts of computer time so that it is not economically feasible to obtain annual energy output from a very large number of combinations of turbogenerators and refilling sluice areas. Accordingly, the authors, from experience gained during previous studies, selected a suitable range of numbers of turbogenerators and refilling sluice areas. It was assumed that the hydraulic machinery would be fixed-blade, straight-flow turbines with rim generators of 30 ft. turbine-wheel diameter. This particular type of machine has been described and its characteristics established in detail by Braikevitch [6].

In this way, the number of variables is reduced and the computing time brought to a reasonable figure.

TYPE OF CONSTRUCTION PROPOSED

Tidal power construction methods have received considerable attention in the past few years and the concept of prefabricating the power station and refilling gates in floatable precast cells has become accepted as likely to lead to greatest economy. The kind of power house structure proposed is illustrated in Fig. 5. Further examples exist in the literature.

The refilling capacity would be provided through submerged Venturi sluices and in secondary crest-gate sluices above the turbines. The open-top sluices have the disadvantage that during the winter they would be subject to severe icing. The submerged sluice gates would suffer to a much lesser extent. In view of the icing problem the possibility of using syphon sluices was also considered. Having no moving parts a syphon might be cheaper and more reliable than a gate structure with equivalent capacity unless the depth at the chosen site is sufficient to accommodate deep gates of the order of 40' or more, in which case the syphon's economic advantages might disappear.

The actual barrage structure proposed is simple and consists of a few main components and many identical units. The technique of estuary closing in high

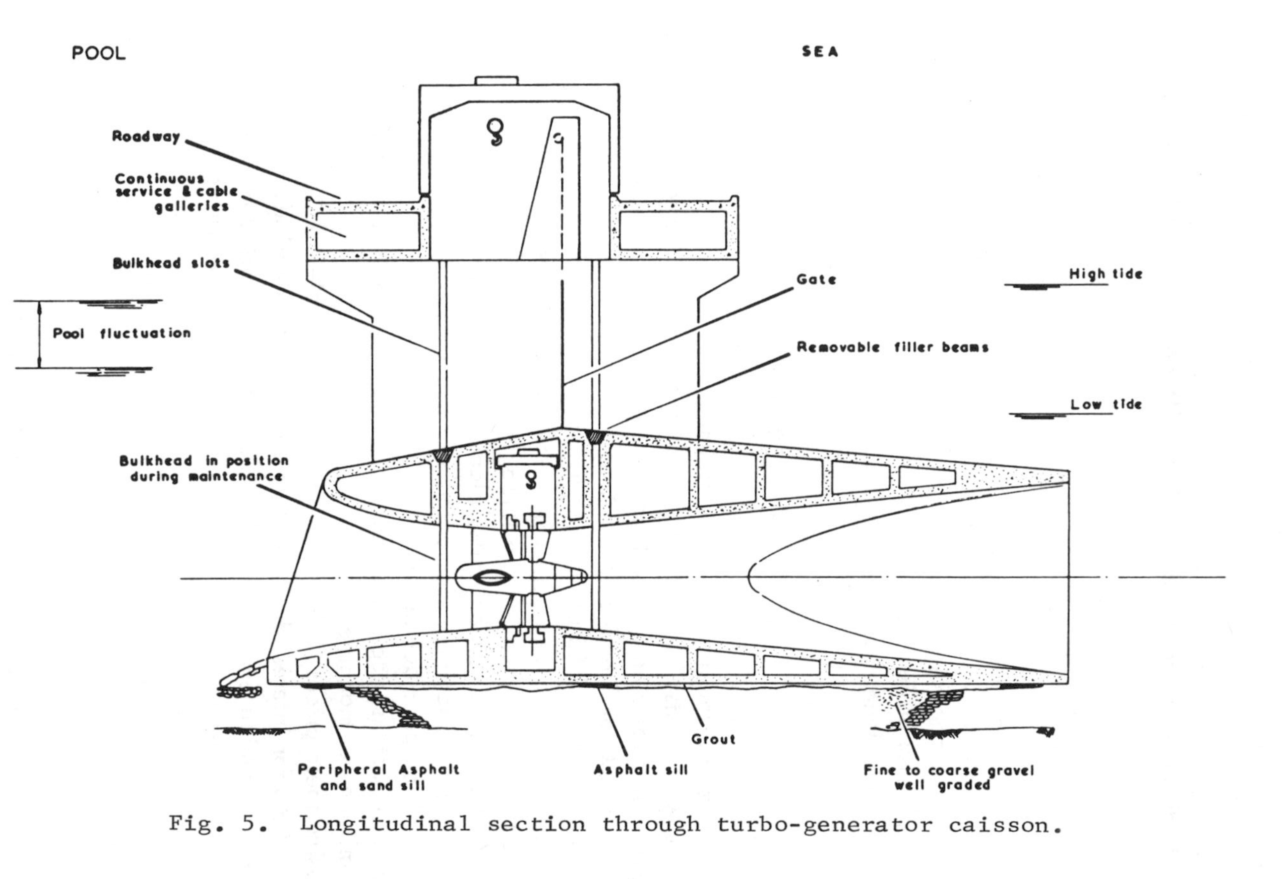

Fig. 5. Longitudinal section through turbo-generator caisson.

tidal ranges has been discussed extensively elsewhere [2][7]. The proposed construction methods provide for this critical operation. The barrage components are:

1. Rockfill base placed by barge dumping, cableways and end tipping;
2. Caisson units of the power station each to contain two (or possibly four) 30 ft. diameter straight-flow turbines with crest-gate sluices 40 ft. deep (see Fig. 5);
3. Refilling units proposed as 40 ft. square Venturi units for the present analysis (see Fig. 6);
4. Hydraulic machinery installed in the caissons to be 30 ft. diameter straight-flow turbines each with rim generators rated at about 35 MW.

There would of course be many more items in an actual structure but the great bulk of the capital cost is contained in these four items and a good estimation of total project cost may be arrived at by considering these four only plus a percentage based on MW capacity.

COST OF ENERGY

Optimization implies energy at minimum cost per unit so it is necessary now to deal with costs. Through other studies recently made fairly accurate cost figures are available for the type of development described. These are summarized below for Site D of Knik Arm:

1. for placing rockfill $5 per cubic yd.
2. for power station caisson units in place: (260 ft. long by 150 ft. high by 120 ft. wide, each for two turbogenerators): $3.6 million each
3. for Venturi sluice caisson units, placed: $1.0 million each
4. for hydraulic machinery (including electrical ancillaries) installed $2.2 million per machine of 35 MW
5. for other costs (sluice gates and machinery, switchgear, transformers, etc.) $1 million per machine.

Using these prices, allowing 15% for contingencies, 7 1/2% for engineering and 6% for interest during a 4 year construction period, the optimum combination for minimum energy cost is 80 turbogenerators with over-sluices and 20 Venturi sluices yielding an annual output from the barrage of 6,130 GWh for a total capital cost of $640 million. This gives a cost per KWh generated of $0.106 for each 1% of capital interest rate. Eg. 6.4 mills/kWh for a 6% interest rate.

ELECTRICAL DEMAND IN ALASKA

The forecast electrical demand in Alaska for 1975 [8] is 312 MW and 1,220 GWh energy. It will be clear from these figures that the available local

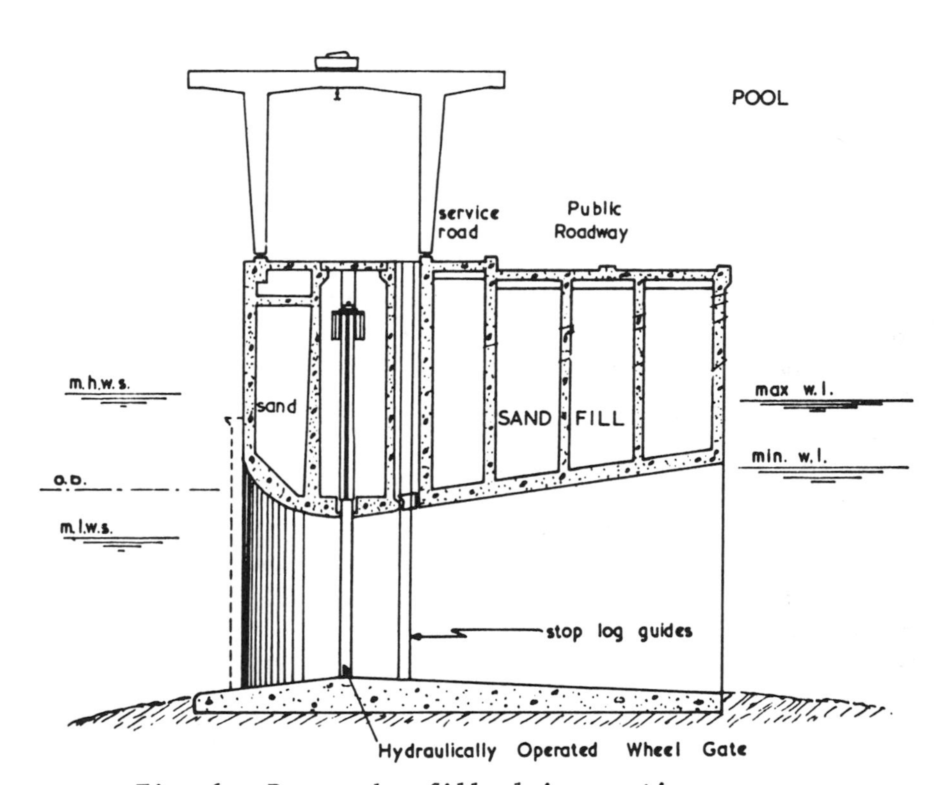

Fig. 6. Proposed refill sluice section.

market could not absorb the full development of Knik Arm for many years. Accordingly, the authors did not think the prospects merited the detailed investigation of pumped-storage schemes to integrate the tidal energy into the system on this scale, though there is little doubt that excellent sites exist for such schemes; eg. at Bradley Lake (see Fig. 1). There did seem to be a case for partial development however and so a request was made to the U.S. Corps of Engineers at Anchorage who kindly supplied the forecast daily demand curves for the integrated Alaskan system in 1975, for each calendar month of the year. These are shown in Fig. 7.

Since no information was available about communication and navigation benefit, it follows that optimization of partial development was not possible here. Nevertheless, it was clearly desirable to carry out a preliminary integration analysis of such partial development energy in the context of forecast system demand. Accordingly, an arbitrary choice was made of a partial development scheme which appeared to be of the correct order of magnitude to supply peaking energy and power throughout the year on the combined South Central and Interior Alaskan system for 1975.

The scheme chosen was for 10 turbogenerators of 30 ft. wheel diameter housed in five double caisson units with over-sluices and an additional effective refilling sluice area of 40,000 square feet in Venturi units. Taking a discharge coefficient of 1.6 for these, this requires 25 such units with 40 ft. square throat openings. These installations, in a Site D barrage, will produce approximately 916 GWh of electrical energy annually.

If the costs of Section 7 are used and only that portion of the whole capital cost directly attributable to energy generation (i.e. caissons, turbogenerators, sluices, etc.) is apportioned to the energy, this gives a unit cost for energy of 1.16 mills/kWh for each 1% interest charge. The total capital cost of the energy producing part of the barrage is $106 million, including contingencies, engineering and interest during construction.

THE INTEGRATION OF PARTIAL DEVELOPMENT ENERGY

It was assumed that the tidal energy, occurring regularly and predictably in magnitude and time would be used through pumped-storage to meet all peak demands on the system above certain load levels, thus allowing the thermal plant to operate on base-load, continuous duty with benefit to its load-factor.

Accordingly, a computation was required to evaluate the levels of demand at which this would occur. The programming for a precise evaluation is complex and was not merited in this case since optimization was not possible.

A close approximation may be obtained fairly simply however, as follows. The total tidal energy available in the year if all the tides were of constant range and less than average is obtained by multiplying a particular range's output, from Fig. 8, by 705. The average daily output is then this

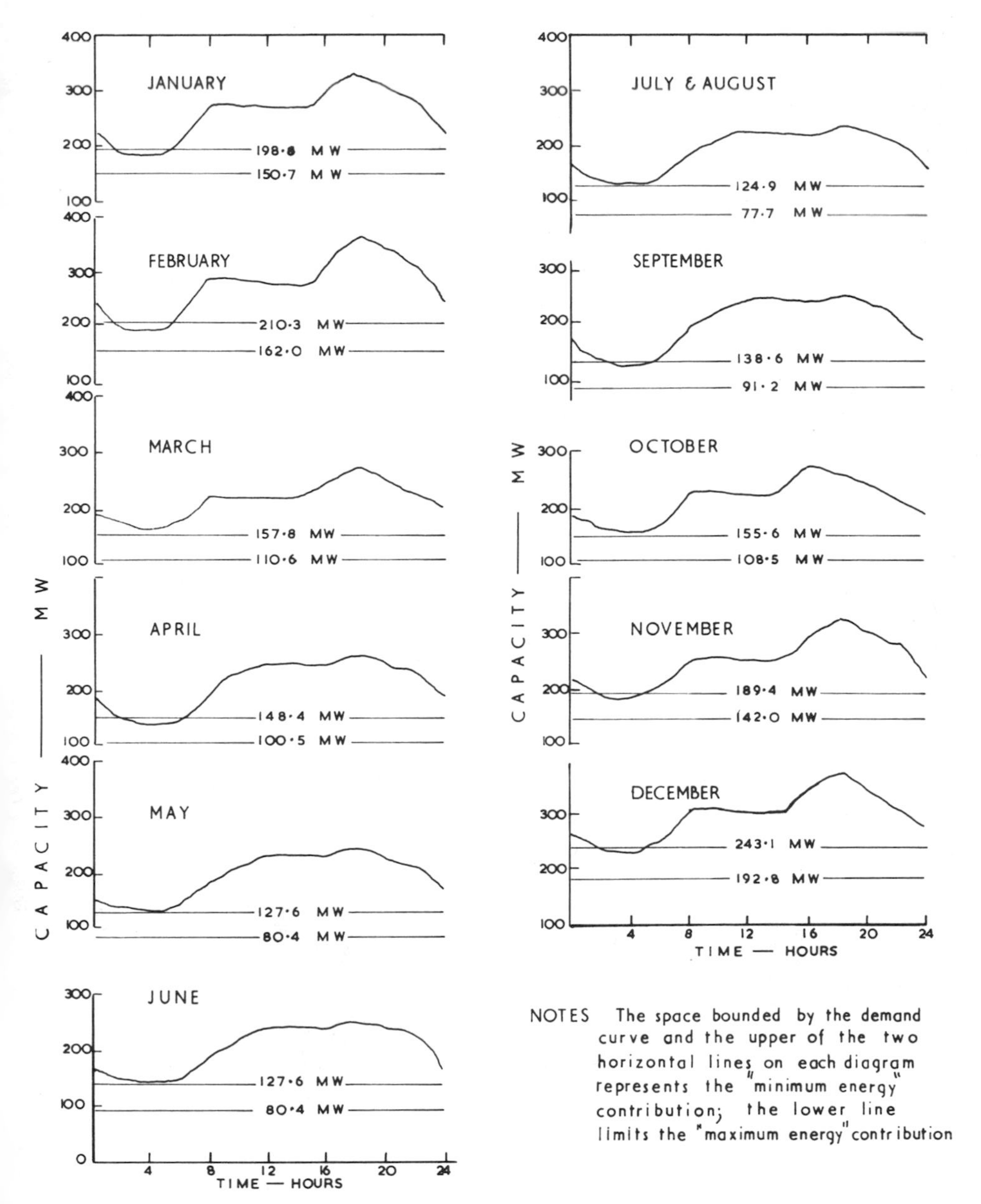

Fig. 7. Forecast daily demand curves for south central and interior system, Alaska, for 1975.

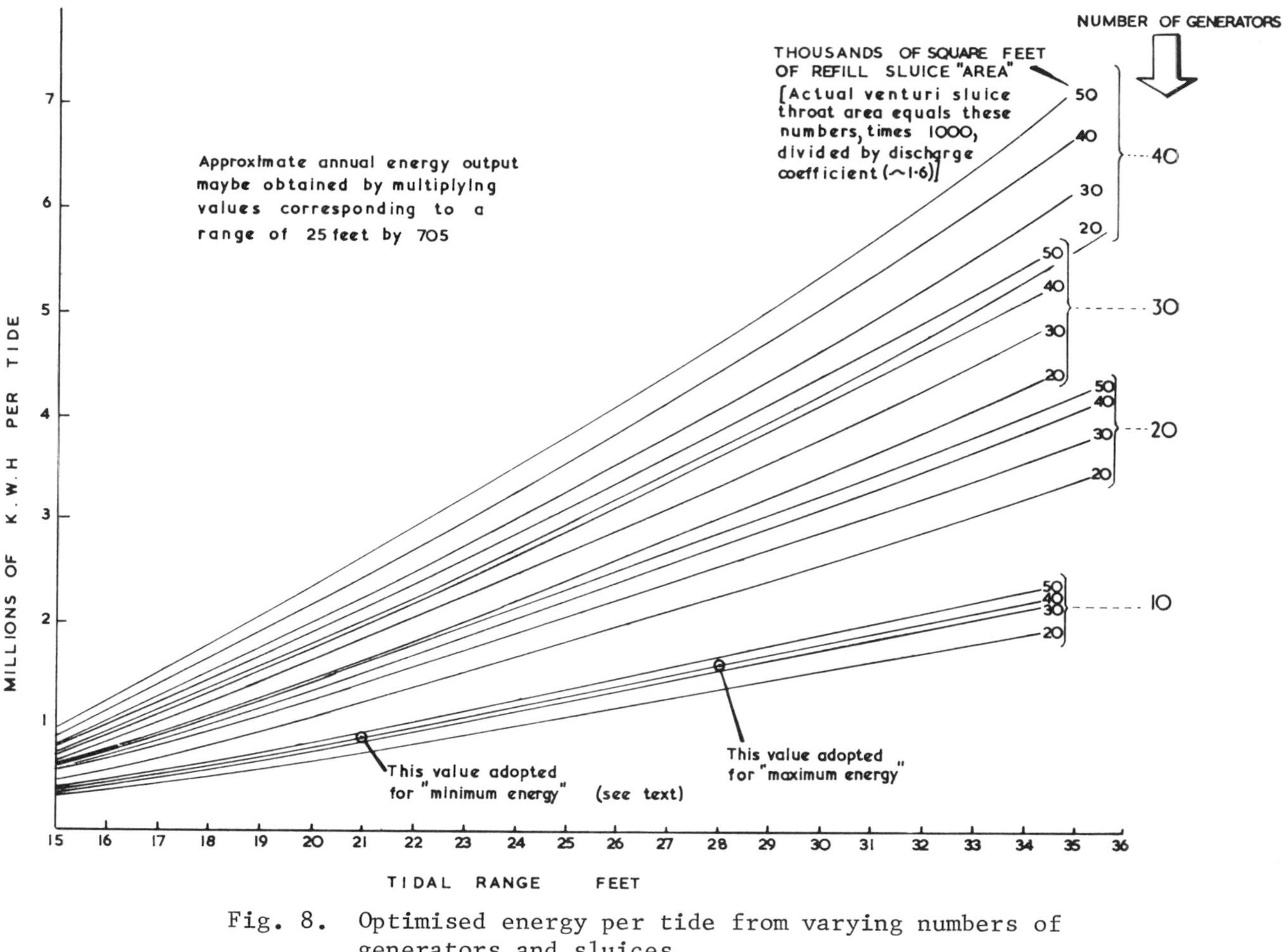

Fig. 8. Optimised energy per tide from varying numbers of generators and sluices.

value divided by 365. Of this daily output one-third is directly injected into the system and two-thirds is pumped-stored and regenerated. (These proportions have been established elsewhere [9]).

A trial and error calculation is then made to find the horizontal line (demand level) above which this net energy will just fit into the space between it and the demand curve for the day. This calculation is made for each month of the year.

A similar process is gone through for the case of greater than average tide, and so two levels of demand are shown on each month's typical daily load curve of Fig. 7 corresponding to the energy levels from uniform tidal ranges, "minimum energy" on the assumption of 21 ft. range and "maximum energy" for a 28 ft. range. Provided adequate storage is available in the pumped-storage reservoir, the true level of demand above which tidal energy may take over is between these two values and probably not far from their mean.

The partial development scheme chosen may be seen from Fig. 7 to be of the right order for the 1975 demand pattern. The energy could be integrated by an addition to the present Eklutna hydroelectric scheme of a pumped-storage facility.

The existing Eklutna reservoir (Fig. 9) has working storage capacity of 7.6×10^9 cubic ft. with an average head above station level of 840 ft. This represents a gross pumping input of around 180 GWh which is much greater than required to allow for output differences between springs and neap tides. Accordingly, the partial development energy could be completely integrated into any conceivable future demand pattern by the installation of about 250 MW pumping capacity and 200 MW generating capacity at Eklutna. Such equipment would of course be available for the developing system's thermal plant also, and so would work at higher load-factors than is customary for such plant.

The tidal energy and integrated pumped-storage would remove between 155 and 135 MW of capacity (depending on the month of the year) from that part of system demand being met by thermal plant, and could also provide spinning reserve capacity of at least 95 MW, rising to 250 MW much of the time.

Obviously, this analysis can only be preliminary, but it may serve to indicate the order of magnitude of energy and cost arising from development of Knik Arm.

BENEFITS ADDITIONAL TO COMMUNICATIONS AND POWER SUPPLY

Improvements of Load Factor in Other Generating Plant

Since the system would now have complete flexibility the future plant

Fig. 9. Eklunta H. E. Project reservoir seen from 5000' looking N.W. (Note complete ice cover.)

could all be designed for the highest load factor possible. This cannot be defined without detailed knowledge of the system but it would be better than existing load factors and so unit costs of conventional generation would fall.

The Stabilization of Water Levels in Knik Arm

With only 10 machines operating the fluctuation of water level on ebb tide would be reduced to about 2 feet and overall fluctuation to about 6 feet. This would have two important effects:

(a) it would considerably reduce the hazard from drift ice within the barrage since less ice would form and that which did form would not be subject to wide oscillations up and down the Arm;

(b) it would have the effect of maintaining at the docks of the Port of Anchorage, a depth within about 6 ft. of the mean spring tide level at all times.

If locks were provided in the causeway, Anchorage would become capable of receiving and discharging shipping at all times. This aspect of the matter deserves further study but there is a prima facie case for benefit.

CONCLUSIONS

The two Arms at the northern end of Cook Inlet, Knik and Turnagain, could be developed to provide large quantities of intermittently produced energy. Turnagain would provide about 12,500 GWh annually at full development and Knik, about 6,000 GWh. The likely costs of such energy would be, at 1970 prices, around 1.2 mills per kWh per 1% interest rate.

The present state of the Anchorage area power market does not suggest that full development of either Arm is worthwhile but does merit a closer look at partial development of Knik Arm as part of a multi-purpose causeway project. Such a project would have the following aims:

1. To provide a permanent road/rail/cable/pipeline crossing of Knik Arm near Anchorage leading to development of the western side of the Arm.
2. To provide the port of Anchorage with permanent deep water, very much reduced currents and probably greatly reduced ice hazard in winter.
3. To provide about 900 GWh of tidal energy per annum, which after integration and using an Eklutna pumped-storage scheme seems likely to provide 750 GWh of completely firm power per annum at a unit cost which may be competitive with alternative sources and with no subsequent increase in price over the working life of the scheme.

Finally, the authors believe that it would be worth investigating the effect of barraging the Inlet between West and East Forelands by a numerical model. If the barrage increased the tidal amplitude through resonance effects, this site might possibly become economically attractive with an energy producing potential that would exceed even the Ramparts Project in Central Alaska.

ACKNOWLEDGMENTS

The authors gratefully acknowledge grants from the University of Washington, Seattle, the University of Sheffield and the University of Salford towards the costs of the investigation. Their thanks are also due for help received and interest shown in the work by the Chief of Engineering Division, Alaska District Corps of Engineers and members of his staff; The Superintendent, Eklutna Hydro-electric Scheme; The Alaska Field Director, U.S. Coast and Geodetic Survey and the Chief Engineers of the electric power utilities in Anchorage.

REFERENCES

1. Bernstein, L. B., Tidal Energy from Electric Power Plants, Israel Prog. for Scientific Translation, Jerusalem (1963).

2. La Rance: A Tidal Power Scheme, Rev. Fr. de l'Energie (1966).

3. Wilson, E. M., "Tidal Energy Development", Handbook of Applied Hydraulics, Editors: C. V. Davis and K. Sorenson, McGraw Hill, New York (1969).

4. Swales, M. C., "Optimization of Ebb-Flow Tidal Power Generation", Ph.D. thesis, University of Sheffield (1968).

5. Swales, M. C., and E. M. Wilson, "Optimization of Tidal Power Generation", Water Power, 20, 109-114 (1968).

6. Braikevitch, M., Gwynn, J. D., and E. M. Wilson, "Tidal Power Development with Special Reference to Plant, Construction Techniques and the Integration of the Energy into Existing Electrical Systems", Proc. 7th Wld. Pwr. Conf., Moscow (1968).

7. Lingsma, J. S., Holland and the Delta Plan, Van Nigh and Ditmar, Rotterdam (1963).

8. Report of the Hydro Resources Sub-Committee of the Alaska Advisory Committee of the Federal Power Commission, Alaska Power Survey (1966).

9. Wilson, E. M., "The Solway Firth Tidal Power Project", Water Power 17, 431-439 (1965).

THE TIDAL POWER PLANT "SAN JOSE" - ARGENTINE

H. E. Fentzloff *

The extraordinary proportions of the very large basin area combined with a medium tidelift indicates a possible utilization of the tidal power of the San José Gulf in Argentine and an initial investigation as to its economy is therefore suggested. The Gulf's basin area of 780 sq. m. and its tidal range of 3.5 - 7.8 m. with corresponding values for Q and H in the energy equation present factors which on first sight one might have every reason to doubt. Almost the greatest factor influencing the economy of any tidal power plant is the condition of the sea bed at the dam site in the basin estuary. This will be a decisive factor for almost all construction work and for the generating capacity to be installed.

From the basin and dam conditions illustrated in Figures 1 and 2, one may easily recognize the remarkable topographical and current advantages offered for the location of the dam and the power plant axis near the 7 km wide estuary. The proposed power plant constructed of lamilarly joined reinforced concrete caissons stretches in a seaward-led concave arch beginning at small dammings towards the shore. The foundations near the shore extend down to the excavated sea bed which consists of tertiary sediments of remarkable strength. For the major parts the power plant would be founded on the dam fill indicated in Figure 2. This fill is obtained from the material excavated from construction pits near the shore and from the enlargement of the present basin estuary.

In the case of single cycle generation of the tidal power plant the length of the structure would be about 8 km. This study, furthermore, deals with double-effect generation (or dual cycle) which appears to be the most favourable scheme here. This would require a length of 16.8 km for the structure.

The main sections (Figures 3 and 4) show the reinforced concrete caissons with the hydroelectric units for single and dual cycle respectively. The generating units consist of tubular turbines with rim generators. Figure 4

---

* Consulting Engineer, c/o Hochtief AG, 43 Essen, Postfach 670, West Germany.

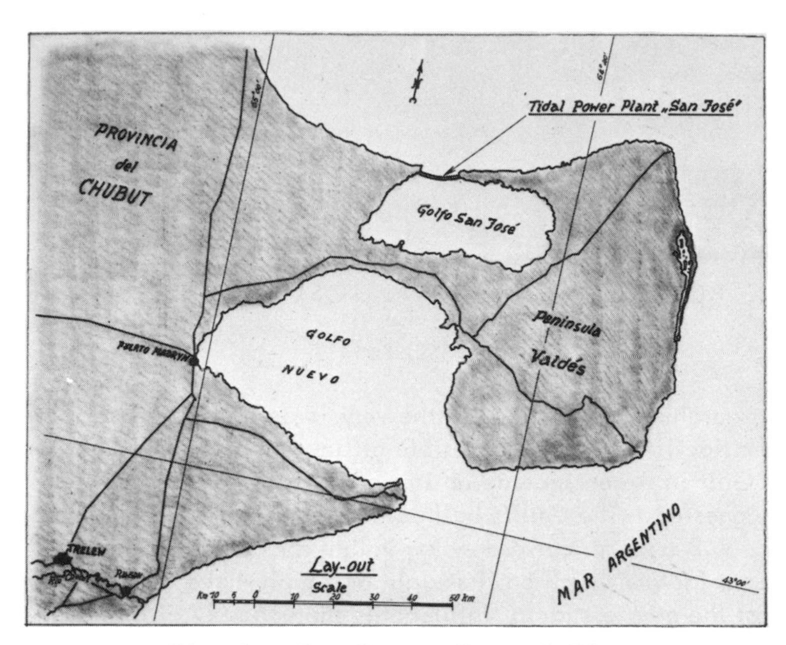

Fig. 1. San Jose - General Plan.

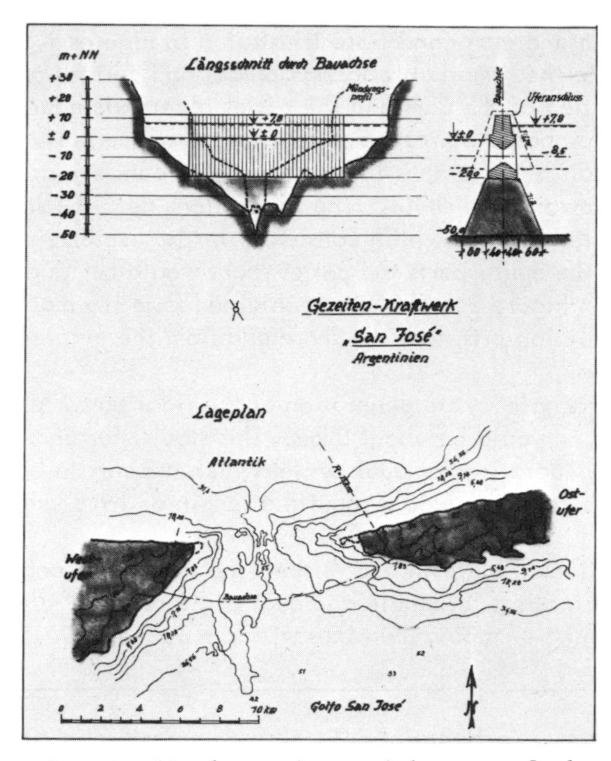

Fig. 2. Longitudinal section and layout of the estuary.

also shows the sluice caissons which in case of the dual cycle system are required for short-time filling and discharge of the basin. The number of these sluice caissons is the same as that of the turbines.

In the case of the single-effect scheme, there will be no need for particular sluice caissons to operate during the filling and balancing period between the working periods. The filling period will have a relatively small flow which the tubular turbines will allow when running without load and at reduced speed.

The selection of the correct runner diameter of the turbine and thus the number of units to be installed will have a great influence on the favourable utilization of a tidal power plant. The maximum runner diameter feasible today for tubular turbines with rim generator is between 7.5 and 9.0 m. Comparing cost estimates will determine the optimum result between the reduction in construction cost and the increase in installation cost with growing runner diameter.

The concept of the San José tidal power plant demonstrated in Figures 2, 3, and 4 corresponds, without compromise, to the postulates for optimum tide utilization which the author prepared in 1955(1). It may be pointed out that the development and realization of the tubular turbine in which the author participated decisively(2) holds the key for an economic tide utilization.

Table 1 shows the main data for the San José tidal power plant for single and for dual cycle operations. It should be mentioned that the final project may entail alterations in the runner diameters and thus the number of units. This may also result from pending considerations of the combination with a pumped storage scheme which is indispensable for optimum power distribution. Only this will allow the discrete blocks of power available from the tidal generation to be integrated in an inexpensive way with the costly peak demand periods in the energy supply system, partly as directly offered peak power, partly as pumped energy.

The technically favourable design of the San José dual cycle plant provides almost 1.6 times the yearly output of the single cycle. From cost estimates for both types of tide utilization, it is expected that the specific cost per kW installed in the dual cycle schemes will be only 1.5 times the cost for single cycle.

For a first value comparison between different tidal power plants, the author proposes two criteria which may be called –

(1) the "construction criterion" as the ratio $\left(\frac{\text{basin area}}{\text{dam length}}\right) \frac{\text{sq.km}}{\text{km}}$

(2) the "energy criterion" as the ratio $\left(\frac{\text{yearly output}}{\text{dam length}}\right) \frac{\text{Gwh}}{\text{km}}$

Among the known projects of tide utilization in Table 2, San José does

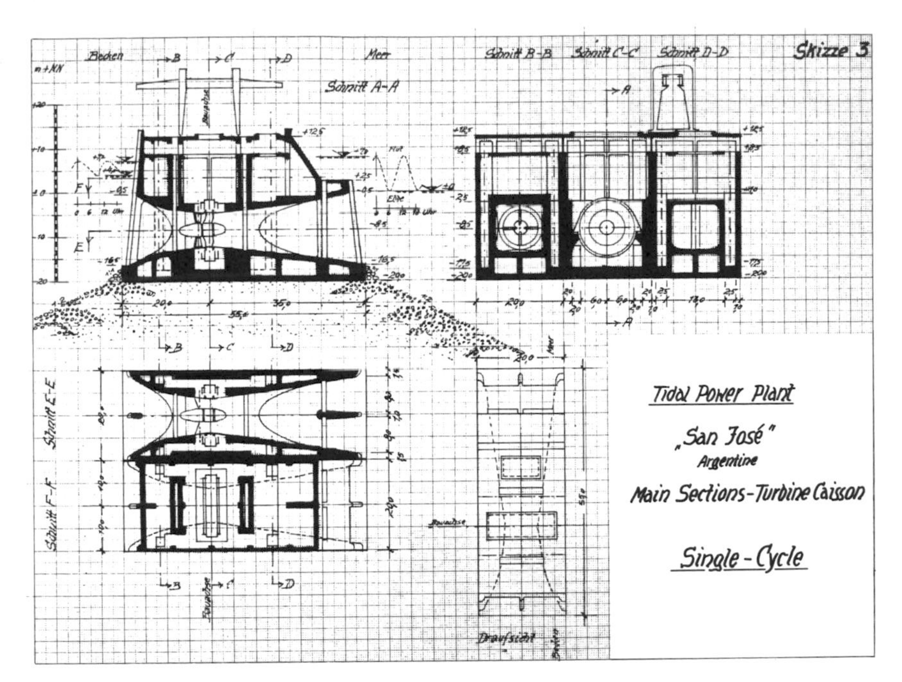

Fig. 3. Main sections, reinforced concrete caissons, single cycle.

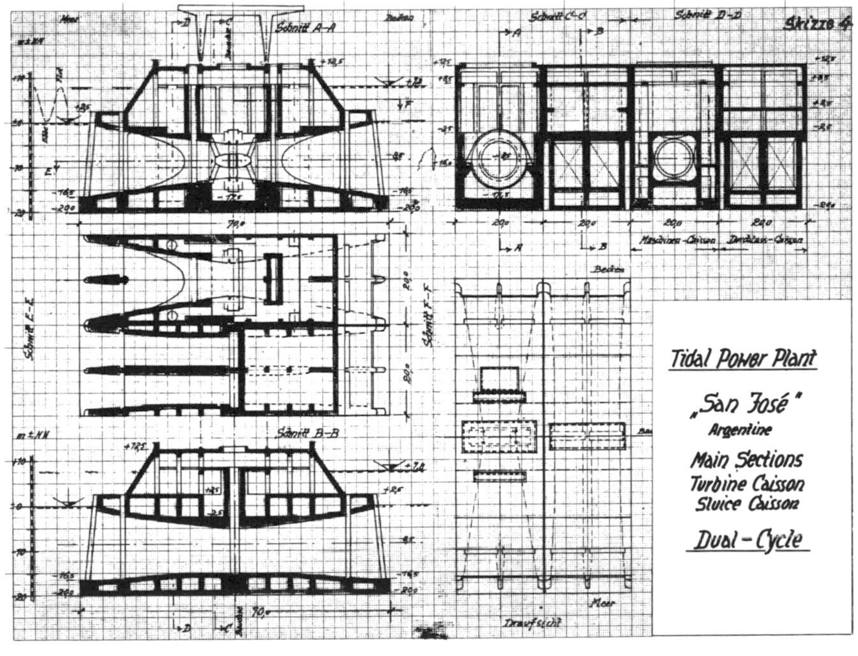

Fig. 4. Main sections, reinforced concrete caissons, dual cycle.

Table 1
Tidal Power Plant, "San José"- Argentine

Main Data

Area Golfo San José	flood	780 sq km
	ebb	725 sq km
Tidal Range	maximum	7,80 metres
	medium	5,65 metres
	minimum	3,50 metres

Power Station

Single Cycle		Dual Cycle
	Installed Capacity	
4·965	MW	6·820
	Yearly Output	
9·500	GWh	75·000
	Installed Units	
400	number	400
	Turbine Runner Diametres	
7,50	metres	9,70
	Capacity each Unit	
4,0 - 72,5	MW	6,6 - 77,0

Table 2
2 Criteria for Value Comparison Between Tidal Power Plants

		Construction	Energy
		Criterion	
San José:	single cycle	1190	98
	dual cycle	890	46,5
Knick Arm - U.S.A.		1120	74
Bristol Channel - U.K.		1038	42
Minas Basin - Canada		876	37
La Rance - France		756	28
Solway Firth - U.K.		248	22
		sq km/km basin area	GWH/km yearly output
		for each km dam length	

very well despite the fact that its tidal range being 3.5 metres minimum and 7.8 metres maximum cannot be said to be very high.

The following photographs are intended to give an impression of the landscape. Figure 5 shows the western bank of the Gulf at low tide.

Fig. 5. Western bank of the Gulf at low tide.

Fig. 6. Width of the Gulf's estuary - 7 km wide.

Fig. 7. Air photo of the site of "La Rance."

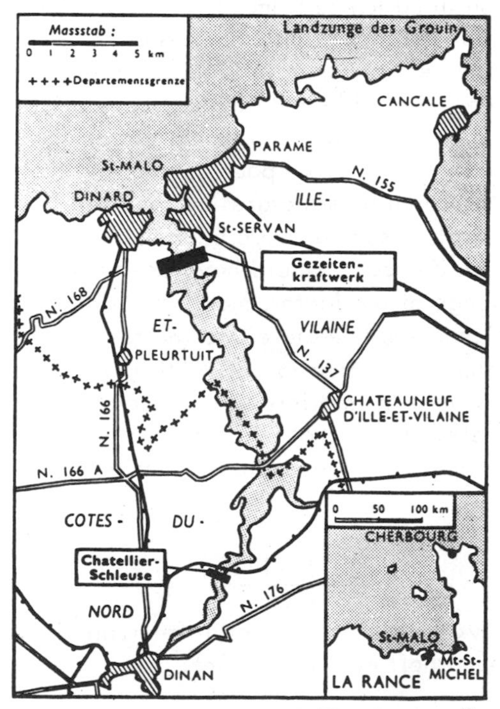

Fig. 8. Layout of "La Rance."

Figure 6 shows the whole width of the Gulf's estuary - 7 km wide - the rocky eastern bank in the background consists mainly of shell lime.

From the point of view of the author's fundamental research, it is necessary to review the details of the costly construction of La Rance. This research led to the proposal by the author of a "LAW OF ANTITHESIS" between river and tide utilization[3]. This proposed law is based on the antithesis in the energy sources: the downward-acting terrestical gravitation causes the water to flow on the earth surface; the upward-acting lunar gravitation causes the water to oscillate in the ocean bays. The inevitable consequence is that in tidal power schemes the concept of hydraulics and of machine and structure design and construction must be different from those of river schemes. The failure to recognize these factors would lead to uneconomical tidal plants.

The air photograph (Figure 7) shows that the La Rance site resembles the conventional type of construction of a river power plant; that means a dry construction pit between cofferdams with dewatering system for the excavation and placement of formwork, reinforcement and in-situ concrete, etc. Contrary to this, the postulate suggests that for the economical utilization of tidal power, the floating and sinking of prefabricated concrete caissons with all preparatory underwater works for their foundation, which consist mainly of fill material, is more appropriate.

The grave deviation in La Rance from the regularity of tide utilization was the inevitable result of the topographical facts of the site. The clear river character of the Rance estuary seen in Figure 8 with the 4 km inland location of the tidal power plant did not allow any type of construction other than the one performed. Another price increasing factor in La Rance was the bulb type tubular turbine selected. The space saving and the corresponding reduction in volume of the concrete structures by the use of a tubular type turbine with annular generator, such as have been working in Germany most satisfactorily for more than 30 years, would have been much more economical.

The lack of knowledge of the factors mentioned above has, unfortunately, led to misleading polemics in the literature and the feeling that one should doubt the economy and thus the utilization of tidal power en bloc!

While no one has a right to interfere with the pioneer work of our French colleagues, it is nevertheless a fact that with the present advances in technology the economic development of the immense power reserves of the tides cannot be stopped.

REFERENCES

1. Fentzloff, H. E., "Systematik der Wasserkraftnutzung", Verein Deutscher Ingenieure, VDJ-Forschungsheft Nr. 453 (1956).

2. Fentzloff, H.E., "Principes Fondamenteaux de la Construction des

Centrales Submersibles", La Houille Blanche No. 5 (1949).

3. Fentzloff, H. E., "A Fundamental Approach to Tidal Power", Water Power No. 8 (1967).

SOME CONSIDERATIONS OF A POSSIBLE NEW ROLE FOR TIDAL POWER

T. L. Shaw*

INTRODUCTION

The tides offer an abundant natural source of energy, yet remain virtually unharnessed. This is not for lack of technical understanding, but rather for a sufficient case to convince supply authorities to invest in this field. The task is made the more difficult by an increasing reliance on familiar and proven generating systems.

The experience of recent years indicates that the pattern of power generation is hardening in favour of nuclear plant. Sites for new conventional hydro stations become increasingly difficult to justify, so despite the attractions of this energy source, its position is declining with many major suppliers.

The emerging pattern is necessarily based on economics. In general terms, as far as possible base-load is met by nuclear plant, with other less efficient older themal sources taking lesser positions; hydro, with certain particular exceptions, is favoured for peaking.

Tidal sources are excluded mainly for well known operational reasons, yet as a large hydro source they could perform base-load and/or peaking operations. However, despite advantages offered by considerable ingenuity in overcoming the vagaries of tidal period and amplitude, much of the available potential is lost in having to overcome lunar/solar cyclic inequalities.

In this paper the author develops what may be regarded as a new case for tidal power, in that the function of this energy source is seen in the part that it could play in an integrated power system. This approach markedly changes the operating characteristics demanded of the tidal component, with resulting efficiency benefits to it as a single component.

Ancillary operating features are introduced as appropriate, along with comments on other aspects of this type of project.

*Department of Civil Engineering, University of Bristol, Bristol, England.

TIDAL ENERGY AND ENERGY NETWORKS

For the purpose of this paper it will be assumed that energies made available from tidal sources are neither trivial nor predominant in the receiving network, i.e. some 3–15% of installed capacity might be of this type. Lesser developments become financially less attractive, whereas larger schemes, of themselves, tend to impose severe restrictions upon the manner of operation of other plant in the network (that is, unless similarly demanding controls are built into the tidal scheme). In either extreme case, competitive energy sources will be preferred for other reasons, although whatever the size of scheme there is still to be overcome the keying of this output with that of other plant to meet instantaneous loads.

The characteristics of coastal tides and the random factors that may exert some influence may be listed as follows:-

1. The tidal period is 12.4 hours, approximately;
2. Cycles of variation of tidal range occur over 2 and 4 weeks approximately and longer;
3. Tidal variations may only vaguely resemble a sine curve;
4. Amplification of tidal curves is largely controlled by estuarial geometry;
5. Atmospheric, barometric and river discharge conditions may all modify the basic wave motion;
6. Barrage construction will lead to a much modified regime, particularly close to the barrage.

Together these factors ensure not only that each estuary experiences unique tidal conditions but also that these conditions vary in a quasi-regular manner, but with random overtones from tide to tide.

Too often proposals to exploit this irregular energy source have been mooted without due regard to the function that such a scheme could perform. This must surely present a regrettable situation. However, those that have worked for such features as a constant power output have only demonstrated how unattractive such a proposition would be in practice. Indeed, a sound case for harnessing tidal energy in any but quite exceptional circumstances is not known to the author.

Mention was made in the "Introduction" of the present emphasis on thermal installations of the "base-load" (constant output) type. Such plant offers almost every advantage, lacking only in flexibility to meet daytime excesses over minimum night-time requirements. Experience does not indicate that this maximum-minimum gap is being closed substantially by preferential off-peak tariffs, its residual magnitude increasing as the total load increases. This combined trend underlines the growing case for energy storage not so much to provide a source of peaking by day but rather to give an additional source of energy operating at a more constant output according to the way in which the excess of peaking over base-load is best supplied. Thus, 2000 MW not being used by night could become 3500 MW of constant output by day according to the periods over which power for storage is available and the combined output is required by day.

But to find sufficient and suitable sites for conventional pumped-storage may not always reasonably be possible. Acceptable schemes are normally reckoned to include a high-head cycle, minimum storage difficulties, and minimum transmission connections. Experience in England, by no means untypical, lends support for the view that it may well not be possible to provide fully for storage needs by this technique.

An alternative possibility, but of lesser immediate attraction, is to accept lower recirculating heads, although the greater discharge then necessary to achieve the same capacity of installation increases the water storage problem. Only at the comparatively low heads available with estuarial storages may it reasonably be possible to realize the required energy storage, yet such schemes for pumped-storage promise to be both comparatively expensive and inefficient when set against high-head 'alternatives'.

Two factors emerge from these considerations. Firstly, there is the need to consider how best pumped-storage facilities are to be provided if these are deemed to be desirable; and, secondly, that in considering estuarial storage there is produced a scheme closely resembling that which would follow for tidal power exploitation. The possibility of absorbing this natural energy source whilst achieving storage of thermal energy in principle offers considerable attractions, the scope for development of which will be discussed in following sections.

THE FUNCTIONING OF AN INTEGRATED BARRAGE SCHEME

Reference has already been made to the form of output that may be sought from tidal schemes. A large supply authority will only be receptive to this constituent if it may be keyed into the generating schedule with no restrictions on its usage. Judging by the present policies typically adopted, the most useful form of output from an estuarial source will involve either a constant or a quasi-constant characteristic of repeated shape and magnitude (with exceptions such as at weekends).

The studies conducted at the University of Bristol and discussed here have concentrated on a repeated and constant output over 12 hours by day (0800 - 2000 hours) although this comparatively straightforward form does not present particularly favourable programming circumstances. Other characteristic outputs could be produced as required with no more trouble.

Besides the variable tidal input to the estuarial scheme, there will be a constant storage component similar from night to night except at weekends. Failure of this input will, of course, influence the output available during the following day, but this contingency will not be pursued here. Also, the situation arising at weekends will not be considered for it is not of prime

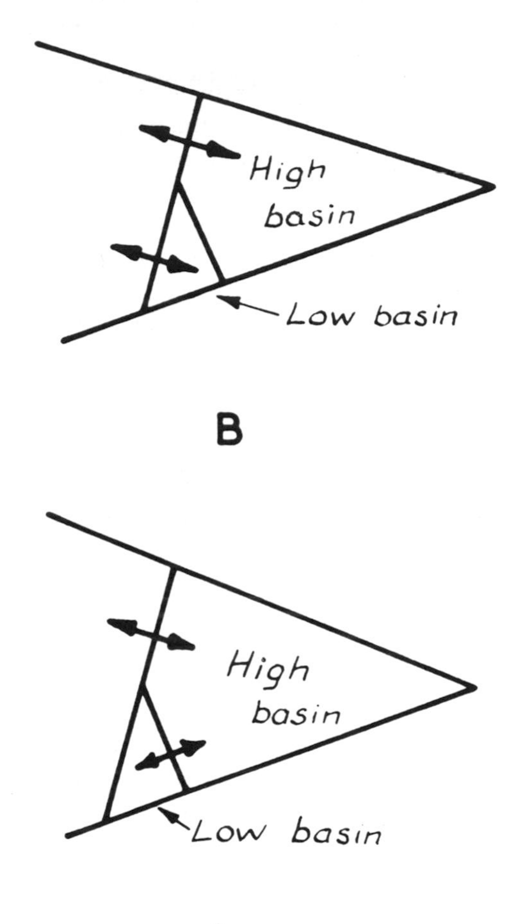

Fig. 1.

importance. What matters basically is that the guaranteed output should be assured irrespective of tidal conditions on any day.

The successful functioning of the concept for integrating tidal power and pumped-storage is fundamentally dependent on the fact that the tidal period is only about 12 hours, i.e. of an amount similar to the pumping and generating periods. This similarity ensures that periods of high and low sea level will occur during every phase of pumping and generating, although these levels may not be as much as the extremes reached in any cycle, being dependent on the duration of each phase.

Hence, in a two-basin scheme there will exist the possibilities that:

(a) Pumping

1. When the tide is high, one basin is filled (by both pumping and gravity if possible);
2. When the tide is low, the other basin is lowered.

The state of the tide when pumping is to commence will dictate which operation will come first, and for how long it will continue before the other becomes preferable. It may be that conditions will favour resuming the first operation towards the end of the pumping phase.

(b) Generation

1. When the tide is high, discharge occurs either between the basins or between the sea and the low basin;
2. When the tide is low, the upper basin discharges to the sea.

As with the pumping phase, there must again be taken into account the state of the tide at the beginning of the generating phase. This will determine the timing and extent of each operation within this phase; optimization techniques must be employed to ensure that these factors are correctly assembled in the programme for any day. A similar procedure must follow for combining the pumping options.

The above alternatives for pumping and generating provoke the following questions:-

1. In addition to installing machines between the sea and the upper basin, is it preferable to link the lower basin with the sea or with the upper basin? These options are illustrated diagrammatically in Fig. 1.
2. If it is physically possible to do so, what advantages would result from arranging for the lower basin to be lowered by pumping to much less than low-tide level? This factor introduces the question of the relative surface areas of the basins that will be required.
3. What consequence, both hydraulic and ecological, may result from the new estuarial environment; in particular, should these considerations impose restraints on the degree of change resulting from a barrage project?

These three factors will be considered in subsequent sections.

THE WORKINGS OF AN INTEGRATED POWER PROJECT

The magnitude of the power output sought of an estuarial project, and the prevailing geological and geographical features, will probably determine within fairly close limits the siting of barrage structures. The extent of variations of water level within the scheme will also be a factor for if these are to be restricted it will be necessary to look to either larger 'impounded' areas or to a lesser output.

Having decided at least the approximate barrage layout, it will be necessary to conduct at least detailed numerical analyses of the modified tidal behavior in the estuary to give a more firm basis upon which to assess the energy potential of the project. Earlier studies may have been performed to consider how various possible barrage sites are influenced in this way. Physical model studies in support of the detailed analytical work may also be deemed essential to study aspects such as the significance of modified estuarial resonances and the possible consequence of Corioli's forces on the new regime (the latter is often of little importance).

It is significant that those sites to which this form of analysis has been applied have all been shown to undergo appreciable tidal reductions by barrage construction [(1)(2)(3)]. Such changes, whilst no doubt being welcomed from certain aspects, only represent a loss of output potential from a power project so need to be assessed accurately.

Generalized analytical techniques for the integration of energy storage with a modified tidal environment have been performed at the University of Bristol. In particular, consideration has been given to the possible methods of development, and the list given below summarizes the main finding of these studies [(4)(3)(5)].

1. The desired constant daytime output, or other generating characteristics, may be developed from a two-basin scheme integrating tidal energy with off-peak energy storage.
2. The variable tidal output is absorbed into the operation by allowing the 'extreme' levels reached in either (or both) basins in any one daily cycle to be a variable throughout the cycle of tides. The extent of this variation will probably be small (1 or 2 feet) in a large scheme taking in more pumped-storage than tidal energy.
3. The scheme outlined in Fig. 1b could well be preferable to Fig. 1a on both operational and efficiency counts (although this order is reversed for non-energy considerations). In either case the schemes show distinct advantages if the lower basin is drawn well down by nightly pumping; more efficient functioning of machinery results, and requiring a smaller basin. An areal ratio of 10:1 for the basins may be possible depending on the extent to which the smaller basin may be emptied and the ratio of the magnitudes of the energy sources to be stored.
4. By the optimum selection of system operation, it should be possible to

develop outputs well in excess of those resulting from the separate and similar development of the tidal power and pumped-storage components (if this were possible). Thus, 100 units of tidal power input could produce an output of up to 80 units; and 200 units of pumped-storage could give a return of 150 units. Together, 300 units of input could give a return in excess of 300 units, i.e. an efficiency over the separate schemes of 130%. The magnified return from the combined scheme is achieved by the timely exploitation of the tide, implying that a greater amount of energy from this source has been harnessed by the integrated scheme than by a tidal power plant acting alone.

5. Peak operation would presumably occur in the winter when maximum energy storage would be required together with maximum daytime output. The scheme could operate at a reduced level in summer still absorbing the full tidal component but with a lesser pumped-storage component. Weekend operation could be included if an energy reserve in the form of more extreme water levels could temporarily be held within the estuary to supplement the day-by-day output during weekdays.
6. Any complete analysis, whether analytical or experimental, should attempt to ascertain the influence that machine operation (i.e. a discharge through the barrage) has on the tidal regime. Certain authors have maintained that, when passing a discharge, a barrage will influence the tidal motion to a degree intermediate between the natural and fully-barraged conditions. The author believes that this view may be misleading, maintaining that this is not necessarily a 'chicken-and-egg' problem. However, substantiating evidence on this possibly important matter is not known to the author although the matter is being given some consideration in Bristol.
7. Operation of the overall scheme would need to be accurately programmed with some account being taken of prevailing natural conditions if an optimum output is to be achieved.
8. Peaking stand-by could be included in the system operation if thought judicial.

ANCILLARY ASPECTS OF AN INTEGRATED POWER PROJECT

Because of the effects that the type of project under discussion could have on an estuarial regime, it could well be that specific restraints may have to be built into the operation from other than energy production considerations.

1. The environmentally most undesirable feature of the schemes set out in Fig. 1 is probably that of having a much amplified tidal range along that section of the coastline bordering the low basin. Whilst economic factors may support including the coast as one side of this basin (despite shallowing waters and hence reduced storage/depth ratio), it may well be deemed necessary to site this whole basin totally off-shore. Naturally deeper

waters in which to site machines could then be more readily available.

2. It is envisaged that the form of power development proposed will be favoured particularly in estuaries having a substantial tidal range and that in these situations the range of oscillation subsequently occurring in the high-level basin could be much less than over the same area in the natural state. The choice of limits between which the modified range will occur, including not only the daily range but also the longer term fluctuation to absorb tidal irregularities, will be influenced by several factors. From plant-efficiency considerations alone, the higher the levels that may be retained the higher the operating efficiency. However, if pre-barrage tidal flooding and/or land drainage difficulties occur, upper limits will automatically be applied. In this respect, there is the opportunity to improve an environment by removing at least the exceptional water levels, although in certain circumstances the occasional inundation of saltings may not be rated as undesirable but necessary to the ecology of those areas. Raised mean estuary levels will almost certainly cause noticeable rises in local water table levels, possibly affecting urban and land drainage, well discharges, and some buildings.
3. Navigational interests are likely to benefit from a generally higher level throughout the upper basin. Since this basin will have a direct connection with the sea (Fig. 1) there will be created the possibility of access to more favourable harbouring facilities for ships of at least the size that previously could be accommodated, but with a reduced restraint on ship movements (tidal/shift working of ports on the high basin would no longer be necessary). Indeed it may well be possible to arrange for unimpeded access to the high basin for all vessels, save only through a lock in the main barrage. Such comparatively sheltered harbouring facilities would become available at little cost extra over that associated with power generation.
4. Coastal amenities, in particular access to beaches (not only above the barrage), may or may not be improved by the new environment, although the chances are that only the upper part of many beaches are, recreationally, particularly attractive. A progression from sand or shingle through to estuarial muds is typical of beaches formed by a high tidal range so the exclusion of the lower levels from the exposed beach profile is not likely to be unacceptable. In this connection, the changed period of oscillation within the upper basin takes on a practical significance. Highest water levels will occur when the power system is fully primed towards the end of the nightly pumping period (0600 hours). From 0700 - 0800 hours onwards this level will progressively fall exposing more of the beaches as the public demand for them increases. Of course, the nightly recovering of the beaches would have to achieve the necessary cleansing.
5. The previous point provokes the question of the stability of solids previously

held in suspension or scoured and accreted with the tidal cycle in the modified estuarial regime. In opting not to pursue this point at length, the author is in no way wishing to detract attention from only one aspect of a much more diverse pollution problem that could be sparked off by disturbing a naturally balanced regime. Whatever the potential difficulties in this respect, whether from sewage effluent, formation of river deltas, thermal power station cooling water recirculation, fertilizers, etc., there is ample evidence from the disturbing experience with some existing enclosed bodies of water to make this a feature for the closest attention.

6. The point has not previously been made that a more horizontal water level will exist over the area enclosed by the main basin than existed previously. In particular, the modified regime will cause greater reductions in level higher up the estuary than at the barrage site. Thus, the maximum permissible level in the upper basin may well be determined by flooding, drainage, shipping and recreational considerations at or close to the barrage site, the modified regime producing more favourable circumstances for the first two of these factors higher up the estuary, but perhaps less so for the latter two (this may well be in keeping with what would be desired in the upper reaches).

CONCLUSIONS

In this paper, the author has attempted to present some general considerations relating to the concept of combining the not uncomplementary energy sources of pumped-storage and the tides to produce at high efficiency the form of output sought of any form of energy reserve.

In avoiding detailed reference to any specific estuary or project, there have arisen points of over-generalization, but these are not regarded as misleading. Each situation needs to be considered on its own merits, yet there remain common denominators for general discussion. Indeed, from the references cited for which the proposed techniques have been studied in some detail, there emerge certain conclusions that apply in general to other sites and thereby give some indication of what might be expected of other schemes [3][5].

Reference has been avoided in the present paper to the economic considerations upon which the viability of any particular scheme may be deemed to depend. This should not be taken to imply that the author is reticent to acknowledge this primary factor quite the opposite is, in fact, the case, these detailed studies to date of the principles outlined herein supporting the broad conclusion that in terms of the energy aspect alone any scheme could be attractive; the additional benefits then accrue from minimal outlay.

As with many large civil engineering projects, estuarial barrage construction involves considerable capital outlay over several years before any return is forthcoming. The power scheme discussed involves two main items, the barrages and the generating and its ancillary equipment. Some return could be forthcoming once the former is substantially complete (shipping, communications, etc.), but an energy usage will have to await overall completion. There then arises the question as to how the earliest possible phasing of the full scheme may be reconciled with the distinct step in pumped-storage and generating capacity that becomes available, and whether the maximum size of any particular scheme needs to be restricted for such operational (economic) reasons.

Such considerations, along with the many others to which reference has been made will call for the fullest planning and programming of any project; yet from an estimation of the possible needs of supply authorities by the time this type of project could be put into service it does appear to offer many distinct benefits as well as a more realistic case for harnessing tidal power.

REFERENCES

1. Ailleret, P., J. Institute of Fuel, 8, 353-356 (1966).

2. Heaps, N. S., Proc. I.C.E., 40, 495-509 (1968).

3. Shaw, T. L. and G. R. Thorpe, "Integration of Pumped-Storage with Tidal Power", (In Press).

4. Shaw, T. L. and S. W. Huntington, Water Power, 4 (1970).

5. Huntington, S. W., Shaw, T. L., Thorpe, G. R., and I. J. Westwood, "Optimized Functioning of Estuarial Storage Projects", (In Press).

PUMPED-STORAGE TIDAL POWER

K. E. Sorenson*

INTRODUCTION

In this paper, a method is presented whereby tidal energy, developed on pumped-storage principles, could be converted to on-peak use with only minor variations in daily, monthly, and annual capacity. In combination with underground pumped-storage and nuclear plants, large generating centers in tidal basins could produce electric energy at high daily load factors. The type of tidal development suggested could also bring many infrastructural benefits to the inner basin.

THE TIDAL POWER PROBLEM

Most of the large, interconnected, power systems of the world are expanding on the basis of fossil and nuclear fuelled generating plants in sizes of 1000 MW or more. These systems have or will soon have excess generating capacity during off-peak hours which can produce energy at a relatively low incremental cost. In the U.S., the practice of capitalizing nuclear fuels results in very low off-peak energy costs.

Any tidal generating scheme that would produce, for the larger systems, intermittent energy on a lunar cycle, and in variable daily amounts would be faced with a market that is overly supplied with cheap off-peak energy.

On the other hand, the capital costs of fossil and nuclear fuelled plants are escalating at an unexpectedly high annual rate. Also, the air pollution, radiation emissions, and heat generation of these plants are causing concern in many parts of the world. A tidal power development that would produce on-peak power and energy in uniform and dependable daily amounts would have a substantial capacity value and would generate the cleanest form of energy. This could be achieved through pumped-storage.

---

*Vice President and Chief Planning Engineer, Harza Engineering Company, Chicago, Illinois.

THE ROLE OF PUMPED-STORAGE

Undoubtedly, the readers are familiar with the conventional application of pumped-storage. These increasingly popular plants convert off-peak energy to on-peak power by pumping from lower reservoirs to upper reservoirs and by generating on the reverse cycle. For economy, most of these plants have relatively high heads and small reservoirs. They are negative energy producers, for which every ten kilowatt-hours expended in pumping only seven kilowatt-hours are regained in generation. Thus, they increase the total energy production from fossil and nuclear fuelled plants above the amounts required by the consumers.

Tidal power pumped-storage can only be at low head with large reservoirs. As single-purpose, negative energy producers, such developments would be uneconomical. However, as suggested in this paper, tidal power pumped-storage can have multi-purpose benefits, be a positive energy producer, and be an essential element in an integrated generating center.

In most major interconnected systems the various types of generation are usually stacked in the ascending order of incremental fuel costs. At the base are the nuclear plants, next are large fossil-fuelled plants followed by conventional pumped-storage, and finally by gas turbines for extreme peaks and reserve capacity. Because conventional pumped-storage is a negative energy producer, the total capacity is limited to the thermal generation available for pumping -- usually from ten to no more than fifteen percent of peak demand.

The large nuclear and coal-fired plants do not lend themselves readily to rapid changes in load. Conventional pumped storage and gas turbines do. Therefore, the latter types of plants are more and more being used for quick response.

The conventional pumped-storage plants have a unique capability of being able to store energy, and are in fact the largest artificial storage battery. This capability is being used in several utility systems to conserve otherwise unusable off-peak generation. Tidal power pumped-storage can serve a similar function. However, instead of conserving only man-made energy, this scheme could also conserve nature's off-peak energy for use when man needs electric power.

CONVERSION OF TIDAL ENERGY

Many schemes have been suggested for continuous generation from the tides, and for supplemental pumping into tidal basins. Dr. Bernstein, in his most valuable compendium, has given a thoroughly comprehensive review of the many possible developments*. The Rance Project has demonstrated that

---

*Tidal Energy for Electric Power Plants, Dr. L. B. Bernstein.

both generating and pumping capabilities can be combined in dual-function equipment.

An apparently attractive scheme for continuous or on-peak generation is the double-basin scheme. Some natural basins could lend themselves to this type of development; for example, Chignecto Bay. However, there are serious disadvantages to a double-basin plan. Only one-half of the area isolated from the ocean is available for daily power generation. The lowering of the upper basin concurrent with raising of the lower basin causes a reduction of head and even greater reduction in power output during the generating cycle. Also, the lower basin suffers from reduced volume of useful storage at the minimum level.

A greater use of the ocean barrier could be achieved if the total volume of water enclosed at the highest level were made available for on-peak power generation. To accomplish this, a modified form of the ebb-tide scheme is proposed.

TIDAL POWER PUMPED-STORAGE

A suggested scheme for tidal power pumped-storage is shown in the plan on Figure 1 and in the section on Figure 2.

Many functional and construction factors have influenced the elements shown for this scheme. The principal factors are:

1. Maximum use of the area enclosed by the ocean barrier, including the fathermost economical oceanward location and the highest feasible elevation of the inner basin.
2. Generation under nearly constant net head.
3. Power house with controlled headwater and tailwater.
4. Power house not exposed to ocean storms.
5. Power house constructed in calm water at constant level.

The major features of the scheme are described below.

The Ocean Barrier

This barrier must rise above extreme maximum tides plus wave effects. The gated spillway would be built first, preferably on a shelf or through a ridge that can be isolated from the currents of the basin for easier construction. During closure of the main barrier (earth and rock fill), the spillway would operate as a diversion channel to minimize the differential water levels within and outside the basin.

The spillway would have deep tainter gates designed for differential pressures in either direction, but with greatest pressure from the inner basin. Temporary navigation passage would be needed through a specially designed section of the spillway, to be used in slack water periods during construction.

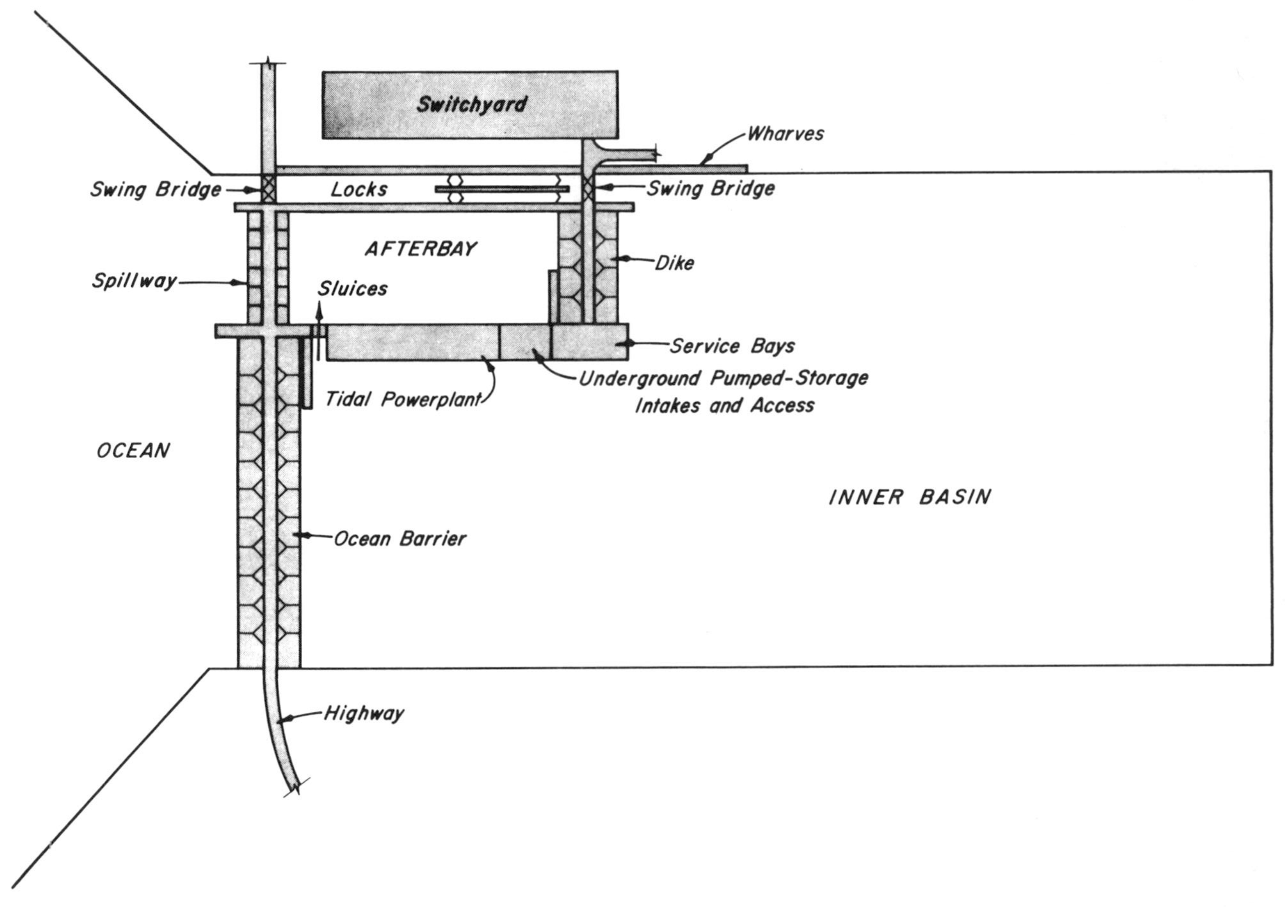

Fig. 1. Tidal power pumped-storage plan.

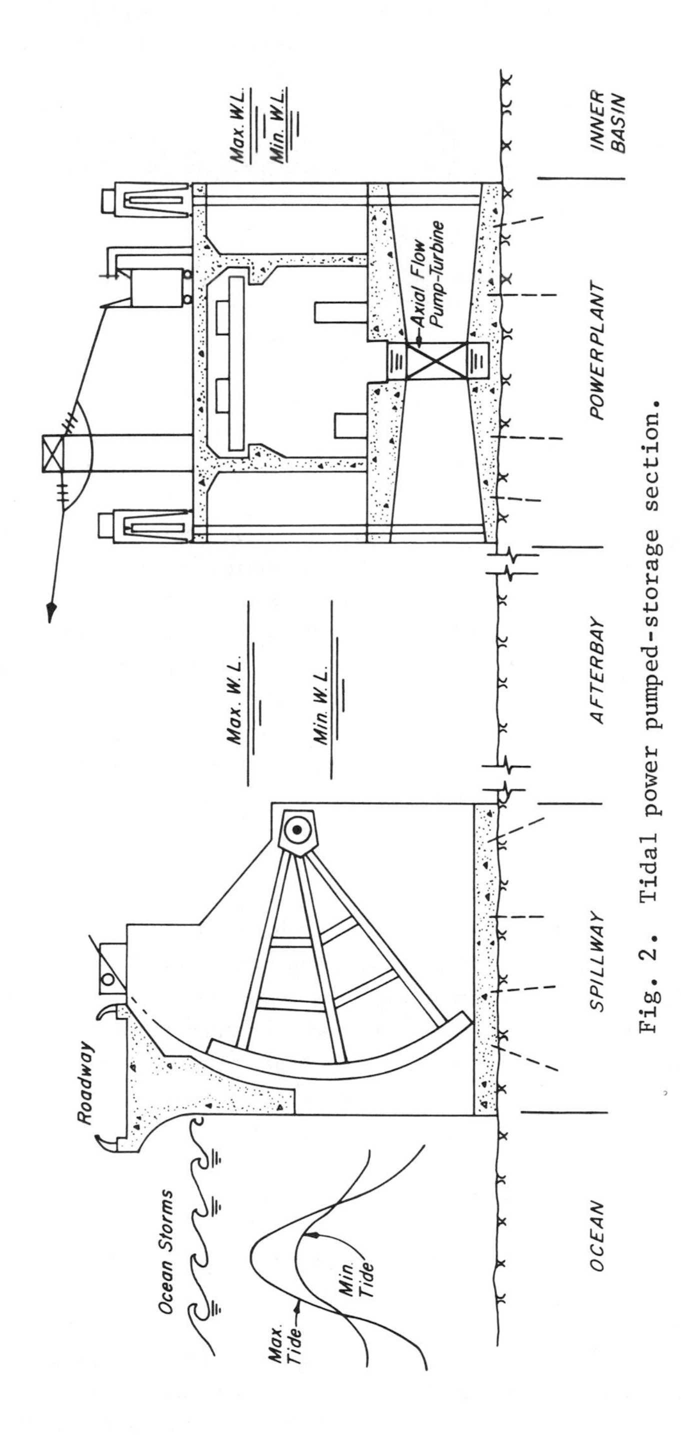

Fig. 2. Tidal power pumped-storage section.

When the barrier is completed, the spillway gates could be closed during a medium tide to permit construction of the powerhouse and permanent locks in calm and non-fluctuating water inside the basin.

Construction of the barrier before commencing the power plant implies a delay in tidal power benefits. This could be offset by earlier transit across a major estuary and by other power installations using the tidal basin.

The Tidal Power Plant

With controlled headwater and tailwater, and without the extreme fluctuations and storm impact of the ocean, the tidal power plant could have a higher draft tube setting and a lower deck level. Construction behind the tidal barrier would permit cofferdamming or 24-hour caisson placement.

Use of axial-flow, rim-generator, pump turbines would shorten the longitudinal length of the power plant, its upstream-downstream width, and the cost of machinery.

Equalizing sluices attached to the power house, together with use of the spillway gates could quickly fill the relatively small afterbay during reversal of operation between generation and pumping. Individual service gates at the pump-generating units could be eliminated. Only maintenance bulkheads and gantry-handled emergency gates for multiples of units might be required.

While generation would be at nearly constant head, pumping heads would be less, ranging from positive to negative static values. Variable pitch pump-turbine blades would most likely be required. To achieve the necessary pumping capacities, dual speed motor-generators might be used.

The Permanent Navigation Locks

Any tidal basin that has present or potential ocean traffic should be improved in its use with the proposed tidal power development. Existing and new wharves would have more constantly deep water levels. However, access from and egress to the ocean must be provided. Permanent locks, as well as power plants, are difficult to build in the face of rapid currents and fluctuating water levels. Under the proposed scheme, permanent locks could be built behind the ocean barrier, to whatever size is needed.

Power and Energy Generation

The generating capabilities and the pumping requirements of the suggested tidal power scheme during minimum tides are shown on Figures 3 and 4, for two different conditions.

Inner basin levels would be controlled by the releases through the power plant during the generating cycle and by the pumping capacities of the units in their reverse mode. During periods of neither generation nor pumping, the

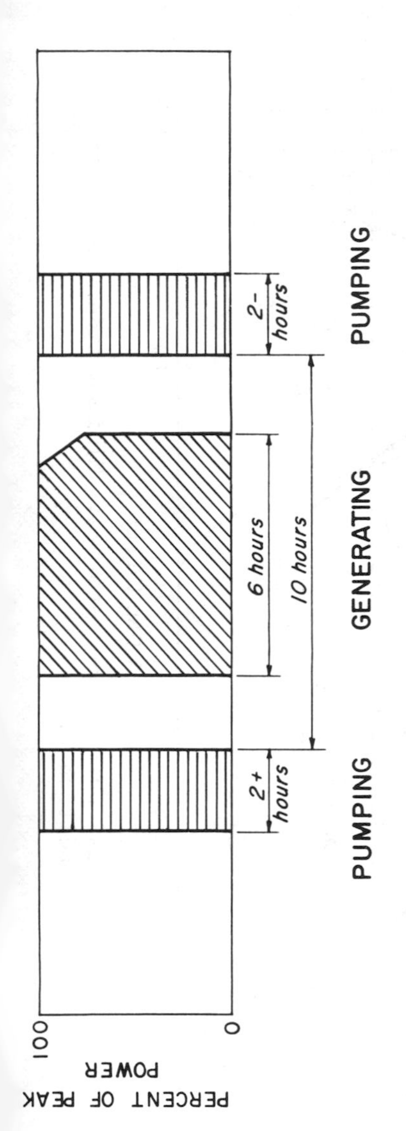

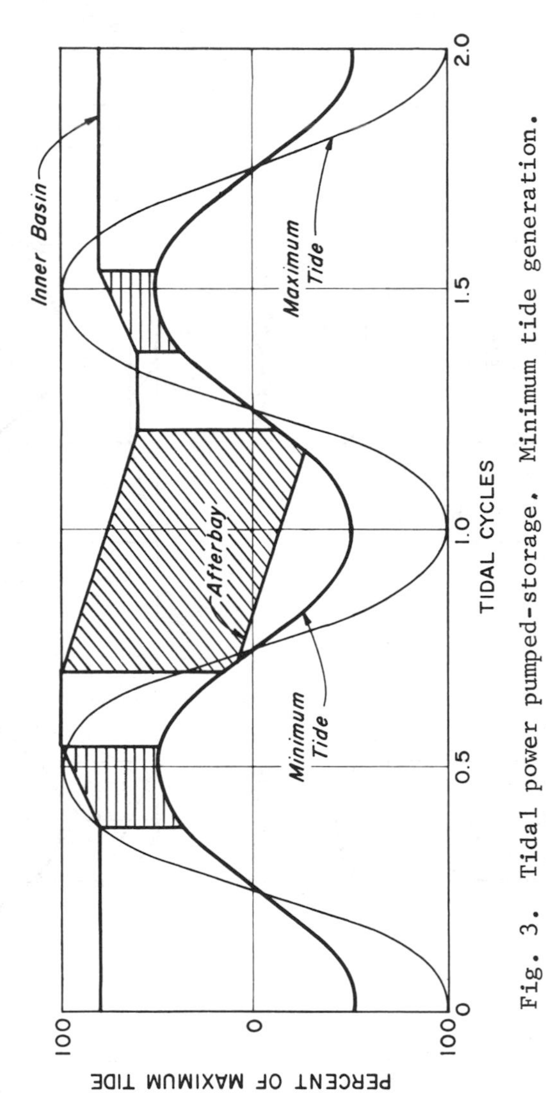

Fig. 3. Tidal power pumped-storage. Minimum tide generation. Condition A.

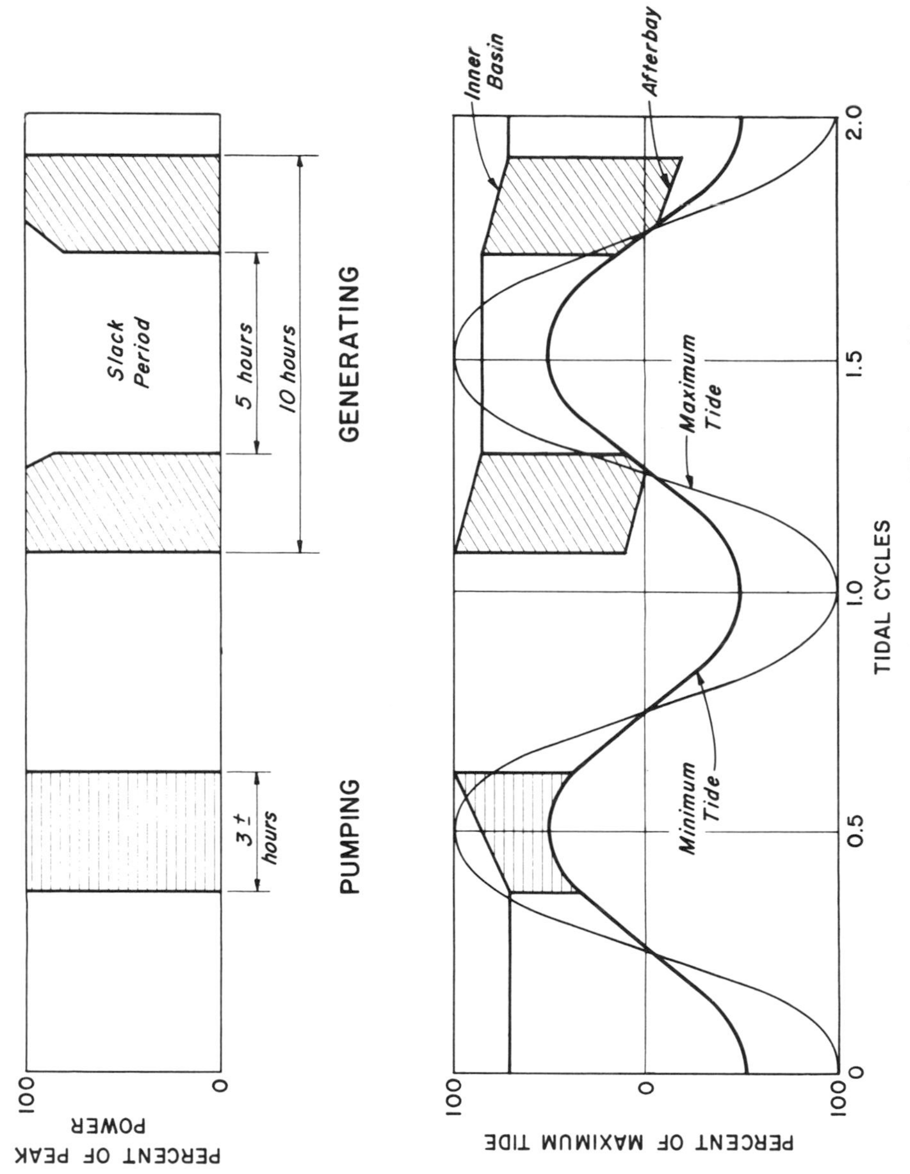

Fig. 4. Tidal power pumped-storage. Minimum tide generation. Condition B.

level of the inner basin would be held static.

Oceanside levels on the power plant would be controlled by the spillway tainter gates. During minimum tide levels, these gates would be partially closed to maintain a constant net head on the units when generating, or fully closed with the afterbay filled when no operation is underway. During high tide levels, the gates would be wide open when pumping, or fully closed with the afterbay filled whenever high tides coincide with on-peak hours of daily generation.

The method of operation shown on Figures 3 and 4 would permit five to six hours of full tidal generation at almost constant capacity within any ten hour on-peak period of a calendar day. Pumping to refill the inner basin could be restricted to off-peak periods of the same calendar day.

Pumping energy input through the tidal plant would be less than the generating output. Minimum tides would require the greatest input, and extreme high tides the least input. At any tide level, the peak motor capacity in the pumping cycle could approximate the generator capacity, but the daily hours of pumping would vary with tidal magnitudes.

The most significant aspect of this tidal power pumped-storage scheme is its independence of tidal variations during the generating cycle. Extreme high or low tides would not change the generating capacity appreciably, nor affect the operation of the power plant.

PUMPED STORAGE COMBINATIONS

The four to five hour gap in tidal power pumped-storage generation must be filled with low cost power to achieve an economic product. Underground pumped-storage projects offer the most favorable combination.

These underground plants have difficulty in competing economically when ten to twelve hours of daily storage are required. The fewer hours of storage needed, the more economical they become. In combination with tidal power, underground pumped-storage can be a most valuable increment as well as a complementary adjunct to underground nuclear installations.

As shown on Figure 5, intakes and outlets of the underground pumped-storage plants can be incorporated as an extension of the tidal power plant. Depths to the pump-turbines, either single step or cascade, would depend upon site geology and the economics of power generation. Water could be withdrawn from the afterbay and discharged to the inner basin, the reverse, or withdrawn and discharged at the same side.

The combined generating output of tidal power pumped-storage and underground pumped-storage would permit ten hours of full generation during any desired period of a calendar day, regardless of the tidal cycle or magnitude. The underground pumped-storage plant would need between five and six hours of generation at a capacity equal to the peak tidal power output.

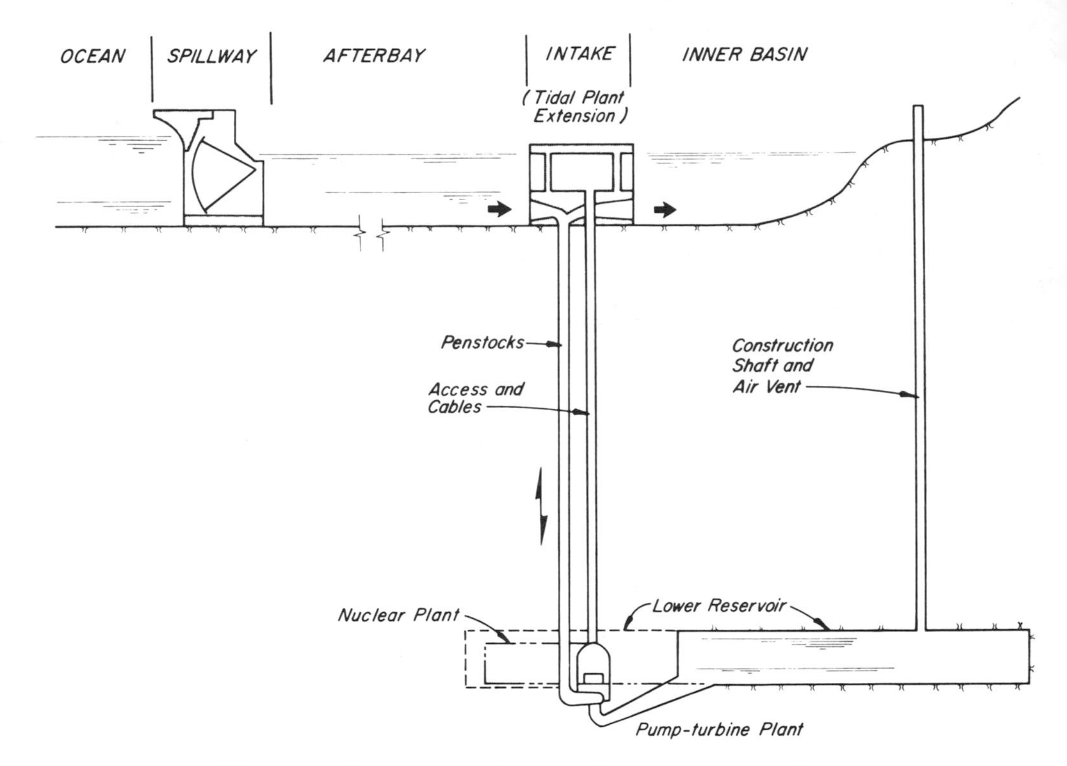

Fig. 5. Underground pumped-storage section.

An underground pumped-storage project, without tidal power, would need very expensive weekly storage to generate ten hours daily, and the availability of off-peak pumping energy could severely limit the installed generating capacity.

TIDAL BASIN POWER CENTER

Power output from the tidal and underground plants must be combined with base load generation for system use. This base load generation could be near the load centers or at the tidal basin. Transmission economics are an important factor, but other considerations could weigh heavily in favor of location at the tidal basin.

Large amounts of cooling water from acceptable sources are not always available near load centers. Opposition is mounting against surface siting of nuclear plants in urban areas.

Tidal basins by their nature offer large volumes of water with daily replenishment, and usually at low temperatures. Major basins, such as in the Bay of Fundy, could provide locations for nuclear plants at considerable distance from permanently inhabited shoreline communities.

Nuclear plants, if incorporated into a tidal development, could be conventionally designed for surface construction. However, significant advantages could be achieved if the nuclear plants were incorporated into the underground pumped-storage plants. Some of these advantages are:

1. Removal from atmospheric disturbances, such as hurricanes; and from accidental impact, such as falling airplanes.
2. Massive natural protection against radiation emissions.
3. Minimum exposure to sabotage.
4. Easier isolation during accidents or future decommissioning.
5. Common use of facilities with the pumped-storage plant such as access cranes, control room, cable ducts, auxiliary equipment, etc.

The combination of nuclear plants with underground pumped-storage is shown in section on Figure 5, and in lower level plan on Figure 6.

Cooling water would be circulated through the underground pumped-storage reservoir. Depending upon depth, nuclear capacity equal to four times the pumped-storage generation could be installed with no more than a double circulation of cooling water from the lower reservoir. A two-times installation would use the cooling water only once.

The inner basin of any significant tidal development would be large. In terms of daily discharges, nuclear cooling water would be much less than the discharge through the tidal units. Nevertheless, pumping into the inner basin from the underground reservoir would result in a mixing of cooler and warmer water. Over a period of years, the inner basin could be warmed a few degrees. An ecological study might show this to be a benefit to fish and

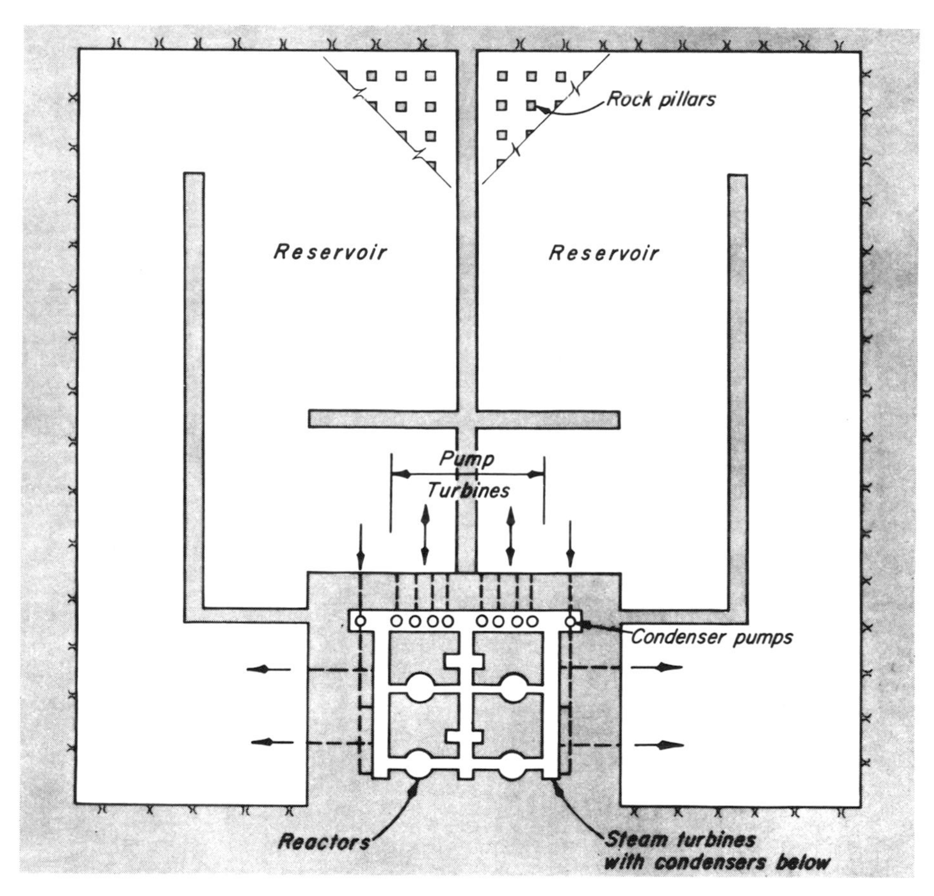

Fig. 6. Pumped storage-nuclear plants. Lower level plan.

wildlife, as well as to inner basin shipping.

Two possible combinations of power output from a tidal basin power center are shown on Figures 7 and 8. With 80 percent base load installation and a minimum of ten hours peak generation, the combinations could produce daily load factors of at least 78 percent.

To minimize transmission costs and achieve quick response to load changes, extreme system peaks and hourly variations in load should probably be carried by plants near the metropolitan centers. The tidal basin power center could give the bulk supply at pre-scheduled block loadings.

MINAS BASIN APPLICATION

Conditions exist at Minas Basin that are generally favorable for the type of tidal basin power center described above. Within precedented transmission distance, there are very large power markets. Physical conditions at the basin would permit a large tidal power development as well as underground construction. In addition, the potential benefits to the inner basin from a tidal development are great.

Power Markets

Within a 1000-mile radius of Minas Basin, present electric power use exceeds 50 million kilowatts. The projected load growth of this area is shown on Figure 9. It is estimated that between 1980 and 2000 the increase in power demand will be over 180 million kilowatts. The siting of generating plants, mostly fossil or nuclear fuelled, presents a formidable problem.

A tidal basin power center in Minas Basin of up to 25 million kilowatts would represent only a fraction of the projected power needs within practical transmission distance.

Interest rates are high at present. However, interest costs affect all types and sources of generation nearly equally, perhaps excepting gas turbines. Most important to the power industry is the identification of new power sources that do not raise exceptional opposition nor entail uncertain delays.

Minas Basin Site Advantages

A barrier across Minas Basin at or near Cape Split, with nine feet of inner basin fluctuation, could support five million kilowatts of tidal power. The tides are high, of sinusoidal shape, and of nearly equal semi-diurnal magnitude. Good quality construction materials are available within reasonable distance. Surface geologic conditions indicate that suitable rock for deep underground construction probably exists near Cape Split.

Minas Basin offers large quantities of cold water for nuclear fuelled

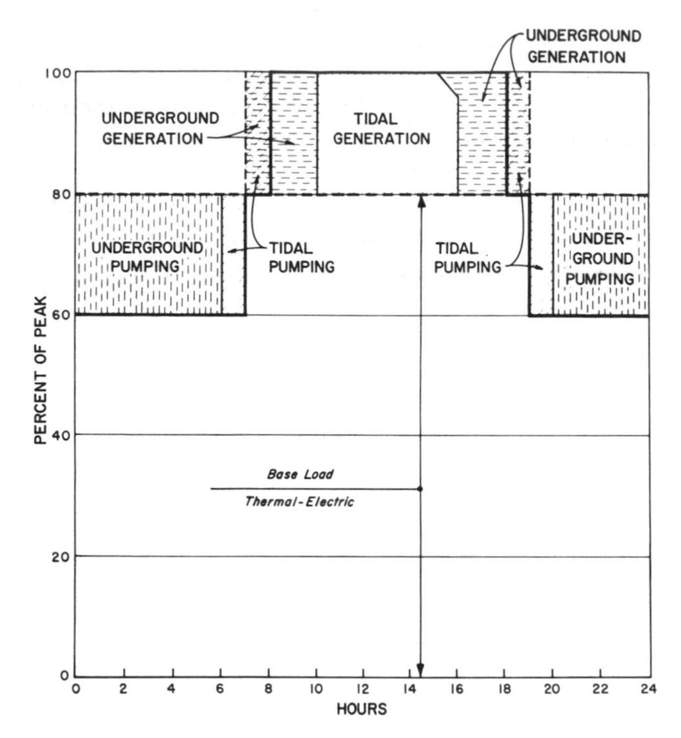

Fig. 7. Tidal basin power center. Minimum tide generation. Condition A. 78% daily load factor to system

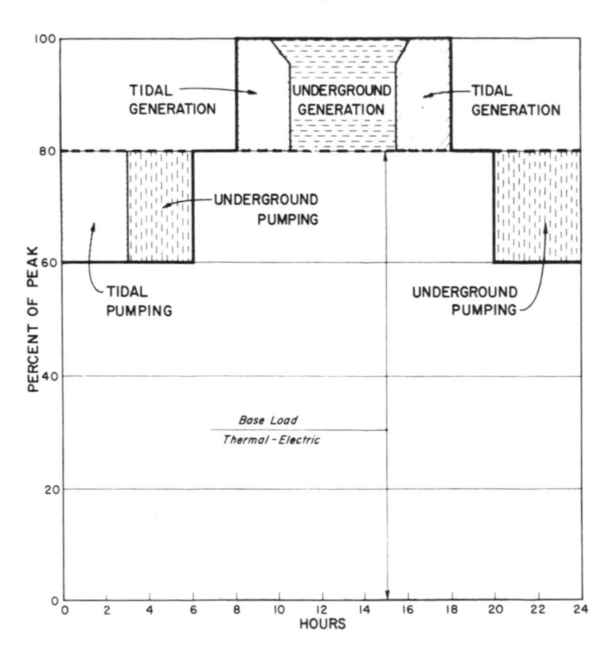

Fig. 8. Tidal basin power center. Minimum tide generation. Condition B. 80% daily load factor to system.

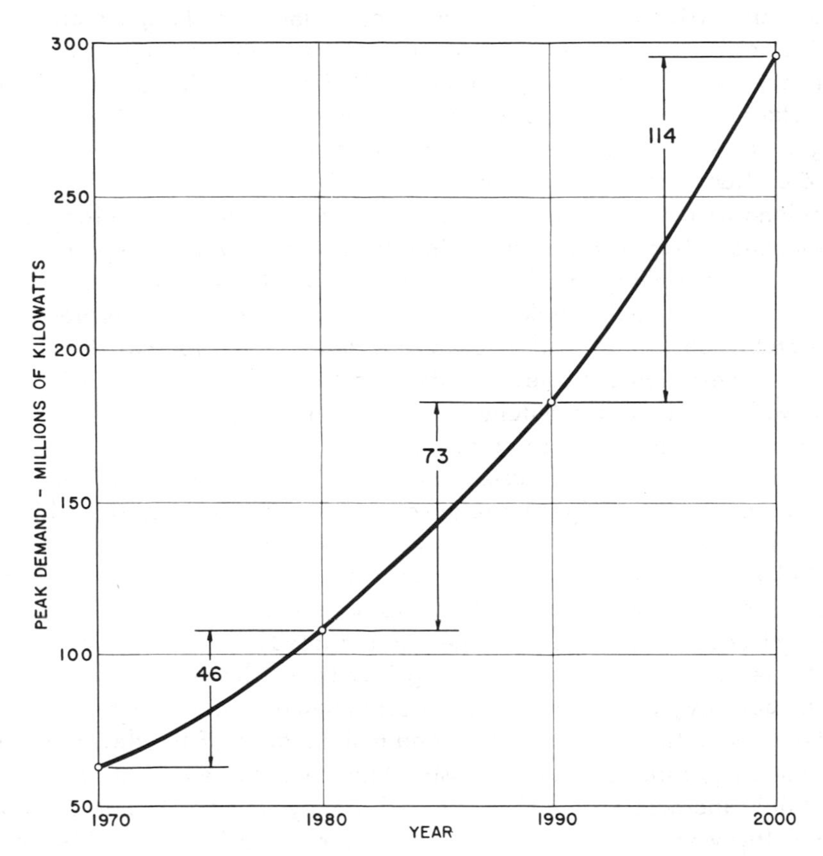

Fig. 9. Projected peak power demand within 1000 miles of Minas Basin.

power plants. While this is also true of the New England and Bay of Fundy coast, the tidal barrier would make this water available within a basin protected from extreme tidal variations and ocean storms.

Inner Basin Benefits

Many substantial benefits, other than power generation, would accrue to Minas Basin with the proposed scheme.

The tidal barrier would offer a causeway connecting the north and south banks of the basin near its mouth, obviating a long detour through Truro. Important ports such as Hantsport, Parrsboro, Walton and Windsor would have nearly constant high water levels twenty-four hours a day for loading and unloading of ships. Navigation within the basin would be less dangerous and unimpeded by tidal levels and currents.

Farmland within the basin must now be protected by dikes which have been overtopped during rare tide and wind conditions. While the proposed scheme would create high inner basin levels, the maximum level would be controlled, and dikes could be designed with confidence in their protection.

The effects on fish and wildlife would require an ecological study. However, the controlled extremes of water level could permit the establishment of spawning grounds and hatcheries that might make this salt water lake a great commercial and sport fishing area.

Municipal and industrial wastes would receive lesser flushing action and treatment of effluents might be required. The costs incurred should be more than offset by the increased value of waterfront property.

Last, but not least, the economic and financial benefits to the Minas Basin and the Province of Nova Scotia could be tremendous.

A major development, such as the one proposed, would imply ten years or more of intensive construction, creating jobs for all levels of skill, accelerated business activity, and a demand for housing and personal services.

After construction, the operation and maintenance of the development, improved shipping, other uses of the basin, and more stable agriculture would certainly increase the present level of employment and business activity.

No matter what sources of financing are used, the Province of Nova Scotia should gradually acquire equity of some degree in a revenue producing project. This, coupled with a greater tax base, could only improve the total income of the Province.

SUMMARY

Tidal power projects by themselves, have little chance of being economical as part of large utility systems. However, in combination with other types of power generation and as part of multipurpose developments they can be an essential element in a unique and valuable use of natural resources.

Obviously, the basic concepts presented in this paper would need extensive study for their application to a particular basin. The Minas Basin, in Nova Scotia, does appear to have many favorable conditions for such a development.

THE TOTAL CONTRIBUTION OF TIDAL ENERGY TO THE SYSTEM

J. G. Warnock* and J. A. M. Wilson**

INTRODUCTION

The economic growth of Canada and its impact on the total picture in North America will be dependent, to a great extent, on the efficiency with which our vast natural resources are employed in the future.

Raw materials and energy resources are widely dispersed throughout the country. The mineral wealth of the Canadian Shield, the forest reserves of Eastern and Western Canada, the power potential and the agricultural and industrial value of our natural water resources, the coal fields of the Maritimes and Western Canada, the petroleum and natural gas fields of the Prairie Provinces, and the uranium deposits of Northern Ontario; all contribute handsomely to the wealth of the Nation.

All this is provided in a portion of the global surface where so far the density of activity has produced a minimum of environmental problems. In fact, Canada can still be viewed in total as a source beneficial to the natural vital needs of humanity. We have adequate resources which are envied by others.

In our resource inventory must be included tidal power -- a source of energy brought closer to reality by advancing technology, by the growing capability of the system to absorb it, and by its minimal environmental effect.

The large blocks of energy and capacity available from our potential tidal power sites cannot be lightly overlooked in the critically demanding atmosphere of the Eastern Seaboard of the USA and Canada. The need for a methodical reassessment of the resources available has been recognized and fulfilled by the studies undertaken by the Atlantic Tidal Power Programming Board.

The conclusion that "economic development of tidal power in the Bay of Fundy is not feasible under prevailing circumstances" is presented together

---

*Vice-President, Acres Limited, 20 Victoria St., Toronto, Ontario, Canada.
**Executive Engineer, H.G. Acres Ltd., Niagara Falls, Ontario, Canada.

with the Board's recommendation that further detailed studies be authorized when (a) the interest rate on money drops sufficiently to suggest the possibility of an economic tidal power development in the Bay of Fundy; (b) a major breakthrough in construction costs or in the cost of generating equipment suggests the possibility of designing an economic tidal power development in the Bay of Fundy; (c) pollution abatement requirements magnify, substantially, the cost of using alternative sources of power; or (d) alternative sources of a more economic power supply become exhausted [1].

Despite these findings which are based on an entirely rational assessment of the Bay of Fundy tidal power resources, we incline to the stubborn belief that this source of continuously renewable, pollution-free energy should be developed to meet the ever-growing energy needs of the Maritimes and the Northeastern United States and provide an escape from the powerful restraints being applied to the other "more economic" sources.

While recent studies indicate that tidal power does not appear, at present, to compete on a cost per kWh basis with large nuclear or conventional thermal power plants, pollution control legislation now contemplated and under active consideration may substantially inflate the capital and operating cost of these plants to the point where pollution-free tidal power can provide a competitive alternative.

The question is then before us -- what should we be doing about this issue -- NOW! It certainly appears that the future may hold a distinct time span during which tidal energy resources would have their place in filling a need -- certainly until alternatives both economic <u>and</u> totally beneficial to man's environment are available.

THE PERSPECTIVE TIDAL POWER POTENTIAL RELATED TO THE MARITIME AND NEW ENGLAND SYSTEMS

In order to put tidal power from the Bay of Fundy into perspective, we might consider the forecast power situation in the Maritimes and New England in the year 1990, twenty years from now.

System Demand

Considering firstly the forecast peak demand, recently published estimates are as follows: [1, 2] (Fig. 1).

	1968 (actual)	1980	1990
Maritime Power Pool	1,161 MW	2,700 MW	5,500 MW
New England	10,045 MW	22,000 MW	41,000 MW
Total	11,206 MW	24,700 MW	46,500 MW

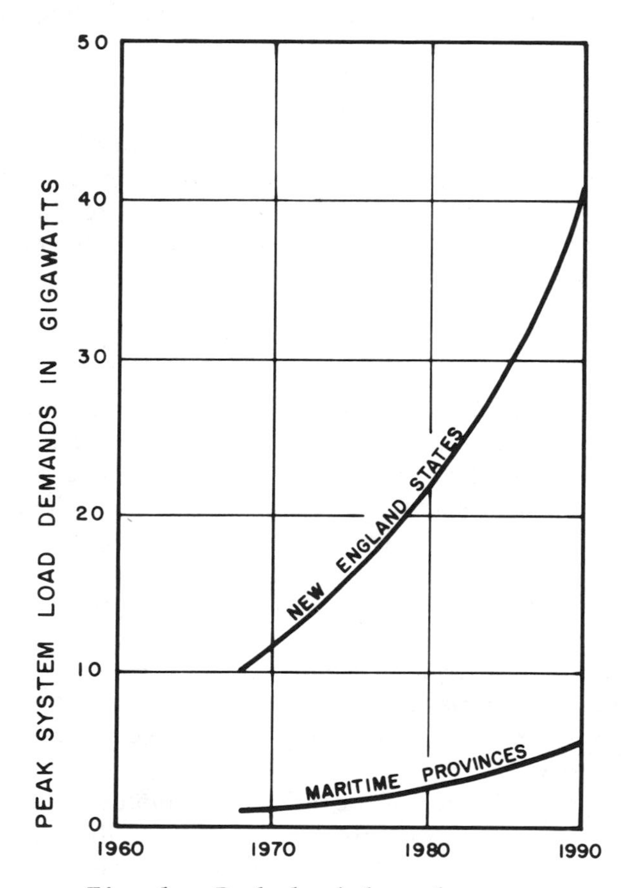

Fig. 1. Peak load demand curves.

Relationship to Tidal Potential

The study initiated in 1967 of potential sites in the Chignecto Bay and Minas Basin areas at the head of the Bay of Fundy led to the selection of a short list of sites at which installed capacities of from 2,000 to 3,500 MW were feasible for single-effect and double-effect operation, respectively. It should be noted that these sites were selected as giving potentially the lowest cost per kWh, and not the largest output. Single basin schemes based on the three sites identified in the A.T.P.P.B. Report as offering the best possibilities for economic power development represent a potential total installed capacity up to 9,000 MW. Other sites permitting an installed capacity of up to 14,500 MW have been identified [(3)].

It is evident therefore that tidal power when introduced to an interconnected system capable of absorbing the deviations of the lunar cycle from daily load variations could make a substantial contribution to the combined installed capacity required in 1990 and more particularly to the growth of 21,800 MW from 1980 to 1990.

Relationship to Conventional Sources of Power Generation

Current planning indicates that the 1990 demand may be met by existing generation facilities plus 25,000 MW of nuclear plants in 8 to 10 locations, 10,000 MW of coal-and oil-fired plants in 7 to 8 locations and 4,000 MW of pumped-storage hydroelectric plants in 2 locations in New England.

The process of locating such plants will meet varying degrees of resistance -- 17 to 20 public confrontations for utilities. No doubt there will be growing pressures for complete preservation of environmental qualities. This can only be achieved at substantial costs.

Applying coarse estimates and some crystal ball licence, we suggest that a 50,000 MW system in this region carries penalties of cost to preserve the environment -- penalties which gradually become heavier and heavier the larger the capacity grows.

Figure 2 gives an indication of the order of magnitude of "penalty" which the generating plant additions may have to carry as the cost of meeting ecological environmental requirements. We suggest that these are minimum values and in all probability are likely to be exceeded in the future when more sophisticated and expensive equipment and facilities, capable of improving emissions and effects, become available.

A nominal value of 1 percent of capital cost has been attributed to tidal power to meet the cost of environmental controls. The differential between this and the value of 33 percent of the capital cost of the final 5,000 MW installed to bring the system capacity up to 50,000 MW goes a substantial way towards meeting the "indicated" higher cost of tidal power.

If tidal power were adopted for this 5,000 MW block of capacity, it has

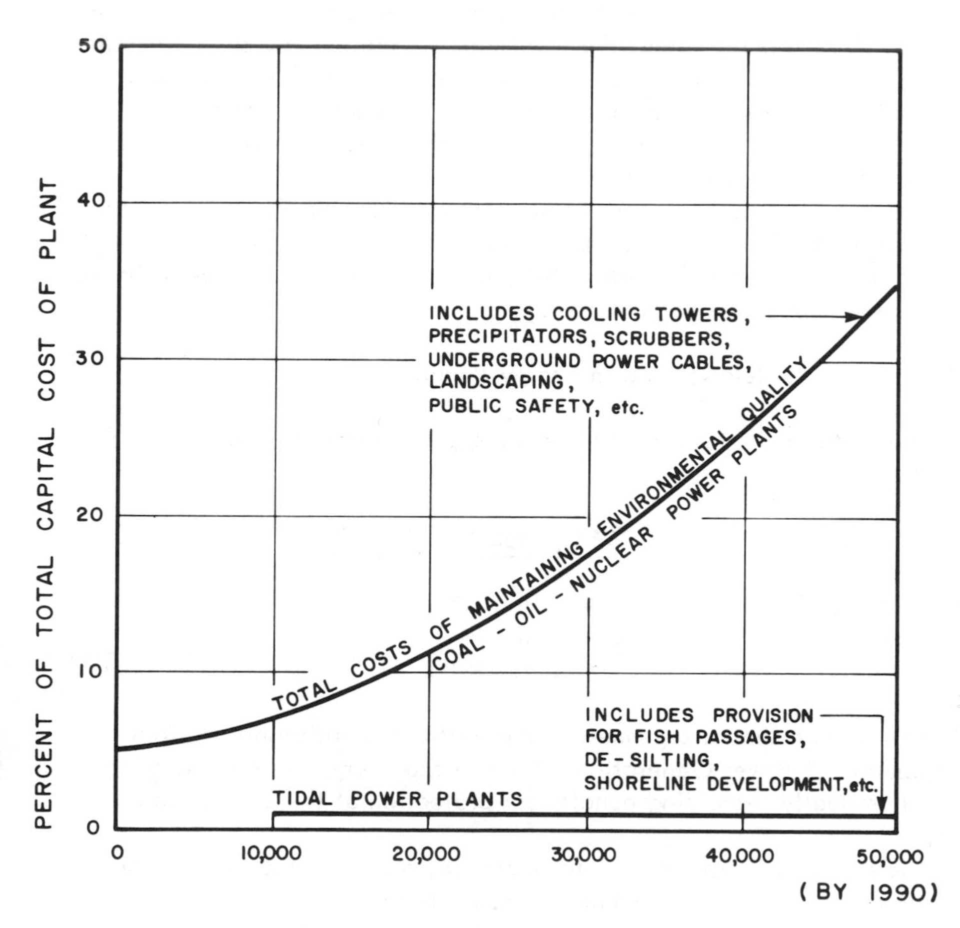

Fig. 2. System capacity - megawatts.

been estimated that two double-effect schemes at the A.T.P.P.B. sites on Shepody Bay and Cumberland Basin (Sites 7.1 and 7.2 respectively), could contribute 5,292 MW of capacity, of which 2,282 MW is considered firm peak capability available 95 percent of the time, two hours per day, 60 days per year, at an estimated total capital cost of $1,360,000,000 [1].

We suggest that the cost of an equivalent 5,300 MW thermal plant added to bring the system to 50,000 MW including a penalty of 33 percent of capital cost for environmental protection, might be as high as $160 per kW, of which $40 per kW represents the investment to protect the environmental qualities of the region. This gives a capital investment of $848 million, including $212 million "write-off" to preserve our ecology. Even after expending this amount, we still would not have total protection and "hidden loss" to environment and economy no doubt would remain.

Certainly the added cost of $500 million to provide the virtually complete protection afforded by tidal power is too great a penalty to be accepted but we feel that it is most important that tidal power costs are viewed in the proper perspective of the total system.

Relationship to Energy Requirements

Energy forecasts for the corresponding periods are at present: [1, 2] (Fig. 3).

	1968 (actual)	1980	1990
Maritime Power Pool	6.4 billion kWh	15	31
New England	52.3 billion kWh	110	210
Combined Total	58.7 billion kWh	125	241 billion kWh

Following a similar argument to the section "Relationship to Conventional Sources of Power Generation", we see the energy demand carrying with it a gradually increasing penalty for protection of the environmental quality.

The energy contribution of, for example, a proposed 2,000 MW single-effect tidal plant has been estimated at 6.5 billion kWh/yr.

The three preferred sites selected in the A.T.P.P.B. Report could provide a total of 13 billion kWh/yr. [1] while the total tidal power potential of the entire Bay of Fundy has been estimated at 175 billion kWh/yr. [4].

From this it can be seen that the total tidal power plant contribution even from a number of sites to the total energy production of the area, due to its cyclic characteristic, is more modest than its contribution to the installed capacity, although large in relation to the Maritime system. We do not therefore, attempt to evaluate the saving in "penalty" for environmental protection.

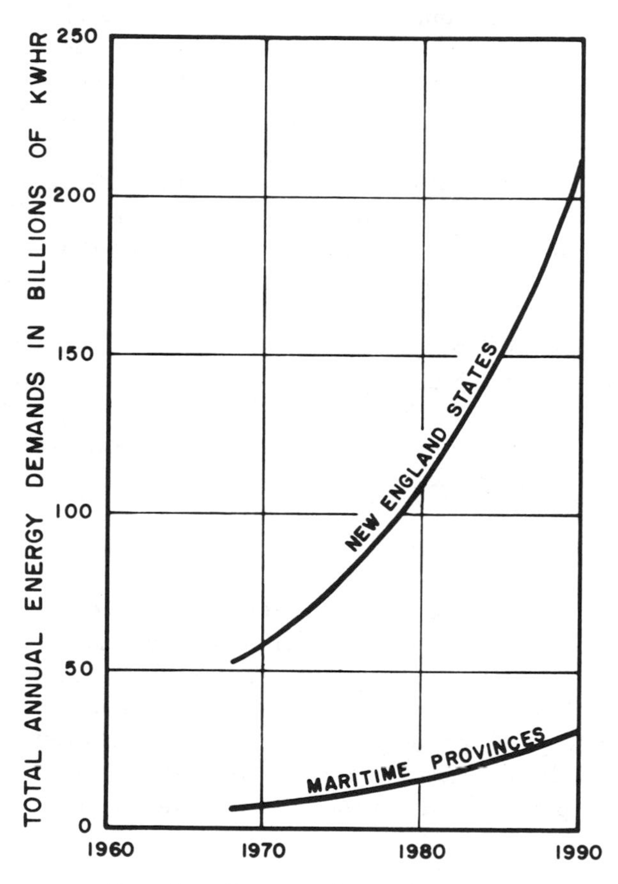

Fig. 3. Energy demand curves.

COMPARATIVE COSTS

Operating Costs (with interest charges)

The energy costs for an optimized hydroelectric single-effect tidal power plant of 2,000 MW has been estimated at 5.6 mills/kWh [1] while the equivalent incremental energy cost of conventional thermal power today is taken as being of the order of 3 to 4 mills/kWh depending on plant size and fuel cost. There is at present an upward trend in these figures and several authorities are now quoting figures of 5 mills/kWh. Nuclear plants by the 1980's may lower this figure but expectations for such cost reduction have, to date, not been realized and energy costs of less than 3 mills/kWh are certainly not assured.

On the basis of energy cost alone, and assuming 3.5 mills/kWh, the penalty for using tidal power in this instance is approximately 2 mills/kWh or $13 million per annum. This should be viewed in relationship to the costs involved in providing thermal stations with the means for protecting environmental quality.

Intangible benefits of tidal power which nevertheless may have a long-term influence on the operating costs are its freedom from inflationary pressures on fuel costs, and the predictability and absolute dependability of the energy source. The movement and height of the tide can be accurately forecast and tides are not subject to unexpected floods, droughts or ice jams.

Capital Costs

The capital cost of a 2,000 MW single-effect tidal plant has been estimated at $475 million while the equivalent capacity provided by a conventional thermal station with predicted costs for a high degree of environmental quality control is of the order of $320 million. The calculation of the capital cost of nuclear plant for the year 1990 is highly speculative, and there is still doubt that it will ever match that of conventional thermal. The penalty in capital cost of using tidal power in this case is therefore approximately $155 million.

THE OPTIONS

It is clear from these figures that on the basis of costs alone, assessed by our present standards, tidal power is scarcely competitive. The question now raised is whether the addition of a pollution-free element of capacity of 2,000 MW is worth $13 million per year to the overall economy of the region.

In looking at the alternatives, the closest to the near perfection of the tidal power projects from the overall pollution standpoint is the nuclear power plant. For the purpose of this survey, we have assumed that the cost per kW installed and the cost per kWh of nuclear plant and conventional thermal plant coincide in 1990.

Nuclear power plants, with proper safety precautions, do not cause atmospheric pollution to any appreciable degree, although appreciable cost may often be incurred to find an acceptable site at which the public will accept even limited hazards, nor do nuclear power plants pollute the cooling water in the normal sense. However, because of their inferior thermal cycle efficiency, they do release larger quantities of heat per kWh generated to the cooling water than conventional plants. Specifically, at the present state of the art, a typical figure for conventional thermal plant rejection is 3,900 Btu/kWh while nuclear plants may reject 6,700 Btu/kWh.

To provide our hypothetical tidal power plant's annual output of 6.500 million kWh by nuclear means signifies a rejection or 44,000 billion Btu per year, or an average of 5 billion Btu per hour. This figure is similar to the present average heat rejection to Lake Ontario in the Toronto area from conventional thermal plants and cannot be disregarded (5).

Heat rejection of this magnitude virtually precludes the use of river water cooling and compels the siting of the nuclear power plants to the sea coast. While it can be argued that this rejection to the Atlantic Ocean is of negligible importance, the configuration of the Maritime and New England coast with its deep inlets and landlocked bays must be taken into account and the influence of such heat rejection locally on fishing and marine life is cause for considerable concern.

The recent examples of thermal pollution of Biscayne Bay in Florida which created in less than two years what has been described as "a barren lifeless zone" in the area of the cooling water discharge from the Turkey Point generating station (6) and the incident in the Cape Cod Canal in the summer of 1968 when a large number of fish were trapped in effluent water at a temperature of 93 to 95 degrees (7) constitute eloquent warnings that thermal pollution is not confined to sluggish inland rivers and lakes but can imperil marine life on the Atlantic seaboard.

CONCLUSIONS

Public concern with environmental pollution has become deep and widespread in the last few years and is manifesting itself in bitter and often blind opposition to any form of technical progress which threatens to increase noise, atmospheric pollution, water pollution, to affect the aesthetics of our surroundings, or to encroach upon undeveloped or agricultural land.

This wave of public consciousness reflected by the elected representatives has already forced a number of hard-pressed utilities to abandon some of their long-range planning and substitute gas-turbine plant for hydroelectric projects for which approval has been long delayed or withheld.

At this time, it is hard to foresee the depth of public feeling on environmental preservation which is likely to prevail in the 1990's, but it is clear that in the absence of major hydroelectric potential in the area, the bulk of the generating capacity to meet the forecast increase in demand of 35,000 MW in the Maritimes and New England by 1990 will be met by an ever increasing number of sea-water cooled nuclear, coal, or oil-fired thermal stations.

In these circumstances, the question in the 1980's may not be whether tidal power can compete economically, but rather whether any alternative to tidal power can be developed to compete with its freedom from any form of environmental pollution. In view of the rapidly growing public pressures seeking assurance that technology is being applied to the ultimate total benefit of humanity, there seems to be a need to look beyond the strict economics of alternative power sources.

While it must be conceded that tidal power will not be entirely without effect on the ecological life of the bays and estuaries within its area of influence, these effects are minimal and far less damaging than the demonstrated damage to marine life caused by other forms of power generation.

The total study effort made to date on the Bay of Fundy tidal power potential, although well conceived and effectively executed, is but a fraction of what could be justified to find some relief from the ever-tightening restraints being applied to power system growth. We have a firm starting point for more determinate studies. It would appear reasonable to expect that each step now taken to acquire further precise knowledge by more detailed study would erode the cost differential and bring tidal power into a more attractive perspective.

Perhaps only then will tidal power come to be recognized, as a benevolent eternal source of energy offering a unique freedom from ecological hazards, to the ultimate benefit of all forms of life on this planet.

REFERENCES

1. "Tidal Power Development in the Bay of Fundy", Report of the Atlantic Tidal Power Programming Board (1969).

2. Load and Capacity Report, Northeast Power Co-ordinating Council (1969).

3. "Tidal Power Studies Chignecto Bay", Report of the Atlantic Tidal Power Programming Board (1969).

4. "Canadian Energy Prospects to 1985, Energy Sources for Water Power", Fournier, R., Can. Inst. MME (1965).

5. "Thermal Inputs to the Great Lakes 1968-2000", Can. Dept. EM & R (1970).

6. "Environment", Time Magazine (1970).

7. "Thermal Pollution and Aquatic Life", Scientific American, 220 (1969).

INTEGRATION OF TIDAL ENERGY INTO PUBLIC ELECTRICITY SUPPLY

E. M. Wilson* and B. Severn**

INTRODUCTION

Arrangements of dams, sluices and turbines, and principles of operation for various simple or complex power generation cycles have been extensively described (1, 2, 3). In these introductory paragraphs it is sufficient to summarize the most significant features of those few systems which at present offer any prospect of commercial importance.

Single-pool arrangements can find fairly wide application among natural bays and inlets. They may be of the simplest type where power is generated during the same part of each tidal cycle by flow in one direction only, or of the reversible-flow type possibly incorporating pumping capability. There can be advantage in electrically interconnecting two or more power plants of the single-pool type, if they are arranged to generate at different times. By the use of adjacent pools hydraulically linked, power can be made available at any time, or continuously. Such systems require unusually suitable topography, if the length and cost of dams are not to be prohibitive.

Selection of the appropriate type and scale of tidal power installations for any particular instance depends upon the relative values which may be set upon various characteristics of the generated output, notably the total amounts of energy produced, and the amount and availability of firm power. Each of these characteristics influences the operation of all the other components in the system, including the volume and unit cost of energy produced at each plant. Accordingly, the tidal power scheme cannot satisfactorily be designed independently: it must be tailored to suit the rest of the particular system in which it is to work, so that total system costs may be minimized. In other words,

---

*Dept. of Civil Engineering, University of Salford, Salford 5, Lancs, England.
**Balfour, Beatty & Co., Limited, Randolph House, Croydon, CR9 3QD, England.

the planning of a tidal power project is to be viewed as one element in total system planning involving the interaction of all elements present and in prospect.

SINGLE-POOL, ONE-WAY SCHEMES

The most convenient standard for comparison among tidal power systems is that which is simplest and most easily understood: the single-pool, ebb-generation system. Water admitted to the pool while the tide is high, is after an interval discharged through turbines to generate electricity. In order to generate most of the energy theoretically available, the pool should be emptied to nearly low water level at about the time of low tide, and almost completely refilled by the next high tide. However, such rapid discharge and full replenishment imply a very large turbine installation and very large sluice capacity, such that each turbine or sluice would only perform a fraction of the duty each would perform if their numbers were much fewer. It is thus necessary to seek an economic compromise exploiting only that part of the total potential which can be developed relatively cheaply. Referring to Fig 1, these economic considerations transform an idealized regime portrayed by the continuous line into a more practical regime depicted by the broken line. It can be seen that in each tidal cycle the period of generation may last six hours or so, and always occurs during the same part of the cycle. In this way, roughly half of the energy theoretically available would be extracted in practice. The artificially maintained water levels may be advantageous for the operation or development of ports and of navigation within the enclosed pool.

The converse system, with power flow from the sea into a previously drained basin, differs principally in the timing of generation periods, which occur about six hours earlier and later. There are disadvantages, in that the useful basin area is less due to exposure of beaches and tidal flats: thus the power and energy capability is always less for this mode of operation. Navigation inside the basin may become impracticable.

Electrical interconnection of two neighbouring one-way schemes of roughly equivalent sizes, one to generate during ebb tide and the other during flood, might in unusually favourable cases offer some prospect of continuous availability of power -- should that be desired.

SINGLE-POOL, TWO-WAY SCHEMES

In this system power is generated by both inflow and outflow through the same powerhouse. For many years past this would have implied switching the flow paths by means of a complex H-form arrangement of sluices controlling a forebay and tailbay. Major simplification of layout became possible with the advent of reversible-flow tubular turbines. Two examples with reversible-flow bulb turbines are now in operation at La Rance and at Kislaya Gulf.

The energy theoretically available by this system is obviously greater than for the one-way cycle. Practically however, this system involves discharging a much larger volume of water in each direction, in every cycle, than does one-way operation. More turbines are required, and it is necessary to curtail the generation periods in order to allow time for a large volume of emptying or refilling, in preparation for the reverse-flow generation phase. This necessitates large emptying and refilling capacity, a portion of which may be provided by free discharge via the turbines. Just after stopping generation the head is still large enough to give a fairly high initial sluice discharge. Nevertheless, there will be rather less complete refilling (or emptying) than in a comparable one-way system, except possibly in cases where it is practicable to use the turbines as low-head pumps at about high tide or low tide. A typical operating regime may be as illustrated by the continuous line in Fig. 2. If this is compared with Fig. 1, it may be seen that there are twice as many generation periods, and assuming application to the same pool, each of them is rather shorter and at lesser average head. There are more turbines and the total sluice area is larger.

This method may yield rather more energy than a comparable one-way scheme. It involves higher cost, generally more than proportionate to the increased energy production. Since the individual blocks of energy are smaller, they may possibly be more conveniently absorbed in the interconnected electricity supply system, in which case a higher unit cost of energy would be acceptable.

The added refinement of low-head pumping capability for boosting the emptying or refilling process at the turn of the tide may confer a worthwhile increase in energy output at modest additional cost. This is of value mainly during below-average tides: it would not generally be practicable to increase the outputs from large tides to the same degree because of limited installed capacity. The broken line in Fig. 2 illustrates how pumping can be employed to increase energy output. A more important aspect is that by such pumping, usually an hour or more before the demand peak, the timing of the next generating period can be advanced (or conversely, retarded).

Such shifts in timing, involving importation of pumping energy and departure from the optimum times for maximum energy production, reduce the net energy output but allow part of the output to be sent out when it is most valuable. Thus a part of the plant capacity can be designated "firm", in the sense that some power can be available whenever desired (provided it is ordered in advance). Since the pumping is generally at lesser average head than the ensuing generation flow, there can be some energy bonus. The Rance scheme is fully versatile in this way.

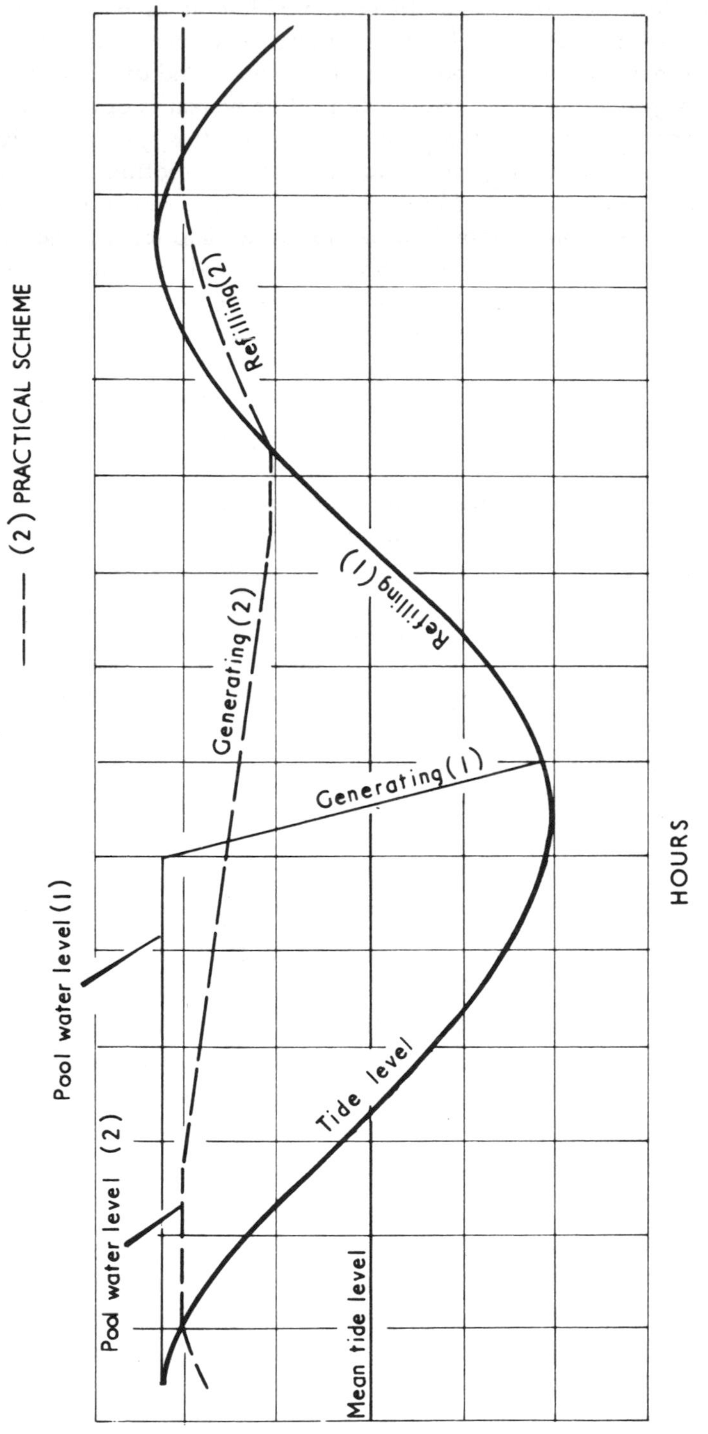

Fig. 1. Single-pool one-way system.

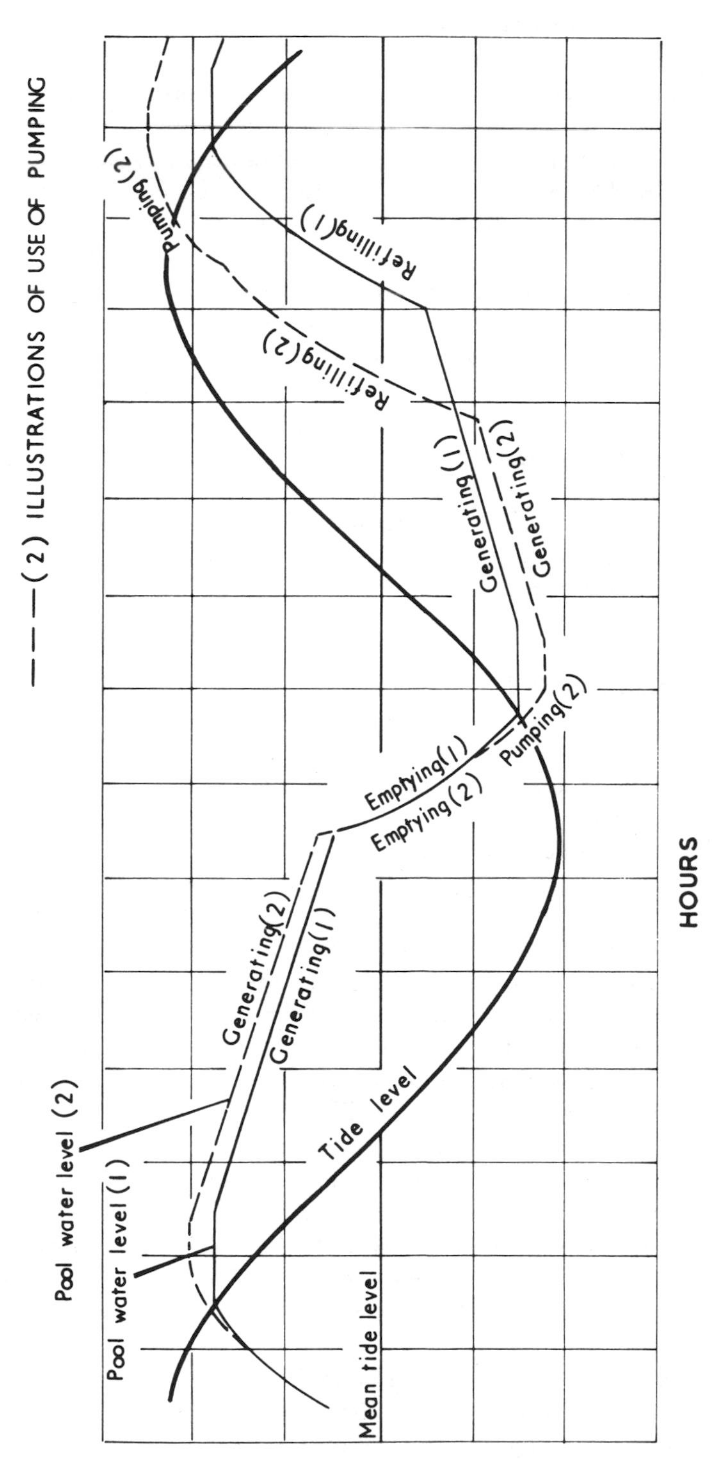

Fig. 2. Single-pool two-way system.

MULTIPLE POOL SYSTEMS

The classical two-pool system (the DeCoeur system) is inherently capable of generating at any time or continuously. The turbines are set to work between two pools, of which one is regularly replenished by each high tide and the other drained by each low tide. Output may be concentrated into selected periods up to the limits of installed turbine capacity. These options are illustrated in Fig. 3. There are numerous other two- or three-pool systems, some of which like the Defour schemes involve turbines working between the sea and alternative pools, by means of dams and sluices arranged for switching flows. These methods afford varying degrees of flexibility in operation. Apart from the complexity and cost of the various dams and sluices, another unfavourable feature is that turbine operation or pumping must generally be simultaneous with, and in series with, flow through at least one set of sluices involving extra head loses. Although philosophically interesting and often highly ingenious, none of these complicated schemes seems likely to find an application in the economically demanding electricity systems of the present day or the foreseeable future.

COMPARISON OF DIFFERENT TIDAL POWER SYSTEMS

The characteristics of the principal systems outlined may be compared by supposing each to be applied to a similar representative tidal inlet. This might be one of the branches of the Bay of Fundy. The several operating regimes are those shown qualitatively in Figs. 1 to 3. Table 1 illustrates the possible distribution of cost among the three main components -- dams, powerhouse and sluices -- for each type of project. The Table also shows corresponding annual energy outputs and energy costs on a comparative scale where values for the single-pool, one-way system are taken as 100. It should be noted that these illustrations have not been optimized for minimum energy cost, and if this were done their relative merits in this respect might vary slightly. Of course, practical scheme optimizations would not be solely in terms of this criterion, but must be influenced by the electrical system characteristics and by the peculiarities of individual project sites.

The dominant elements in project costs are clearly seen to be the powerhouse, in all the single-pool schemes, and the dams, in multiple-pool schemes. The small energy production of the latter virtually rules them out of consideration unless the cost of dams can be unusually low, e.g., due to exceptionally favourable site conformation, or if much of the dams will in any case be built for some other purpose.

Comparative installed generating capacities and firm capabilities during least tides are listed in Table 2. These firm capabilities are minima: the pumped single-pool system has a higher output when the chosen generation

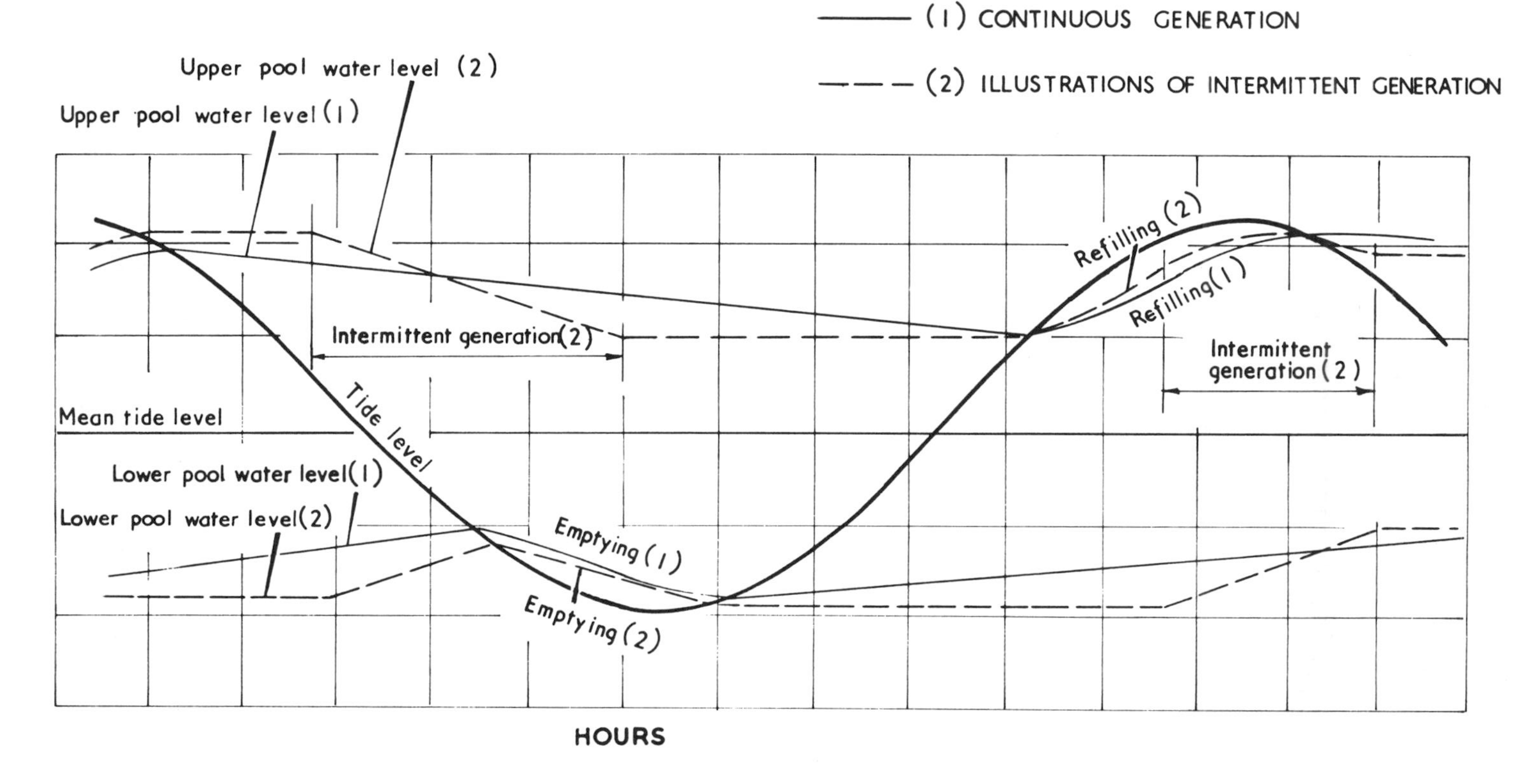

Fig. 3. Two-pool (DeCoeur) system.

Table 1
Comparative Tidal Power Project Costs and Energy Outputs

	Costs				Annual	Energy
	Dam	P.H.	Sluices	Total	Energy	Cost
1. Single Pool Ebb	28	57	15	100	100	100
2.1 Single Pool 2-Way	28	102	31	161	124	129
2.2 Single Pool 2-Way with Pumping	28	103	31	163	142*	110*
3.1 Two Pool Continuous	63	12	23	98	53	184
3.2 Two Pool Intermittent	63	35	24	122	53	230

Table 2
Comparative Output Characteristics

	Generating Capacity		
	Installed	Firm**	Notes
1. Single Pool Ebb	100	-	generates 4–7 hr. in each 12 1/2 hr.
2.1 Single Pool 2-Way	128	-	generates 3–4 hr. in each 6 1/4 hr.
2.2 Single-Pool 2-Way with Pumping	140	22	for at least 2 hr. whenever demanded
3.1 Two Pool Continuous	27	11	base-load supply
3.2 Two Pool Intermittent	80	34	for at least 3 hr. whenever demanded

*If worked for maximum energy yield.
**During least tides.

period is not at the least convenient stage in the tidal cycle. However, it is obvious that for each scheme the output must be well below average during least tides (generally neap and apogee tides). For single-pool schemes there must also be substantial variation in capability during any one tidal cycle.

These disadvantages can only be partially compensated within a tidal power scheme by foregoing part of the energy potentially available and by incorporation of an element of pumped storage. The only other alternative to simply accepting the disadvantages is external compensation for which the principal auxiliary is interconnected pumped storage (and in some systems, river hydroelectric plants with large storage). It will be clear that complete compensation for tidal output variations would be uneconomic even where practicable: any real solution will be a compromise involving the costs and the convenience of operation of the tidal plant and of the other plants interconnected with it. Some of these possible solutions are discussed further in the following sections.

PUMPED STORAGE IN A SINGLE POOL SCHEME

Since the tidal cycle is not in general in phase with the cycle of electrical demand, some means is required for shifting the time of availability of the generated electrical energy from a tidal power scheme. This can be done very effectively in a single-pool project by the use of a reversible pump-turbine as outlined previously. For example, reference to Fig. 4 will show that electrical energy may be produced at any required time by arranging that the pool level is kept up or down so that power may be generated when requested, the water moving either into or out of the pool, depending on whether the sea is higher or lower than the pool. It is not suggested that a regime like that of Fig. 4 is necessarily a likely one, but it illustrates the remarkable adaptability of such a scheme for which La Rance is the prototype.

The imported energy would be off-peak system energy and using it in this way meets two requirements. It ensures that such energy is available at peak demand periods, and it shifts the availability of at least some tidal energy to coincide with the peaks also. It is thus a particular form of pumped storage. Provided the timing is right, the energy used for pumping can be more than recovered by pumping against a low head and generating with a comparatively large one. Too much should not be made of this however, as the time of peak demand may preclude the use of off-peak energy at the optimum time in the tidal cycle to maximize the energy gain. In other words, flexibility in time can only be won by yielding potential tidal energy.

There are other disadvantages to the system as a form of pumped storage as follows:-

1. The special pump turbines and their containing structures are a good deal more expensive than simple one-way turbines: e.g., a draft tube is necessary at both ends of the water passage.
2. A compromise is necessary between the ideal modes for pumping and for generating, which reduces the efficiency in each mode below that for the corresponding single-purpose machine.
3. The maximum period of generating availability may vary from as little as

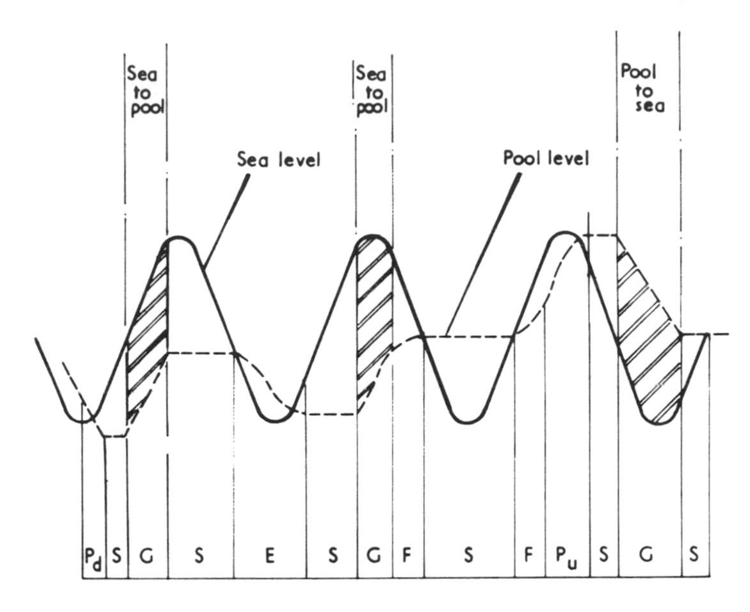

P_d = pumping pool level down with imported energy

S = standstill

G = generating

E = emptying

F = filling

P_u = pumping pool level up with imported energy

Fig. 4.

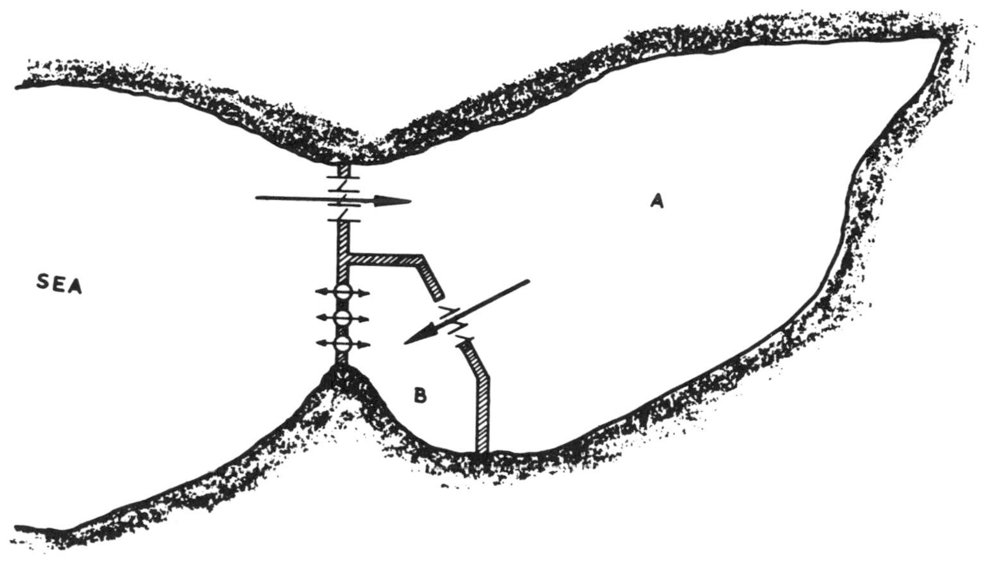

Fig. 5. Layout for tide-assisted pump storage plant. (Barrage built for other main purpose.)

three hours on neap tides to seven hours on springs. Beyond these limits, the variation in sea level precludes generation.

4. The machines are not generally available as spinning reserve on the system, which is increasingly one of the great attractions of pumped-storage plant. This is because the pool is being prepared (by being raised, or lowered) for its role as part of a peak energy producing system some hours later.

Generally, the movement of large quantities of water through a small head rather than small quantities through a large head is an uneconomical way of storing energy.

PUMPED STORAGE IN A DOUBLE-POOL SCHEME

Double-pool schemes have many disadvantages but theoretically they are well suited to integral pumped storage. In a two-pool scheme of the DeCoeur type, two alternatives are available. If pumps are sited in the two sea/pool barrages, then off-peak pumping power can be usefully absorbed at any time either by pumping the upper pool up or the lower pool down. If the pumping head to or from the sea is less than the pool to pool generating head an energy gain is possible.

Alternatively the pumps can be in the pool/pool barrage, either separately or more probably as reversible pump-turbines. Pumping power can still be absorbed in any non-generating period but without prospects of energy gain. The former method is the more expensive but also more flexible since pump capacity is independent of generating capacity. A number of variations on these alternatives is possible.

In any of these systems the generators can be used as reserve capacity except where there is insufficient head immediately following a generating phase. Generally however, double-pool schemes are too expensive because of extensive barrage construction. Provision of pumps as well as turbines compounds this disadvantage, whatever the elegance of the theoretical operation, and of course the general observation at the end of the section entitled Pumped Storage in a Single Pool Scheme still applies.

TIDE-BOOSTED PUMPED STORAGE

It is also possible to design a reservoir system in an estuary with the primary function of a pumped storage scheme. Such a scheme may be termed a tide-assisted or tide-boosted pumped storage scheme. The two pools need not now be of approximately the same surface area as is desirable for the classical double-pool tidal power schemes, nor need the level of the larger pool necessarily fluctuate appreciably with the tide.

Consider Fig. 5 which shows a possible layout for such a scheme, and Fig. 6 which shows the relative movement of pool and sea levels with time in a seven-tide period. The smaller pool B is regulated up or down to ensure that there is always a difference in levels between it and the sea at a forecast time of peak demand. This is achieved by having sluices in the main dam, sluices between the basins A and B and pump turbines between basin B and the sea. Off-peak energy is used to raise or lower the level of B by pumping between it and the sea according to the tidal levels expected and the time of the next peak demand. This energy may be recovered, often with a bonus, during subsequent generation. The system can be a very attractive proposition if the main barrage is being provided for some other principal purpose.

Even so, on certain occasions it is still very difficult to provide generating capacity and recourse must be made to very unusual and disadvantageous methods to make it possible,as the first two cycles of Fig. 6 illustrate. Of course these difficulties may be overcome by another machinery installation between pools A and B, but this kind of proposal is usually much too expensive to be attractive.

EXTERNAL PUMPED STORAGE

Although this is the simplest and most obvious method of integrating tidal energy it lacks elegance as a solution. It is frequently criticized as making the cost of tidal energy prohibitive through the provision of two sets of plant, one in the tidal station and the other in the pumped storage station. This would be valid only if there were no other generating plant in an electrical system. Normally other base-load thermal plants, fossil-fired or nuclear, already exist, and in recent years numerous pumped storage installations have been built to complement them in meeting peak demand. If the pumped storage plant exists already, its load factor may be improved by integrating tidal energy while still performing much of its original function with the thermal plant. Similarly, new pumped storage plant would be designed for multiple duties,handling off-peak output from both thermal and tidal generating stations. In the U.K. electrical system the function of the existing pumped storage plant in acting as spinning reserve is of increasing importance, and it is difficult to envisage new pumped storage installations which do not have this function.

The foregoing combination of duties is not satisfactorily fulfilled solely by incorporation of pumped storage capability in tidal power schemes, nor by tide-assisted pumped storage. Thus in most large interconnected systems the obvious answer will be the right one, enabling a comparatively simple installation to be used to harness the maximum practicable tidal energy for generation of electricity, of which some 35% can be consumed as generated, the remaining 65% being converted to peaking energy via external pumped storage. In

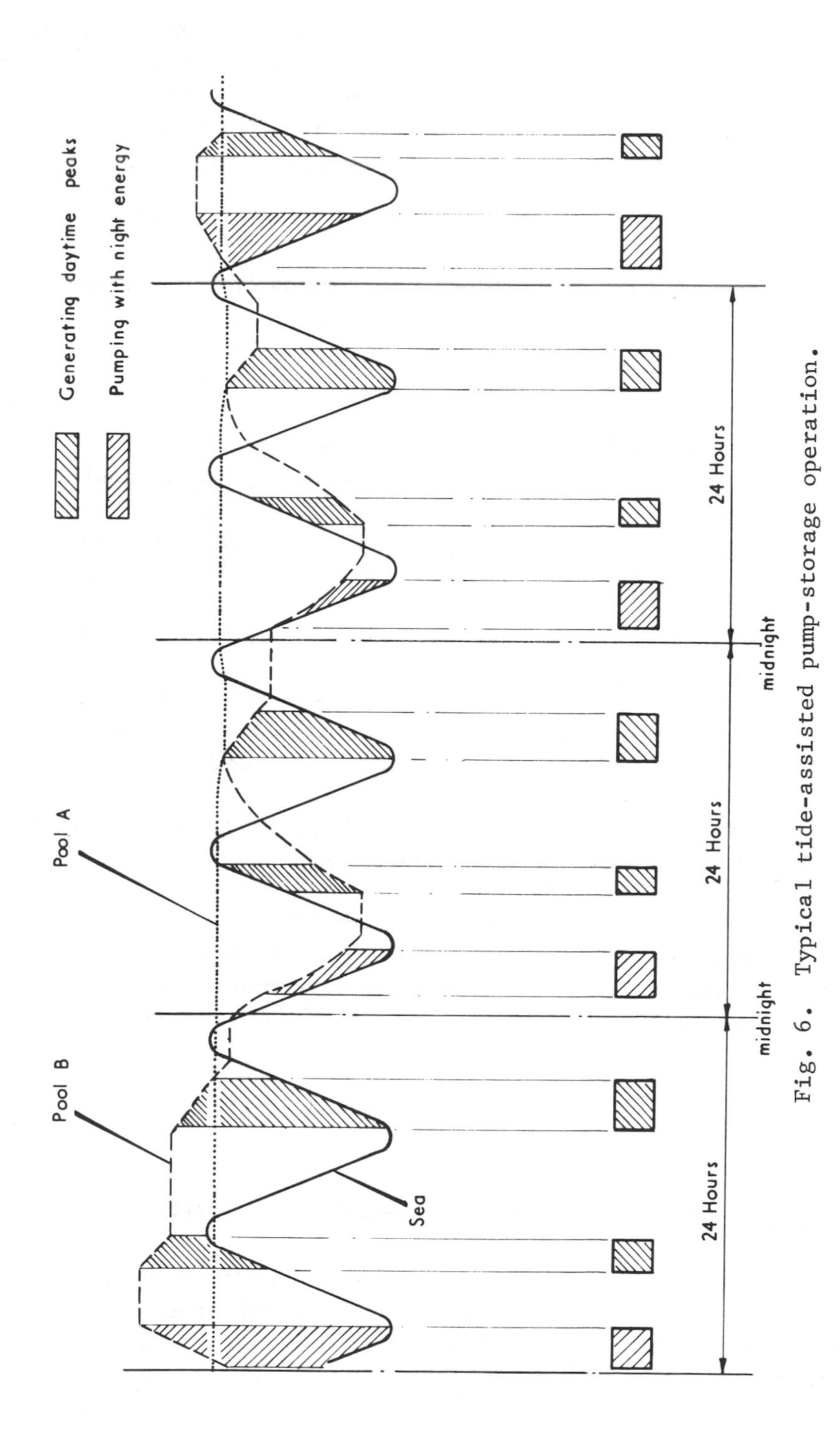

Fig. 6. Typical tide-assisted pump-storage operation.

most cases this will give best economy since the pumped storage schemes may be sited to exploit suitable topography, existing reservoirs, etc., and designed to appropriate capacities for the system. The tidal plant energy production may at the same time be simplified in comparison with other solutions.

RELATIVE MERITS OF THE MOST FAVOURED SYSTEMS

Many studies of actual possibilities, with reasonably detailed construction cost and energy production estimates, have led the authors to conclude that the choice for future development is likely to be between single-pool ebb-flow schemes with external pumped storage, and single-pool integral pumped storage schemes similar to La Rance (with or without external pumped storage in addition). The former have the following advantages:-

1. Simple, and hence relatively cheap, tidal machinery may be used since the ability to pump and generate in two opposite directions is not required.
2. All the energy available to such an installation may be extracted since no manipulation of water levels to meet forecast peak electrical demand is required.
3. The external pumped storage plant is at all times available to act as reserve capacity in the system.

To the extent that two-way, single-pool schemes with integral pumped storage do always have firm on-peak capability (see section entitled, Single Pool, Two-Way Schemes, and Table 2) they can be regarded as fully flexible, independent producers of power and energy. However, most of the potential output is over and above this firm component and could only be exploited by compensating operation of other plants in the system -- most probably, independent pumped storage installations.

In the final analysis, the optimized solution adopted will depend on the relative costs of the power and energy delivered, and their value in terms of overall system cost over a period of years. The same solution will not necessarily emerge for the same estuary at different stages of development of the interconnected electricity supply system.

REFERENCES

1. Studies in Tidal Power, N. Davey, Constable, London (1923).

2. Bernstein, L. B., "Tidal Energy for Electric Power Plants", Gidroenergoproekt (1961). Israel Program for Scientific Translations, Jerusalem (1965). U.S. Dept. Interior and Nat. Sci. Foundation, Washington (1965).

3. Wilson, E. M., "Tidal Energy Development", Handbook of Applied Hydraulics, Eds. C. V. Davis and K. E. Sorensen, McGraw-Hill, New York (1969).

SOME ELECTRICAL DESIGN AND OPERATIONAL ASPECTS OF A LARGE TIDAL POWER DEVELOPMENT

H. A. Erith*

INTRODUCTION

Considerable study has been devoted of late to the development of the better tidal power generating sites in the Bay of Fundy. Aspects of power transmission to potential market areas, however, have not received the same attention. The power developments are unique in many ways: in their power capability, as a function of the lunar cycle; in their siting, some miles from land; in the multiplicity of units and in their characteristics (e.g. low speed, low inertia, submerged bulb-type with an inherent sensitivity to wave action). The associated power plant operation and transmission is equally unique. If the simplest of the many available generating patterns be adopted -- namely, one-way generation during emptying of the storage basin -- without pumping operating regimes, the following requirements must be satisfied at all times:

1. Fast, controlled, unit start-up and connection to the system twice a day.
2. Fast loading of machines at tidal development and equally fast and co-ordinated unloading of system generation.
3. Controlled reactive loading of the machines with associated reactive adjustment of system generation, to obtain effective voltage control.
4. Some 6 hours later the reverse of the above operations and the fast, controlled shutdown of all units.

To illustrate and emphasize the various parameters involved, the paper considers a typical high-power tidal project employing one-way generation, and connected to a power system providing limited local power supplies and a major power supply to a distant market, this latter incorporating E.H.V. long distance transmission. The typical power project and transmission is described and the variation of certain key quantities during unit start-up and loading to full generator capability is investigated by establishing load flows at intervals

*Senior Staff Engineer, Power Systems, The Shawinigan Engineering Co. Ltd., P. O. Box 3010, Station B, Montreal 110, P. Q., Canada.

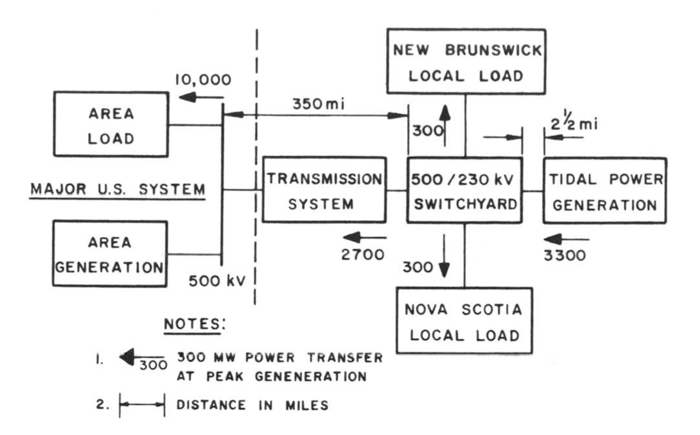

Fig. 1. Basic system.

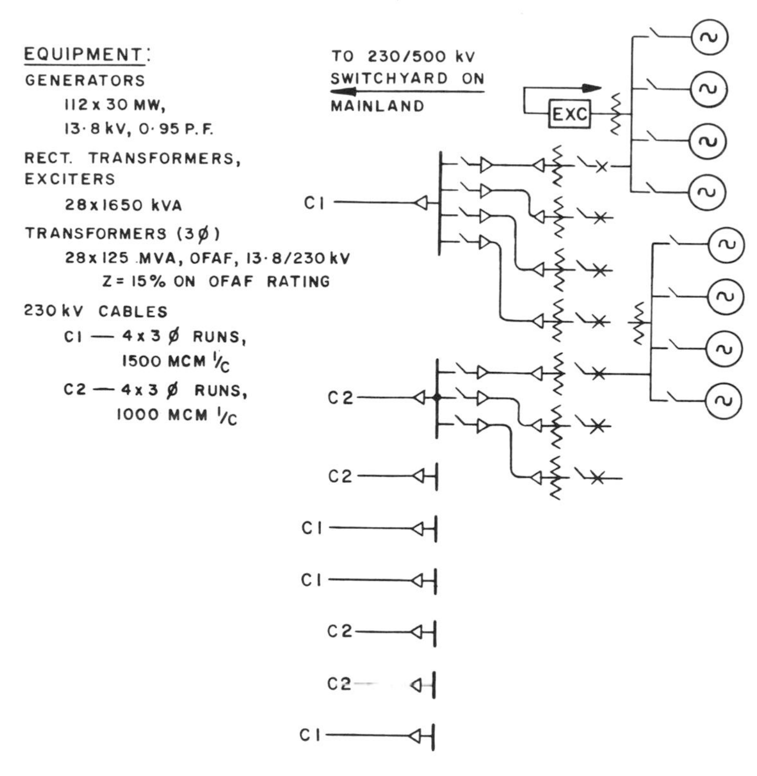

Fig. 2. Generation and transformation to 230 kV.

during the start-up process.

CO-ORDINATION OF A MAJOR TIDAL POWER DEVELOPMENT WITH LONG-DISTANCE TRANSMISSION

The typical power system, the performance of which is investigated in succeeding sections is shown in Fig. 1, 2 and 3. A logical sub-division of the basic system (see Fig. 1) is:

1. Generation
2. Transformation
3. Transmission
4. Load

Generation

The tidal power station is assumed to comprise one hundred and twelve 30 MW bulb-type generating units. The units are arranged in groups of four and bussed at 13.8 kV, and a single static exciter is employed for the group; the four machines self-synchronize as a group upon unit start-up. In turn, the group is synchronized conventionally at 13.8 kV with the system via a 125 MVA, 13.8/230 kV transformer. The twenty-eight 125 MVA transformers are bussed in four groups of four (each group representing 500 MVA total capacity) and four groups of three (each group representing 375 MVA. The generating units are located some 2 1/2 miles from a mainland switchyard facility, and 230 kV single core cable is utilized for power transmission along the barrage to the main switchyard.

Transformation

Transformation to 500 kV is achieved by four 900 MVA, ONAF, auto transformer banks. Unscheduled outage of one transformer bank would also remove one 230 kV cable and 12 units from service. After isolation of the faulty bank the 12 units could be restored to service and the remaining three banks would be adequate to handle the full power transfer.

Transmission

The 2700 MW maximum transmission requirement assumed establishes transmission at 500 kV via three circuits and using a 4-conductor bundle per phase. To minimize the difference angle between the bus voltages at the receiving- and sending-end 500 kV busses, 30% series compensation has been provided at the midpoint of each 350 mile line. Without the series capacitors the angle between the bus voltages at maximum transmission is some 37° and

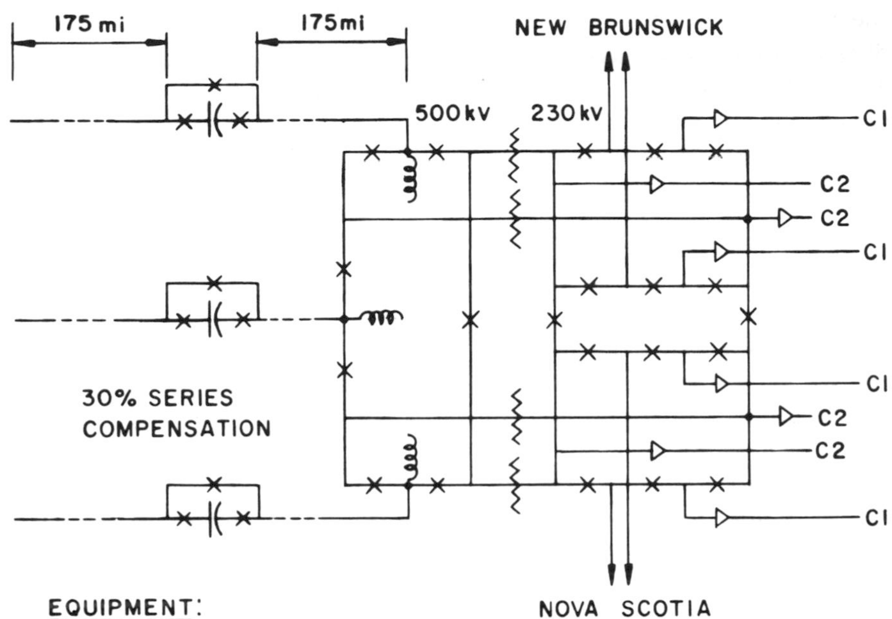

EQUIPMENT:

230 kV CABLES

C1 — 4 x 3 ϕ RUNS, 1500 MCM 1/C
C2 — 4 x 3 ϕ RUNS, 1000 MCM 1/C

TRANSFORMER BANKS

4 x 900 MVA, ONAF, 230/500kV, AUTO
Z = 10% ON ONAF RATING

TRANSMISSION LINES

3x500 kV — 4x0·9" O.D. ACSR/ϕ, S.I.L. = 903 MW AT 500kV

SERIES CAPACITORS

3x255 MVA — 1200A CONT. CURRENT RATING, 59Ω/ϕ

SHUNT REACTORS

3x200 MVA

Fig. 3. Transformation to 500 kV and transmission.

would, under more normal circumstances, be considered a satisfactory stable operating value. However, in view of the low inertia and H-constant of the tidal power units and the possible fluctuating power output of the units due to wave action, it is considered good practice to reduce this angular deviation to less than 30° and thus obtain a higher stability margin.

To provide improved system voltage control and reactive loading, a 200 MVAR, 500 kV shunt reactor is connected directly to each incoming line circuit at the 230/500 kV switchyard.

Load

The provincial area utilities are assumed capable of absorbing a total of 600 MW (300 MW in Nova Scotia and 300 MW in New Brunswick) of the tidal power generation, at 0.975 lagging p.f.

The remaining 2700 MW peak is assumed transmitted to a major U.S. network at a time when the U.S. area load is steady at 10,000 MW and 0.95 lagging p.f.

PLANT OPERATION AND PERFORMANCE

A typical operating cycle for the tidal power plant is shown in Fig. 4. For the assumed tidal conditions and installed generating capacity, the period of generation is some 6 1/2 hours per tide or a little over 50%. Each generating cycle may be broken down into five stages, viz:

1. Unit start-up, assumed complete in 15 minutes, at which time all units are operating at 100% gate opening, but the generator output is only some 40% of nameplate rating because of the minimal operating head on the turbines.
2. Relatively slow load build-up to generator nameplate rating, due to increased operating head on the turbine as the sea or tailwater elevation drops away faster than the slow draw-down of the storage basin.
3. Constant power generation at nameplate capacity where, due to the high available operating head, turbine output must be cut back to avoid generator overloading.
4. Relatively slow load decrease, as low tide conditions are reached and the turbine tailwater elevation begins to rise.
5. Relatively fast load decrease, over a period of say 15 minutes, during which all units are shut down due to insufficient head.

Subsequently, a further operating sequence is initiated, involving:

(a) the opening of refilling sluices when the basin and sea level elevations are equalized;

(b) the closing of the refilling sluices when the basin and sea level elevations are again equalized following "topping up" of the basin.

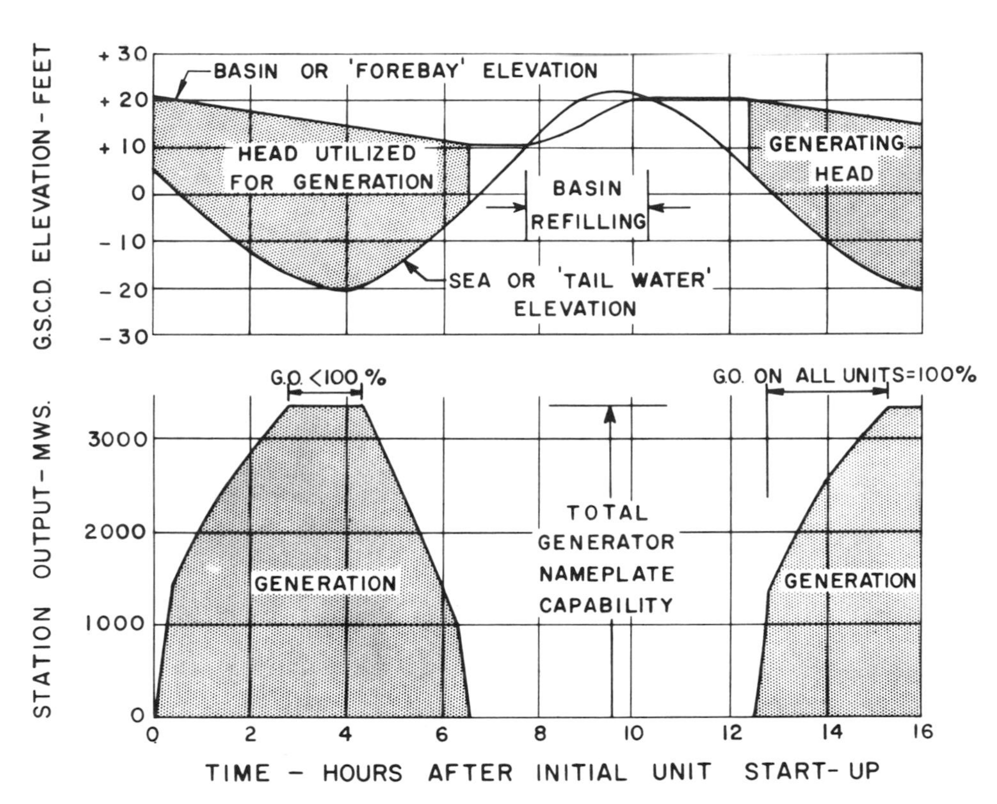

Fig. 4. Typical operating cycle.

TABLE 1

SUMMARY OF LOAD FLOW DATA FOR TYPICAL PROGRESSIVE LOADING OF TIDAL POWER DEVELOPMENT

Time (hrs. after start of initial machine groups)	Total Station Generation			Loading on each 125 MVA transfr.		Total Local Load (N.S. + N.B.)		Loading of distant generation (to meet 10000 + j 3300 area load)			Major Bus Voltages			
	P (MW)	Q (MVAR)	P.f. (lag)	P (MW)	Q (MVAR)	P (MW)	Q (MVAR)	P (MW)	Q (MVAR)	P.f. (lag)	500 kV rec.-end bus (% on 500 kV)	500 kV send.-end bus (% on 500 kV)	230 kV switch-yard bus (% on 230 kV)	generator terminals (% on 13.8 kV)
0	0	0	-	0	0	0	0	10006.6	1720	0.986	95	105.6	104.4	101.8
1/4	1320	137	0.995	47.2	4.9	240	55	8945.2	1888.2	0.977	"	105.9	104.7	102.5
3/4	1850	285	0.989	66.0	10.1	336	77	8527	2043	0.973	"	105.5	104.4	102.7
1 1/4	2310	457	0.981	82.5	16.3	420	96	8170	2209	0.965	"	105.1	104.1	103.0
1 3/4	2700	630	0.974	96.5	22.5	490	112	7876	2391	0.957	"	104.7	103.9	103.4
2 1/4	3020	840	0.963	108.0	30	550	125	7633	2520.6	0.950	"	104.6	104.1	104.2
2 3/4	3300	1020	0.956	118.0	36.5	600	138	7426	2643.3	0.942	"	104.5	104.2	104.5

Notes

1. Full load output of each unit is 29.5 MW net, the 0.5 MW reduction covering double-transformation losses and auxiliary power.

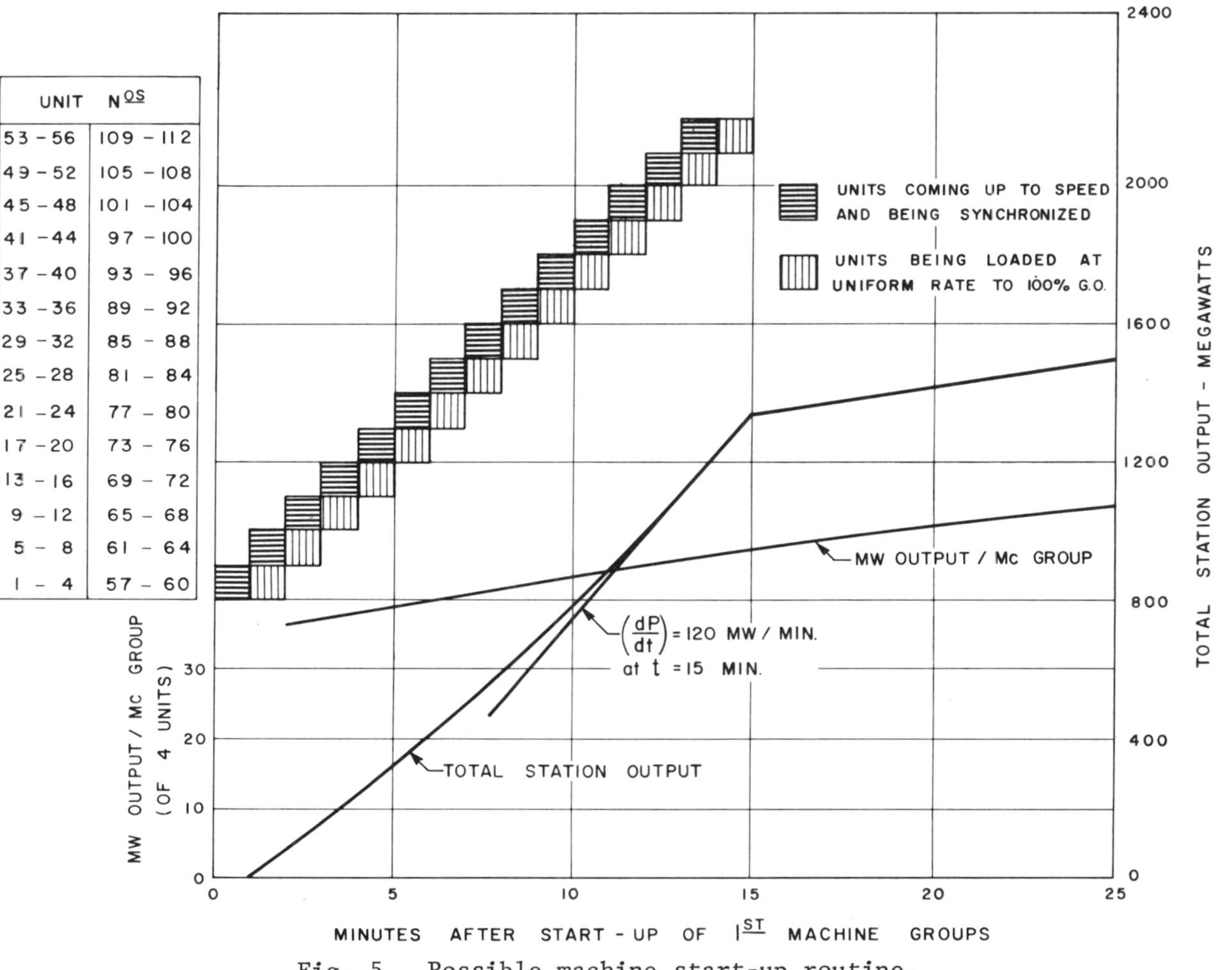

Fig. 5. Possible machine start-up routine.

Stage (1) above -- namely, unit start-up -- is of particular concern insofar as operation of the generating plant is concerned. The development of a start-up routine is shown in Fig. 5. The major steps are:

(a) the start-up of two groups of four machines each, every minute;
(b) synchronizing of machine groups one minute after "start" initiation;
(c) opening of wicket gates to 100% G.O. in a time of one minute at a rate such that power output increases linearly with time.

By the 15th minute after "start" initiation of the first two machine groups, the station output is increasing at the rate of 120 MW/min.

The 15 minute time interval for complete station start and loading has been selected arbitrarily. A shorter start-up time would gain a little in energy generation per tidal cycle but would call for a corresponding greater rate of generation reduction on the systems utilizing this tidal energy. We believe the 15 minute start-up to be a reasonable value to adopt at this stage for analytical purposes.

SYSTEM OPERATION AND PERFORMANCE

The system performance corresponding to the first three stages of plant operation and performance is summarized in Table 1 and the time variation of major power and reactive quantities is developed in Fig. 9.

Three key points in the loading cycle have been selected and load flow diagrams prepared (see Fig. 6, 7 and 8). Fig. 6 is applicable for the condition of complete plant shutdown, Fig. 7 for all units operational at reduced load, 15 minutes after start-up, and Fig. 8 for nameplate generation on the entire plant.

Table 1 shows that, even adopting single-ratio transformers for the 13.8/230 kV (nominal) and 230/500 kV (nominal) transformations, it is possible to obtain very satisfactory voltage control of all busses and equipment by reasonable adjustment of reactive generation. Table 1 shows the receiving-end 500 kV bus held at 95% voltage for the entire load variation. The voltage range at the sending-end 500 kV bus is 104.5-105.9%; the 230 kV bus has an even smaller voltage variation, whilst the generator terminal voltage increases with station output from 101.8% to 104.5%.

Fig. 7 shows the very large effect of the 500 kV line charging on the distant U.S. system reactive generation. In fact, as the tidal project is loaded up, the power loading of the distant generation is reduced from 10006.6 to 7426 MW (i.e. a reduction of 2580.6 MW) whilst the reactive loading of the distant generation is increased from 1720 to 2643.3 MVAR (i.e., an increase of 923.3 MVAR).

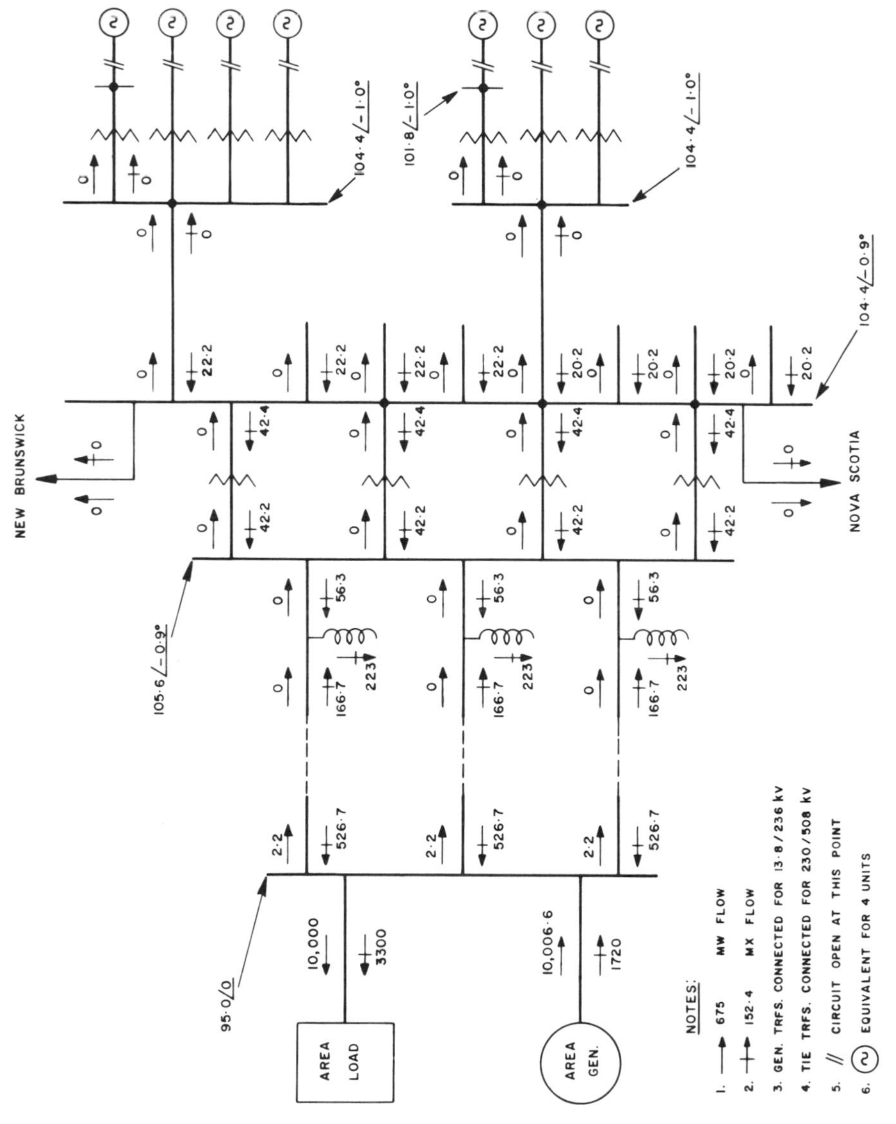

Fig. 6. Typical load flow. I. Zero generation.

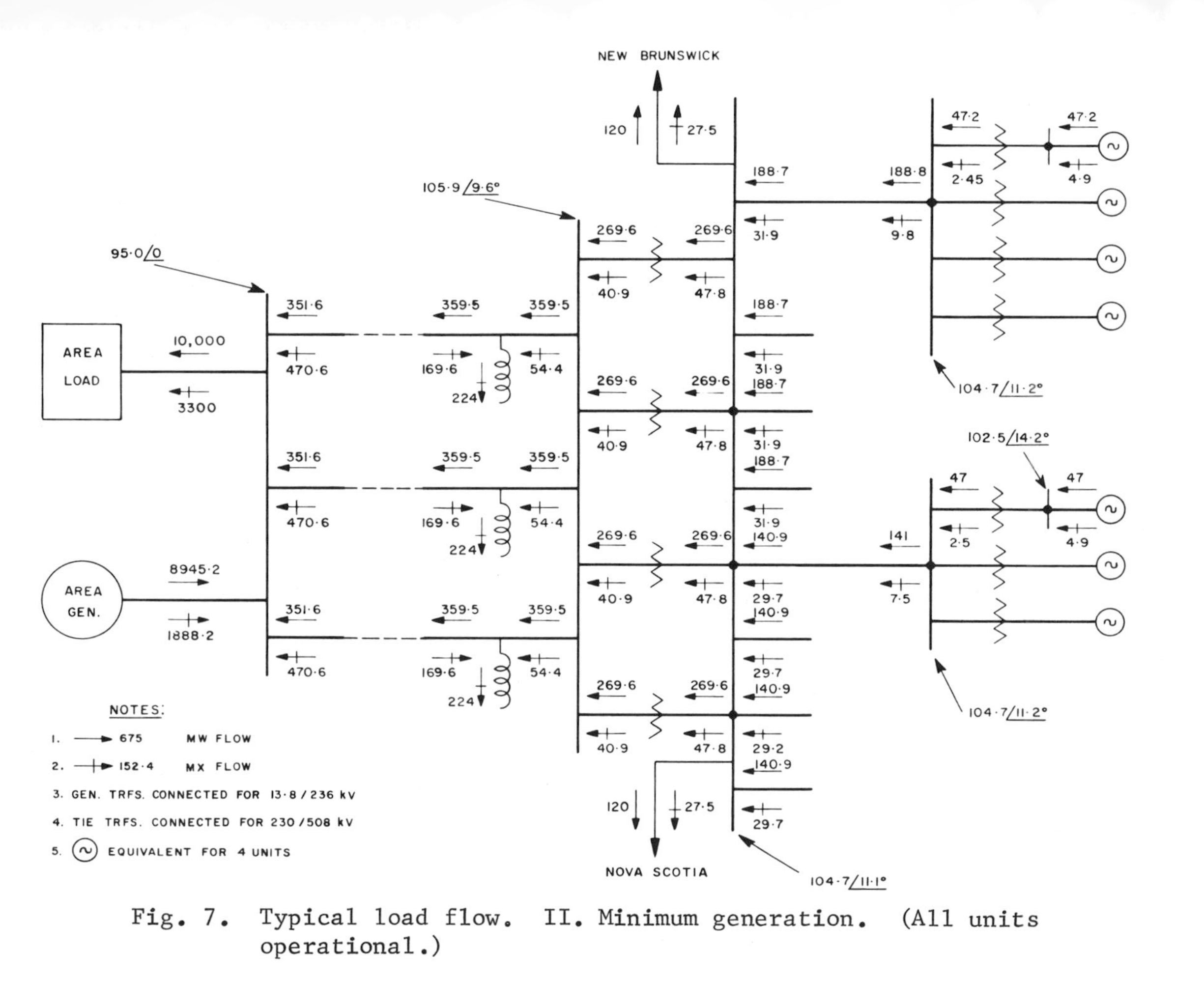

Fig. 7. Typical load flow. II. Minimum generation. (All units operational.)

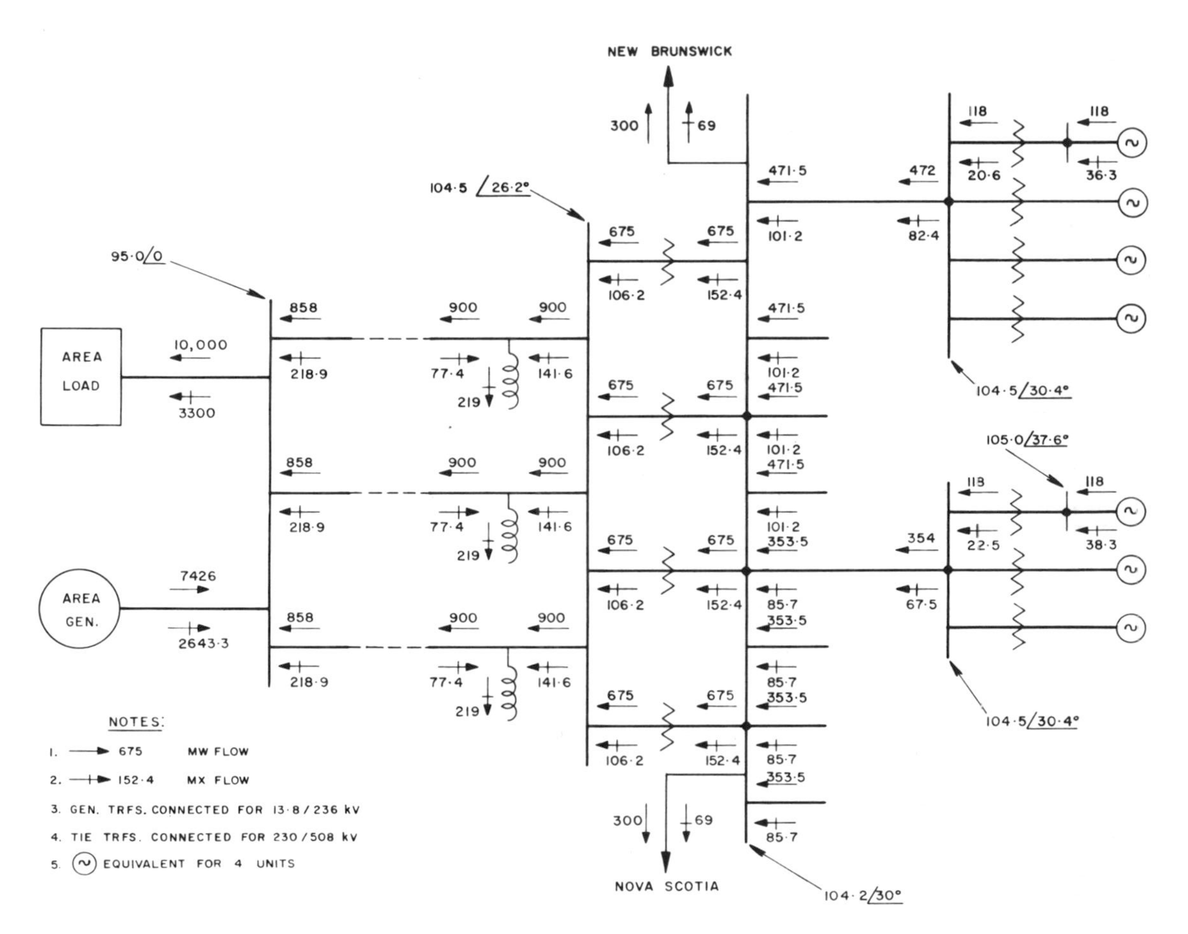

Fig. 8. Typical load flow. III. Maximum generation.

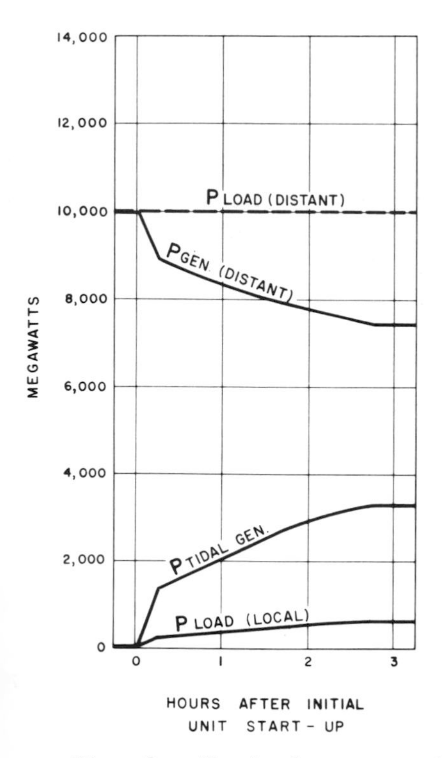

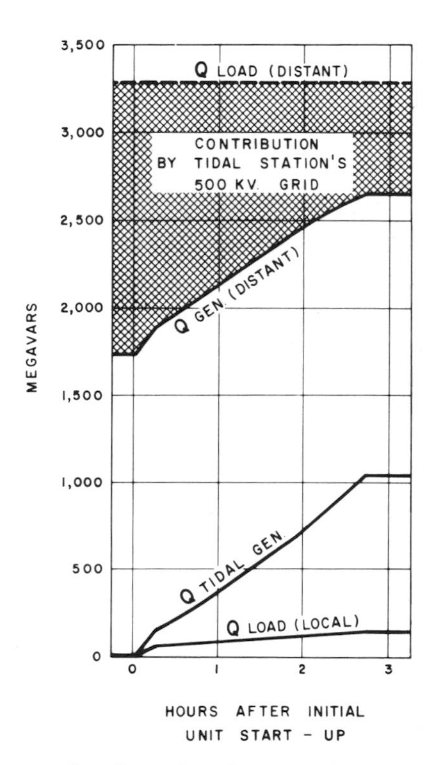

Fig. 9. Typical power and reactive variation for increasing tidal generation.

CONCLUDING REMARKS

The paper represents an attempt to cover some of the rather unusual concepts of power generation and transmission inherent in a large tidal power project. The simplifications and assumptions have been many -- probably the most important has been the assumption of constant power and reactive at the distant U.S. system. It is realized that large variations in the power and reactive coinciding with start-up could significantly affect the plant loading. Certainly we believe that the problems of power and reactive loading and control could only be solved and handled on a day-by-day basis using a high degree of automation.

ACKNOWLEDGMENTS

The author acknowledges with gratitude the significant assistance of Mr. B. J. Gevay and other colleagues in The Shawinigan Engineering Co. Ltd. for facilitating preparation of this paper.

Much of the background for this paper resulted from the author's involvement as power systems consultant to Tidal Power Consultants Ltd. in their study of the Minas Basin sites on behalf of the Atlantic Tidal Power Programming Board in 1967.

MODERN TECHNIQUES FOR BARRAGE CONSTRUCTION

J. D. Gwynn*, F. Spaargaren** and A. J. Woestenenk***

INTRODUCTION

The timing and content of this Conference is dominated by the recently issued Fundy report and there is some danger of concentrating too much on the circumstances peculiar to Fundy. However, other potentially important tidal power sites exist which indicates the need for a more general approach. This paper applies certain engineering and economic considerations towards the solution of the problems associated with barrage construction in tidal conditions.

The greatest experience in this field of recent years is being obtained in the Delta Plan works in the Netherlands where seven major tidal inlets have to be dammed. Five of these have already been completed, the sixth is planned for completion in 1971, and the seventh and largest, on which work began in 1968, is to be finished in 1978. This whole project was put in hand as a result of the disasterous flood of 1953, so the period of construction spans 25 years. It had been preceded by years of study which made it possible to begin promptly when the decision was made to go ahead.

The total cost will be about 1000 million dollars.

This effort has required extensive research which has already yielded important improvements and economies in construction techniques.

It seems advisable to examine their applicability to tidal power construction with a view to reducing costs.

There was and is a need to update the study of the Minas Basin prepared for the Atlantic Tidal Power Planning Board in 1967 in the light of additional data arising from the activities on the Delta Plan.

---

*Engineering and Power Development Consultants, Marlowe House, 109 Station Road, Sidcup, Kent, England.
**Waterloopkundige Afdeling van de Deltadienst, Van Alkemadelaan 400, The Hague, Netherlands.
***Bitumarin, P. O. Box 42, Zaltbommel, Netherlands.

There are two main systems of construction:

1. Dry enclosure - requiring cofferdams.
2. Directly in the sea.

To compare the merits of these two methods, consider the foundation possibilities:

a. Structure directly on a rock bed.
   Associated problems: Bed preparation generally involving difficult excavation.
b. Structure on a rock bed overlain by sediments which must be removed.
   Associated problems: As in a., plus the removal of the sediments either by dredging or by dry enclosure.
c. Structure on sediments, i.e. sand, consolidated silt or clay.
   Associated problems: Protection against erosion under and on both sides of the power house and refill structures caused by head differences up and downstream in either direction.

Of these three possibilities, rock is of rare occurrence, rock overlain by sediments is more common, while sediments is the most usual case.

Returning to the two techniques of construction and considering the most usual case, sediments, first, protection must be applied to the sediments before the structure is added irrespective of whether this is done by dry enclosure or directly in the sea. Proven techniques now exist to work directly in the sea with the various sediments already referred to. This avoids the heavy expense and risks associated with deep cofferdams in variable tide conditions. The extent of the protection is dictated by the requirements of the permanent structure which, in general, are enough for construction conditions so that no extra protection costs are incurred.

An important advantage of avoiding cofferdams is the saving in time which may amount to 50% on the foundations. Speed of construction is particularly desirable, notably to minimize the duration of temporary risks due to scouring and imperfectly controlled tides and also to minimize interest costs during construction.

Apart from this, the construction of caissons can be put in hand at the same time as the preparation of the bed. In this respect, the adoption of caisson construction without use of cofferdams confers a decisive advantage. (Fig. 1).

A cofferdam enclosure would have to be considerably larger than the corresponding portion of the barrage to be constructed. For two or three years it would have to withstand heads possibly three times as great as the finished barrage, and the tolerable leakage would be considerably less. Such temporary works would be very expensive and time consuming.

Cases do occur where it has been decided to work in situ in a dewatered enclosure. This procedure, as exemplified by the Rance project, has been well described.[1] Underwater construction of fill embankments

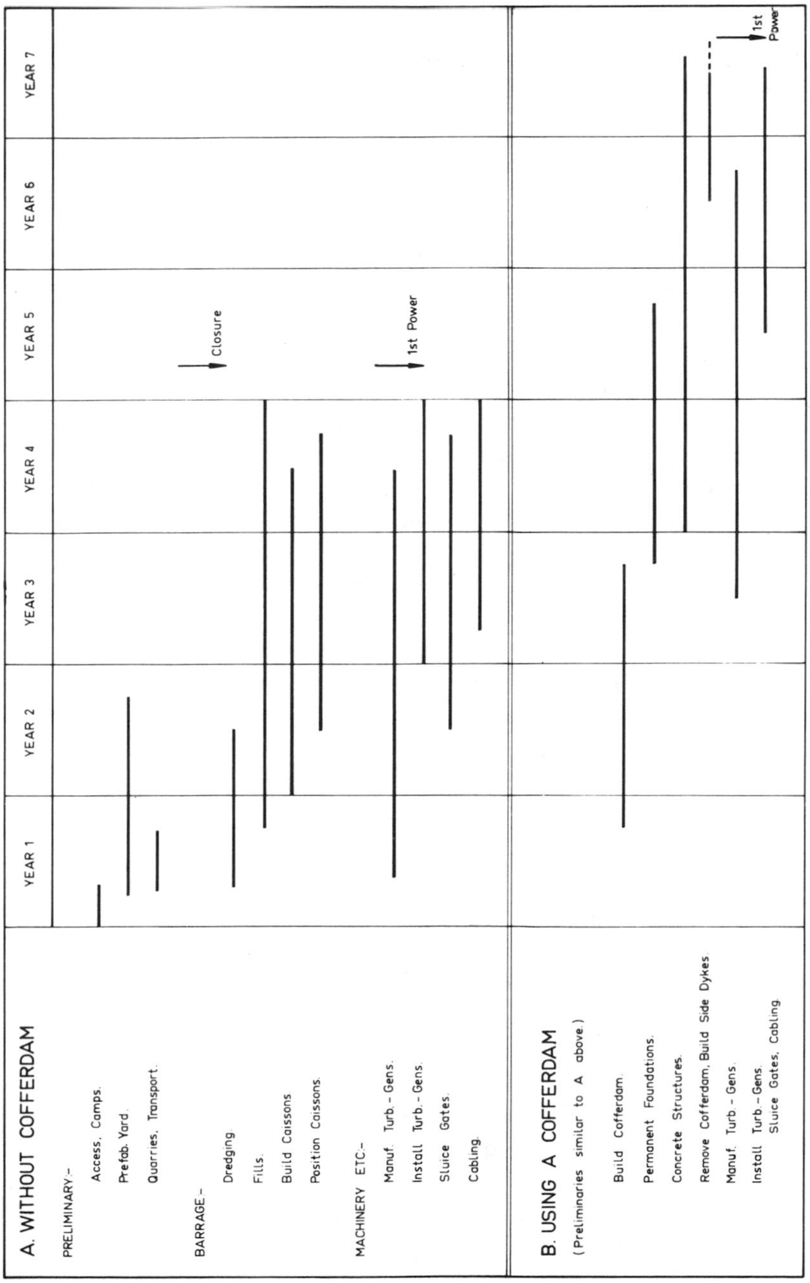

Fig. 1. Typical construction schedules.

has been,and continues to be, the subject of extensive research and field experience. Reference may be made to the published accounts.(2)(3)(4)

The cost of prefabricated caisson construction on shore is lower than in situ, and the cost of positioning the caisson (based on Dutch experience) is only 2 or 3% of the construction cost. Hence construction directly in the sea is always preferable.

In the case with rock overlain by sediments which must be removed, there is again the choice of excavating by dredger or in the dry.

If dredging is adopted there is a risk of the area excavated being refilled by siltation. In Holland, this problem is generally solved by using the dredged material to build part of the barrage and sufficiently narrowing the gap so that the velocity of the current is kept high enough to avoid siltation.

On the other hand, the cost of excavation in cofferdam construction is higher than dredging, in addition to which there is the cost and time loss of constructing and removing the cofferdam.

Moreover, the existence of a cofferdam gives rise to problems of protection of the adjacent areas. Hence, in this case also, construction directly in the sea is preferable.

In case of rock, it will also always be cheaper and quicker to work directly in the sea for similar reasons.

It may be remarked in passing that this conclusion is at variance with Mr. Lawton's address with respect to the Minas Basin (under heading Principal Findings).

Referring again to Figure 1 (Typical Construction Schedules), if cofferdams are adopted, the building of side dikes would usually have to follow removal of cofferdam and commissioning of the sluices to avoid creating excessive velocities. This would further postpone the date of first power possibly by years.

SCHEDULE CONSIDERATIONS

In the execution of a large and complex scheme such as a tidal power project, there are likely to be several operations demanding extensive investigation and planning, or great thoroughness and care, or - by virtue of simple magnitude - application of large resources of materials and skill. This situation has obvious implications, from the scheduling point of view, and considerable ingenuity and energy are called for to secure and maintain control of the linked parameters of time and cost.

It is usually desirable, other things being equal, to complete a project as soon as practicable after the necessary plans and financial arrangements are completed. The paramount reason is that once heavy expenditure is under way, it should be brought to a productive stage with the minimum of accumulation of interest costs. All the available time prior to starting heavy expenditure should be employed in refinement of project plans, so as to maximize the beneficial aspects and minimize costs.

The particular constraints limiting acceleration of the schedule for a tidal power project are:

1. Necessary preparatory operations before permanent construction works can begin on site.
2. The large scale of structural concrete work necessary and inconvenient access (along the barrage) for materials and work force.
3. Restricting the degree of obstruction of the tidal inlet until operations or works vulnerable to increasing currents have been accomplished or protected.
4. Protracted manufacturing programme for large numbers of machinery items, especially the turbine generators.

Where major portions of the project works are to be built in situ, in a dewatered enclosure, the task of constructing that enclosure will dominate the first half of the construction period. On the other hand, if reliance is placed on extensive use of prefabricated caissons for the permanent structures, large prefabrication and fitting-out facilities must first be prepared at some suitable location. This latter method permits these major concrete works to be started at least two or three years earlier, and centralizes most of this concrete work at a shore location convenient for highly mechanized operation. It must be noted that this faster progress with civil works necessitates correspondingly earlier delivery of machinery, and it will usually be necessary to order the turbine-generators by the time that the main civil engineering contract is let. Moreover, until these machines are defined, civil engineering design and detail construction planning cannot be finalized. Since the decision to proceed with a tidal power project may precede the start of construction by two years or more, such procurement considerations can be foreseen and allowed for.

As the project nears completion, the date for first production of power will be determined by the stage reached in building the barrage. If the main works are being built in situ, within an encircling temporary dam, all the work which is to be immersed must be completed, or separately enclosed, before the main enclosure can be flooded. After this, the temporary dam, or most of it, must be removed before power can be generated.

When the barrage is principally constituted of prefabricated caissons, first power can be generated as soon as the whole barrage is sufficiently complete to withstand full working heads. This requires that all caissons be in position before the closure operation is effected by completing any rockfill embankments and by shutting off all gate openings and temporary sluices.

Using either of the above procedures, commissioning of installed plant would proceed in stages and the work on the later stage would be completed from within the barrage.

Alternative construction schedules for a typical project are illustrated

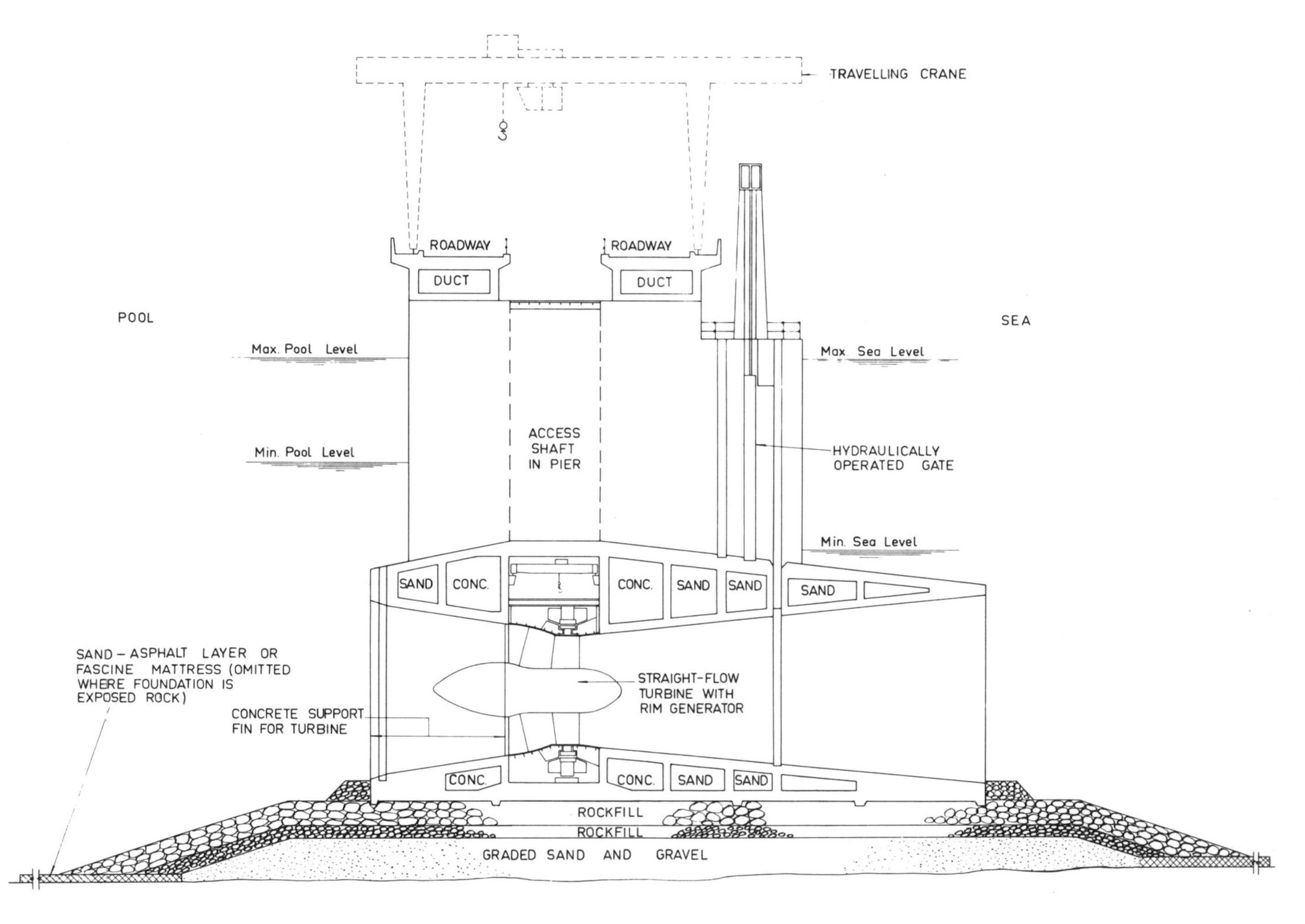

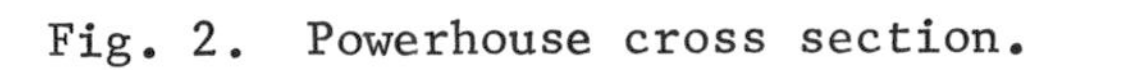

Fig. 2. Powerhouse cross section.

in Fig. 1. The first of these shows positioning of caissons spread out over a period of two years or so. This means that it may be practicable to schedule the mechanical and electrical installation work quite conveniently, and there is not much need for an intermediate storage area for the completed caissons.

However, wherever it is possible to position all the caissons in a single season, there could be much saving in idle time for the large tugs and much less risk of dangerous scouring developing.

USE OF CONCRETE CAISSONS

Construction in the form of prefabricated concrete caissons has three principal attractions which are of general application:

1. Much of the work can be concentrated in a conveniently located "factory wharf" area rather than strung out along the barrage possibly miles from shore. Working conditions, mechanization, and productivity can thus be akin to those of a factory - in contrast to those of a relatively exposed marine construction site.
2. There can be a net saving in construction cost by obviating the need for a very large deep-water cofferdam.
3. The total period of construction can be minimized, not only for the reasons stated in 1. above but also because the major concrete structures can be begun independently of the work along the barrage alignment.

At some sites, there may be other even more compelling factors. Where the sea-bed rock is deeply fissured or lies at great depth below permeable sediments, establishment of a secure cofferdam to withstand heads of 100 feet or more may be difficult or even, for practical purposes, impossible.

Types of Caisson

The main elements of a tidal power barrage are the power house, the sluices, the remainder of the dam, and possibly navigation locks. Prefabricated caissons may be used in all of these.

The power house consists of numerous parallel water passages and housings for the turbines, which are axial-flow machines with axes horizontal or near-horizontal. The superstructure accommodates sundry auxiliaries, connections and means of access, and possibly gates and a highway. An example is illustrated in Fig. 2. The fundamental caisson unit is based on one turbine, but for stability of flotation, as well as for manufacturing convenience or to facilitate grouping of auxiliary services, multiple units to house two or even four turbines will generally be preferable. According to consideration of buoyancy, stability, and ease of construction, parts of the

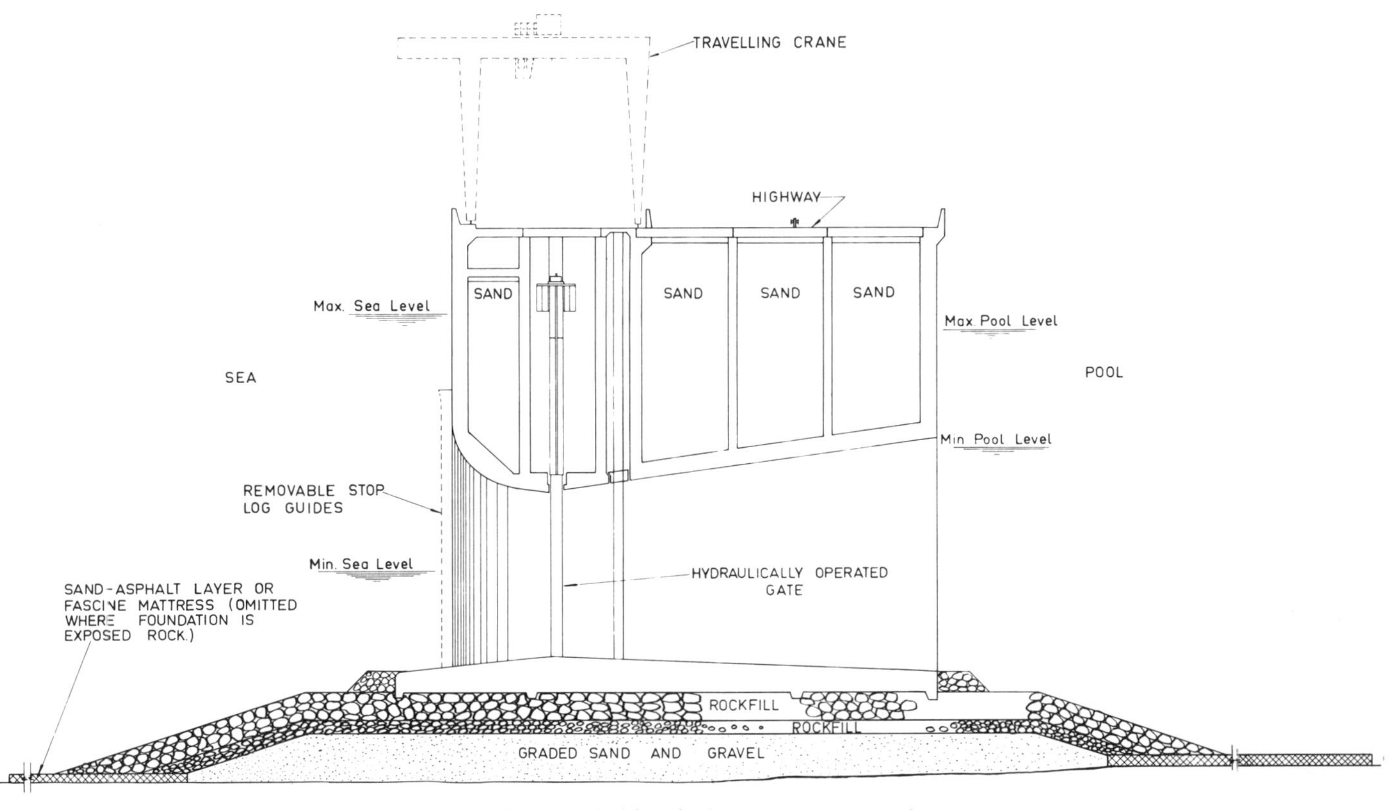

Fig. 3. Refill sluice cross section.

superstructure and much of the mechanical and electrical equipment may be added after the caissons have been positioned.

Various studies made into the performance characteristics of sluices and gates combine to indicate that the most efficient and economical form, where independent of other structures, will be of the submerged venturi type.[2] A caisson form is illustrated in Fig. 3. These caissons will probably be in multiples of two or four gates rather than single units .

Depending on the proportions of the project and the possible severity of tidal currents at certain stages in construction, it may be necessary to provide temporary sluices in addition, to maintain sufficient waterway opening. Their function would be essentially the same as that of the Dutch "culvert caissons" of which two or three patterns have been used very successfully in the Delta Plan.[4][5] It may also be convenient to form other portions of the barrage, especially where constituting abutments between dissimilar sections, by the use of units similar to the Dutch "box caissons".

Navigation locks would not be numerous, if any, and so repetitive caisson prefabrication would be possible only to the extent that several caissons might be used in one lock. A notable example of this method of subassemblies made up into one very large floating caisson is the No. 5 dry dock at Genoa. The base of this dry dock was built of 15 caissons stressed together (Fig. 4), and then the walls were built on this floating raft until the dock was complete. It was then towed into position (Fig. 5) and sunk on a prepared fill foundation. The finished monolith was 855 feet long, 185 feet wide, and 70 feet high, and weighed some 140,000 metric tons excluding ballast.[6]

Of the many examples of concrete caissons, the best known is the "Mulberry" harbour, units weighing up to 4000 tons each were towed across the English Channel and used in the invasion of Normandy in 1944. Since then, the idea has been further applied in harbor and sea defence works in Holland and Libya, and at Montreal, Dublin and Tilbury. Caissons have also been used for numerous underwater tunnels, including examples at Montreal, San Francisco, Antwerp, Cairo and Rotterdam.

Caisson Construction and Handling

The large concrete caissons for the Dutch Veersegat, Volkerat and Lauwerszee works were built as monolithic structures by more or less conventional in situ concrete construction techniques. This was done in a conveniently located construction dock prepared by dredging and diking, and was large enough to contain several caissons which were all completed before the pit dike was breached to float them out.[4] Fig. 6 shows this process in the Volkerat.

In some instances, it may not be practicable to complete the caissons before launching them. The depths of water available in or near the construction dock may be insufficient, or it may be preferred at a stage of partial

Fig. 4. Joining two caissons, Genoa dry dock construction.

Fig. 5. Genoa dry dock under tow.

completion to float the substructure to a quayside or jetty for further building and fitting-out, while the vacated dock bed is used for another substructure. A variation on this general principle was used to build numerous caissons for quay wall elements at Dublin. In that instance, the caisson bases were constructed one after another on a temporary lifting platform which could be lowered into the water to launch them. Permanent "Syncrolift" installations of the type exist at Halifax, at North Vancouver, and elsewhere. An alternative, practicable for small caissons, would be to construct above water level and launch by a slipway.

Another method follows the example of industrialized building, the caissons being constituted of an assemblage of pre-cast elements, mostly panels. By this procedure, construction in the assembly bay can be rapid and factory methods can be used for repetitive production of the constituent elements.

A major tidal power project of the future will require large numbers of turbines and gates. An appropriate form of construction for the concrete works is likely to be caissons of which the cellular substructure parts (at least) are assembled from "factory made" structural elements of high quality concrete, with extensive use of pre-stressing techniques. If the caissons are not completed in this way, they may be floated out at an earlier stage and finished off while afloat, including installation of parts of the machinery and equipment.

Manoeuvring the completed caissons will not, in general, be easy. Open sea towing in the conventional way by tugs is straightforward and can be satisfactory for relatively small units. This was the method used at the pilot tidal power station at Kislaya Gulf, for which the power house caisson was towed some 60 miles from a construction dock at Murmansk. The main tow occupied about 18 hours, using two 2000 h.p. tugs.

Large units require the application of several powerful tugs, some of which will be more effective if strapped alongside or pushing. Caissons are not designed primarily as seaworthy vessels and may be apt to roll or yaw in wind and waves. As with the Kislaya Gulf caisson, stability may be improved and the attitude in the water adjusted by attachment of buoyancy pontoons or by provision of ballast in suitable positions. Push-towing will help to maintain directional control during the tow: pushing alongside will be more effective than from astern for caissons floating deep in the water.

During positioning operations it is possible for pusher tugs to vary the direction of thrust on the caisson much more rapidly than for tugs towing by means of hawsers. Handling of the Dutch caissons at this stage has generally involved a combination of pushing and towing, where possible employing the adjacent portion of the works as an abutment or pivot for one end of the caisson (Fig. 7).

Less conventional equipment and methods could also be used in these operations. Pre-cast segments for the LaFontaine road tunnel at Montreal

Fig. 6. Completed caissons ready to tow at Volkerak.

Fig. 7. Positioning the last Volkerak caisson by tugs.

were winched into position from a casting yard in the bank. For some sites, this procedure might be developed to suit a tidal power house.

Equipment of a type much used for ready and accurate manoeuvring of floating craft is made by the Schottel Company. This is available in package-type units, comprising engine and propulsion unit, suitable for mounting temporarily or permanently on pontoons, barges, etc. The thrust may be turned in any direction while running at full power: these are in effect "outboard motors" of 1000 or even 2000 h.p. Such units could be employed about the works, pushing bottom-dump barges and in other operations besides handling caissons: thus their use might prove economical. Arrangements adopted for handling and positioning caisson units for a road tunnel under the Parana river in Argentina are illustrated in Fig. 8.

The positioning operations will generally be carried out at about the time of slack water, at high or low tide. Deep-draught units will be more conveniently brought over their prepared foundations at high water, and the falling water level thereafter will increase the stability after grounding, during the early stages of ballasting. In order to minimize the amount of vertical movement required after each caisson is correctly located above its foundation, a preliminary ballast adjustment should be made to allow only the necessary bottom clearance while being brought into this position. Until the grounded position is finally checked and approved, ballasting should be with water only, so that re-floating remains possible. The greater weight of sand or even concrete ballast may, however, be necessary before the structure is sufficiently secure as a dam to withstand full design heads.

Extensive investigations will be required into the patterns of water movement about the partially completed works, in order to plan and schedule the necessary manoeuvres. As far as possible a "fail-safe" principle is to be followed, e.g., approaches should be timed and made in such a way that if tugs or winches should fail the resulting drift should be away from possible collision. The likelihood of hawsers breaking is less with tugs or limited-force winches than with ordinary fixed winches.

Foundations and Scour

The stability of the barrage to be built will depend mainly on the reliability of its foundation, the type of which is principally determined by the currents, sea-bed conditions, and constructional and economic factors.

When the sea-bed is composed of solid rock, the current erosion resistance and bearing capacity will hardly present any difficulties although close attention should be paid to the flatness of the bottom. Where the bottom is very irregular, the strength calculations must allow for the less favourable or more problematic support conditions for the founded caissons. The bending and torsional strength and stiffness must be adequate and must meet the tolerance requirements of gates and other moving parts. Con-

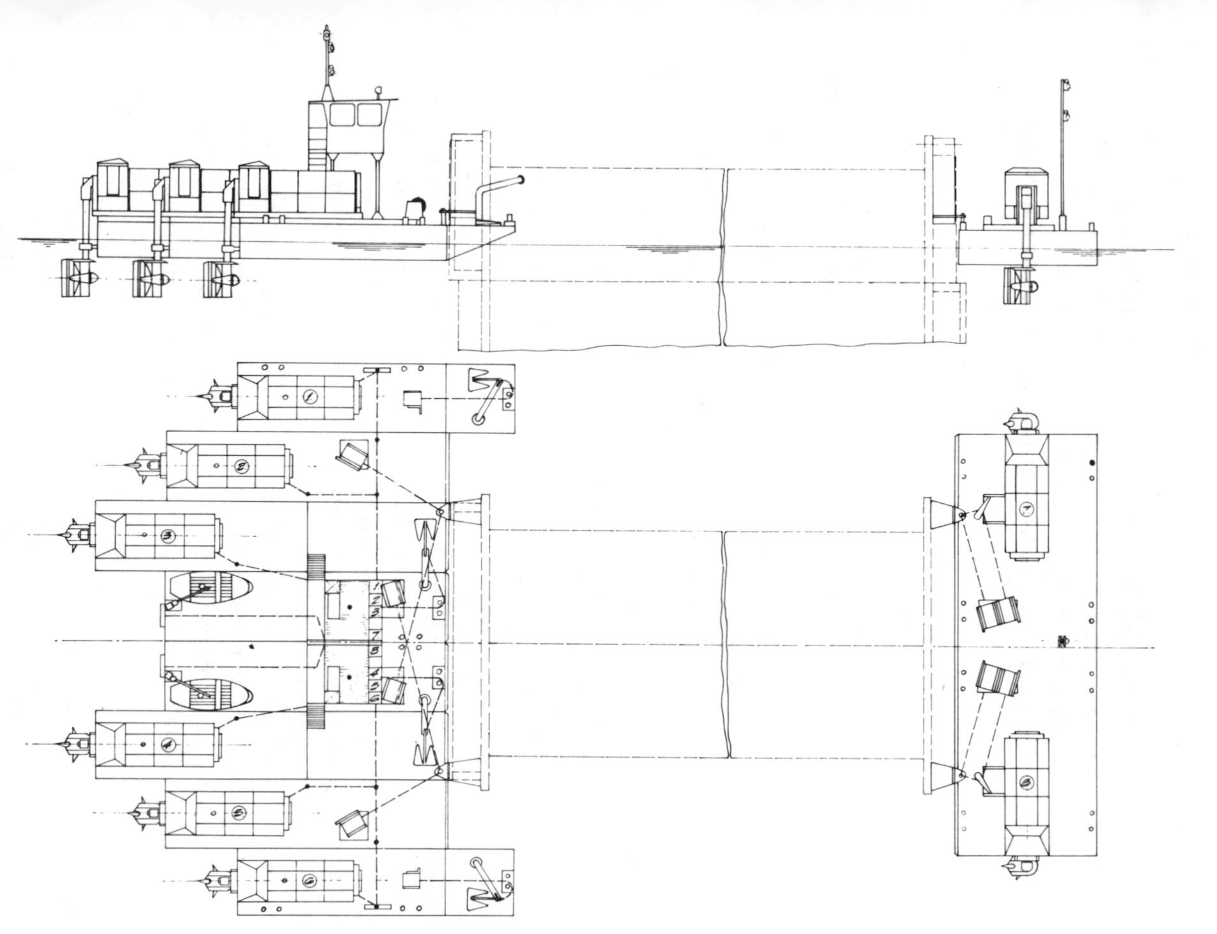

Fig. 8. Tunnel section steered by eight schottel units.

sequently, a bottom as flat as possible is preferable.

When the bottom consists of weaker material such as clay or sand, both the resistance to erosion and the bearing strength must be considered, although a sandy bottom will usually present no difficulty in respect of bearing strength.

The dynamic equilibrium between the current and sand transport is generally disturbed by the placement of caissons. The cross-sectional flow area is decreased, thereby increasing the current velocities and turbulence. Erosion so caused can threaten the stability of the caissons. In the Netherlands, three types of bottom protection are used; fascine mattresses, a filter construction or asphalt mastic.

Protection for a sandy sea-bed underneath the caissons should be sand-tight under the influence of the flows and pressures resulting from difference in head between the upstream and downstream sides of the caissons. Moreover, the normal and shear forces must be transferred from the caissons to the sea-bed.

All these requirements can be met by a filter construction and fascine mattresses. In the fascine mattress sandtightness is achieved by incorporating a layer of woven fabrics. The fabrics are permeable, so as to preclude local hydrostatic overpressures which could start a concentrated leakage flow at the seams and cause severe local erosion. The mattresses should furthermore be flexible enough to follow the bottom configuration very closely, otherwise small gullies could be formed underneath.

A filter built up of granular materials affords a highly satisfactory solution, as illustrated in Fig. 9. The construction consists of layers such that each successive layer protects the underlying one against erosion, the top layer being sufficiently stable without further covering. The sill should be flat, and so the stones used in the upper layer of the sill should not be too big, to facilitate levelling. This may conflict with the hydrodynamic stability of the discrete stones. A solution can be found by grouting with mastic asphalt, discussed in the next section.

Adjoining the sill on which the caissons are to be placed, the sea-bed may require protection against scouring, in order to ensure the stability of the sill. The possible scouring effects near the end of the protected area must be considered. The slope of scour holes can approach the natural slope of sand or even exceed it, which may cause undermining at the toe of the bottom protection. Thus a stone filter construction is not suitable near the toe because no internal cohesion exists and the structure could quickly deteriorate. Fascines and mastic asphalt in this position are more satisfactory, inasmuch as they can follow the slope of a scour hole thus giving partial protection which ultimately leads to a condition of equilibrium.

Vortex streets generated at the ends of the abutments and of caissons already placed in position may determine the intensity of erosive action. By choosing the proper hydraulic shape of the final closure gap, the genera-

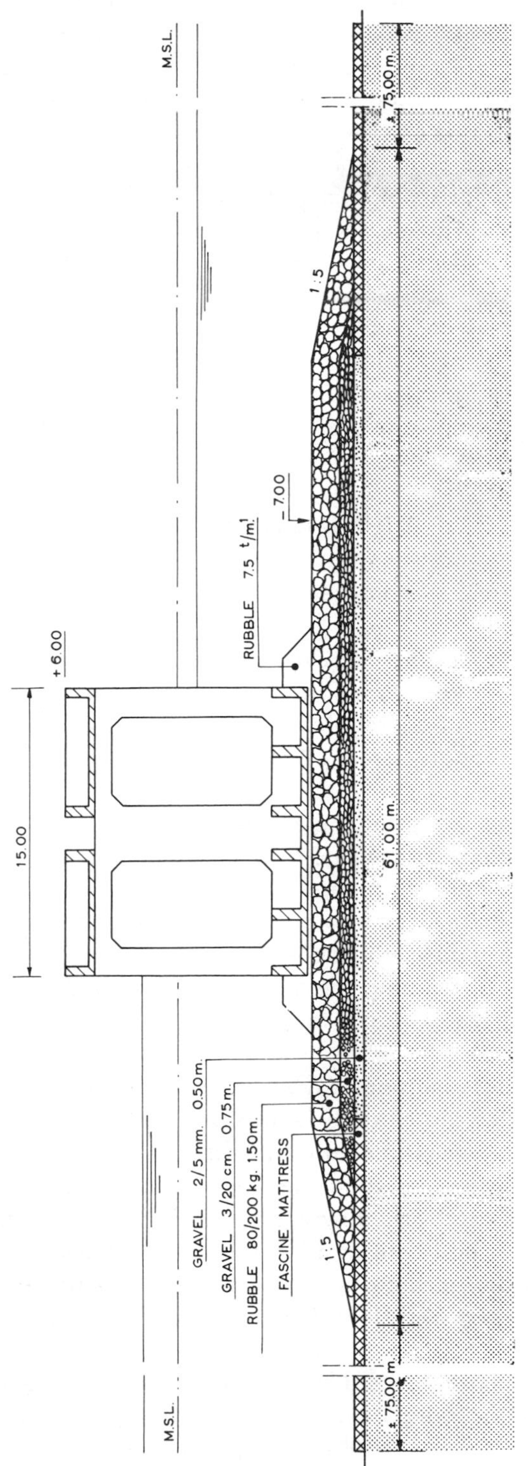

Fig. 9. Cross-section of sill under Volkerak caisson.

tion of large speed gradients, which in turn cause vortex streets, can be avoided. When placing caissons, vortex streets cannot be avoided entirely and any scouring should be limited by placing the caissons in sufficiently rapid succession. Where this cannot be done, more extensive and consequently expensive bottom protection must be applied.

When the bottom consists of compact sand, the bottom protection should extend far enough to avoid sharp pressure gradients in the sand, which might cause slides towards any scour hole and endanger the stability of the caissons. The same requirements apply to the sides of the closure gap. If the sand is insufficiently compact, quicksand can occur due to quick pressure loading (founding of a caisson) or relief (erosion). When this possibility exists, fuller investigations are necessary and some form of consolidation treatment may be required before building can proceed upon such an area.

USE OF BITUMINOUS MATERIALS

Characteristics

Mastic asphalt is a mixture of bitumen, filler, sand and, for some applications, gravel or broken stone, in which the bitumen content is higher than the voids in the mineral aggregates. For example, a typical composition could be: 66% locally dredged sand, 17% limestone filler, 17% bitumen.

The mechanical properties of the mix are determined by the extent to which the voids in the aggregates are overfilled by bitumen and by the viscosity of the binder, which in turn depends on the grade of the bitumen and the temperature. Thus the mix can be poured as a hot phase and is highly viscous after being cooled down to ambient temperature. Design methods have been described for optimizing the mixture of sand and filler in order to achieve minimum voids, using graphs to design mixtures of the desired viscosity.[7]

More recently, these methods have been further refined. It has been found that the viscosity of the mixture is also influenced by the surrounding pressure, due to the small amount of air entrained in the mastic asphalt by the mixing process. In other research the mechanical properties in the cooled phase have been established by measuring stiffness as a function of time under load. It must be borne in mind that mastic asphalt remains a viscous building material, acting as a stiff material in response to sudden loads but showing plasticity and flexibility in following gradual changes of shape.

Stone asphalt is characterized by a double mixing procedure. Big stones (up to 60 kg), after being oven dried, are mixed with pre-mixed mortar to form a mix which is readily pourable in the hot phase but which, due to its high stone content, is highly stable when cold.

Both of the preceding materials are suitable for grouting rock: mastic asphalt for grouting stone up to say 300 kg, and light stone asphalt for grouting rock of bigger categories. By massive grouting all the stones are fixed in position, thus achieving maximum stability against waves and currents. Having attained a stable skeleton before grouting, this type of construction is able to withstand pressure and shear forces comparable with non-grouted rock. It is a relatively impermeable construction, which makes it unsuitable for some applications, e.g., where large uplift pressures are likely.

Pattern grouting avoids this problem by leaving the revetment permeable. Patches of grouting material are applied on the revetment in a fixed pattern in such quantities that approximately 60% of the interstices between the stones are filled. The grouting material is designed according to the size of stone to be pattern grouted, so that excessive flow in the hot phase is prevented. From tests executed by the Delft Hydraulics Laboratory it can be concluded that the stability of pattern grouted slopes under wave attack is comparable with that of rubble mounds covered with non-grouted stones at least five times heavier. These investigations continue.(8)

Asphaltic concrete, the well known road building material, is widely used in the construction of dike revetments above water. In contrast to the preceding materials it has to be compacted artificially after laying, which results in very low voids content (say 4%).

Lean sand-asphalt consists of sand bound by a small amount of bitumen (3 - 6%) and is suitable for bulk application, in general only compacted by its own weight. In contrast with the preceding materials it has only limited durability, due to its higher microvoid content, but it is fairly resistant to the action of waves and currents (up to 3 m/sec.). The permeability equals that of the sand used to make it.

Design of Bottom Protection

In order to increase the hydrodynamic stability of the sill, massive (or nearly massive) grouting could be applied. The grouting should not impair the shear strength of the sill. Tests have shown that as the shear strength is developed, some vertical settlement and horizontal displacement take place. Therefore, some of the interstices among the stones should remain open to give room for the grout, avoiding an internal buildup of pressure which would result in loss of shear strength.

Where the sill rests on an erodible granular bed, scour protection may be provided by a mat of mastic asphalt. Because of its impermeability, hydrostatic uplift pressures may occur, especially towards the downstream toe. In some situations, occasional lifting and release of accumulated water may be tolerated but generally it will be more satisfactory to ballast vulnerable areas with rock. A limited amount of rock may in any case be dumped over the mat to roughen the surface; otherwise velocities near the bed and beyond the toe might be excessive. It is normal to thicken or ballast the toe to counteract any possibility of it being turned over by reverse currents. Large-scale tests have shown that a mastic asphalt slab, even if ballasted by stone, will follow any changes of bottom profile without being cracked. The overlap of joints between adjacent strips or sheets is not always watertight; this is usually harmless and may permit local pressure relief.

Placing in Undersea Locations

A satisfactory method of placing mastic asphalt under water is decisive for its feasibility as a sea-bed protection. Methods have been described for applying mastic asphalt in situ, using a pouring-pipe provided with a distributor, and mounted on a vessel carrying the mixing plant.[7][9] Development was started in 1956 by Bitumarin, in close cooperation with the Netherlands Rijkswaterstaat and Royal/Shell. As a consequence of the tests run with the pilot plant "Dorus Heijmans" and the successful construction of approximately 80,000 m^2 of asphalt mastic bottom protection in the Grevelingen sluicing gaps, after full consideration Rijkswaterstaat commissioned Bitumarin to build a specially designed asphalt ship, the "Jan Heijmans", (Fig. 10), which started work in the Brouwershavense Gat in 1968.

The "Jan Heijmans" is designed to show negligible movement in waves up to a metre high, with a period of 5 seconds. Her asphalt drying and mixing plant has a maximum capacity of 250 tons per hour. The mastic asphalt is fed continuously via an agitator tank to the deposition apparatus, which essentially consists of a vertical feed pipe and a distributor tube 5m wide, fitted with eight pairs of orifices. Its working range reaches to 25m below the surface, but can easily be increased by elongating the feedpipe. (The low viscosity at pouring temperature (approx. 800 poises at 140° C) and the density of 2 tons/m^3 permit gravity feed.) The viscous resistance is concentrated in the orifices, serving the double purpose of maintaining even distribution and keeping the water out, thus preventing steam formation inside the apparatus. A feeler is attached to the distributor, operating a hydraulic height compensator which ensures that the nozzles are always at the right distance from the sea-bed.

Uniform thickness of the deposited carpet results from a constant uniform discharge of mastic asphalt and constant speed following a prescribed track. For this purpose, six centrally controlled electric anchor-warping winches are installed. The position and speed of the ship are continuously measured and recorded by tellurometer and speedometer, checked periodically by line measurement. The control panels for asphalt production and deposition, and for all movements of the ship, are installed in the control room high amidships.

To minimize any unavoidable variations in thickness of the mat,

Fig. 10. The 'Jan Heijmans' at work.

generally the tracks are laid tile-wise, in two or three lanes. For checking the thickness after completion a special penetrometer apparatus has been developed.

Slope Protection

Where sand is abundant, as in the Netherlands, it can be used at relatively low cost in hydraulic fill for dikes and estuary barrages. Bituminous materials have proved to be very suitable for the protection of these works against waves and currents.

Most of the applications are well illustrated in the Brouwershavense Gat dam (one of the Dutch Delta dams). (Fig. 11). Generally on the flats in the tidal zone a so-called "pancake" is made. The hydraulic fill is retained within dikes which give temporary protection. For these dikes any conveniently-handled material of fair resistance to wave action is suitable. In the Netherlands, mine waste is often used but, of late, lean sand-asphalt has come into use, being relatively inexpensive, more resistant than mine-waste, and easy to handle.

For toe protection the mastic asphalt slab, poured in-place, is now generally in use, varying in thickness from 0.1 m to 0.2 m. Two types of slope protection are generally in use: grouted stone and rolled asphaltic concrete. Both types are impervious and therefore susceptible to seepage pressure, which can be reduced by the provision of a permeable bottom strip such as gabions, between the asphalted surface and the sheetpiling at the foot of the slope. Alternatively, use of lean sand-asphalt rather than mine-waste decreases the permeability of the substructure and can result in savings in the required thickness of the main protection. The thickness required depends on the severity of wave action, but is generally dictated by the required weight for balancing seepage pressures.

In the tidal zone, grouted stone is generally employed because of the ease of construction in a temporarily wet environment. Above the tidal zone, where the required thickness is only small, asphaltic concrete is more suitable. Where tidal ranges are large, which will be the case in tidal power schemes, uplift pressures under an impermeable blanket could necessitate excessive thickness or extra rock ballast. Use of a conventional filter construction would avoid this. The uppermost stones must be stable against waves and currents: if this would normally require very large rocks, considerable savings could be made by using smaller stone and pattern-grouting.

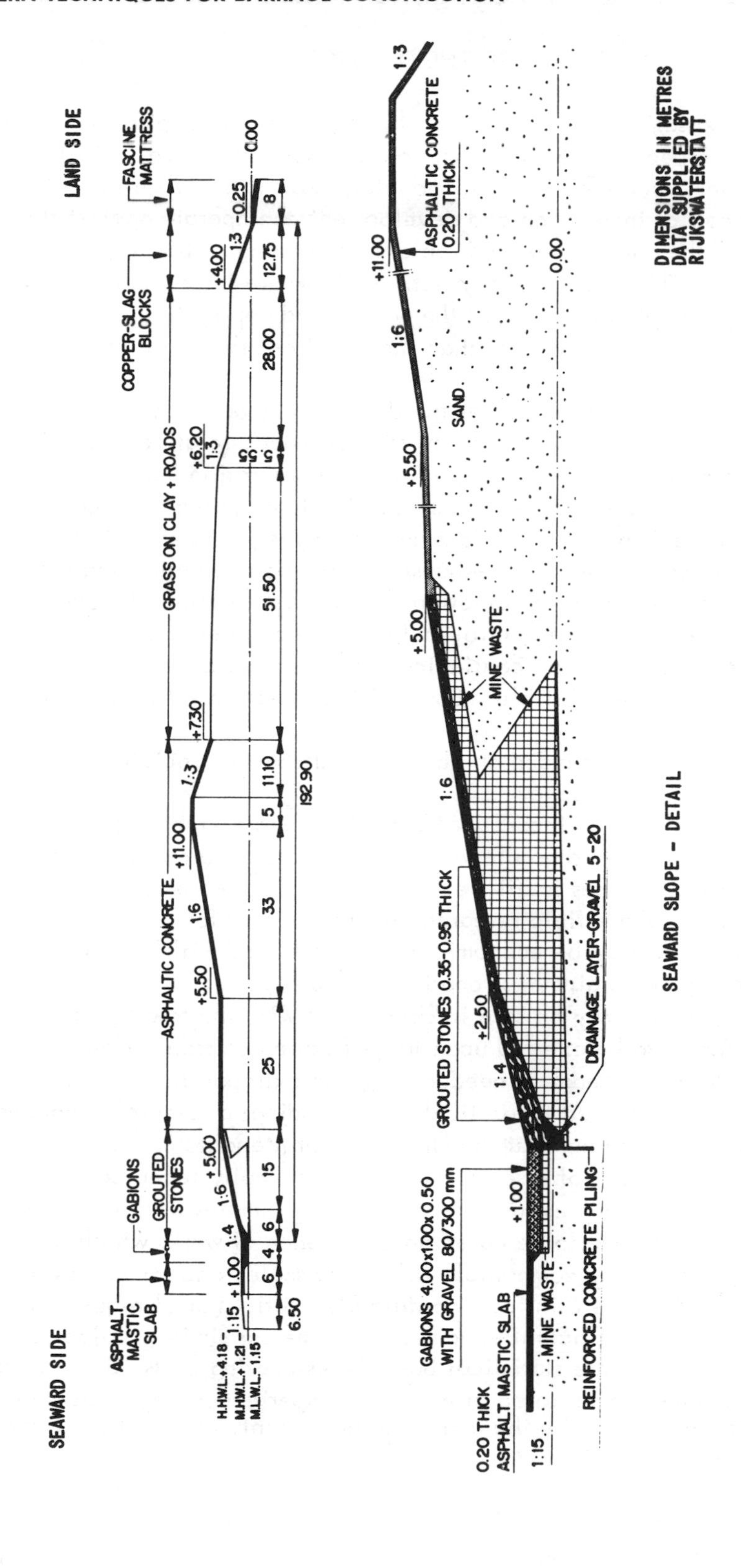

LAND SIDE
SEAWARD SIDE
FASCINE MATTRESS
COPPER-SLAG BLOCKS
GRASS ON CLAY + ROADS
ASPHALTIC CONCRETE
GROUTED STONES
GABIONS
ASPHALT MASTIC SLAB
H.H.W.L.+4.18
M.H.W.L.+1.21
M.L.W.L.-1.15
0.00
-0.25
+4.00
+6.20
+7.30
+11.00
+5.50
+5.00
+1.00
1:3
1:6
1:4
1:15
8
12.75
28.00
5
51.50
11.10
33
25
15
6
4
6.50
192.90
ASPHALTIC CONCRETE 0.20 THICK
SAND
MINE WASTE
GROUTED STONES 0.35-0.95 THICK
+2.50
DRAINAGE LAYER-GRAVEL 5-20
GABIONS 4.00x1.00x0.50 WITH GRAVEL 80/300 mm
REINFORCED CONCRETE PILING
0.20 THICK ASPHALT MASTIC SLAB
SEAWARD SLOPE - DETAIL
DIMENSIONS IN METRES
DATA SUPPLIED BY RIJKSWATERSTATT

WHAT OF THE FUTURE?

With the passage of time, inflation tends to increase capital and operating costs, especially for items which are labour-intensive. On the average, however, such increase is in monetary costs rather than in real costs. The process of innovation and development can operate against this increase and, in the case of new and rapidly developing fields, may more than outweigh it. This is particularly noticeable where large scale mechanized production is applicable, e.g., the widespread application of plastics in recent years, often at lower cost than the metal or other articles previously used.

There may be several aspects of the design and execution of tidal power projects where real economies may still be made, relative to comparable cost levels hitherto accepted. Confining consideration here to construction techniques and related matters, it should firstly be noted that scope for more intensively mechanized operations and factory-style manufacture in civil engineering works is far from exhausted. The possible advantages are more likely to be realized in combination with the application of modern techniques of project schedule control and of value engineering.

With further research and continuing technological development, other substantial progress is also possible. Some of the directions of interest are as follows:

1. More information is needed on abnormal tides with associated maximum levels and heads frequency.
2. Amid great volumes of theoretical consideration and laboratory experimentation, established usable knowledge on the occurrence and magnitudes of wave forces is still tentative and rule-of-thumb. There is a need for large scale field research, so that designs for permanent and temporary works may be economically proportioned with sufficient but not extravagant margins of strength and stability.
3. Something similar may be said in connection with the towing and mooring forces to be exerted upon large floating caissons of awkward shape. More knowledge is needed to guide extrapolation from past experience and limited theoretical understanding: a research programme on this subject is in hand at the University of Manchester.
4. A much-needed form of research which is not wholly technical is in the field of prospective construction costs. The estimates which can be made at present have to be based on experience of works which are generally not strictly comparable in type or scale, and sometimes on past prices which were in fact inordinately cautious or else quite unprofitable. Short of actual execution of numerous similar projects, there cannot be a true statistical basis for estimating costs, but it does seem that more concentrated attention by experienced engineers and contractors would reduce the degree of uncertainty about likely work

costs.

5. Bituminous grouts and asphaltic concrete have found application where they can be used with less expense or more effect than alternative materials and methods. As well as gradual refinement of current technology and more advanced machinery for producing and placing such materials, it may be anticipated that new types of low-cost matrix materials will be introduced from time to time, probably originating from the petroleum and petro-chemicals industries.
6. Industry has already furnished useful protective materials for marine use, for example nylon mesh used in asphaltic mats. Other materials may find further application in barrages, e.g., in repetitive forming of complex shapes.
7. As a possible alternative or supplement to the use of tugs in areas where these are not readily available, the Schottel screw deserves consideration. Fig. 8 shows an arrangement which would enable 10,000 h.p. or more to be concentrated on one floating caisson unit.
8. The placing and manipulation of underwater fill and excavation or levelling of sea-bed areas may be considerably aided by introduction of remotely controlled underwater machinery. For example, an underwater bulldozer is being developed in Japan. Various forms of free-standing or moored working platforms might be used: a very effective moored platform was used for grading operation for the San Francisco Bay subway tunnel sections. Special vessels have already been developed for placing fill underwater. These may be increased in capacity and refined in the precision of their control.
9. An important aspect in the security and cost of barrage works is the permeability and strength of sea-bed material underneath. While these properties can be modified by compaction or grouting, the effects in poorer natural materials are not at present very reliable. New and more powerful compaction equipment is constantly being developed for use ashore: relatively little effort has been directed to its use underwater, and there is need for research on its effectiveness.
10. New chemical grouts suggest the possibility of much better control of foundation properties in future, thus providing a worthwhile cost savings.
11. Where very large future excavation and fill operations are contemplated, it is nowadays the fashion to suggest nuclear explosives as a solution. There may be some remote locations, unpopulated and undeveloped, where such explosions could be released harmlessly to shift huge volumes of excavation or to block an inlet. However, such sites do not include places where there are towns and ports in the vicinity, subject to damage by blast and waves, hence excluding the Bay of Fundy or the Bristol Channel. The use of nuclear energy does not appear to offer substantial advantages in connection with tidal

power barrages.

CONCLUSION

It has been shown how some recently developed techniques could be applied to tidal power construction. The current velocities in the case of the Bay of Fundy are closely similar to those found in the Delta Plan. These velocities are among the most important parameters in planning such barrages so that the technical feasibility of the methods outlined is already within the experience obtained in Holland.

In our last section a few research trends have been outlined, some of which may be helpful in the future. Among these, particular attention is drawn to the need for more knowledge of the occurrence and magnitude of wave forces. The fuller understanding of these might result in significant design economies.

ACKNOWLEDGMENTS

The authors gratefully acknowledge the assistance accorded by the management and staff of Engineering and Power Consultants (J. D. Gwynn), Rijkswaterstaat (F. Spaargaren), and Bitumarin (A.J. Woestenenk), and for the information and photographs kindly provided by Schottel-Nederland NV and Schottel England Limited, by Societa Fincosit, Genoa, and by Collen Bros. (Dublin) Limited.

REFERENCES

1. Revue Francaise de l'Energie, Paris (Sept. - Oct. 1966).

2. Straub, L. G., "Hydraulic Design Schemes for Tidal Barriers and Closure for Proposed International Passamaquoddy Tidal Power Project". St. Anthony Falls Hydraulic Laboratory, University of Minnesota (March 1959).

3. Golder, H.Q. and D. J. Bazet, "An Earth Dam Built by Dumping through Water". ICOLD 9th Congress, Istanbul (1967).

4. Lingsma, J. S. , "Holland and the Delta Plan". Nijgh en van Ditmar, Rotterdam, 3rd Ed. (1966).

5. Hydro-Delft No. 15, Delft Hydraulics Laboratory, Delft (1969).

6. Borzani, G. and P. Vian, "La Costruzione del Bacino di Carenaggio No. 5 nel Porto di Genova", 'Costruzioni' No. 86, pp. 181-201, Casa Editrice La Fiaccola, Milan (1963).

7. Kerkhoven, R.E., "Recent Development in Asphalt Techniques for Hydraulic Applications in the Netherlands". Ass'n Asphalt Paving Technologists, Philadelphia (1965).

8. d'Angremond, K., v.d. Weide, J., Span, H.J. Th., and A. J. Woestenenk, "Application of Asphalt in Breakwater Construction". Coastal Engineering Congress, Washington (1970).

9. Meijer, P.C., "Mastic Asphalt as Bottom Protection", International Dredging and Port Construction (July and August 1968).

PRECAST FLOATED IN CONSTRUCTION AS APPLIED TO TIDAL POWER DEVELOPMENTS

P. R. Tozer* and T. J. Sluymer**

INTRODUCTION

Detailed investigations performed to date of the potential power resources of the Bay of Fundy have proven beyond a doubt that the harnessing of this power is quite feasible yet economic considerations appear to preclude the development of this resource at least for the immediate future. Without innovation and development of radically new methods of construction or power generation, there can be little hope of economically unlocking the vast tidal power resources of the Bay of Fundy. In past investigations, numerous design and construction concepts have been advanced, all aimed specifically at reducing construction costs. Although these concepts have in the main been proven, the large scale of the tidal undertaking envisaged is still virtually unsupported by precedent. To date, only a few tidal developments including La Rance and Argentat in France, Kislaya Guba in Russia, have been brought to fruition, and these are comparatively small developments, experimental in nature.

The consideration of conventional, proven methods of construction as applicable to a tidal scheme indicates that the use of cofferdams is certainly feasible and would provide maximum protection for working areas and a guarantee of the quality of workmanship being performed due to direct control. However, it has been estimated that cofferdams alone probably represent some 20 to 25 percent of the total capital cost of a development. That this almost certainly will lead to the exclusion of such a scheme from a tidal development is self-evident, in terms both of direct capital costs and of financing expenses over the very prolonged period of time necessary for this type of construction.

Although the cofferdams associated with a development are only one of

---

*Senior Engineer, H. G. Acres Ltd., Dorchester Road, Niagara Falls, Ontario, Canada.
**H. G. Acres Ltd., Dorchester Road, Niagara Falls, Ontario, Canada.

many aspects requiring refinement, the object of this paper is to give brief consideration to the concept of dispensing with cofferdams and investigating the possibilities of building precast sections in the dry and floating them into position, sinking and anchoring them on prepared foundations.

The prime problems which must be overcome and which are associated with the construction of a tidal development may be roughly placed in two categories, namely "natural", which includes water depth, tidal ranges and current, climate, width of closure, bed profile, foundation material and overburden type and depth if any, scour and silting; and "developed" which is the increase in water flow speeds induced by closure of the water passage during construction. With precast floated-in construction, units of a development may be fabricated in the dry under conditions virtually unaffected by tidal range, current and climate. Apart from performing its required in situ function of power house, service bay or sluiceway, any unit must be designed not only for stability in the final operating location, but must accommodate within its design the requirements for stability during floating and sinking operations. In order to reduce the "developed" problem of increase of water velocity with increase of closure, units should be capable of allowing the passage of water when in their sunken position.

Precedents

Operations with large floated-in structures have been performed, successfully, including the Mulberry Harbour, light houses in the St. Lawrence River and overseas, dike closures in Holland, Deas Island in Vancouver and the Lafontaine and Chesapeake Bay tunnels.

The Chesapeake Bay tunnel utilized floated-in segments approximately 34 feet in diameter and 300 feet long. The tidal range was of the order of two feet, and the current velocity two feet per second with a slack period of approximately two hours during which to sink the tunnel sections in position. The water depth was a maximum of 40 feet and the depth of excavation was a maximum of 100 feet. There was a 24-hour period available after grading the sand bed at the bottom of the trench to place the tunnel segments. If they were not placed within this period it was necessary to regrade the trench bed.

The Lafontaine Tunnel in Montreal comprised tunnel sections 360 feet long, 120 feet wide and 25 feet high. The riverbed excavation depth was 100 feet and the tunnel sections and sinking operations were designed for a current velocity of five feet per second with no tide.

In Holland, extensive construction work has been carried out entailing the placing of large caissons for dike closure under extremely difficult tidal, current and scour conditions.

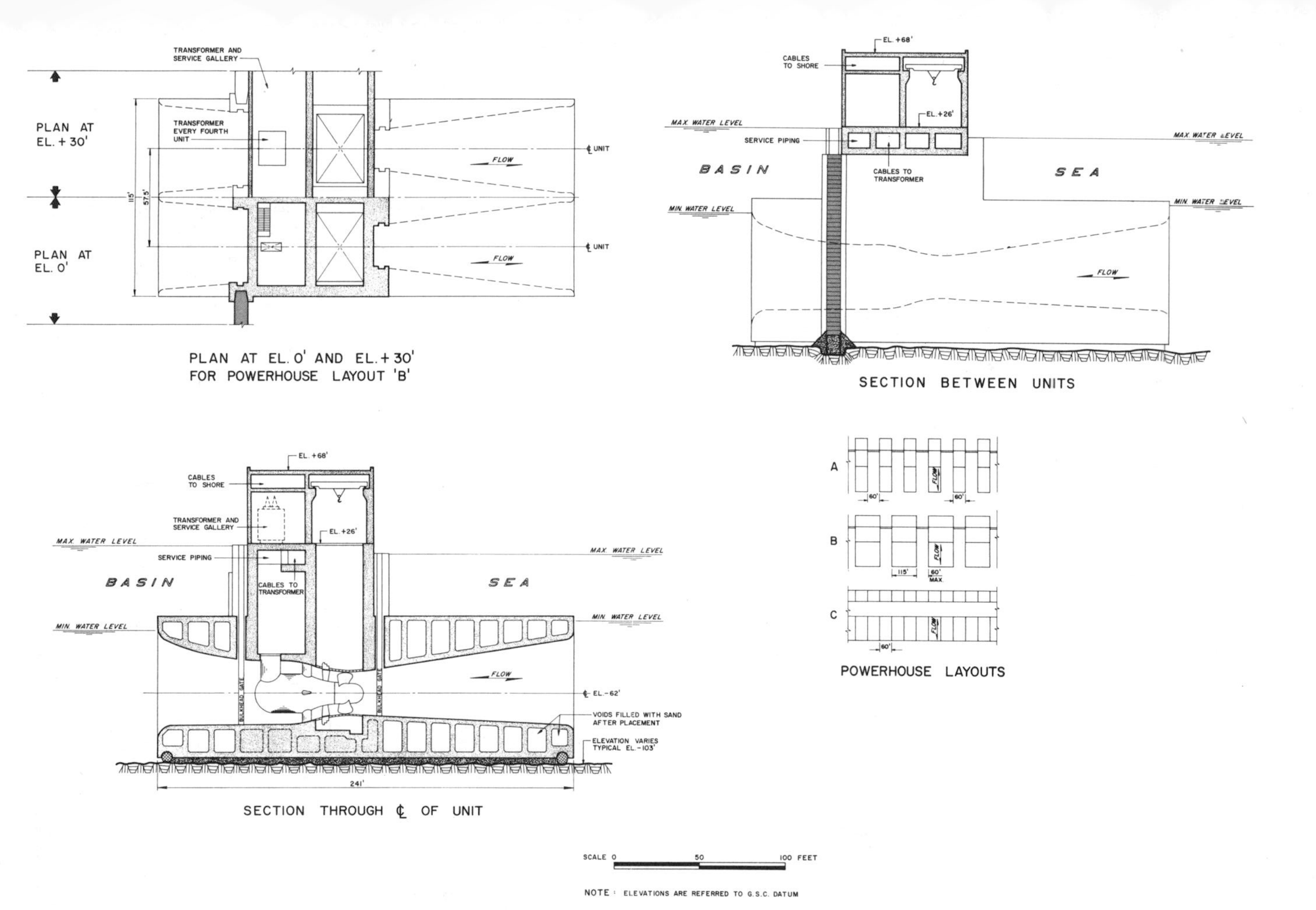

Fig. 1. Double effect generation enclosed powerhouse.

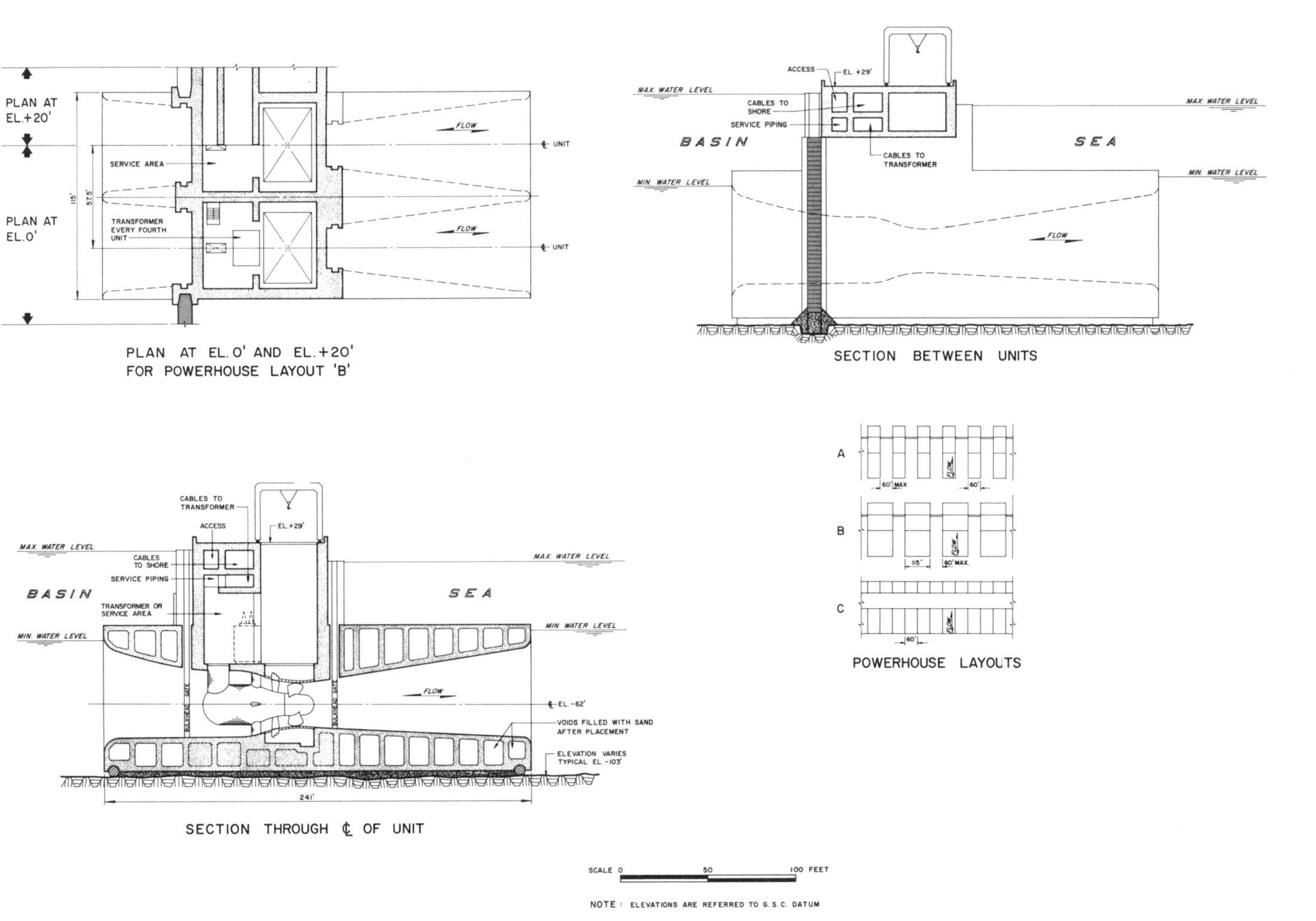

Fig. 2. Double effect generation open type powerhouse.

Alternative Methods

There are basically two types of floating operations, and these may be defined as follows:

1. Tunnels or other structures that will submerge and require barges or other means to maintain stability and control during sinking.
2. Structures that do not completely submerge and can be placed by flooding, the stability of the structure being maintained without assistance from outside sources, such as barges.

The Chesapeake and Lafontaine tunnels were floated using system (1) above, but with a tidal development it is method (2) that will probably predominate.

An indication of the size of single and multiple units possible for a tidal power development is given on Figures 1, 2 and 3. These show typical examples of structures that have been developed to house single and double effect generating units, with enclosed or open type power houses. The magnitude of range of the Bay of Fundy tides increases with passage up the bay, with the result that the spring tide of 30 feet at Saint John, attains a range of 46 feet in Chignecto Bay and 53 feet at Minas Basin in the head of the Bay. The calculated natural flow velocities at maximum tides in Chignecto Bay vary between four and nine feet per second.

It will be seen from the foregoing that precedents for precast floated-in construction have been created, and that the basic problems associated with that type of construction have been solved. However, the large head variation, current velocity and scour inherent with a tidal power development require that considerable investigation and research be undertaken to improve existing and develop new methods of floating, sinking and anchoring units. All investigations for tidal works must be undertaken with extensive resort to model study, prototype development and field testing where warranted.

The tidal current velocity is a major factor in the design and construction of floated-in structures. It affects not only the floating, locating and sinking of precast sections, but also the scour action at the seabed and the silting of foundation excavations, or removal of preplaced bedding material. A suitable anchorage system may be designed to ensure one hundred percent control of floating units from the dry dock to placing.

The major problems of foundation preparation, including silting or scouring, require extensive research and will be commented upon in more detail later.

A general description of the various individual operations entailed with precast floated-in construction follows.

Forming, Casting and Launching

The fabrication and casting of any unit can be undertaken in a coffer-dammed, drained area adjacent to the shore, in a specially constructed dry

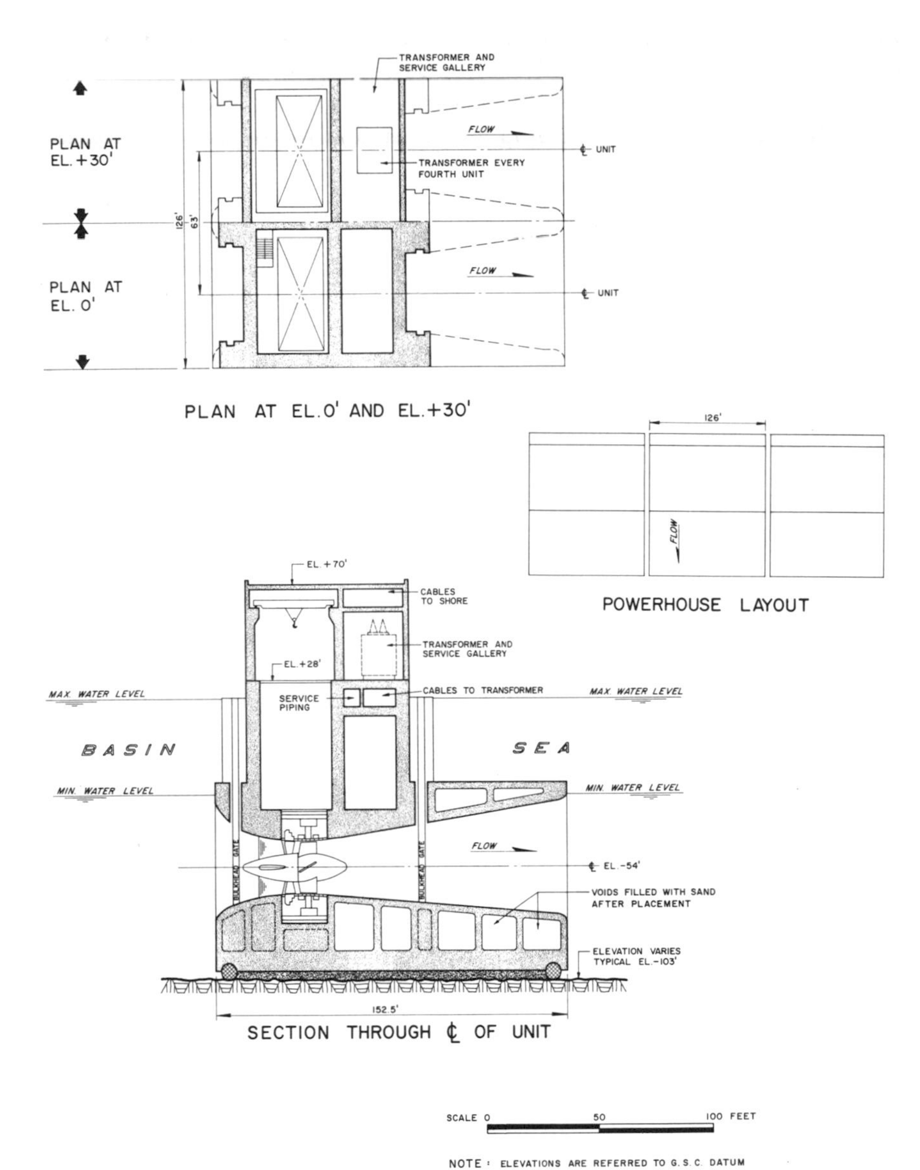

Fig. 3. Single effect generation enclosed powerhouse.

dock, or on shore above high water level. The main advantage of the cofferdam or dry dock casting area is the relatively simple flooding procedures when the unit is ready for launching. The initial cost of the construction facilities will be high, but the number and sizes of units to be cast should result in a reasonable cost of dock per unit.

The physical size of units, if cast on shore above high water level will require very sophisticated handling and slipway systems which may not in fact be practicable.

The weight of sections in both cases must be kept to a minimum to reduce draft and, consequently, the channel dredging necessary to launch and float the sections from shore. It may well be that the natural tidal cycle could be used to advantage, with all launchings taking place at high tide, thus minimizing any requirement for channel dredging.

Should any unit of a development prove to be too unwieldy, consideration could be given to the possibility of casting and floating the unit in two sections. For example, with a power house unit, the draft tube and power house superstructure could be formed as separate items. The base section is then floated out, sunk and anchored in position, and successive sections are floated out and connected to the base.

However, this procedure should only be considered as a last resort for it will more than double the difficulties that would be experienced with one piece operation, including:

1. The base section will be completely submerged requiring barges or similar equipment to maintain control during sinking.
2. The amount of work entailed is doubled by virtue of the number of separate sections to handle being doubled.
3. A complicated seal will have to be developed between the sections.

A further serious problem associated with the concept of floated-in construction for a development of this nature is the interconnection of successive units, which will be discussed later.

Floating

The movement of floating sections can be achieved in a number of ways including:

1. Self-propulsion, with each section being provided with a removable system of propelling motors so located as to permit total maneuverability.
2. Tug operations.
3. A pattern of fixed anchor points upstream and downstream from the development with a cable winching system to move sections hand over hand into position.

The successful system adopted at Chesapeake comprised slinging the tunnel sections between two barges towed into position with tugs. One major point with this method was the filling of the floating section with sufficient

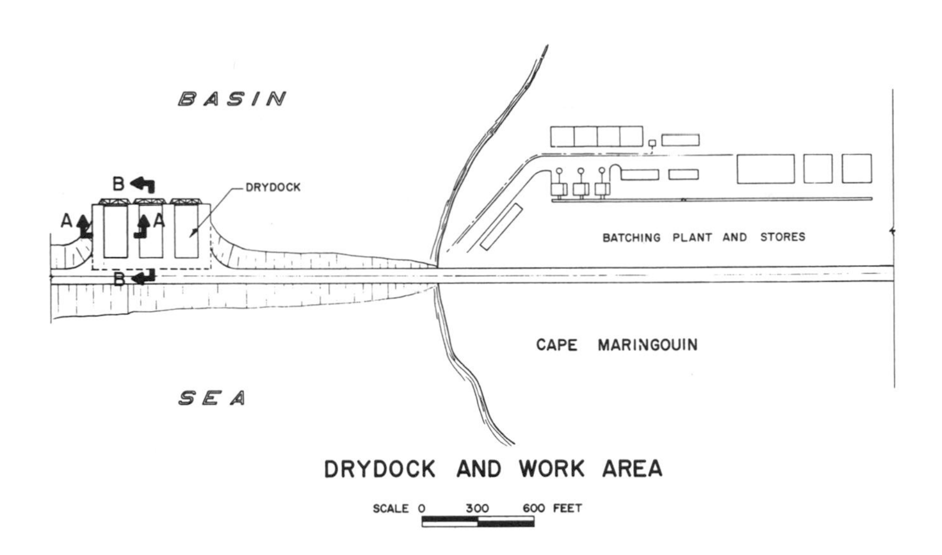

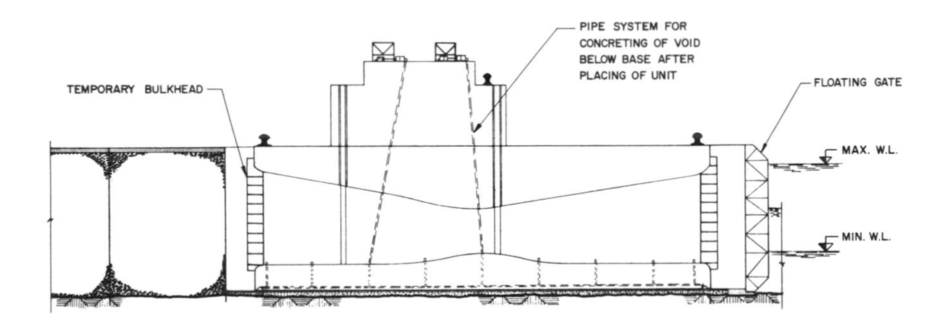

SECTION B-B

CONSTRUCTION OF UNIT IN DRYDOCK

SCALE 0 30 60 FEET

PROCEDURE

1. BUILD POWERHOUSE UNIT IN DRYDOCK
2. ATTACHED UNIT TO TWO BARGES, EACH EQUIPPED WITH TWO WINCHES CONNECTED TO ANCHORS
3. UNIT MOVED LATERALLY BY TENSIONING TWO LEADING CABLES AND SLACKENING TWO TRAILING CABLES
4. DISCONNECT AND RECONNECT CABLES TO OTHER ANCHORS, AS REQUIRED
5. REFLOATABLE CONCRETE ANCHORS USED TO ATTAIN FINAL LOCATION.

Fig. 4(a). Proposed method of construction.

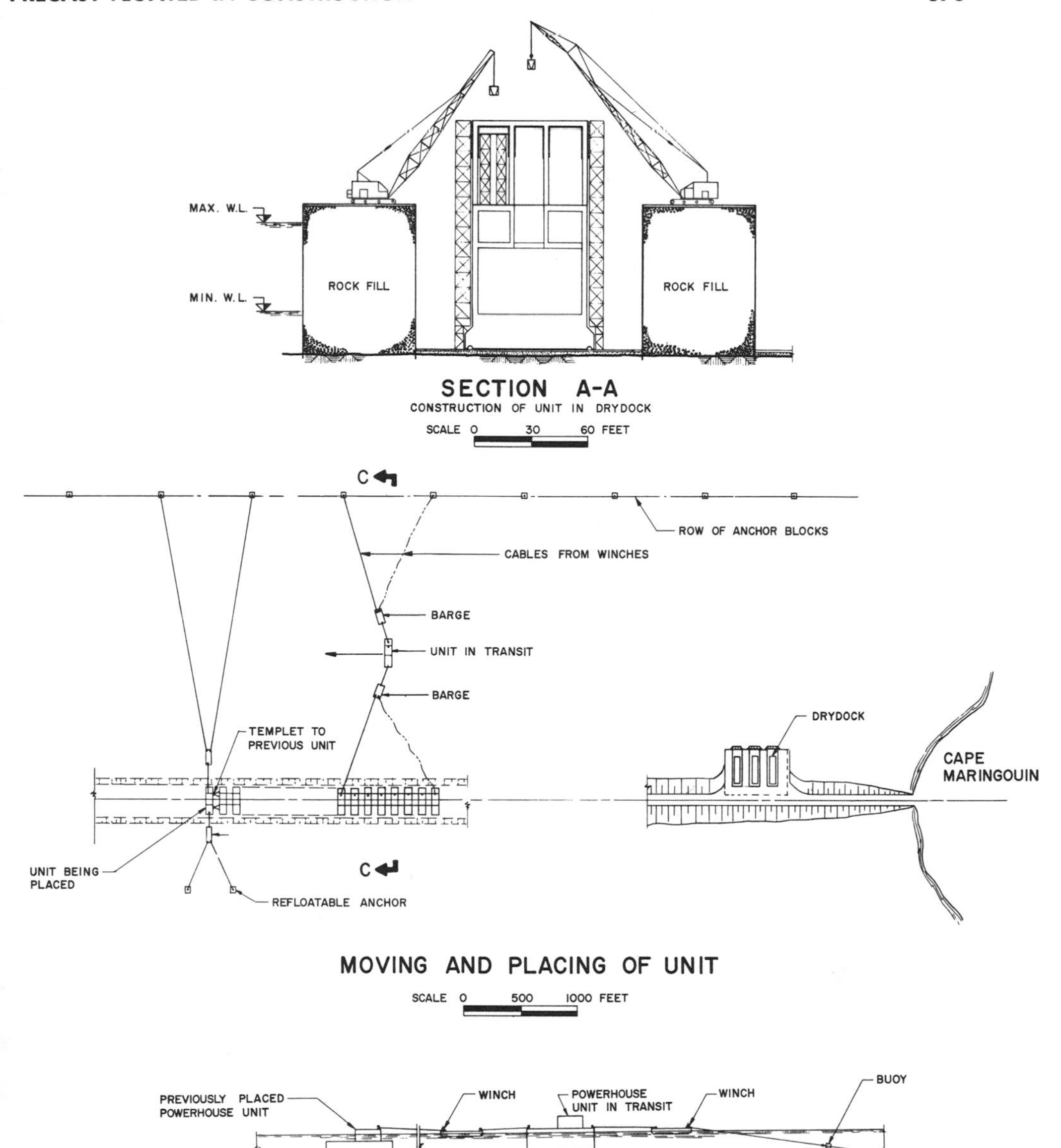

Fig. 4(b). Proposed method of construction.

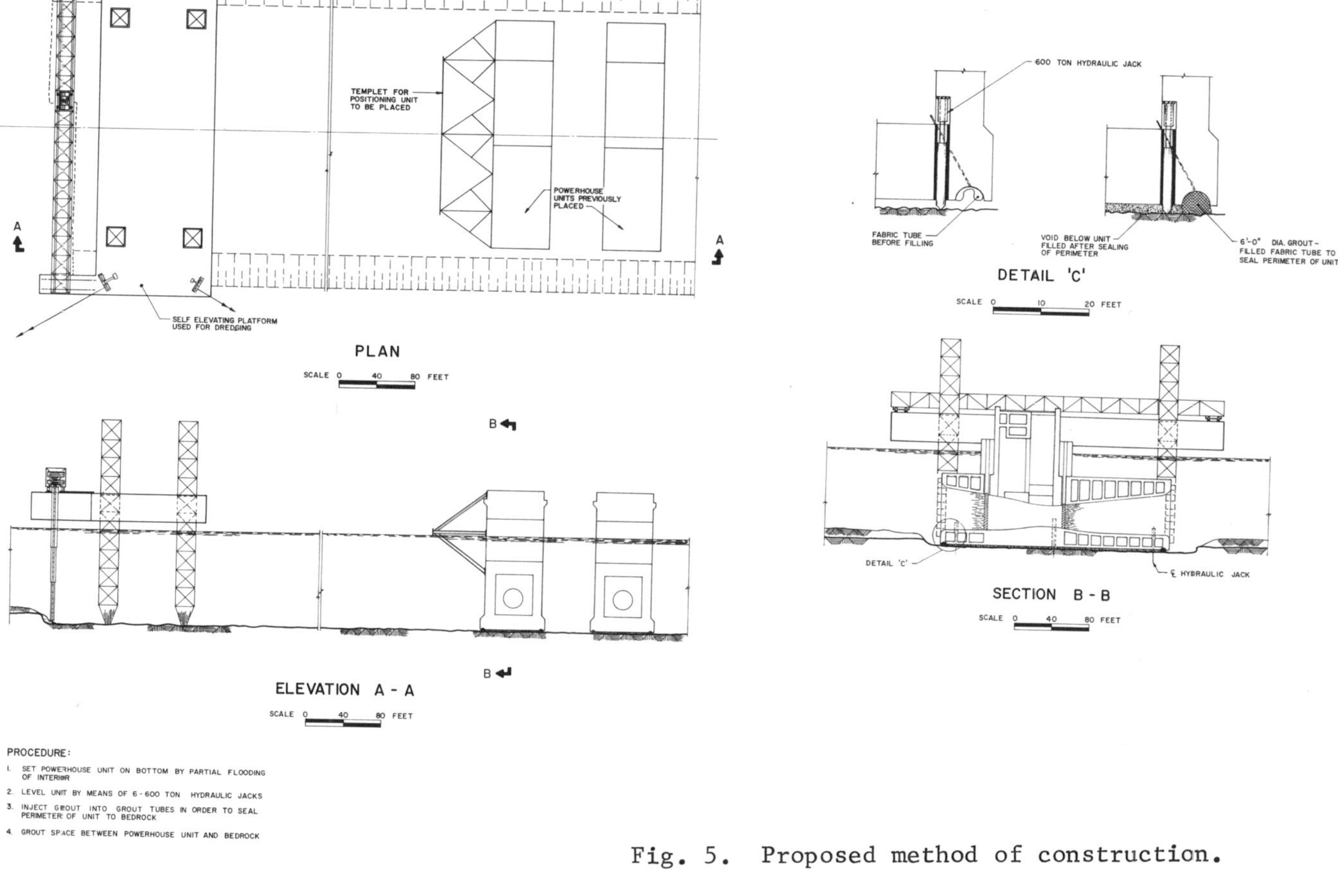

Fig. 5. Proposed method of construction.

ballast to create barely positive buoyancy. During the Chesapeake operations, the tunnel sections were moved with about two feet of freeboard. This resulted in the minimum freeboard being presented to wind but a maximum area to current velocity, which in this case was very small. As a consequence, once a section was in its correct relative position, sinking could be achieved with the addition of a minimum amount of ballast.

With a tidal development, current velocities are high, and the merits of employing tug operation (with or without barges) or self-propulsion for section movement and control would be extremely dubious. Should tugs or self-propulsion units be employed for moving and placing sections, it is probable that a capacity of 10,000 to 20,000 horsepower would be required, dependent upon the current velocity, with guaranteed continuous mechanical performance and precise timing of operations.

An anchorage system, with anchor points upstream and downstream from the development can be designed for a maximum current velocity so that timing of the operation will be of little importance. The possible application of this method to a tidal development is shown on Figures 4 and 5. To give an indication of the magnitude of the force to be controlled by the winch cables, consider a power house unit 60 feet wide, with a draft of 70 feet. With a current velocity of eight feet per second and a shape factor of 1.3, the maximum current force would be of the order of 175 tons.

For a tidal development under natural conditions such as those encountered in the Bay of Fundy, any of the above methods would require considerable analysis involving the use of models.

Sinking and Anchoring

Once a unit is floated to a final location and held in position, the sinking operation would, preferably, be performed at low slack tide by the addition of water ballast. The whole operation, however, will have to be performed during a relatively short slack water period, and positional control and rate of lowering can be critical. Positional control includes holding a section in a horizontal plane at all times within very close tolerances to offset the effect of current forces. This virtually eliminates the use of tug boats due to the risk of mechanical failure. The accuracy that will be achieved during a sinking operation of this magnitude is unlikely to be better than plus or minus 12 inches.

The method by which a section will be anchored and sealed to its foundation will depend upon the type of foundation constructed. The foundation selection will, in turn, depend upon the geology of the site selected. There is no doubt that anchorage can be achieved by gravity loading, shear keys, caissons or some other methods. A serious problem is the sealing of the section to the foundation to avoid piping and foundation erosion under a 40 to 50 ft. reversible head.

Season

The season for floating out and placing units will be limited by virtue of weather conditions. Every consideration should be given to casting sections and stock piling them during the winter season for use in the months suitable for water operations. One possible method of storing sections would be by sinking them in shallow water for recovery at a later date.

DISCUSSION

The above paragraphs describe briefly some of the basic operations and problems associated with a tidal power development using precast floated-in construction. There are many other requirements which are also extremely important, particularly the sealing between units, their interconnection, both mechanically and electrically, anchoring of units to their foundation, and the foundation construction.

A tidal power development will require power houses for the generating equipment, transformer bays, service bays and other ancillary areas. Every attempt should be made to simplify the design of a basic precast unit so that it could, with minor modification, serve any function, whether it be power house or service bay. A further simplification would be to construct groups of possibly two or three units in one section. This would greatly reduce the extent of the floating and sinking operation. However, current forces and handling forces would be increased proportionally.

With large tidal ranges and considerable current velocities to be overcome, it is highly unlikely that adjacent units or groups of units could be placed with a great degree of vertical or horizontal accuracy.

The system of sealing between units would have to be extremely flexible. A possible solution to the problem of sealing would be to deliberately keep successive sections separated by as wide a gap as possible. When closure is required, the gaps could be sealed by means of stoplogs tailor-made to take up any discrepancies of elevation or location between adjacent sections. An added benefit of this form of construction would be a reduction in the amount of dike needed for closure. Consideration could also be given to these gaps being utilized for the installation of sluice gates.

Mechanical and electrical connection between the units could be achieved by a bridge system across the stoplog gaps, containing electrical ducts, walkways, mechanical ducts, and a rail system for gantry cranes to service stoplogs, gates, transformers and turbines.

One phase of the construction which has been mentioned only briefly is the foundation on which the precast sections are to be placed. No matter how sophisticated construction methods may become for fabricating and placing units, all will be to no avail unless the foundations are equally practicable.

Full emphasis should be placed on selecting a site, not only acceptable in terms of the energy available, but also acceptable geotechnically, incorporating as far as possible acceptable water depth, minimum overburden and a regular fault-free rock surface. While the likelihood of finding such a degree of perfection in sites is remote, there will continue to be development of underwater working capabilities. Rapid advances have been made in recent years in Holland with the in situ placing of mastic asphalt under water as a bed protection and founding medium.

Oil drilling platforms are becoming very sophisticated and are used for other types of seabed construction. They may well be adapted to placing, sinking and anchoring units.

Underwater equipment has been envisaged capable of operating on the seabed for dredging and excavating rock. Prototypes, when built and tested, could possibly fulfill a major operation in foundation preparation.

SUMMARY

An attempt has been made in the foregoing to describe briefly some of the possible methods of utilizing a precast floated-in system for a tidal power development. Most of the solutions suggested for the construction problems described are quite feasible in terms of everyday practice. While there are few precedents of comparable magnitude for the construction techniques envisaged, all of the methods indicated have been successfully carried out on a smaller scale.

It could well be that a future need to utilize the abundant supply of tidal energy available, coupled with an innovative approach and technological improvements will prove the precast floated-in system for tidal power development economic as well as technically feasible.

Unfortunately, the ultimate in economy is rarely achieved by applying mundane practice to unique problems. Unique solutions are required, and it is to be hoped that the advancement of technology in the years to come will produce equipment and methods equal to the challenge of placing structures in deep tidal waters.

REFERENCES

1. "Tidal Power Development in the Bay of Fundy", Atlantic Tidal Power Programming Board (1969).

2. International Passamaquoddy Tidal Power Project (1961).

3. "Tidal Power Studies Chignecto Bay", Atlantic Tidal Power Programming

Board (1969).

4. "Bay of Fundy, Tidal Power Development Studies", Atlantic Tidal Power Programming Board (1969).

5. Jellett, J. H., "Docks and Harbour, The Lay-out, Assembly and Behavior of the Breakwaters at Arromanches Harbour (Mulberry "B")", The Civil Engineers in War, Vol. 2.

6. "La Rance Tidal Power Scheme", Revue Francaise de L'Energie (1966).

7. Bernstein, L. B., "Tidal Energy for Electric Power Plants" (1961). (Trans. by Israel Program for Scientific Translations, Jerusalem, 1965).

8. Bernstein, L. B., "Hydrotechnical Construction: Kislaya Guba Tidal Power Plant", Trans. for ASCE (1969).

9. Wilson, E. M., "Solway Firth Tidal Power Project", Water Power, 17 (1965).

10. Wilson, E. M., "Tidal Power from Loughs Strangford and Carlinford with Pumped Storage at Rostrevor", Proc. I.C.E. (1965).

POWER UNIT AND SLUICE GATE DESIGN FOR TIDAL INSTALLATIONS

E. Ruus*

INTRODUCTION

Tidal power plants are characterized by low heads and large discharges. They are economically feasible only as high capacity installations where the tidal range is large and where large bays are naturally available or can be created at low cost.

For economical handling of large discharges under low head it is necessary that water passages are large and streamlined. This applies to turbine water passages as well as sluice gates. Great care must be exercised in hydraulic design to keep the head losses down. On the other hand the value of water lost due to leakage is relatively inconsequential and therefore the seals, usually required at the gates to prevent leakage, can be omitted.

Low head, large flows, streamlined water passages are the main hydraulic considerations for the turbine and sluice gate design. For the hypothetical installation the availability of two pools in a reasonably ice-free and protected location is assumed.

TURBINES AND GENERATORS

Maximum Output of the Unit

Recent developments in turbine manufacturing make it feasible to use large units, which results in reduction in cost of the power plant substructure. For a given head the turbine speed is reduced as the output of the unit increases. This results in an increase in the cost of a direct-connected generator. The increasing generator cost limits the economical maximum output of a unit without a step-up gear.

---

*Department of Civil Engineering, The University of British Columbia, Vancouver 8, B. C., Canada.

Specific Speed

For a tidal power development [2, 5, 6] only high specific speed runners, fixed or movable blade propeller turbines are considered. The movable blade runner is normally preferred for its high efficiency at partial loads. In comparison to a fixed blade runner a Kaplan runner offers somewhat larger output while working under a head other than the design head. In a tidal installation however, because of the large number of units, the turbines can always be operated at the best gate opening, i.e. very nearly at the point of best efficiency. Therefore the fixed blade propeller, which for a given output is smaller than the Kaplan turbine and cheaper (particularly when the corrosion protection is considered) is suggested for a tidal plant.

For a given output, a higher specific speed results in a faster and smaller turbine. However, the setting of the turbine must be deeper to avoid cavitation. Recent recommendations show a substantial rate of increase of the required cavitation coefficient σ at high values of specific speed. For an arbitrarily chosen design head of 20 feet for example, the turbine setting must be lowered by 13 feet for an increase of specific speed from 200 to 225. It therefore seems to be advisable not to exceed the presently used maximum specific speed which for propeller turbine is about 200.

Turbine Shaft Arrangement

A vertical shaft arrangement requires deep excavation for the draft tube and a large distance between the units. Substantial head losses occur in the draft tube bend. These losses are critically dependent on the depth of the draft tube bend, measured vertically from the runner elevation.

In modern designs for low head plants [4] with horizontal shaft arrangement the water passage is made more or less straight and tubular. It resembles a horizontal or slightly sloping venturi, with the runner at its throat section. There is sometimes a sweeping bend in the upstream portion of the water passage. Because of the absence of the sharp bend in the downstream portion of the water passage, draft tube losses are smaller than those at vertical shaft arrangement.

Turbine Generator Arrangement

In recent designs for low head plants the generator with its appurtenances is built into a bulb that is located in the water passage. There is an access shaft to the bulb from the main floor. This arrangement is quite satisfactory for small and medium size units. For large units it becomes impractical because of the size of the bulb.

For a given capacity the size, weight and cost of the generator depend much on its speed. Speeds obtainable for medium and large size turbines

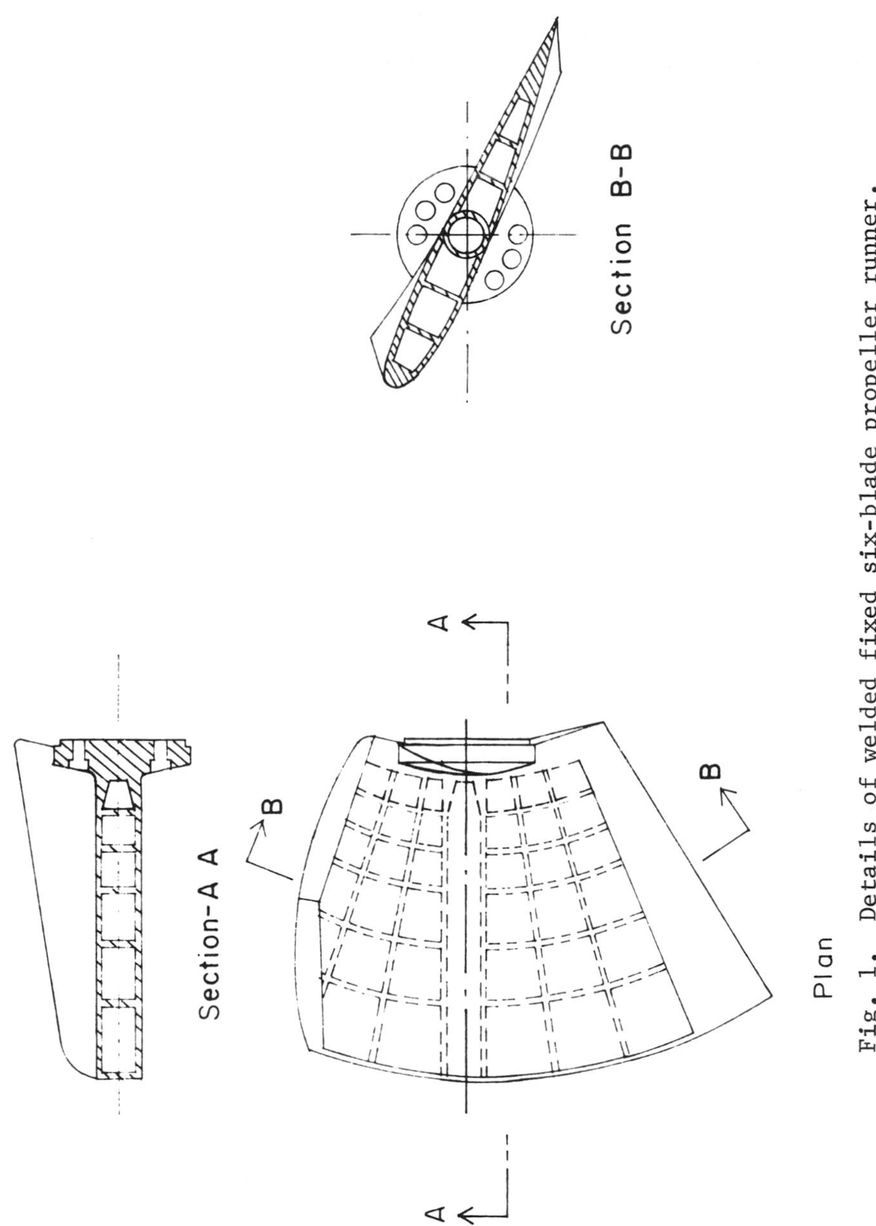

Fig. 1. Details of welded fixed six-blade propeller runner.

working under low head are much lower than those economical for the corres-ponding generators (too many poles are required). The designer is faced with the choice of using a step-up gear between the turbine and the generator or a vertical shaft layout. In the first case the speed of the generator is increased. This results in a smaller and more economical generator, which can still be placed into a bulb in the water passage. The step-up gear is, however, an added cost and reduces the overall efficiency of the unit.

With advances in manufacturing and technology of materials suitable for gears the choice in recent years in Europe has been in favour of gears. The output of such units with step-up gears is still limited to about 10-15,000 kW. According to Kovalev (3) gears for transmitting up to 70,000 kW are technically feasible today. This statement is supported by the author in Appendix I.

For small units, successful attempts have been made to eliminate the step-up gear and the bulb, and still use the horizontal shaft layout by attaching the generator rotor at the periphery of the runner. This design is however critically dependent on the sealing detail at the fast moving runner periphery required to avoid short circuiting the generator by water leakage. The design holds good promise for medium size units.

In this paper the author proposes for medium and large size turbines a peripheral gear rather than a peripheral generator. This design would combine the advantages of large units, horizontal tubular water passages, elimination of the bulb, economical generator speed and elimination of possible water leakage into the generator because of the failure of the seal. It does require a large and expensive gear wheel. For successful operation of the gears, the manufacturing tolerances are limited to less than a thousand of an inch. The author believes that such a gear would work satisfactorily when attached to a large turbine runner.

Turbine and Gear

Fig. 1 shows the details of the suggested six blade turbine runner. Normally a Kaplan runner working under a low head has only four blades. The number of blades is kept to a minimum to alleviate blade operating mechanism problems in the hub. (For fixed blade propeller these problems do not exist.)

Blades are of a welded type with ribs in the radial and the circumferential direction between the top and bottom surfaces. This construction reduces the weight and increases the stiffness of the blade in comparison with one of solid casting. Surfaces in contact with water should be covered with stainless steel overlay. Each blade is fastened to a radial shaft, which is bolted through a flange to the hub. Blade shafts act as spokes for the gear.

To reduce the axial deflection of the blade tip, six tension rods are provided between the collar at the upstream face of the hub and the free end

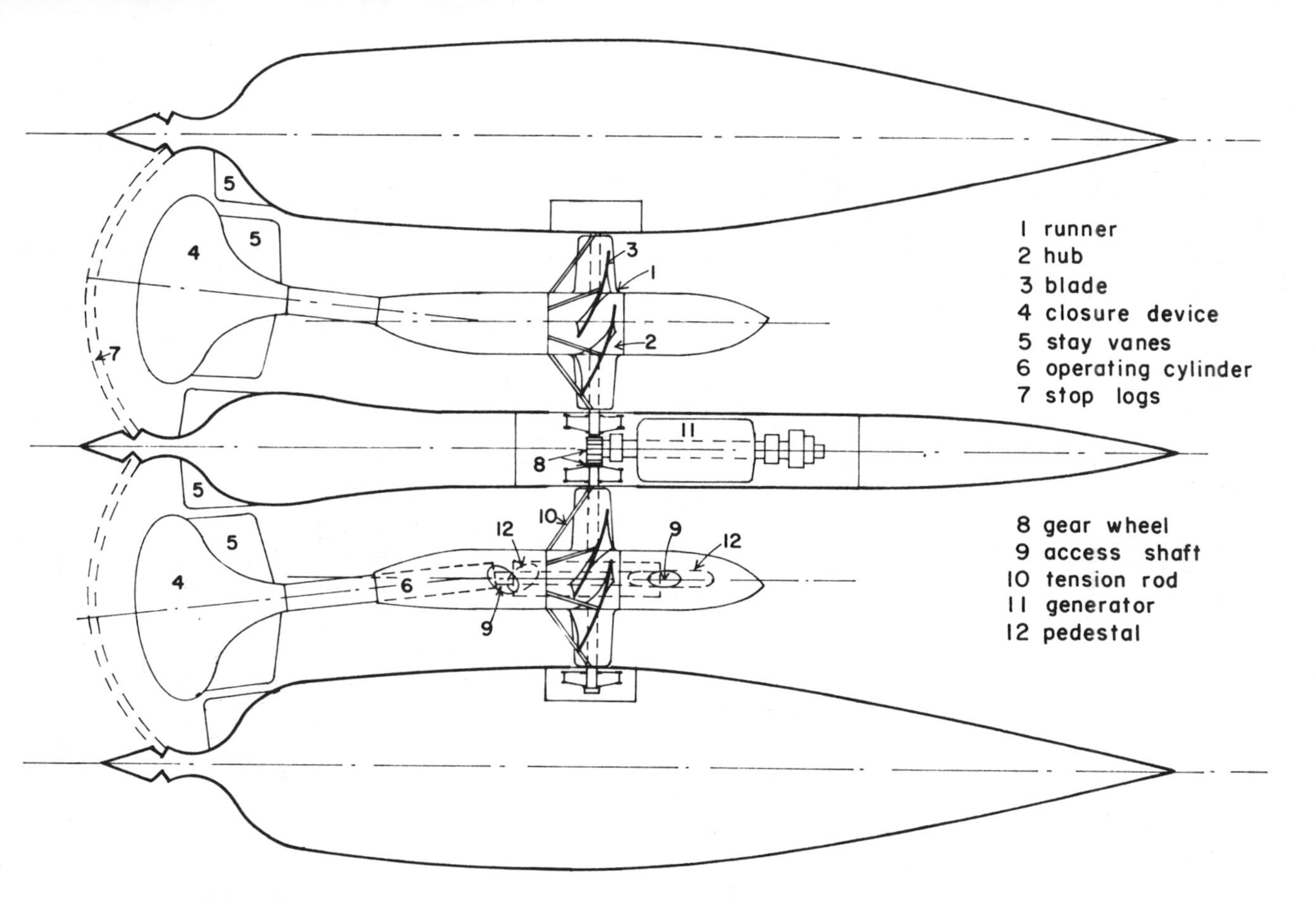

Fig. 2. Section of power plant at runner elevation, high speed generator with step-up gear.

of the blade shaft (see Fig. 2).

Tests made in the Hydraulics Laboratory of The University of British Columbia with a model runner of 10 inches in diameter, with such rods, each having a diameter of 0.020 of that of the runner, showed a decrease in efficiency of 2% (at the point of best efficiency) in comparison to the runner without these rods. The L/d ratio for the rods was approximately 1/25.

The proposed runner has a continuous peripheral rim. The width of the inner surface of the rim facing the water is kept to the minimum required by the diameter of the spokes. This arrangement will keep the vibrations initiated by the non-uniform water pressure acting on the rotating rim to a minimum. The rest of the rim is widened to reduce any axial distortion of the gear. The same effect is obtained with positioning rollers. The transfer of the bending deformation of the spokes into the gear is eliminated by the bolt attachment that runs tangentially to the periphery of the runner.

Fig. 3 shows the detail of the rim and the gear. To provide the necessary stiffness, a continuous box girder is welded to the steel casting that constitutes the gear wheel. This should be done before the gear is cut. The rest of the rim is later bolted to the box girder. The box girder keeps the deformations caused by the rotation of the wheel and the tooth load to within a few thousands of an inch.

Fig. 2 shows a schematic plan of the main elements of power generation for alternative (a). Two of the 40 feet diameter turbines are connected through the peripheral gear to a common generator located in the center pier. The water passages leading to and from the turbines are slightly unsymmetrical to provide a width of the center pier just large enough to house the generator. This minimizes the required size for the gear wheel. The water passage upstream of the turbine is circular in cross-section. The cross-section of the passage downstream of the turbine is a transition from a circle at the turbine to a rectangle at the draft tube outlet.

No wicket gates are provided with either of the alternatives. These are omitted because of the high cost resulting from their large size and necessary corrosion protection. The resulting loss in power output is relatively small. Because of the fixed propeller blades and the absence of the wicket gates no load regulation can be done. A mushroom shaped steel shell is provided as a closure device. This is operated by the horizontal pressure cylinder and serves as wicket gates while the unit is brought up to synchronous speed during the start-up procedure. The stay vanes required for whirling motion are provided as shown in Fig. 2.

Two shafts are provided for access to the turbine bearings. The downstream bearing also serves as the thrust bearing. The pedestal supporting thrust bearing is prestressed to the bottom slab of the power house substructure. The horizontal thrust of the mushroom is carried through the hollow thrust bearing of the turbine and to the pedestal. Axially directed rails are provided at the periphery of the mushroom as guides for opening and closing. An air

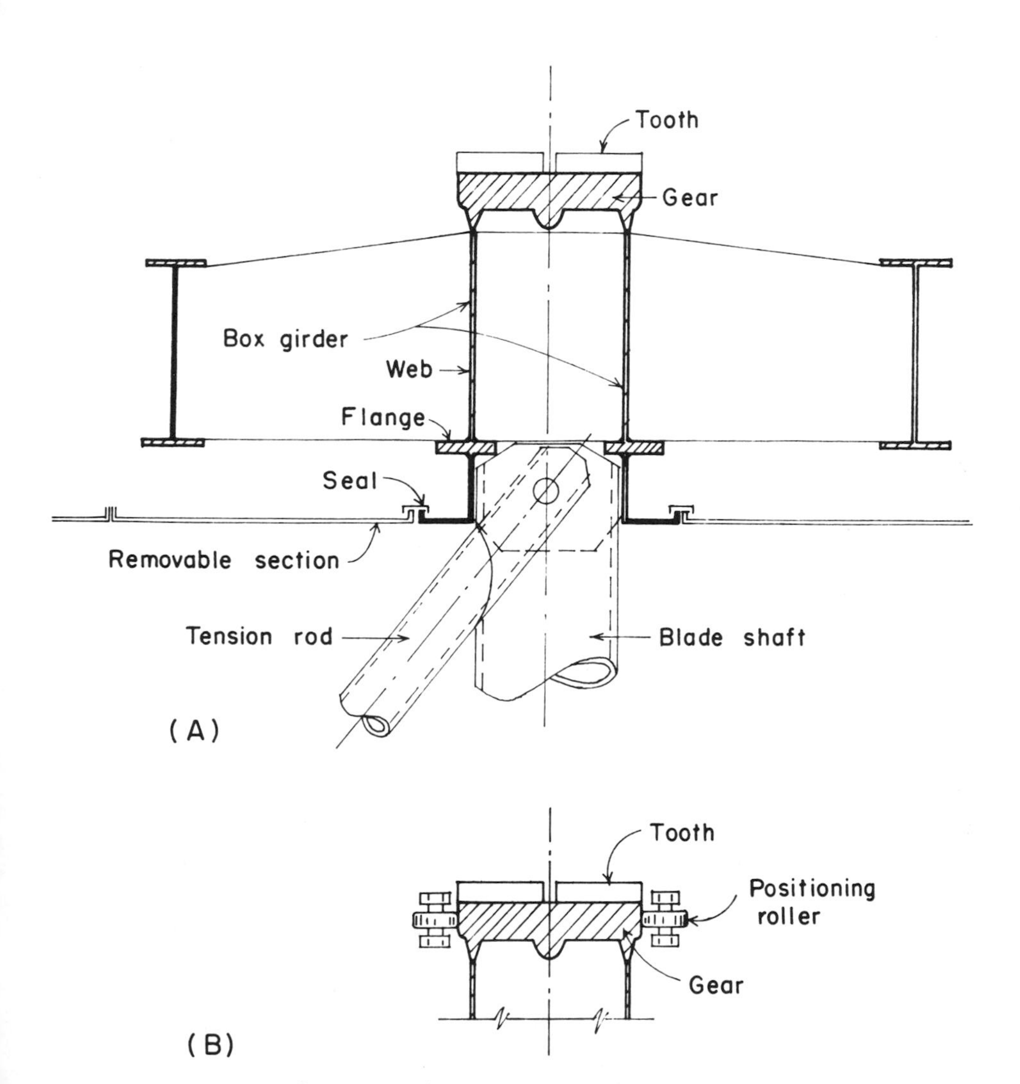

Fig. 3. Peripheral gear, detail of rim.

tank embedded in the mushroom holds the latter buoyant. Circular stop-logs serve for dewatering.

Power House

An outdoor type power house superstructure is proposed with outdoor transformers and heavy gantry cranes to lift the turbines, generators, transformers and stop-logs.

A section for housing two turbines and one generator could be built in a drydock, floated into place and submerged.

Fig. 4 shows a schematic plan of the main elements of the power plant for alternative (b). Note the rotor attached to the runner at its periphery.

SLUICE GATES

Requirements for tidal sluice gates are different from those appropriate for conventional gates. An unusually large water passage is needed to handle large flows. A large number of closings and openings is required during a year of service. Normally the gates need be operated only when the ocean and pool levels are nearly equal. Failure to open one gate during a cycle will cause only a few percent of loss in energy output. If a gate fails to close, the corresponding loss will be somewhat greater. It therefore appears that the gates need not be required to operate (open or close) against the full pressure head. However, the wave action can still cause a substantial intermittent pressure differential that must be coped with during an opening or closing operation. Because of the small head the water tightness of gates is relatively inconsequential.

Fig. 5 shows a steel shell with cylindrical vertical axis used as a gate between the piers. The maximum pressure head acting on the gates and the piers in normal operation is assumed to be about 40 feet, acting in the direction opposite to that of the flow. In this condition the shell is working in tension and the water pressure is carried horizontally to the piers. The maximum pressure head on the gate in the direction of flow would result from the wave action, when the water level in the ocean and in the pool are approximately equal. For this condition, a head of 5-10 feet seems appropriate. The shell is working in compression and needs horizontal stiffeners to prevent buckling. These stiffeners together with the shell serve also as beams carrying the bending moment caused by unevenly distributed pressure originating from wave action. The water pressure acting on the shell is in this case carried through the vertical edge beams and through top and bottom diaphragms to the four corners of the shell which are supported on the rollers. These rest on the tracks located in pier slots. In Fig. 6 the gate is indicated with dotted lines in half-open positions.

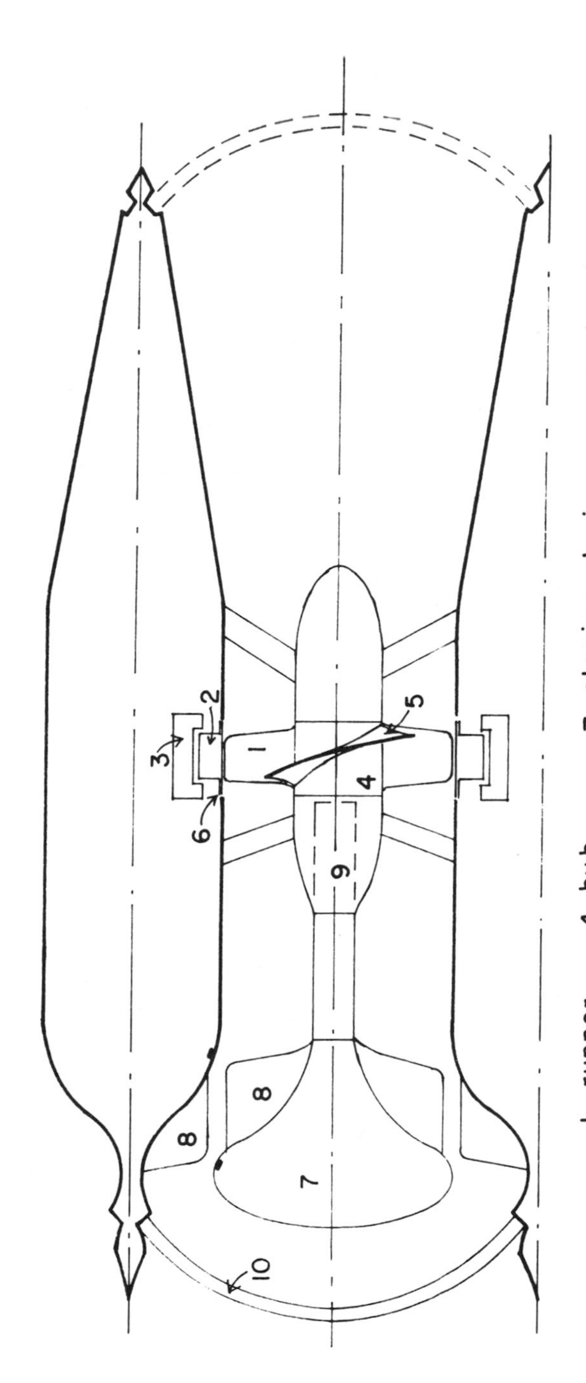

1 runner
2 rotor
3 stator
4 hub
5 blade
6 seal
7 closing device
8 stay vanes
9 operating cylinder
10 stop logs

Fig. 4. Section of power plant at runner elevation, low speed peripheral generator.

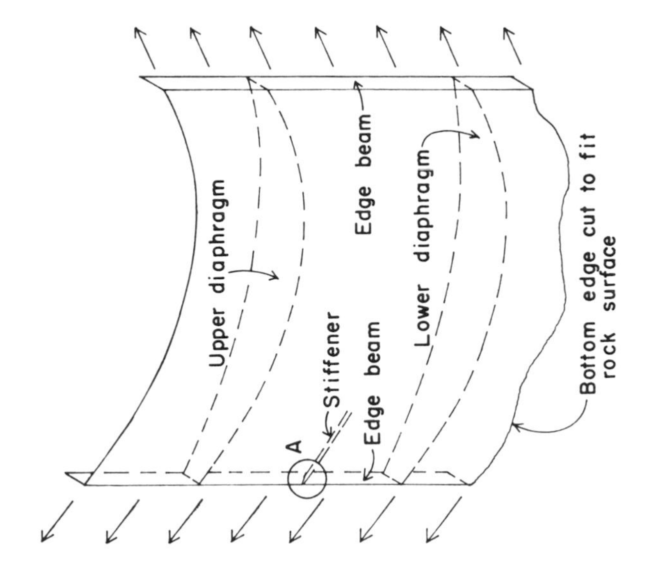

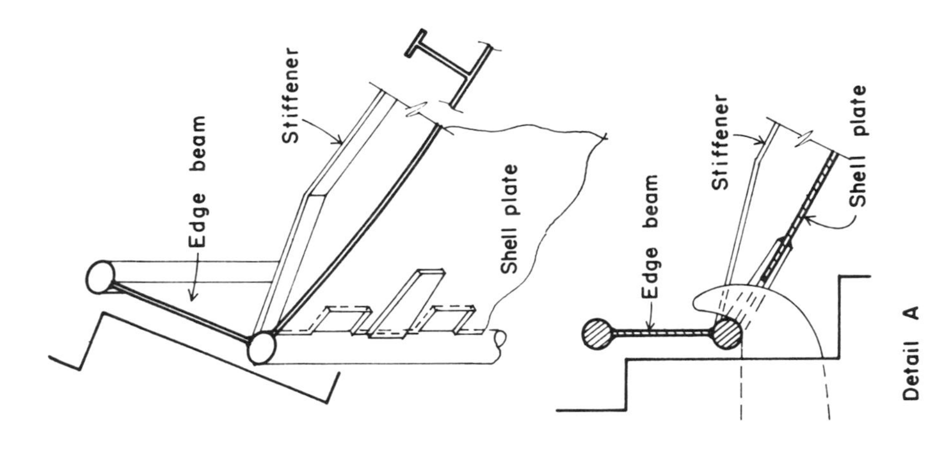

Fig. 5. Sluice gate.

When foundation conditions are favourable (relatively even rock surface at a suitable depth), the shell could be extended directly to the rock surface. The portion of the shell below the lower diaphragm works as a cantilever. Local reinforcing may be necessary where this cantilever length becomes excessive. Accurate rock surface elevations are needed in order to cut the lower edge of the shell to fit the rock to minimize losses through the gap. An average gap of 3 inches for a total of 50 gates at a water depth of 50 feet would result in loss of energy output approximately equal to 25% of that for the case where one gate is left open. To reduce this loss, flexible seals could be used.

The gate is operated by four ropes attached at its corners. Counterweights are provided to balance a part of its weight and to reduce the required hoisting capacity.

During a lowering operation the bottom guide rollers move in vertical slots provided in the pier. When the gate reaches the vertical position before final closing, all its weight is suspended by the vertical ropes attached to the bottom corners of the gate. The final positioning of the gate occurs under its own weight and water pressure. A kink at the bottom of the vertical slot is provided for this purpose. A pressure cylinder operated locking device keeps the gate in position and prevents it from swaying due to wave action. During a raising operation the gate is first unlocked and unhooked by tilting the gate a little and then cracking it open a few inches.

By using post tensioning, with cables anchored in the rock foundation, the cross-sections of the pier below the water level could be kept to a minimum. Above the water level the piers could be widened and lengthened to the dimensions required at the top. Slots are provided for an emergency gate. Fig. 5 shows the detail of the anchorage of the gate at the pier.

DISCUSSION

The design proposals are aimed at obtaining maximum firm capacity rather than the maximum energy output of the plant. First it is assumed that major costs of a tidal installation are related to the power plant and sluice gates. An attempt is made to reduce these costs by installing large horizontal shaft turbines and generators and large sluice gates.

The design proposed for the Passamaquoddy [8] tidal installation shows that more than 50% of the power plant cost of $151 million was allotted to cofferdams, excavation (mainly rock), backfill and concrete in the outdoor type power house substructure. The spacing between units of 10,000 kW nameplate capacity was 78 feet. An installation capacity nearly six times as high for that spacing is suggested in this proposal. This should result in a substantial reduction in these costs. With a step-up gear, the speed of the generator will be close to the optimum. As the turbine with the gear is a major

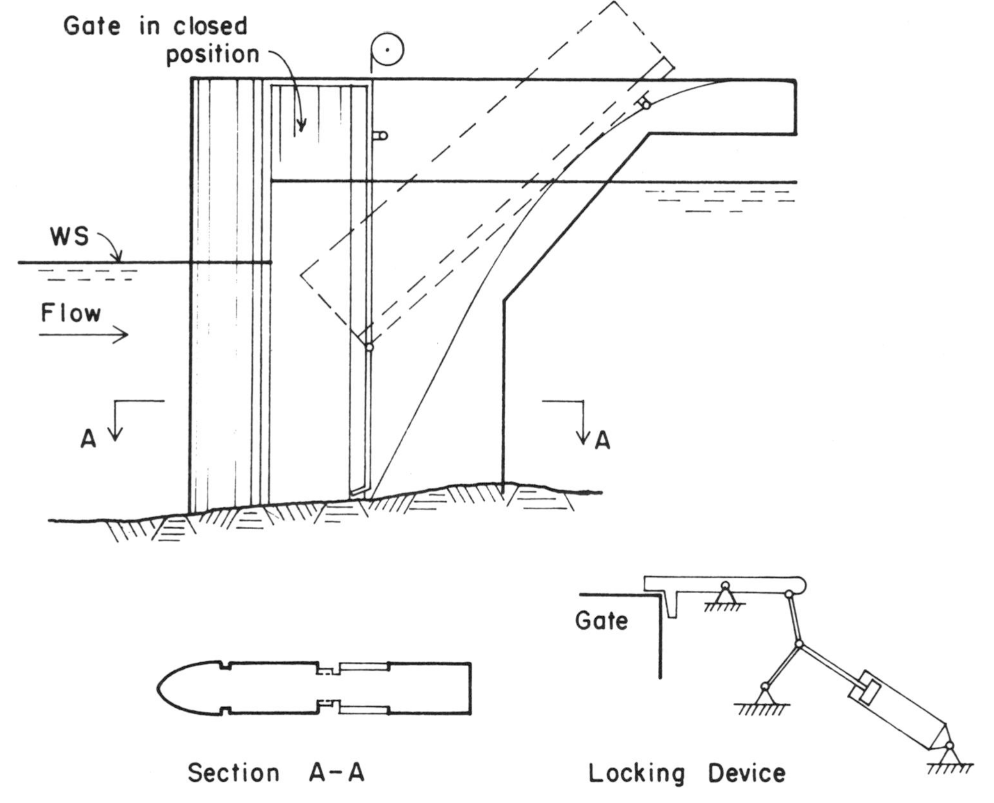

Fig. 6. Sluice gate pier.

cost item, it seems justifiable to omit wicket gates and limit maximum power output to about twice that obtained under minimum head, or about 1.6 times the output obtained at the design head.

The minimum head is based on an assumed minimum tidal range of 25 feet obtainable 92% of the time on the frequency curve. With such a minimum tidal range and with a corresponding assumed sum of upper and lower pool level variations equal to 10 feet an average minimum head of 25 - 5 = 20 feet is obtained, which is available 96% of the time. The design head is made equal to this head.

This limitation in the maximum power output limits the required generator, transformer and transmission line capacity and also limits the maximum force acting on the gear tooth. It is proposed to limit the maximum differential of pool levels to about 35 feet by raising the lower pool level. This reduces cavitation. At extreme Spring tides it is also advisable to lower the level of the high pool. Both pool level adjustments during extreme Spring tides reduce the maximum water pressure on the gates. These level adjustments can be accomplished by closing the gates before the water level variation reaches its extreme. Gates are then subjected to a pressure in the direction of flow. The gates could withstand a maximum head differential of 10 feet (equal to that resulting from an assumed wave action) without any further reinforcement.

The cost of cofferdams, excavation and concrete at the emptying gates of Passamaquoddy amounted to more than 75% of a total of $61 million. Only a minimum sheet piling acting as a cofferdam for the pier, and a minimum amount of concrete below the ocean water level is needed in the suggested design. No rock excavation is planned. The simplified gate design should result in reduction in the cost of gates and machinery relative to that of Passamaquoddy for equal sluicing capacity.

No auxiliary power plant is needed in conjunction with the proposed hypothetical tidal development.

REFERENCES

1. Krueger, R. E., "Selecting Hydraulic Reaction Turbines", U.S. Dept. of Interior, Bureau of Reclamation Engineering Monographs No. 20, Denver, Colorado.

2. Mosonyi, E., "Water Power Development", Hungarian Acad. Sci., Budapest (1963).

3. Kovalev, N. N., "Hydro-turbines, Design and Construction", Moskva-Leningrad (1961).

4. Press, H., "Wasserkraftwerke", Wilhelm Ernst & Sohn, Berlin-Munchen (1967).

5. Creager, W. P., and J. D. Justin, "Hydroelectric Handbook", John Wiley & Sons, New York (1949).

6. Brown, J. G., "Hydro-Electric Engineering Practice", Blackie, London (1958).

7. Shigley, J. E., "Machine Design", McGraw-Hill, New York (1956).

8. "Investigation of the International Passamaquoddy Tidal Power Project", International Passamaquoddy Engineering Board, Washington-Ottawa (1959).

9. Mutch, R. D., "Design of Industrial Gears", Eng. Journal (1960).

10. British Standard Specifications No. 436.

APPENDIX I

Calculations for Propeller Runner and the Gear Wheel (1, 7, 9)

Design head h_d = 20 ft. (available 96% of time).

Choose specific speed n_s = 200 rpm.

Runner diameter D = 40 feet.

Hub diameter D_h = 0.33 D = 13 feet.

Net area of water passage (allow 11% for turbine blades) A $= 0.89 \frac{\pi}{4} (40^2 - 13^2) = 1000 \text{ ft.}^2$

Throat velocity V_{max} = 26 ft./sec.

Turbine full gate efficiency η_o = 0.85

Turbine discharge Q_{max} = 26.0 (1000) = 26,000 cfs.

Turbine maximum output $= \frac{Q_{max}\, h_d\, \eta_o}{8.8} = \frac{26{,}000\,(20)\,0.85}{8.8}$

$= 50{,}000$ HP

Speed n' $= \frac{n_s}{\sqrt{HP}}\, h_d^{5/4} = \frac{200}{\sqrt{50{,}000}}\, 20(2.11) = 37.7$

Use n' = 38.5 to conform to nearest synchronous speed of generator considering step-up gear.

Angular velocity $\omega = \frac{38.5}{60}\, 2\pi = 4.05$ rad./sec.

Torque T $= \frac{550 \text{ HP}}{\omega} = \frac{550\,(50{,}000)}{4.05} = 6{,}800{,}000$ lb. ft.

$= 6{,}800$ k. ft.

Radium of gear R = 26.0 f.

Maximum force on tooth P $= \frac{T}{R} = \frac{6.800}{26.0} = 260$ kips.

Gear diameter = 624 inches, number of teeth = 468

Pinion diameter = 48 inches, number of teeth = 36

Pitch = 0.75

Maximum allowable load on one inch width of gear tooth considering wear [10], P_w for a pitch 0.75 $= \dfrac{X_c\, Z\, S_c}{K}$

Speed factor X_c = 0.38 (Chart 11 [10]).

Zone factor Z = 5.3 (Chart 7 [10]).

Pitch factor K = 0.79 (Chart 12 [10], extended to pitch 0.75).

Surface stress factor S_c = 5100 (A number of chromium and other high quality steels have S_c value equal to or exceeding 5100) [10].

Max. P_{wear} $= \dfrac{0.38\ (5.3)\ 5100}{0.79} = 13.0$ k/in.

Required face width of tooth = 20 inches.

Use face width 30 inches.

Use material with S_c = 9000 for pinion.

The allowable load for strength is considerably greater than that for wear. Therefore the gear could be loaded for short duration to approximately 1.65 times the values shown for the design case.

A name plate generator capacity of 60,000 kW is provided for each of the turbines with a total firm capacity of 120,000 kW for a double unit (two turbines + one generator) at a load factor of 0.60.

Generator speed N_{gen} = 13 (38.5) = 500 rpm.

IDAL POWER: RESEARCH SUBJECTS AND SOME DEAS FOR CONSIDERATION

A. N. T. Varzeliotis*

INTRODUCTION

The power potential of tidal bays has attracted a great deal of attention 1 recent years. The La Rance Project, completed in the mid-sixties, has een a pioneer development in the field of tidal power and is still the only lant of its kind in the world. It has received a great deal of publicity and as caused widespread controversy.

Several tidal bays in the United Kingdom, USSR, Canada and the United tates have been the subjects of feasibility studies; however, only the USSR as gone beyond the study level with construction of the Kislaya Guba experi- ental tidal power plant.

It is not at all surprising that the state of the art of tidal power develop- ent is in its infancy. After all, despite the large number of river hydro lants in existence, we are still learning the art of river hydro development.

Since all the available experience in tidal power design, construction, nd operation has been derived from the La Rance Project, it is understandable hat the task of evaluating the power potential of tidal bays is carried out ainly by river hydro specialists. Thus, the work done is influenced heavily y experience in river hydro development. Components of plants, equipment, achinery and methods of construction are taken from river hydro technology, odified, and applied to tidal power. Modifications, however, are often nadequate as the original concepts are retained. Frequently their availability nd use prevents development of the proper concepts.

Tidal power is very different from river hydro. The similarity between ne two ends abruptly once it has been stated that, in both cases, the fluid otivating the prime movers is water in its liquid state.

Dr. H. E. Fentzloff[1] attempted to clarify the situation by listing the issimilarities between river hydro and tidal power and calling them "the

---

Engineering Division, Inland Waters Branch, Dept. of Energy, Mines & esources, 404-1001 W. Pender St., Vancouver 1, B. C., Canada.

Laws of Antitheses". Actually there are neither "laws" nor antitheses"; there are simply dissimilarities, just as there are dissimilarities between apples and oranges although both are edible.

Tidal power should be recognized as a special source of energy, requiring special technology for its development. Approaching tidal power as salt water river hydro is bound to cripple development of the appropriate technology.

If the substantial power potential of the tides is to be developed, we must recognize the individuality of the subject and concentrate on learning how to utilize this renewable source of energy. The way is through research and development. We must learn how to build tidal plants, which materials to use, and how the plants should be operated. Most important, we have to learn how to assemble components in both the qualitative and quantitative sense in order to produce optimum systems. To acquire the knowledge is going to be expensive, especially if our objective is to make a concentrated systematic effort.

The different feasibility studies mentioned above have shown tidal power to be more expensive than alternative developments. Among the factors which contribute to the relatively high cost of tidal power is the lack of specialized knowledge, which is the subject of this paper.

This paper has been written with a twofold purpose and has, therefore, been divided into two separate parts. First, some of the areas where major efforts of research are needed are outlined; secondly, some specific ideas providing potential solutions to some of the problems are presented.

SUBJECTS FOR RESEARCH

In this section of the paper, some topics in tidal power technology which I think deserve research are outlined. A breakthrough in the development of one or more of these subjects may be decisive in making tidal power economically feasible.

Schemes of Development

The term "scheme" is used to denote the geometric and qualitative arrangement of structures and equipment to comprise the working total of a tidal power generating facility.

Although a tidal bay may lend itself to several schemes of development, one scheme will be the optimum in relation to (a) the bay characteristics; (b) the economics of development; and (c) the economics of power available in the area the plant is to serve. Quite a number of schemes for tidal power development have already been proposed. They range from single-basin unidirectional power production schemes to multi-basin schemes with main and

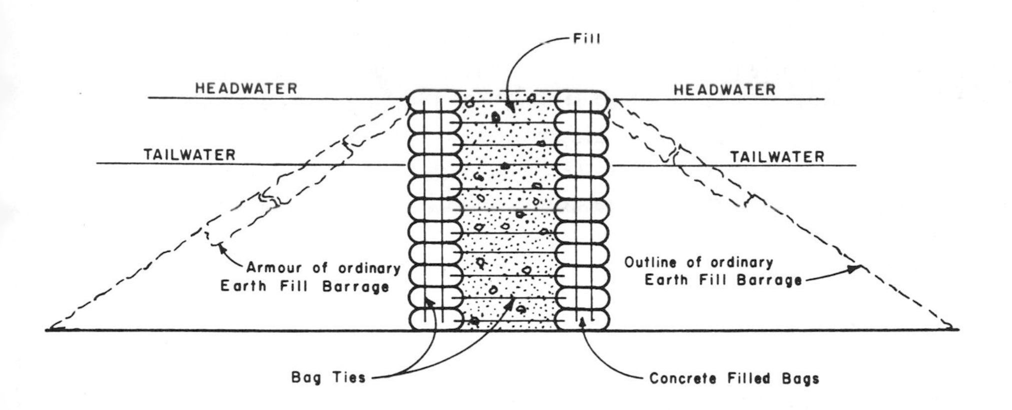

Cross-Section of vertical wall bag barrage

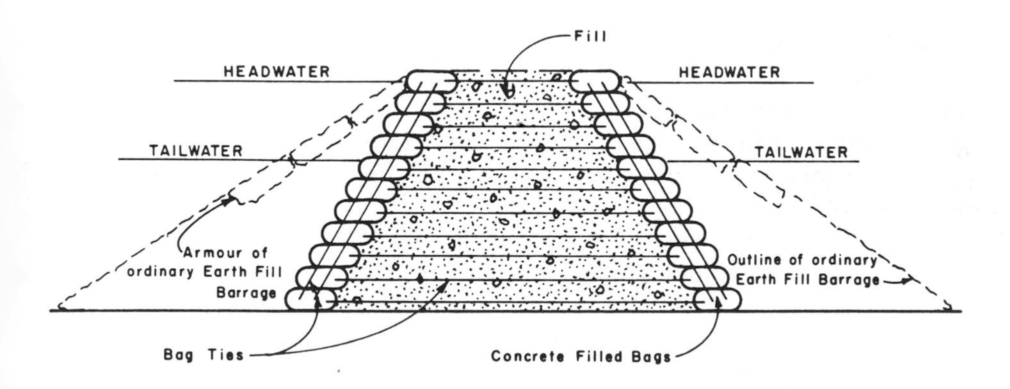

Cross-section of inclined wall bag barrage

Fig. 1.

auxiliary power houses and pumping. The proposed schemes, or combinations, may not have exhausted the possibilities, as indicated by the presentation of the double-basin transconnected scheme in the second part of this paper.

Regardless of whether new schemes can be devised, it is important that we learn how to evaluate alternative schemes with proper recognition of imposed constraints. It is true that it does not take much to exclude certain schemes once the aims for the development of a bay are specified; however, when the work advances to a more detailed stage difficulties arise. In river hydro, through long experience we can eliminate some alternatives by inspection but I question our ability to do so with tidal power. We must,therefore, acquire the knowledge to make the choice through analysis.

Barrages

The term "barrage" is used in order to emphasize that tidal power basin closures are different from dams. Unlike dams, deep water barrages only have to withstand hydrostatic heads which are a fraction of the height of the structure. The pressure side of the barrage alternates continuously and wave action on barrages may be quite severe. Leakage through the barrage is of no importance, providing that it does not endanger the structure. Cofferdamming of barrage sites for construction in the dry may involve prohibitive costs.

These unusual characteristics of barrages point out the need for research and development. Of interest are:

1. types of barrages
2. methods of construction in the wet
3. construction equipment
4. materials
5. protection against wave action
6. means and ways for barrage closure in fast tidal waters.

In the second part of the paper an idea is presented as a potential basis for development.

Tidal Range Compensation

The construction of a tidal plant often affects the tidal range of the bay, the effect usually being adverse.

The tidal amplification which takes place in a bay is a function of the "effective" length of the bay in relation to the wave length of the tide before reaching the bay. As the effective length is a function of both the physical length and shape of the bay it may be altered, within limits, by changing that shape.

The shape of the bay may be affected by spur dams, construction of underwater "sills", and by removal or relocation of natural obstructions, etc. to augment the tidal range. This can be significant as a "foot of head" at a tidal plant has a relatively high value.

It may be worth some efforts to develop further knowledge on the subject.

Gates and Gate Structures

The gate-controlled passages used for filling and emptying tidal power basins constitute a major component of the cost of the development. The cost of providing these facilities, expressed as a percentage of the total capital investment varies with the scheme of development; however, it is always significant. The magnitude of the cost of the gates and passages suggests that research should be undertaken to develop gates and structures suited to tidal power development.

There are special requirements for gates in tidal power projects as these gates work under conditions much different to those used in river basin developments. They are:

1. Tidal Power Gates have no strict sealing requirements as some leakage is quite tolerable.
2. The gates will operate frequently and regularly.
3. The gates may be subjected to a differential head which alternates from one side of the gate to the other.
4. The gates could be totally submerged at all times.
5. The gates should be capable of rapid operation for opening and closing.
6. Power requirement for gate operation should be low, especially for schemes with interrupted power output.
7. The differential heads at the time of opening or closing are very small or negligible.
8. Corrosion problems are quite acute.
9. Ice problems may dictate special designs for installations in some bays.

With these requirements in mind, research should be directed to the development of a relatively cheap gate suited to the purpose. The advent of substantial non-ferrous materials in recent years may provide the answer to some of the related problems.

Generating Units - Power House

Tidal power plants are usually big. They utilize low heads and tremendous volumes of water, and therefore have large power houses with numerous large turbine generator units. These large sizes are reflected in the capital investment in power house and equipment expressed either as a percentage of the total project cost or as dollars per horsepower installed. The power house and its equipment represent an investment of well above 50% of the total for a tidal power project.

At present, the prime movers most suitable for tidal power are hydro-turbines of the propeller type with horizontal shafts. The demand is for large

POWERHOUSE AND GENERATOR
PRIME MOVER ARRANGEMENT FOR TIDAL PLANTS
SCHEMATIC

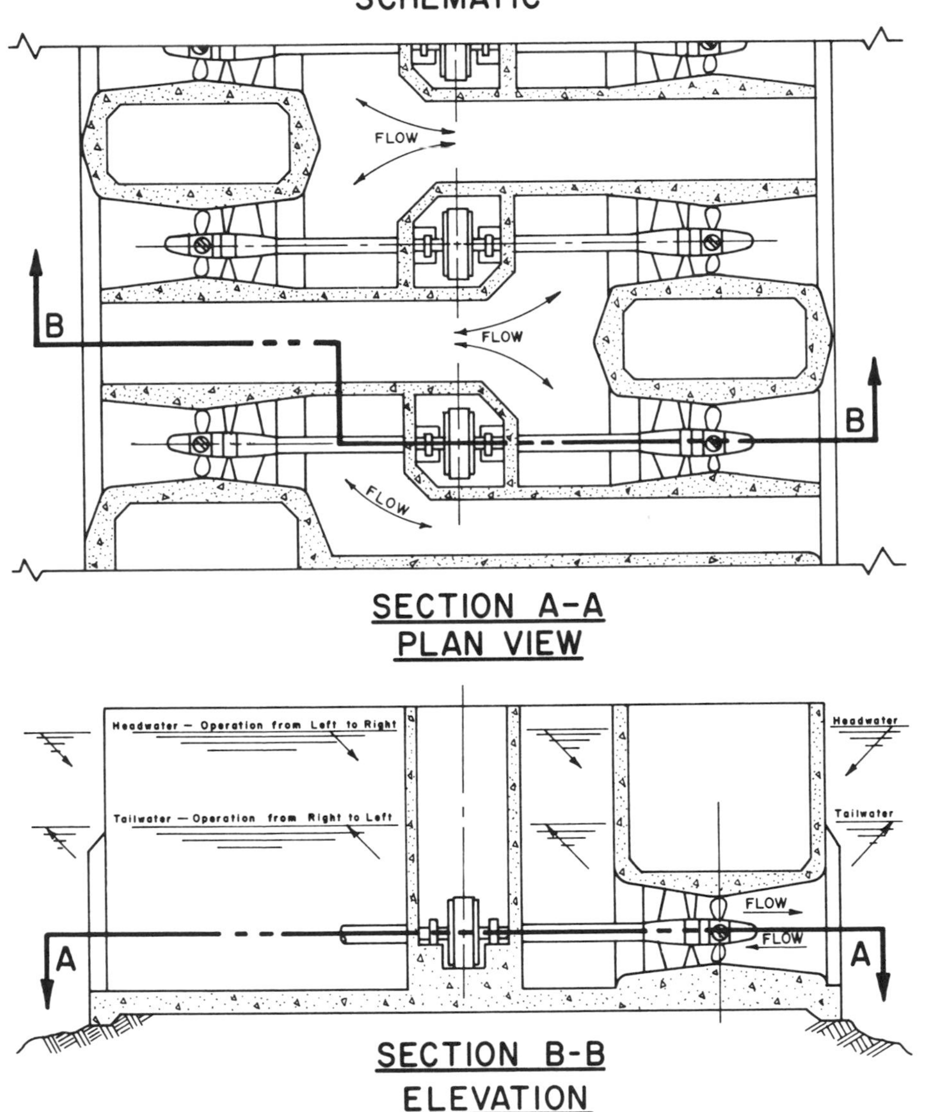

Fig. 2. Schematic.

units. Their dimensions are set by limitations in manufacturing and by limitations imposed by site geometry. The generators are relatively small in comparison to those used in development of river hydro.

Obviously there is substantial ground for research in the field. New types of prime movers may be developed or the existing ones may be improved. Design and construction of power houses may also be improved to provide structures better suited to tidal power at lower cost.

An idea for a power house generating unit arrangement is presented in the second part of the paper.

Optimization of Solutions

The objectives of a project are dictated by the demands for its product. Design of the project attempts, naturally, to achieve the objectives as cheaply as possible. To achieve the goals the designer must, within the framework set out by the objectives, evaluate alternative solutions and also evaluate components in order to optimize the solution.

The factors which affect optimization of design of a tidal power project are greater in number than for other power projects. Furthermore, in tidal power we have neither accumulated experience in the process of evaluation nor have we experience in assessing the relative importance of each component as a cost contributor and revenue producer. The problem of optimization can therefore be very complex.

Research is needed. We should endeavour to develop means of analysis which will allow us to determine the optimum site in a bay and the optimum position with regard to scheme of development, installed generating capacity, type of prime mover generator units, type and capacity of gates, method of operation, etc. Many more factors are involved but enough items have been listed to indicate the tremendous number of combinations from which the optimum solution must be chosen.

SOME IDEAS FOR EVALUATION

In the following section of the paper, some ideas on barrage construction and power house arrangement as well as a scheme for tidal power development are presented for consideration as research subjects.

"Bag Construction" of Barrages

Because the differential head on a tidal power barrage is usually small compared to its height, the normal problems of stability for earth or rock-fill structures are greatly reduced. The size of an earth-fill or rock-fill barrage is governed to a large extent by requirements for the side slopes. For many reasons these slopes cannot be steep and large amounts of fill are placed for reasons other than stability. Moreover, this fill is relatively costly due to the cost of cofferdamming and de-watering, and because of expensive precautions which must be taken if fill is to be placed under water.

In view of the above, I suggest the following idea for evaluation and, if promising, for development.

Modern technology allows the fabrication of large bags, as for example, the so-called "sausages" used for the sea transport of liquids, or the huge rubber water-storage tanks used by the U.S. Army and other agencies.

A type of bag may be developed which can be placed on the barrage site under water and then filled with special concrete from barges, or from the shore, by pipeline. Once a bag has been filled others may be placed on top and alongside, and tied together either by keys, steel rods, or wire rope, etc., which may penetrate or be attached to the bags. Thus the barrage can be built in a way which resembles construction of a brick wall with the bags resembling the bricks. On tidal flats, and when a deep water barrage reaches elevations above low water, the bags can be placed in the dry.

When the water is deep, the barrage may be constructed as a "sandwich" barrage consisting of fill between two walls of either vertical or slightly inclined bags. The bags may be tied together by wire rope or some other means to offset the horizontal pressure of the earthfill, or part of it. Obviously, any ties used must either be corrosion resistant or encased for effective protection against corrosion.

The size and shape of the bags should be designed to suit conditions and methods of construction. For example, one or more sides of the bag may be rigid and could be made of concrete, steel, or similar material. The bags may be designed with a self-levelling upper face and the interfaces of the bags may be provided with seals for watertight construction, etc.

Bags may be fabricated of plastic, canvas or any other suitable material. The strength of the bag material need not be great because the differential pressure will be relatively low during both the filling and concrete hardening stages.

In the case of "sandwich" construction, filling may be carried out while the bag walls are being raised. Because the fill material will be enclosed, it may be possible for fill to be dredged from the bottom of the tidal bay with little need for control and compacting as is required in placing ordinary fill.

Another advantage of this method of construction is that the expensive armour needed to protect an ordinary barrage against wave action should not be required to protect a bag barrage as the concrete-filled bags of the walls would serve this purpose. This is a substantial item as protection against wave action must extend from below low tide to the top of the barrage. Bags might also be utilized in the foundations of other structures such as power houses and gate structures which would be built in place.

A final consideration is that the barrage closure may be facilitated by leaving tunnels in the barrage until construction has been completed, and then closing the tunnels at low tide by filling empty bags anchored in the tunnels.

Figure 1 is a schematic presentation of the idea.

Power House Structure - Generating Units

An idea for tandem connection of prime movers to generators and a power house structure arrangement for these units is shown schematically in Figure 2.

The idea originated from the observation that the generators are relatively small in comparison with the physical dimensions of the prime movers. Although the size of prime movers is limited, the same restrictions do not apply to the size and capacity of generators. Thus, increasing the capacity of the generator is desirable.

Naturally, the units may be unidirectional or two-directional, fixed blade or Kaplan type, and may be capable of pumping, etc.

The main advantages of the arrangement are: (a) it eliminates half the number of generators that would be used otherwise; (b) it eliminates half the number of turbine speed governors; (c) simplicity of equipment as the units may be supplied by any manufacturer of water wheels; (d) a reduction in the length of the power house over some other arrangements but with an increase in width.

The Double-Basin, Transconnected Scheme

This scheme of development incorporates all the advantages of single-basin development with those of a double-basin scheme which has the power house located in the divide. It therefore utilizes most of the available head and produces continuous power output. If night power output is not saleable, the scheme may use its own power, or power from outside sources, to pump in to or out of the basins in order to augment daytime power output.

The scheme and its various methods of operation are illustrated in Figures 3 to 9. In its basic form the scheme consists of a high basin, a low basin, and a "central" complex of structures. This central complex, enclosed by gated waterpassages, is located at the junction of the two basins and the sea. The gate arrangement provides for flow in the following five sequences:

1. sea - power house - low basin,
2. high basin - power house - sea,
3. high basin - power house - low basin,
4. sea - high basin,
5. low basin - sea.

Sequences 4 and 5 mean that some of the gates of the central complex can serve as emptying gates for the low basin and others as filling gates for

DOUBLE BASIN
THREEWAY TIDAL POWER PLANT
SCHEMATIC
DAY OPERATION
LEGEND
BARRAGE
SLUICEWAY
POWERHOUSE
POWERHOUSE - PUMPHOUSE
P = PUMPING
G = GENERATION
F = FILLING
E = EMPTYING
TIDE (Sea)
LOW BASIN
HIGH BASIN
G
E
G
E
SEA

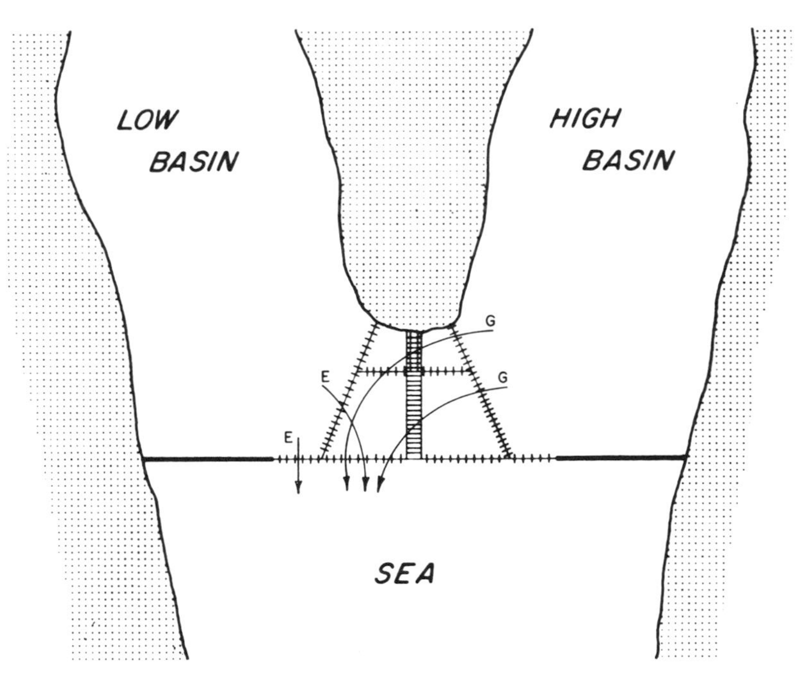

Fig. 3.

DOUBLE BASIN
THREEWAY TIDAL POWER PLANT
SCHEMATIC
DAY OPERATION
LEGEND
BARRAGE
SLUICEWAY
POWERHOUSE
POWERHOUSE - PUMPHOUSE
P = PUMPING
G = GENERATION
F = FILLING
E = EMPTYING
TIDE
(Sea)
LOW
BASIN
HIGH
BASIN
F
G
F
SEA

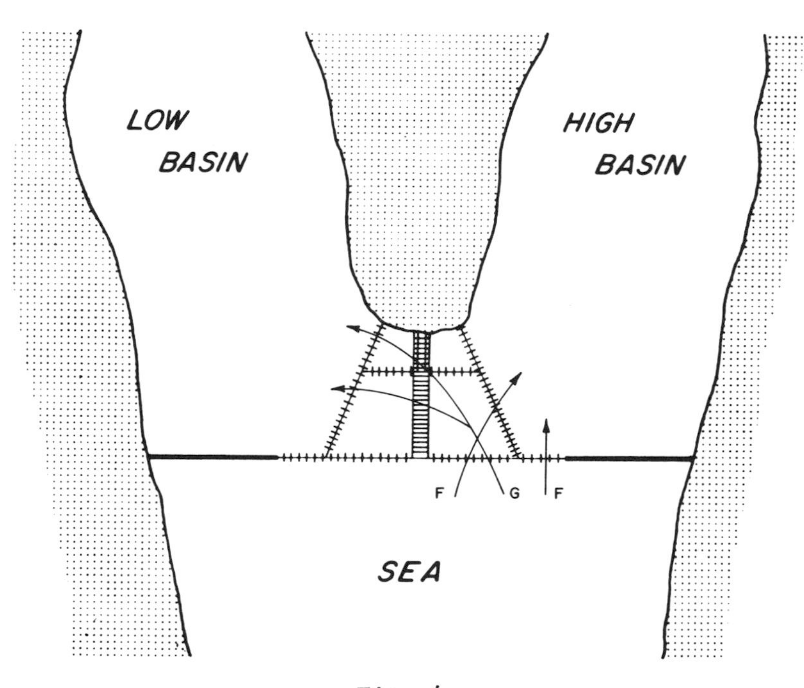

Fig. 4.

DOUBLE BASIN
THREEWAY TIDAL POWER PLANT
SCHEMATIC
DAY OPERATION
LEGEND
BARRAGE
SLUICEWAY
POWERHOUSE
POWERHOUSE - PUMPHOUSE
P = PUMPING
G = GENERATION
F = FILLING
E = EMPTYING
TIDE
(Sea)
LOW
BASIN
HIGH
BASIN
G
G
SEA

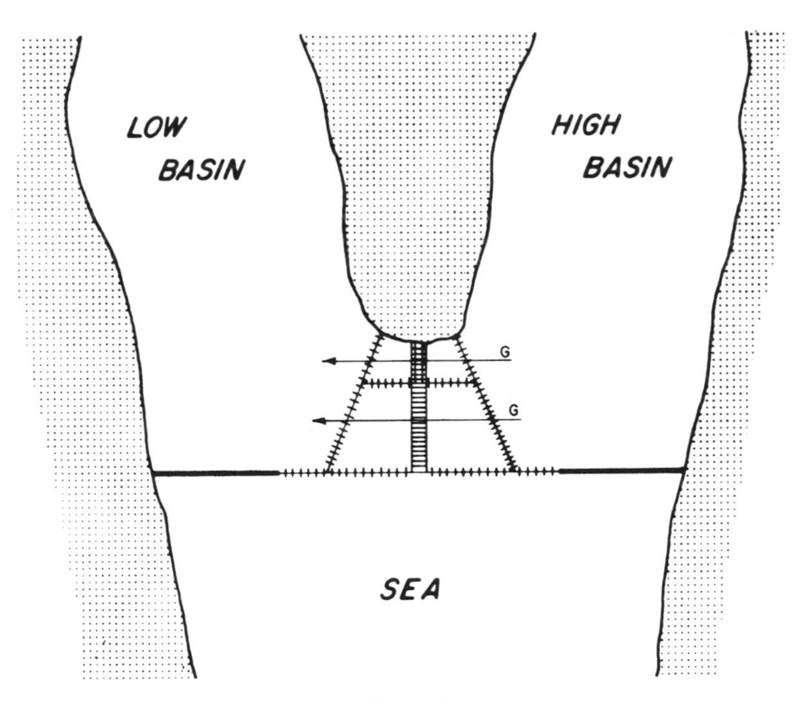

Fig. 5.

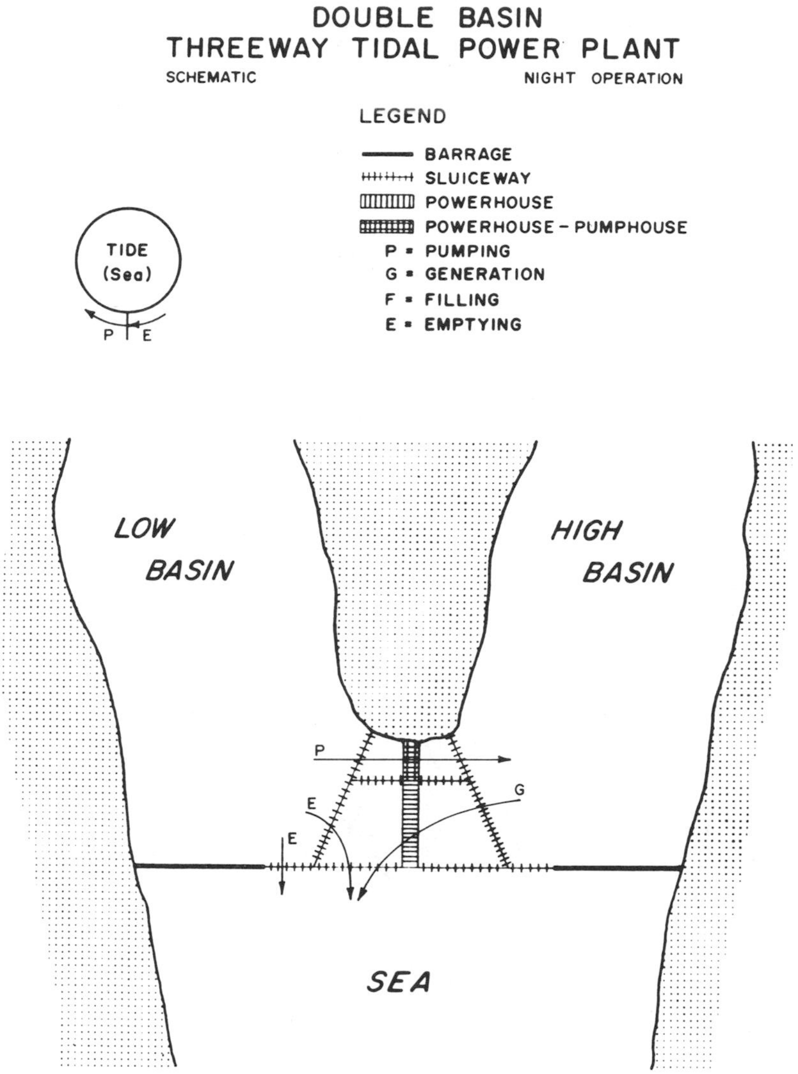

DOUBLE BASIN
THREEWAY TIDAL POWER PLANT
SCHEMATIC
NIGHT OPERATION
LEGEND
BARRAGE
SLUICEWAY
POWERHOUSE
POWERHOUSE - PUMPHOUSE
P = PUMPING
G = GENERATION
F = FILLING
E = EMPTYING
TIDE
(Sea)
P
E
LOW
BASIN
HIGH
BASIN
P
G
E
E
SEA

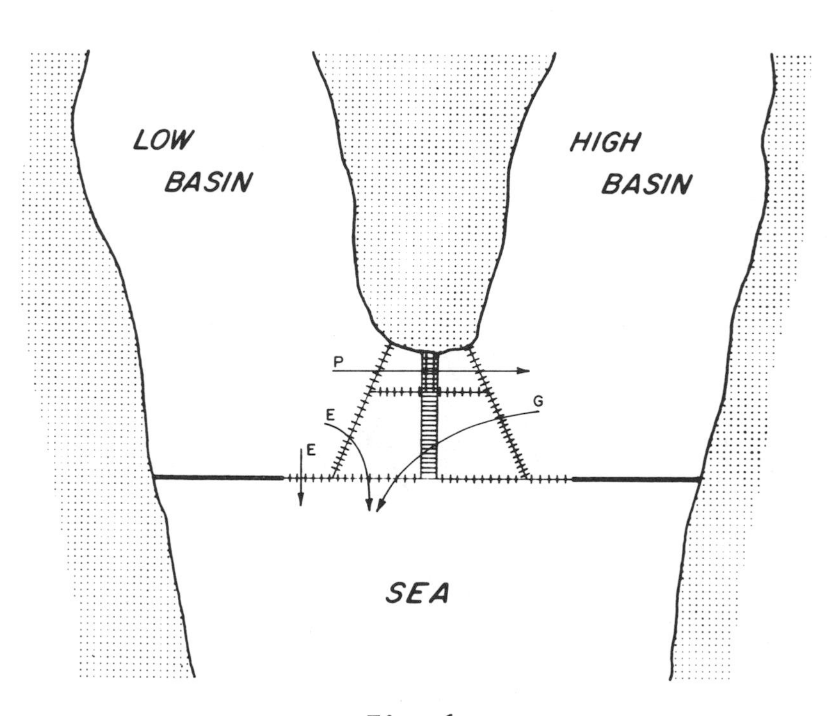

Fig. 6.

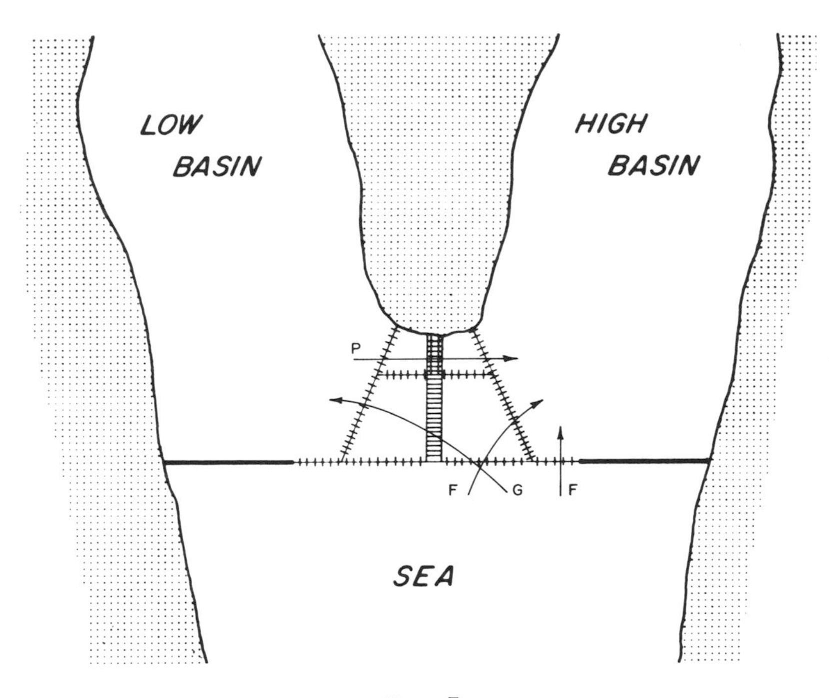
DOUBLE BASIN
THREEWAY TIDAL POWER PLANT
SCHEMATIC
NIGHT OPERATION
LEGEND
BARRAGE
SLUICEWAY
POWERHOUSE
POWERHOUSE - PUMPHOUSE
P = PUMPING
G = GENERATION
F = FILLING
E = EMPTYING
F
P
TIDE
(Sea)
LOW
BASIN
HIGH
BASIN
P
F
G
F
SEA

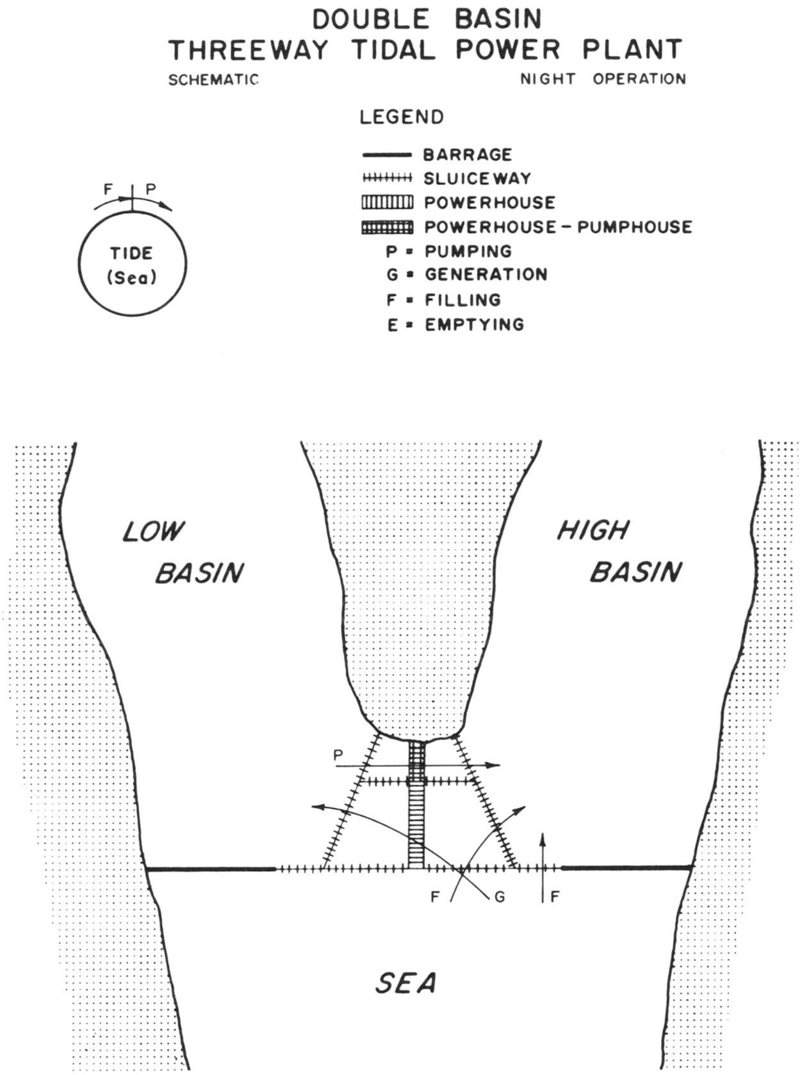

Fig. 7.

DOUBLE BASIN
THREEWAY TIDAL POWER PLANT
SCHEMATIC
NIGHT OPERATION
LEGEND
BARRAGE
SLUICEWAY
POWERHOUSE
POWERHOUSE - PUMPHOUSE
P = PUMPING
G = GENERATION
F = FILLING
E = EMPTYING
TIDE (Sea)
LOW BASIN
HIGH BASIN
P
G
SEA

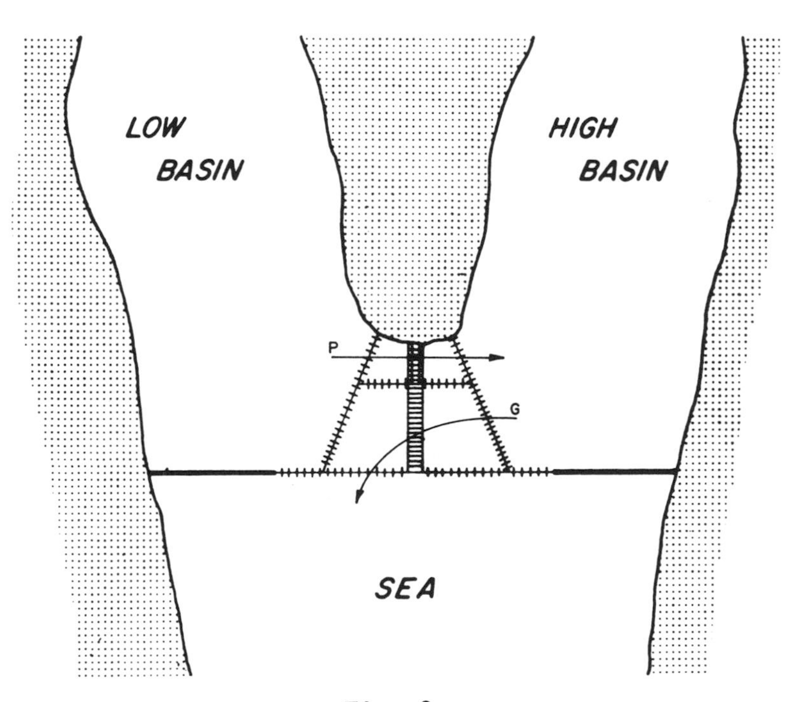

Fig. 8.

DOUBLE BASIN
THREEWAY TIDAL POWER PLANT
SCHEMATIC
NIGHT OPERATION
LEGEND
BARRAGE
SLUICEWAY
POWERHOUSE
POWERHOUSE - PUMPHOUSE
P = PUMPING
G = GENERATION
F = FILLING
E = EMPTYING
TIDE
(Sea)
LOW
BASIN
HIGH
BASIN
P
G
SEA

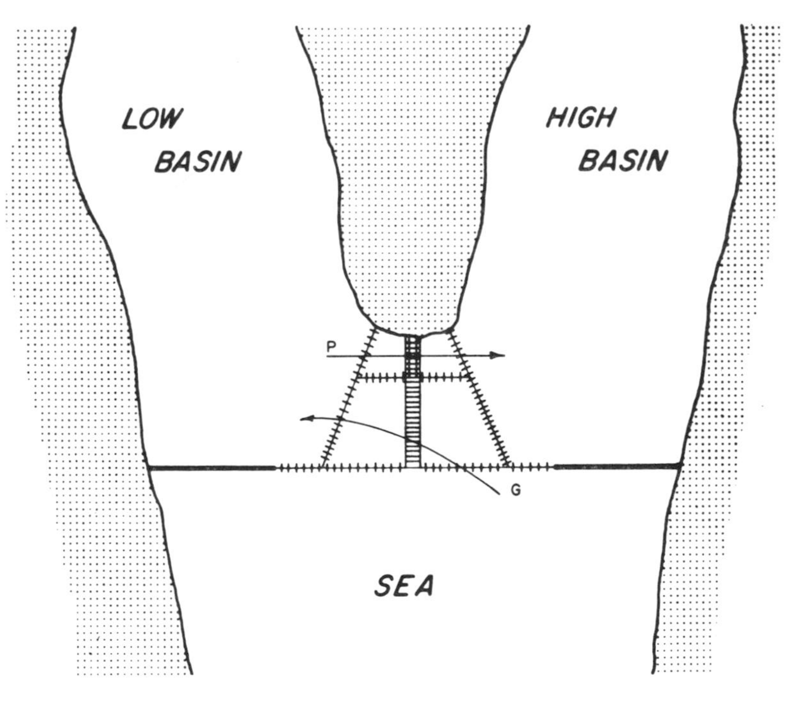

Fig. 9.

the high basin. They will not, of course, provide adequate capacity for this use and additional gates must be provided. Nevertheless, the number of gates required otherwise is reduced by this secondary utilization of the gates.

Now we may look at the operation. If we consider sequences 1 and 2 only, the scheme should be equivalent to what could be achieved by developing the two basins separately as single-basin schemes. The power house and its equipment, however, are shared by the two basins. This means that the power house structure and equipment are roughly half the size that they would be if the two basins were to be developed independently as single-basins.

If we consider sequence 3 only, the scheme is equivalent to the two basins being developed as a double-basin, interconnected scheme.

The single-basin scheme yields the greatest amount of energy and the double-basin interconnected scheme yields continuous power and peaking capacity. The proposed scheme combines all these features.

The proposed scheme lends itself to a variety of modifications. "Auxiliary" power houses may be added to one or both of the barrages. Pumping capability may be added to some units in the "central" power house or in the "auxiliary" power houses, or to both.

In the latter case, where water is pumped in order to augment peaking capacity, the scheme may generate its own power or it may utilize power from outside sources. In Figures 3 to 9 the scheme shown includes pumping with some of the units of the central complex being able to act as pumps.

When compared to the two basins developed as single-basin schemes, the double-basin transconnected scheme requires only about half the power house and generating units. However, the saving is not exactly "net", for the latter scheme requires a much greater number of gates than the first. To give an idea of this trade-off of units for gates it is estimated that, on the basis of 25 MW units and 50' x 50' gates, every generating unit which is eliminated must be replaced by roughly one and a half gates. Gates, however, are cheaper than generating units. This would result in a cost saving and increase the advantages of the scheme.

Some of the advantages of the proposed scheme are tabulated for convenience of the reader:

1. a smaller number of generating units are used in comparison to other methods of developing the basins;
2. simple unidirectional generating units, possibly fixed blade, can be used;
3. the power house can be constructed in the dry with the gates and structures acting as a cofferdam;
4. energy output would be continuous over any predetermined period of time up to 24 hours daily;
5. peaking power capacity would be greater than that obtained from other schemes;
6. pumping power can be produced by the plant itself, if this is desirable.

Finally, we must consider the question of cost comparison between the proposed scheme and the more conventional methods of tidal power development. Although it would be desirable to present a cost analysis so that the reader can readily evaluate the scheme, a meaningful cost comparison is very difficult, if not impossible, and would probably be misleading.

Some of the difficulties in making a cost comparison are:

1. One has to choose (a) a tidal bay, fictitious or real, and (b) a power output, again real or fictitious. In an abstract comparison the choice may be made, intentionally or unintentionally, of a tidal bay and a development which favour a chosen scheme in order to prove that it is the most economical scheme of development. For example, the proposed scheme is well suited to a naturally Y-shaped bay with the two arms having about equal water surface area at the operating level. Also, it is well suited to producing continuous power at one rate during the night and at another rate during the day as well as being capable of meeting a certain one hour duration peaking demand. Obviously the proposed scheme has great advantages over more conventional schemes which could not give an equivalent result. Naturally, the same result could be achieved by a combination of conventional schemes. For example, a single-basin scheme may be constructed in the stem of the Y along with an external storage scheme; or two single-basin schemes may be constructed in the branches of the Y and an interconnected scheme constructed between the branches of the Y. In both cases, the cost of the prime-mover generator units and of the power house structures could be roughly double the cost of these items in the proposed scheme. Therefore, if the replacement of machinery and power house by barrage length and additional gates results in equivalent costs as is probable for a favourable site, the proposed scheme would have a definite cost advantage.
2. In order to compare different schemes of development, the parameters involved must be optimized. This is a tedious process which would require a team of specialists from different fields of engineering working with computers. The situation is not like river hydro where an experienced eye can conceive the entire project upon viewing the site because of the tremendous amount of data available. This task is beyond the resources available for preparation of this paper.
3. The cost of a project is quite sensitive to local conditions. For example, the relationship between the cost of machinery and the cost of construction labour is vastly different between highly developed and undeveloped countries.

In conclusion, I believe that the reader will be best served if I avoid a misleading cost comparison between my scheme of development and the more conventional ones. This way it may be given a fair chance next time the feasibility of a tidal project comes under serious study.

REFERENCES

1. Fentzloff, H. E., "A Fundamental Approach to Tidal Power", Water Power, August (1967).

STRAIGHT FLOW TURBINE

M. Braikevitch*

INTRODUCTION

The simplest arrangement of a low-head hydro-electric unit is that proposed by the well-known American engineer, L. F. Harza in 1919-22; Fig. 1 is taken from his Patent Specification. The generator rotor is mounted on the propeller runner, the water flowing through the generator rim and the turbine casing being sealed against the flanks of the generator rim on either side. The unit has a horizontal shaft so that an inlet bell mouth suffices to lead the water to the runner, and the draft tube can be the straight diffuser, which is the most efficient form of energy recovery.

Thus the machinery is cut down to a minimum, the civil work is cheapened because the waterway is short, and since there is no vertical draft tube leg and bend, excavation is drastically reduced. Yet in spite of the simplicity and economy of the Harza concept, the idea has not found favour in our age, where the accent is always on economy and where the art of engineering is being invaded more and more by the accountant.

The answer is that the very simplicity of the conception makes it difficult because it depends essentially on the ability to seal reliably the turbine casing against the rotating flanks of the rotor rim, and so far this problem has defied a complete solution.

The fundamental reason for failure was the attempt to seal by bringing together the sealing faces completely, which involved great danger of heavy mechanical rubbing and wear, instead of concentrating on trying to reduce the leakage to a very manageable amount by maintaining automatically a constant small gap between seal and rotating rim: in fact, following the principle widely used for mechanical seals today.

The necessity to avoid the sealing problem then led to the development of the bulb turbine unit, which is more conventional in its approach, being fundamentally a vertical shaft unit laid down horizontally so as to obtain the

---

*Consultant, English Electric-AEI Turbine Generators Ltd., G.E.C. Power Engineering Ltd., East Lancashire Road, Liverpool L10 5HB, England.

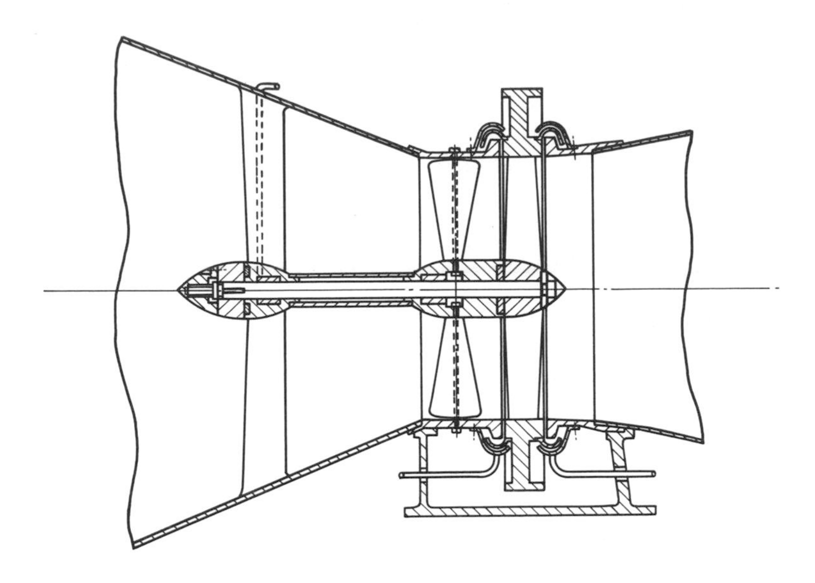

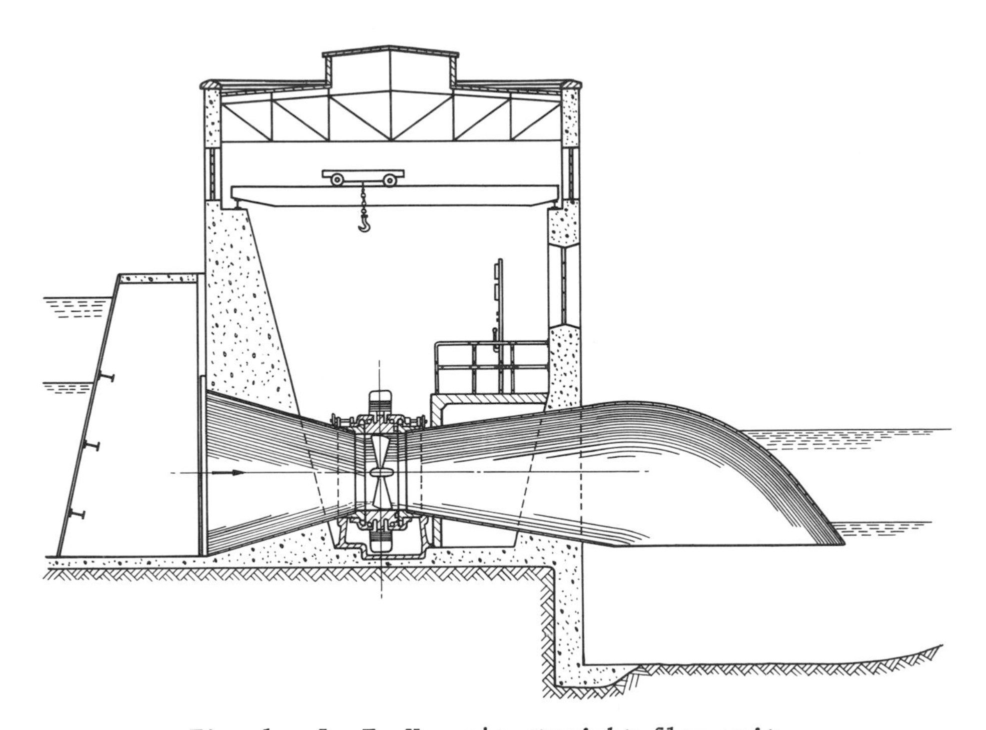

Fig. 1. L. F. Harza's straight flow unit.

straight draft tube very desirable for high specific speed runners.

It also led to a revival of the old horizontal shaft arrangement where the driving shaft comes out of the draft tube bend, with the important difference of a very small angle for the bend and a correspondingly large radius, so as to increase draft tube efficiency, which we see in the slant axis unit.

It will be of interest to review the development of the Harza concept, which could be described as the Straight-flow Unit, the water passing straight through the turbine-generator assembly.

The first time the idea was put into practice was by Arno Fischer in Germany in 1937, the turbines being supplied by a firm of world repute. Unfortunately politics got involved, Fischer emphasising that since the units could be housed in the body of the weir the station could be completely submerged and so would be proof against aerial attack. He also greatly exaggerated the possible savings. In all, seventy units were installed on the rivers Iller and Lech in Germany, and three in Austria. They were all fixed-vane propellers and are very similar, the head ranging from 8 to 9 metres, the output from 1.3 to 1.8 MW, the speed from 214 to 250 rpm, and the wheel being 7 ft. or so in diameter. Fig. 2 shows a typical power station and Fig. 3 a typical turbine section.

It will be seen that the rotor of the unit is carried in two bearings. The vaned structure supporting the downstream bearing is a drawback hydraulically. The turbines being fairly small, access to the bearings in the nacelle requires dismantling. A butterfly valve was placed immediately upstream of the unit so that the flow could be shut off from inside the station without having to use a gantry crane, as would be the case with the more usual sluice gate. The butterfly had an unfavourable effect on the efficiency.

The sealing arrangement in shown in Fig. 4. It consists of rubber sealing rings (1) pressed against the rim by tightening up the metal bands (2). A leakage of 15 litres per second was envisaged to lubricate the faces. Sealing rings (3) were inflated by water pressure to seal completely at standstill. The remainder is a thrower (4) and an auxiliary seal (5); pumps evacuate the leakage.

The trouble was with the wear and failure of the seals (1) which caused flooding on occasion. The correct adjustment of the seals was made more difficult by the slight axial movement of the rotor rim with load, which caused the upstream face to move away from its seal and the downstream face to press on its seal. The stainless steel faces on the rim stood up very well.

Finally, the seal also shown on Fig. 4 was developed and proved successful. This seal (1) has a flexible lip held against the rim by water pressure and so able to follow the slight axial movement of the rim, but most important the lip has a series of radial grooves cut in its sealing surface so that water can enter between seal and rim to lubricate and cool the surface and possibly building up some hydrostatic pressure, and even forming a series of liquid wedges from groove to groove, somewhat on the lines of the liquid wedges which sepa-

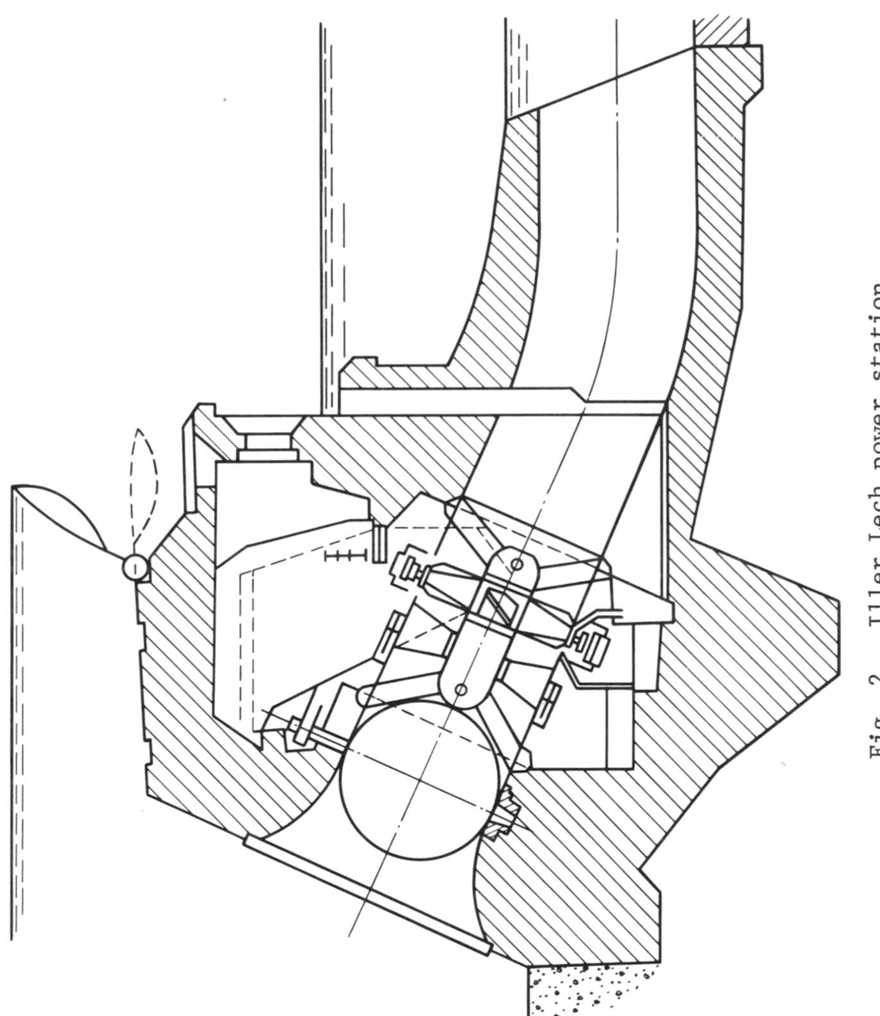

Fig. 2. Iller Lech power station.

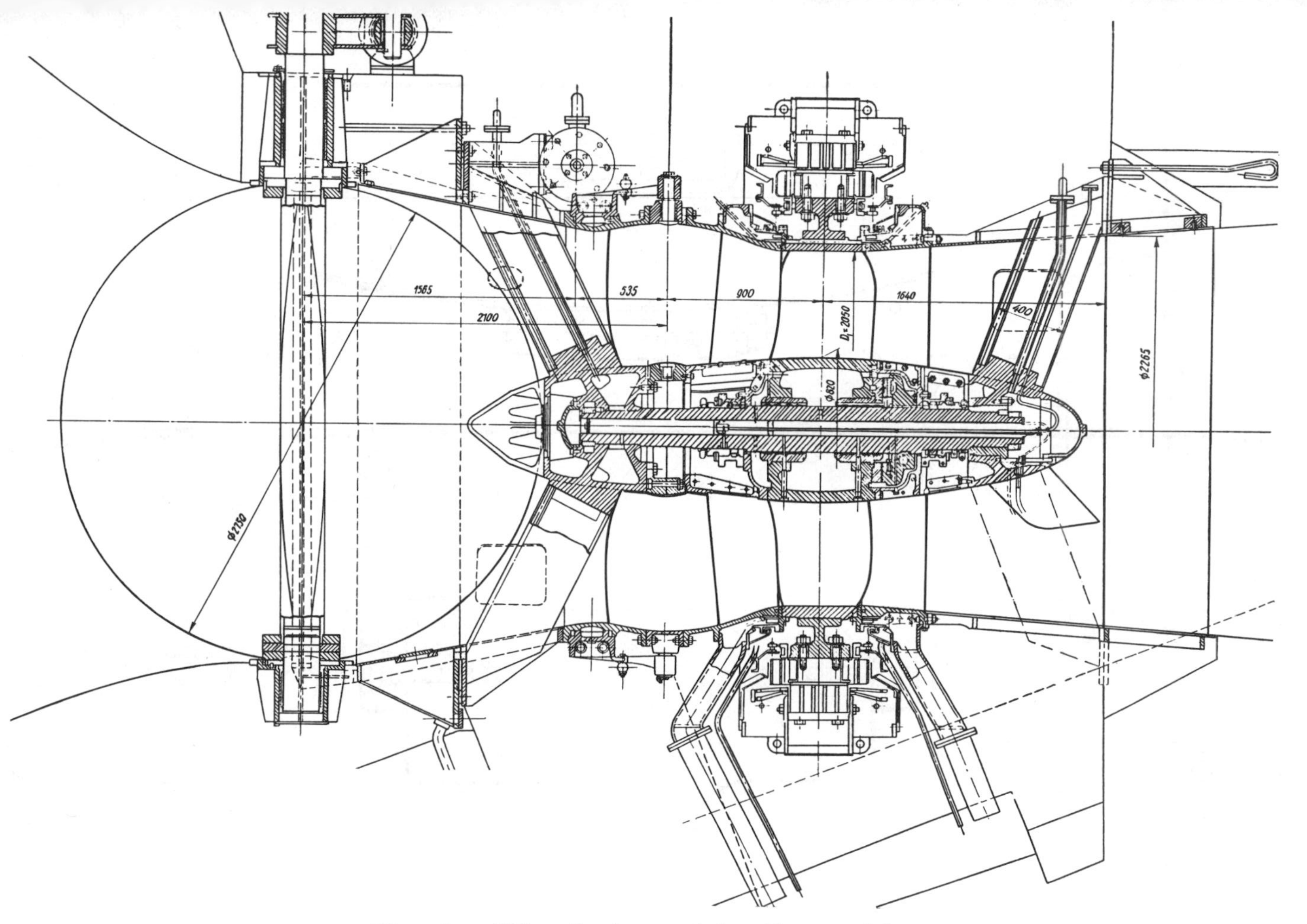

Fig. 3. Iller Lech straight flow turbine.

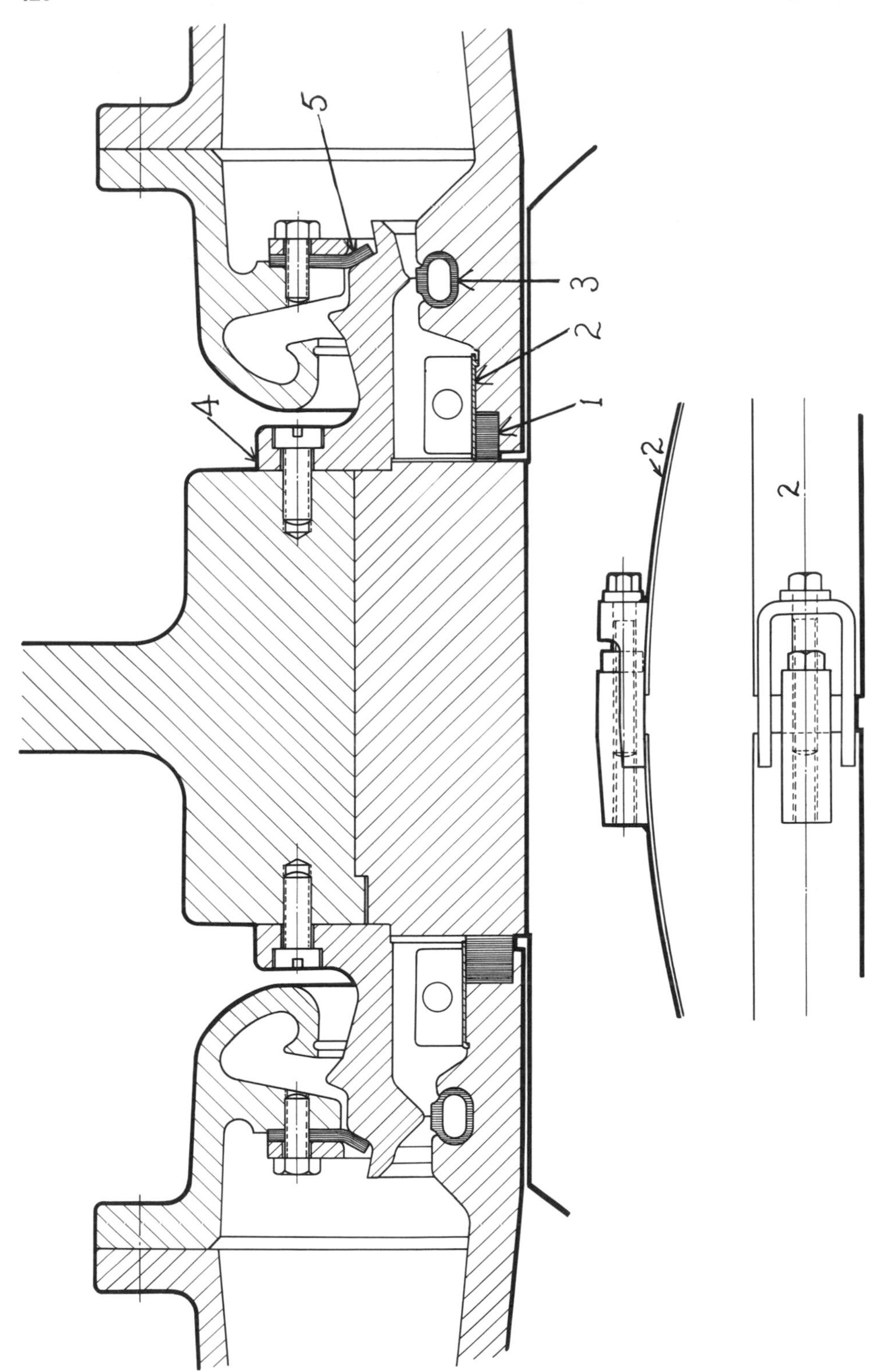
5
3
2
1
4
2
2

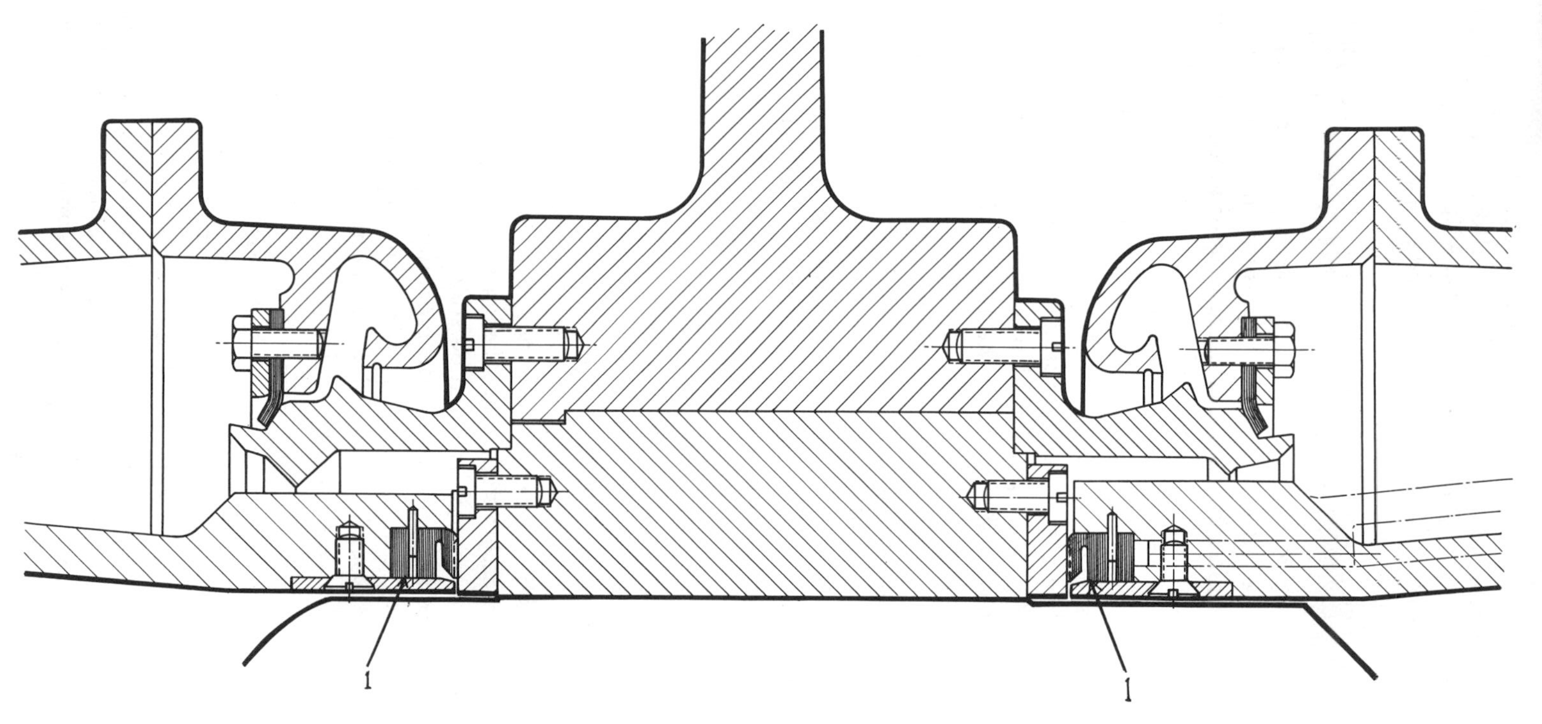

Fig. 4. Iller Lech turbine original and modified rim seal.

rate the pads of a thrust bearing from the thrust collar. The magnitude of the forces is related to the gap between seal and rim face, and so heavy rubbing of the surfaces is avoided. In fact, this modified seal is beginning to apply the principle essential for success stated previously.

A further source of trouble was the mounting of the generator rim. This was shrunk onto a rim cast on the outside of the propeller runner vanes, the idea being that as the generator rim would expand much more than the runner, the difference could be reduced by pre-compressing the runner vanes. This was adequate for normal speed, but could prove insufficient at runaway when the generator rim might tend to separate from the runner. A further method tried was the driving of keys between the two rings.

Nevertheless, in spite of the original difficulties the Iller/Lech units are operating successfully now, 25 years after installation.

The straight-flow design was also studied independently in the U.S.S.R. and tried out soon after the War. Fig. 5 shows the unit. This has a runner with six movable vanes, and again is of the two bearing type. The design head is 10.25 metres, the output 6.3 MW, the speed is 125 rpm and the wheel diameter is 10 ft. 10 in. The unit is a considerable advance on the German machines, but was a failure probably because the water carried a lot of sharp sand in suspension, which ruined the seals and attacked the bearings. The rubber sealing bands were in this case pressed against the rim by a series of hydraulic cylinders which enabled the seals to follow the rim without reduction in or excessive contact pressure. Springs were located between cylinders to retract the seals when the hydraulic pressure was cut off. The pressure contact between seal and rim in the presence of sand proved fatal. There was also difficulty with the mounting of the generator rim. As a result, the straight-flow turbine was abandoned in the U.S.S.R.

Such is the history of Harza's simple conception. The vital question is, can the problems be solved, or must the design of horizontal shaft low-head units always be based on the necessity of avoiding rim seals? The answer, based on an intensive design study and laboratory research on an 8' 4" diameter model seal rig, is that the problems can be solved and so straight-flow units of the largest sizes can be built both with propellers and movable vane runners. The approach used was to study all the experience available, so as to be able to understand the problems fully, and find out the reason for failure, and then to find a solution. Fig. 6 shows the resulting straight-flow unit.

The problems are described below.

Seals

Owing to the large diameter of the generator rim and its high peripheral speed, the seal must have some flexibility and must never press against the rim. What is required is a seal of rubber (or some other elastic material) which will automatically maintain a small gap between sealing face and rim.

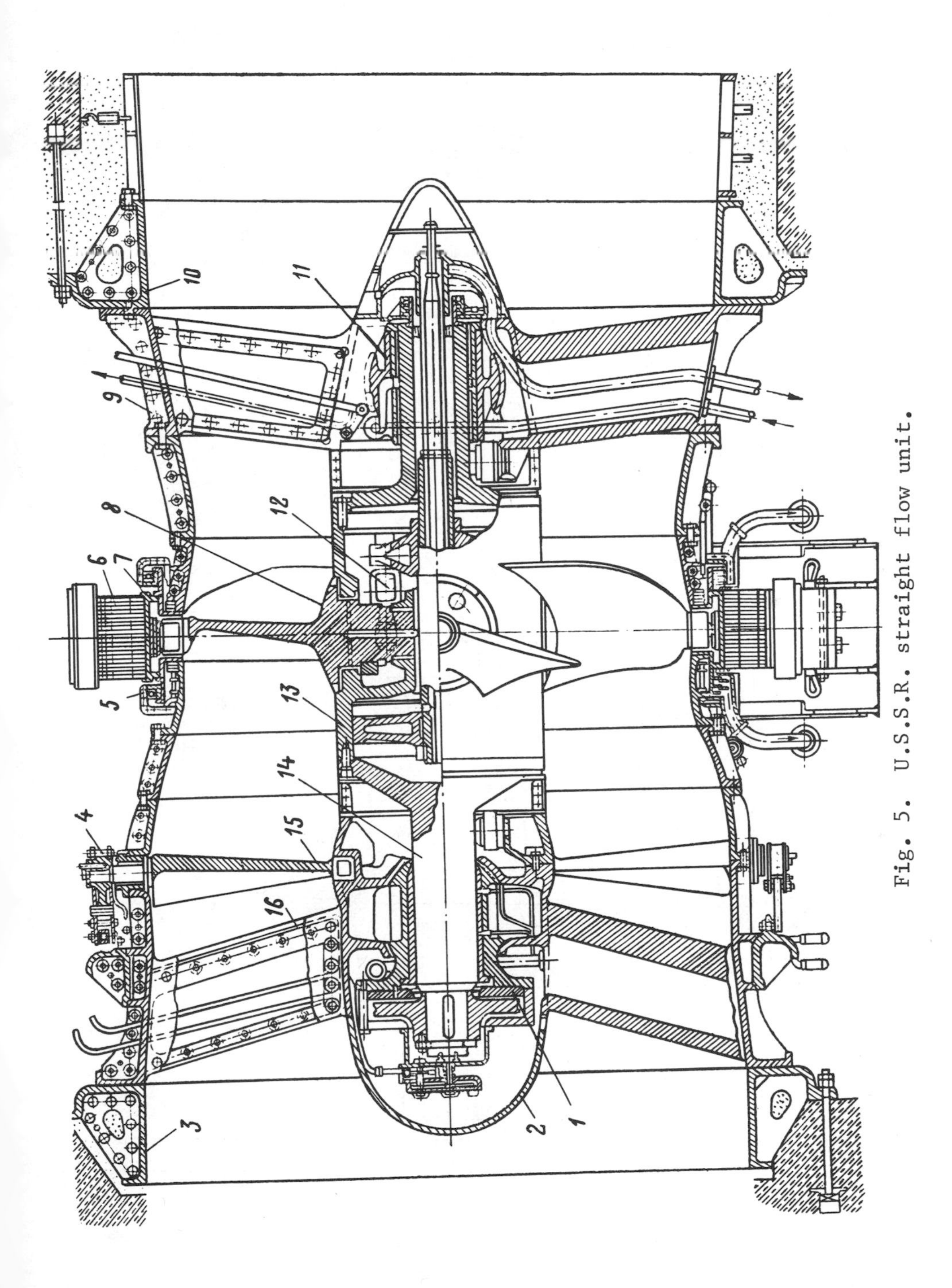

Fig. 5. U.S.S.R. straight flow unit.

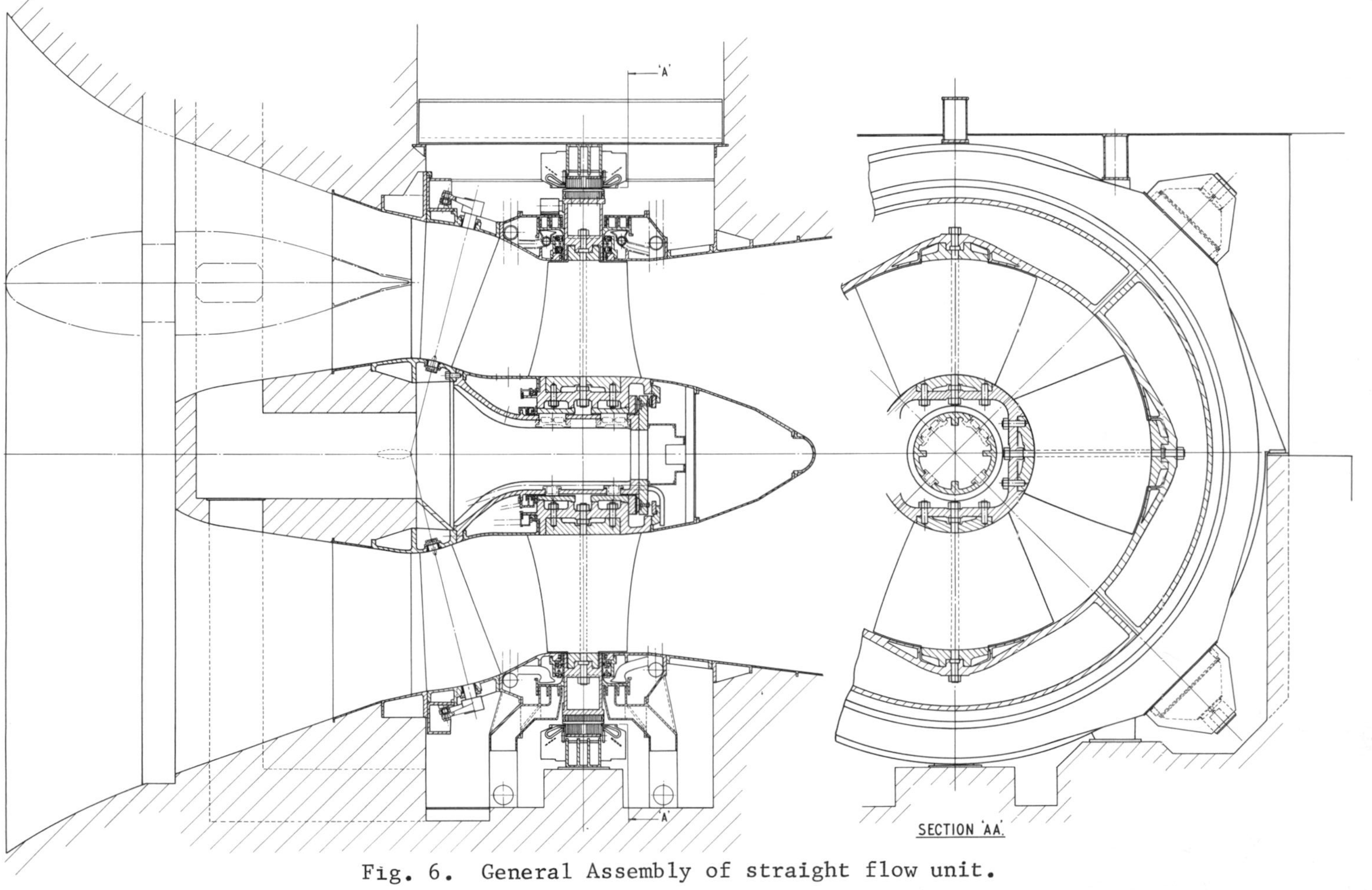

Fig. 6. General Assembly of straight flow unit.

To work, the seal must leak so that a thin water film may be maintained between sealing face and rim flank, and an opening force may be exerted on the sealing face. The magnitude of this opening force must be very definitely related to the magnitude of the gap, and then by applying a constant pressure to the back of the seal in the closing direction, a balance can be achieved which will automatically maintain the gap at the required safe (against touching) small value.

If there is sand in the water, the only answer is two seals with sand-free water introduced between them. One seal then seals, always discharging some clean water, against atmosphere, while the other seals against the turbine, preventing the entry of sand by always discharging some clean water into the turbine.

Fig. 7 shows the seal we have developed and tested, and this is of the twin-type because tidal water is bound to contain sand. For normal river water the inner seal is omitted, clean water injection is not required, and the arrangement is simpler.

(1) is a rubber seal bonded to steel segments, bolted to the seal carrier ring (2). The back of this is sealed by the L-shaped rubber (3) against the escape of operating pressure water to the exhaust side of the seal and forms a servo-motor providing the closing force.

The opening force on the seal face is produced by the water pressure remaining after some of the head has been transformed into velocity and some used to overcome friction as the water flows through the gap. Since the flow varies with the gap, the residual pressure also varies. In addition, there are radial and a circumferential groove to ensure that the seal will separate from the rim face immediately pressure is admitted, and to add a possible hydrodynamic force component.

To seal when the unit is stationary and to allow an inspection of the sealing assembly, inflatable rubber rings are provided.

Fig. 8 shows the seal proving rig with the 8' 4" diameter seal used in our laboratory. It consists of a disc mounted on a shaft in bearings to simulate the generator rotor rim. On each side there is a set of two seals with water admitted in between. Beyond the outside seals there are throwers to guide the leakage, and scoops to collect it so it can be recirculated. The whole is enclosed by a removable casing, which can be replaced by a polythene hood so as to observe the discharge of the leakage water. Exhaustive tests proved that the seals float and that the desired gaps can be maintained. Throughout the tests the seals never got damaged (meaning that dangerous mechanical contact was avoided), even when a certain amount of sand and rust was present in the circulating water. Actual dimensions of the seals correspond to the full sized job.

Fig. 9 shows the relation between opening force on the seal and gap. The shape of the seal settles the force and the relationship and in addition allows for wear. The same figure shows the experimental relationship between

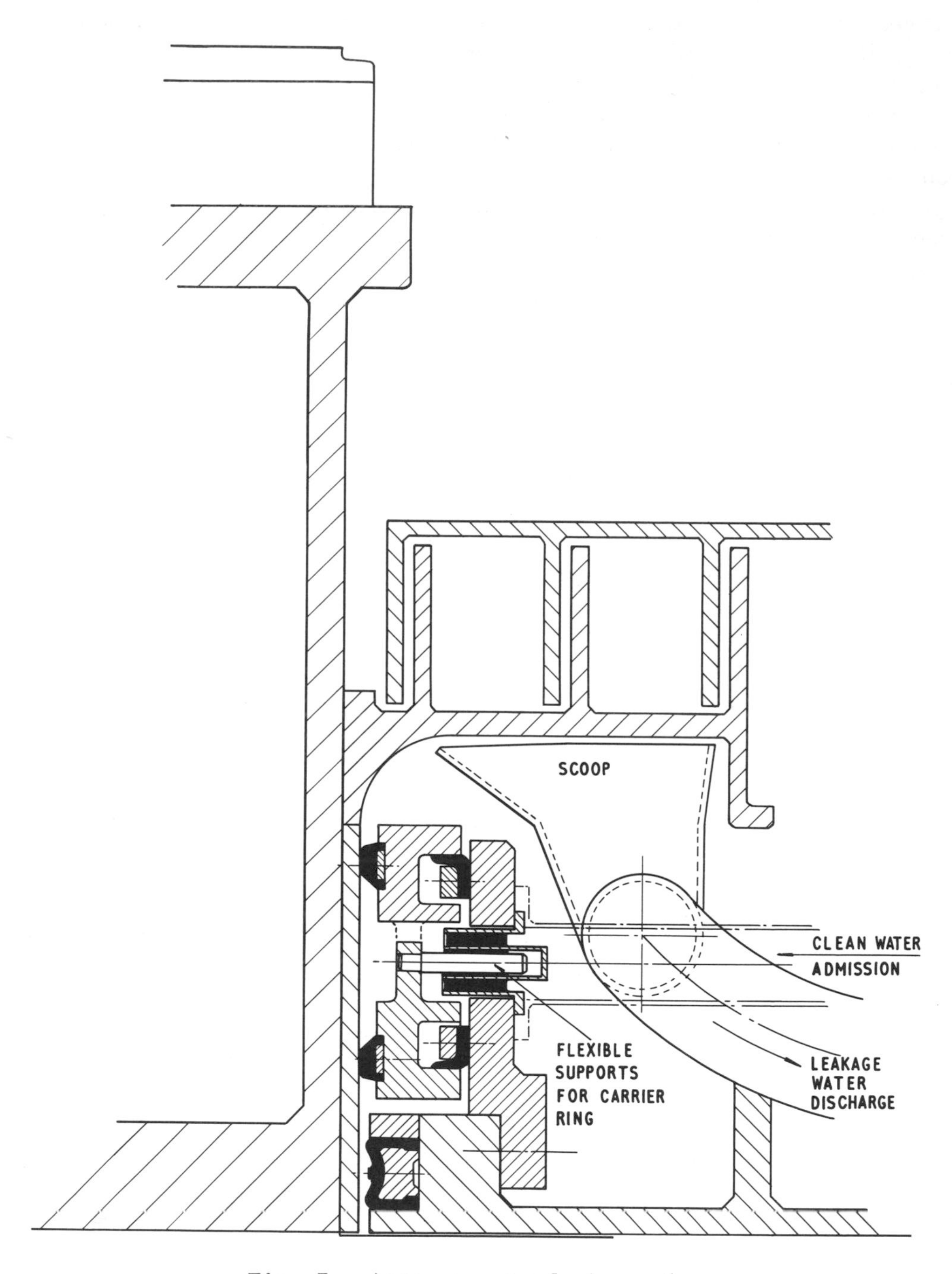

Fig. 7. Arrangement of rim seals.

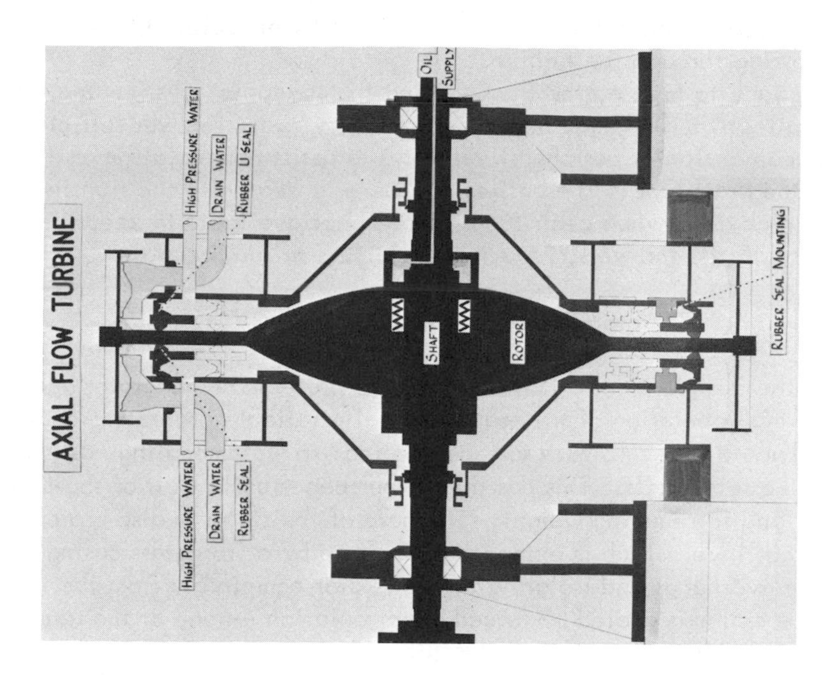
AXIAL FLOW TURBINE
HIGH PRESSURE WATER
DRAIN WATER
RUBBER U SEAL
HIGH PRESSURE WATER
DRAIN WATER
RUBBER SEAL
OIL SUPPLY
SHAFT
ROTOR
RUBBER SEAL MOUNTING

leakage and pressure; the total leakage can be kept well below 1/8 of 1% of the water flowing through the turbine.

Going back to Fig. 6, the throwers and the scoop nozzles feeding into ring mains collect the leakage, and as indicated by tests, recover sufficient pressure (by conversion of peripheral velocity) to assist recirculation or discharge into the draft tube. The baffles are there to deal with the transient conditions which arise when centrifugal force is not available to keep the water spinning in the throwers, i.e., when starting or running down.

Water-Proofing of Generator

Since the design is new, it is advisable to protect the generator against any possible mal-operation of the seals. A really reliable protective coating for the rotor would interfere with surface cooling, so water cooling has been adopted for the rotor coils. This has the further beneficial result of shortening the rotor and so reducing weight. The bore of the stator is also protected by a waterproof layer which is extended either end by a fibreglass casing enclosing the overhangs and isolating the generator completely from the turbine pit. The stator is cooled by forced air circulation -- one of the fans is indicated on the figure.

Attachment of Generator Rotor Rim to Runner

The rotor rim expands due to centrifugal force and heat. The design of the vanes is settled by hydraulic considerations, and being in the water the vanes remain cold. Thus any radial extension of the vanes cannot match the increase in the radius of the rim. The difficulty is overcome by fixing the rim to the hub by four bars, each of these passing through a radial aperture in its respective vane.

The bars when they are above the unit's centre-line carry their share of the rotor weight in compression, and when below, in tension. The bar has a guide bearing in its vane to secure very adequate stiffness for the strut action. As the unit starts rotating and the generator rim gets warmer, so the compression is reduced until all four bars become ties. The stiffness of the vanes in bending will also make a contribution towards carrying the load.

On the outer periphery of each vane there is an integral bridge piece through which the bar passes. Each quadrant of the rim has a boss which fits into a corresponding aperture in a bridge. The torque is transmitted from the vane to the rim through this projecting boss which can slide in and out of the bridge as the rim changes in diameter.

If it is desirable to reduce the thickness of the vanes near the periphery, and there is not enough metal for drilling the vane radially right through, the thinner portion of the vane is slotted and then covered over by thin plates to maintain the contour. Furthermore, the bar can be flattened in this region

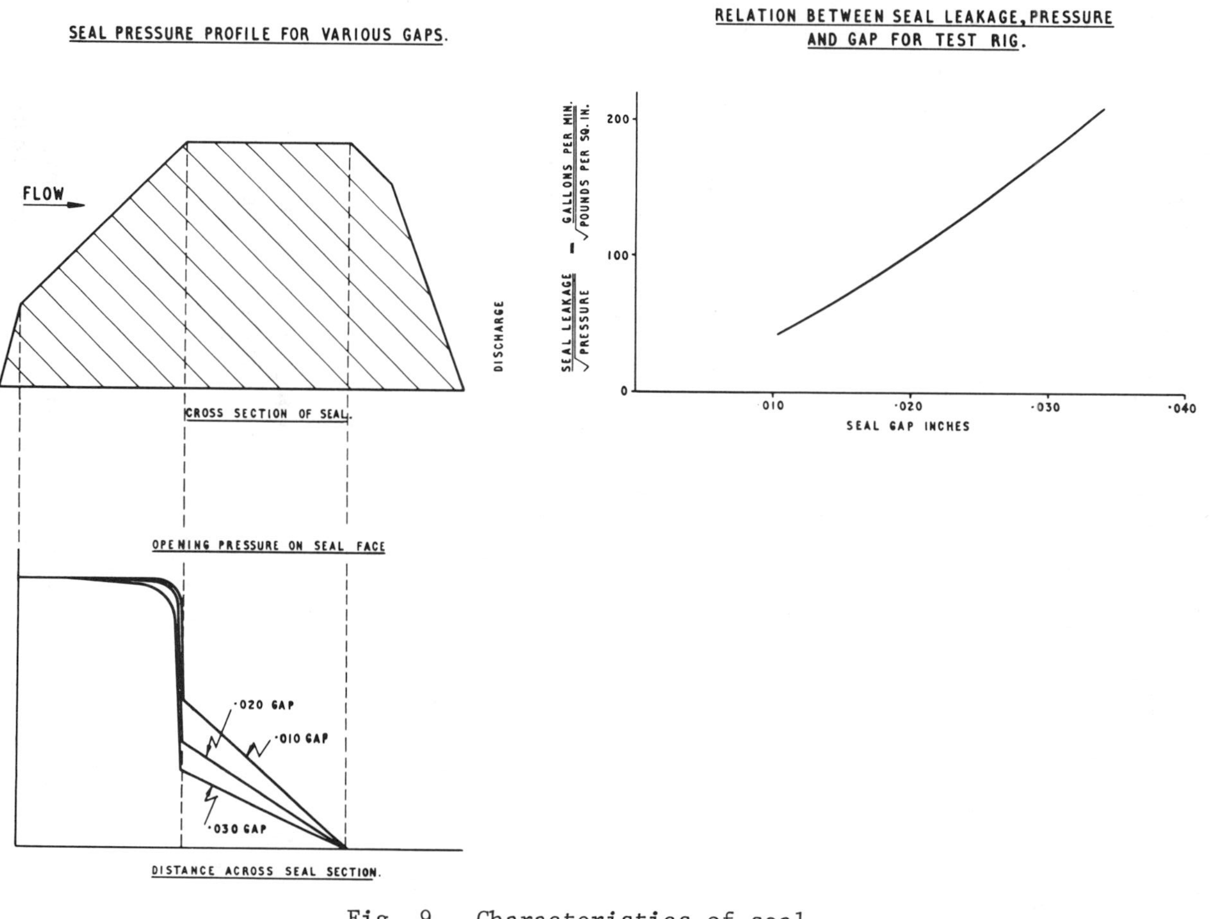

Fig. 9. Characteristics of seal.

and can still be threaded into the vane from the outer periphery provided the diameter of the hole through the bridge matches the larger dimension of the flattened bar. Details of the arrangement will be found on Fig. 6, which also shows the connection to the rim. Since the bar is fixed in the hub and in the rim and so does not move relatively to either but extends and contracts with the rim, it can be drilled to receive the excitation leads and to supply the cooling fluid to the rotor.

Since the vane is not attached to the bar but only guides it, the vane could rotate around the bar, thus making a Kaplan runner possible. In this case, the torque would still pass to the rim by way of the vane bridge and boss, but with the difference that the bridge now rotates around the boss.

The vane can be fixed in the hub by one of the methods developed for Kaplans and can be operated by a servo-motor located in the runner discharge cone.

Support of Rotor

This is achieved by a hydrostatic oil bearing placed on the vertical centre-line of the rotor and running on a fixed shaft overhanging from the nacelle. Thus the bearing load will not exceed the weight. Because of the low peripheral speed and heavy load a hydrostatic bearing using pads fed by high pressure oil was chosen. Similarly, the hydraulic axial thrust is also carried by a hydrostatic pad thrust bearing which is double acting.

The stub shaft is attached to the nacelle, supported by a vertical fin of prestressed concrete. This fin extends right up to the intake and incorporates the gate grooves and the access duct to the inside of the turbine. The large depth of the fin combined with its considerable thickness and the fact that it is completely tied into the dam and its foundations, makes it a rigid beam for carrying the overhung load. The prestressed concrete design reduces deformation under load. The stiffness rate of the support is several times greater than the rate of any forces which can be applied to the rotor, thus preventing vibration. Two horizontal fins near the gate apparatus provide lateral stiffness. For smaller units where the absolute thickness of the concrete may have to be less, the vertical fin may have to be partly of steel, in which case it will extend into the gate apparatus.

Auxiliaries

These are certainly more extensive than with a normal turbine, and their reliability must be beyond doubt, but they are all located outside the machine and can be duplicated as required for safety and will all be readily accessible for inspection and maintenance.

There will be pumps for removing the leakage water, pumps to supply the oil bearings, and a cooling water system for the generator. Some items will

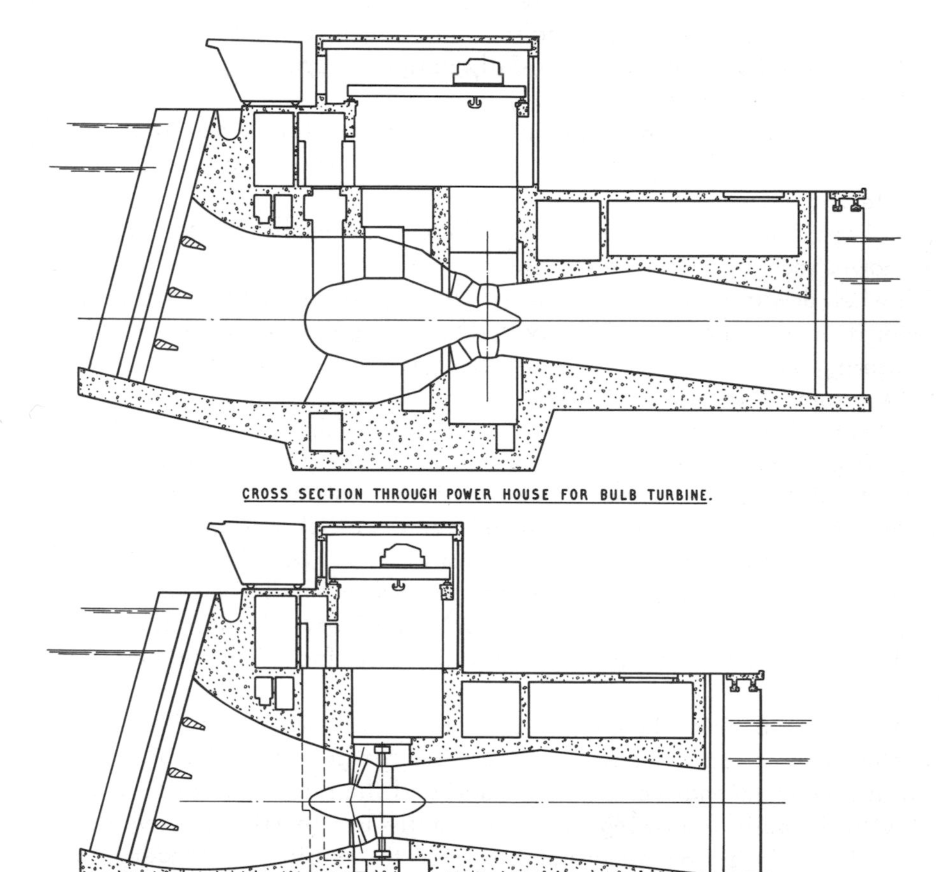

CROSS SECTION THROUGH POWER HOUSE FOR STRAIGHT FLOW TURBINE.

Fig. 10. Comparison between a bulb and a straight flow unit installation.

be duplicated for security, but as there will usually be several straight-flow turbines per installation, the systems can be "bussed up". Undoubtedly, careful attention must be given to these systems.

Overhung Arrangement

This has been chosen because it is cheapest and most efficient hydraulically, since there is no supporting structure downstream of the runner. It requires a heavily reinforced, carefully designed supporting fin, but once the loads to be carried have been established, a safe design can be evolved. The overhung arrangement results in the shortest unit and therefore in the shortest station block. Moreover, it reduces the amount of machinery in the nacelle and improves access.

Regarding access generally, it should be remembered that the upstream and downstream gates are now close together, so the water volume to be dealt with when dewatering is much less than usual and the downstream gate is no longer deeply submerged. Moreover, the leakage pumps are available for dewatering duty, consequently dewatering is simple and so access to the inside of the turbine from outside is easy.

We will now see what the straight-flow design does for the station block. Fig. 10 shows the comparison with a bulb unit, and Fig. 11 with a slant axis unit. In all cases the runner diameter is identical and the draft tubes are equivalent. The straight-flow unit reduces the block to a bell mouth and a draft tube, and therefore shortens it considerably. Both generator and turbine are in the same vertical plane and so can be handled by the same crane, which reduces the area to be covered by cranes very substantially. As a result, the cost of the station block and handling facilities can be cut by up to 40%.

In contrast with the bulb unit, there is no restriction on the diameter of the straight-flow unit's generator, therefore the generator inertia is nearer that of a conventional vertical shaft unit and the regulation difficulties inherent with the bulb unit owing to its abnormally low inertia, are avoided.

The straight-flow generator is outside the turbine (no longer in a steel shell in the water) and so is readily accessible, especially if provision for moving the stator axially so as to uncover the rotor is made as was often done with conventional horizontal shaft sets, -- with the difference that it does not lengthen the units.

Cheapness is essential if low heads are to be developed and the straight-flow unit, whether of the propeller or the Kaplan type, will, we hope, make the development of more of these low-head sites an economic proposition.

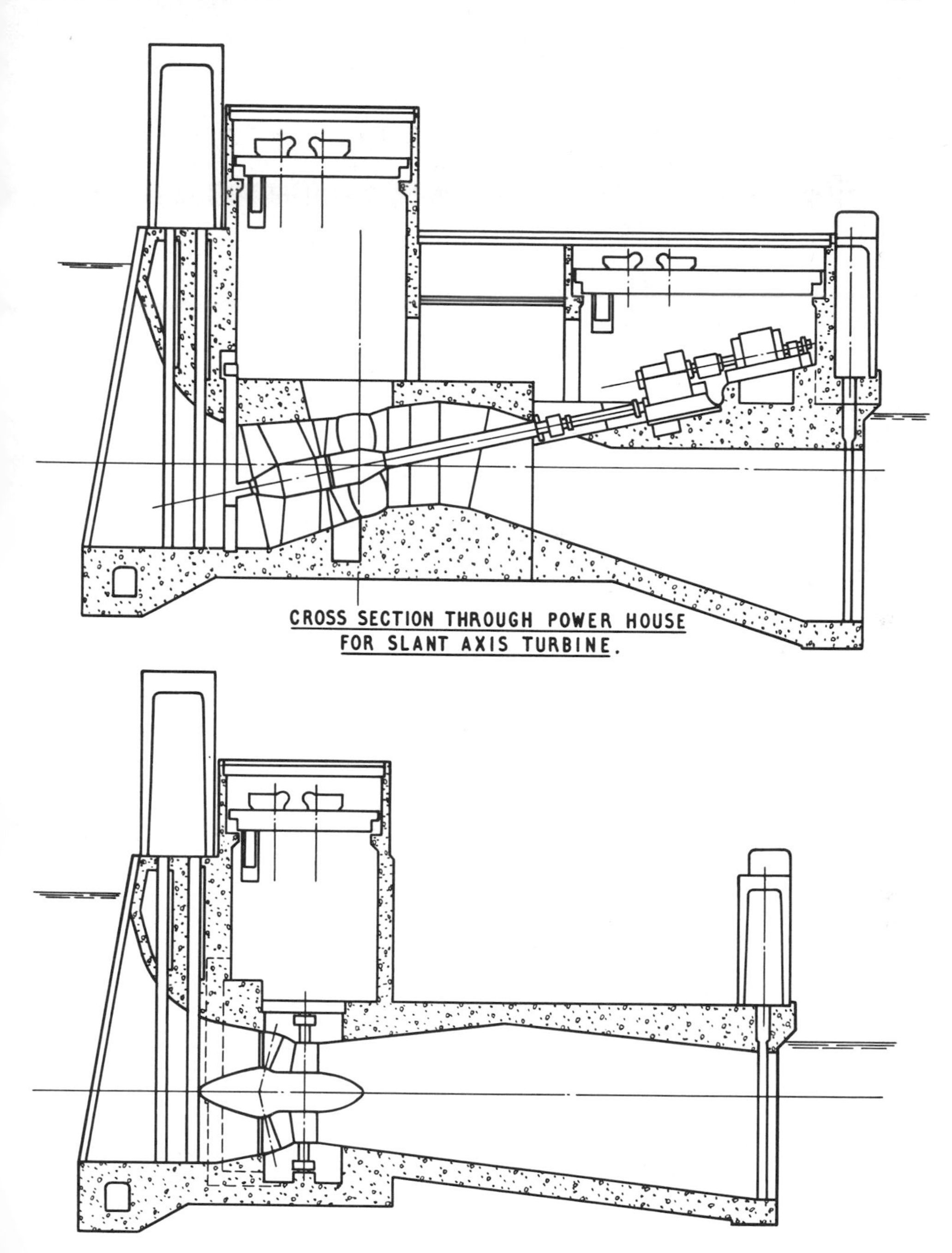

Fig. 11. Comparison between a slant axis and a straight flow unit installation.

BIBLIOGRAPHY

1. Books:

Bernstein, L. B., "PRIAMOTOCHNIE I POGRUJENIE GIDROAGREGATI" (Straight flow and submerged hydro-electric units), CINTIMASH (Central Institute of Scientific & Technical Information on Machine Building); Moscow (1962).

Bernstein, L. B., "KAPSULNIE I SHAHTNIE GIDROAGREGATI" (Bulb and pit hydro-electric units); ITOGI NAUKI ITEHNIKI (Progress of Science & Technology); Moscow (1968).

2. Articles from Journals:

Von Widdern, H. C., "The Tubular Turbine", Escher Wyss News, Vol. XXV/XXVI (1952-53).

Zinter, O., "Operational Experience with Tubular Turbines"; Escher Wyss News, Vol. XXV/XXVI (1952-53).

3. Papers:

Canaan, H. F. (J. M. Voith), "The Underwater Power Station and the Arno Fischer Underwater Turbine", August (1945). (Translated from the original German by English Electric Co. Ltd.)

Braikevitch, M., Gwynn, J. D., and E. M. Wilson "Tidal Power with Special Reference to Plant Construction Techniques and the Integration of the Energy into Existing Electricity Schemes"; 7th World Power Conf., Moscow (1968).

TIDAL EFFECTS DUE TO WATER POWER GENERATION IN THE BRISTOL CHANNEL

N. S. Heaps*

INTRODUCTION

In a previous paper [1] the author estimated the effects on tides in the Bristol Channel of a closed barrage placed across the Channel at various positions. The one-dimensional hydrodynamical equations for the Channel were integrated numerically in the manner described by Proudman [2] using transverse sections 0 to 37 shown in Figure 1. The method yielded tidal elevation at odd-numbered sections and mean horizontal tidal current directed across even-numbered sections. Attention was restricted to a single harmonic constituent, the M_2 tide. Non-linear terms were excluded from the dynamical equations and therefore there was no reproduction of shallow-water tides.

Some of the results obtained from the above-mentioned investigation are presented in Figures 2 and 3. With M_2 tidal elevations given by $H_1 \cos(\sigma t - g_1)$, where σ denotes the speed of M_2, variations of amplitude H_1 and phase-lag g_1 along the length of the Channel are shown for the case representing existing conditions with no barrier and the cases associated with a barrier located at sections 4, 6, 8, 10 and 12 respectively. It is apparent that decreases in tidal amplitude are produced by introducing, across the upper reaches of the estuary, a closure located seaward of section 6.

Continuing the study of barrage effects in the Bristol Channel, the present work was carried out to determine how the tides in the Channel might be further affected by the presence of a permeable rather than an impermeable barrage, the former representing the action of a tidal power station. The postulation of a closed or impermeable barrage in the original paper, while a first step, does not represent the conditions obtaining if a tidal power scheme is operating, since in such circumstances large quantities of water pass the barrage during every tidal cycle.

Specifically, the tidal calculations described here refer to a tidal

---

*Institute of Coastal Oceanography and Tides, Bidston Observatory, Birkenhead, Cheshire, England.

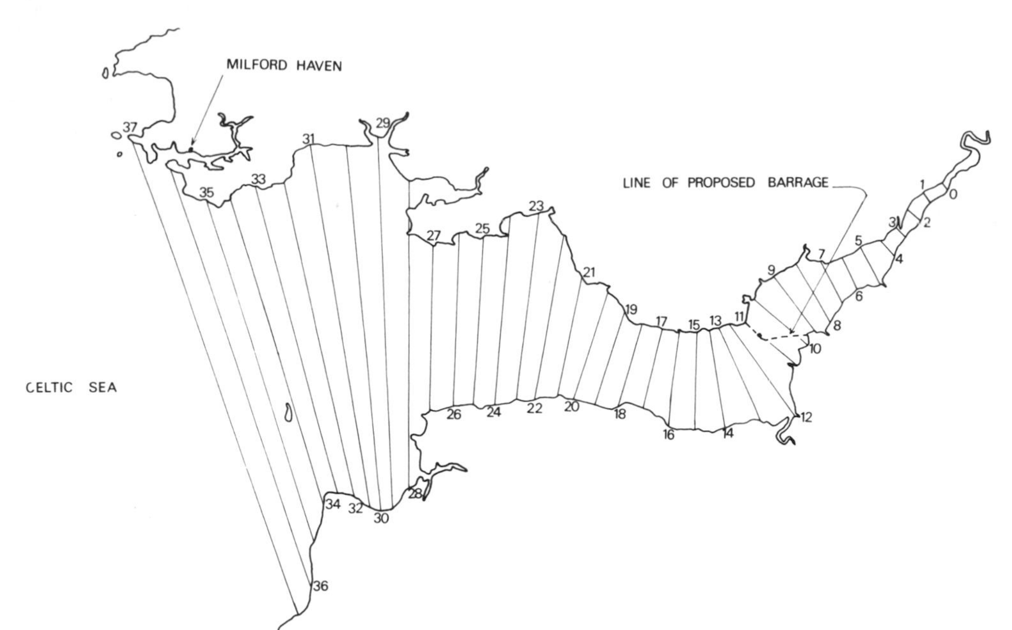

Fig. 1. The Bristol Channel, showing sections used in the calculations.

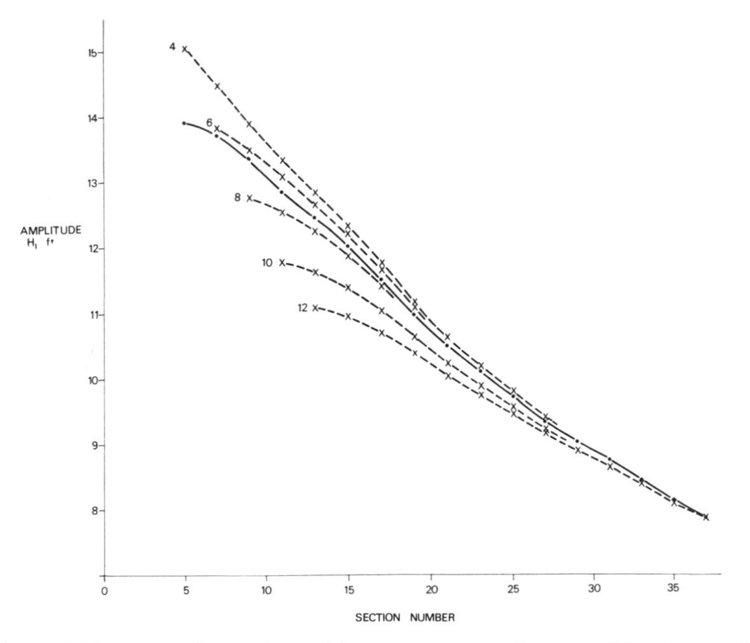

Fig. 2. Effects of a closed barrage on the amplitudes of M_2 tidal elevation (from Heaps 1968):
—•—•—•—•—•—•— without a barrage
n---x---x---x---x--- with a barrage at section n

scheme proposed by Wilson, Severn, Swales and Henery [3], incorporating a barrage situated near to section 10 as indicated in Figure 1. For the purposes of the mathematical analysis the barrage is assumed to coincide with section 10 but formulae are used, giving rates of flow through the turbines and sluices, relating to the design features of the actual proposed system. Because the operation of the scheme involves periods of no flow through the barrage as well as others of seaward discharge and upstream flow, it follows that the tides at points in the estuary cannot be defined in terms of a simple periodic function as in the earlier study. For this reason, an adaptation of the initial value method of computation described by Rossiter and Lennon [4] has been used to determine the tidal variations in the estuary. In this method the tides are built up numerically from an initial state of zero motion to a pattern of steady oscillation which is the required tidal regime. An advantage of the latter approach over the one used previously is that non-linear terms may be retained in the dynamical equations and account thereby taken of the generation of shallow-water tides.

THE TIDAL MODEL

Again the transverse sections 0 to 37 of Figure 1, equally spaced at intervals of 5.11 km, form the basis for a mathematical tidal model. From this, tidal variations along the length of the Channel are determined assuming:

1. a closed end condition at section 0
2. a permeable water power barrage coinciding with section 10
3. a prescribed time-variation of tidal elevation at the mouth, section 37, taking the form of a simple harmonic law $H\cos(2\pi t/T)$; three different cases of such tidal input are considered (each case being defined by an amplitude H and a period T) corresponding to the M_2 tide, conditions at mean springs and conditions at mean neaps respectively.

Thus, behind the barrage there is a reservoir, extending from section 0 to section 10, which connects through the barrage installation with the open part of the Channel extending from section 10 to section 37.

A significant feature of the present theory is the specification of flow rates through the barrage according to the law:

$$
\begin{aligned}
Q_B &= K(\Delta\mathcal{J})^{1/2} && \text{for } \Delta\mathcal{J} > a && \text{(seaward discharge through the turbines)}\\
&= 0 && \text{for } 0 \le \Delta\mathcal{J} \le a && \text{(no flow)}\\
&= -K^1(|\Delta\mathcal{J}|)^{1/2} && \text{for } \Delta\mathcal{J} < 0 && \text{(upstream flow, reservoir filling).}
\end{aligned}
\qquad \ldots\ldots.(1)
$$

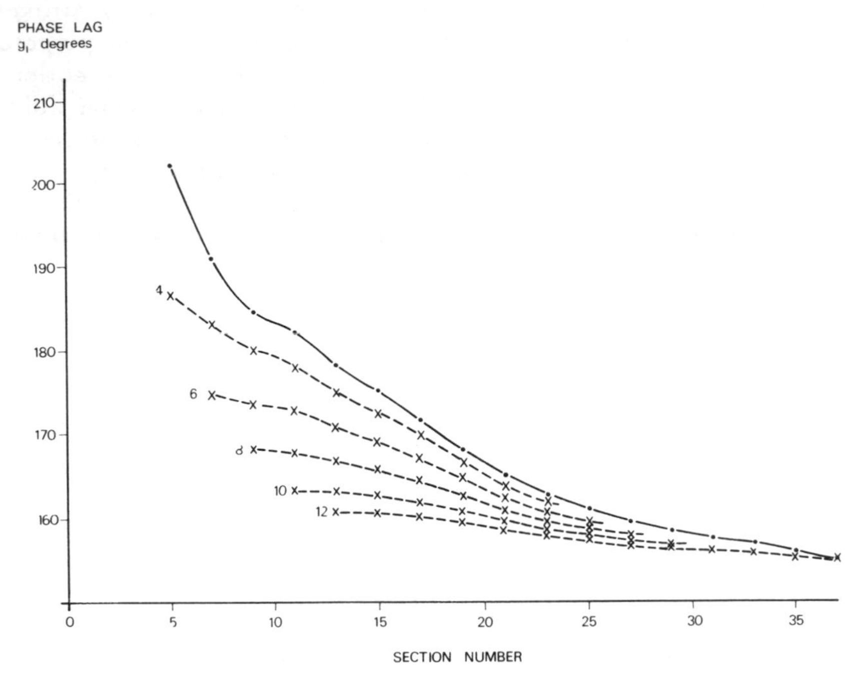

Fig. 3. Effects of a closed barrage on the phases of M_2 tidal elevation (from Heaps 1968):
•—•—•—•—•—•—•— without a barrage
n---x---x---x---x--- with a barrage at section n

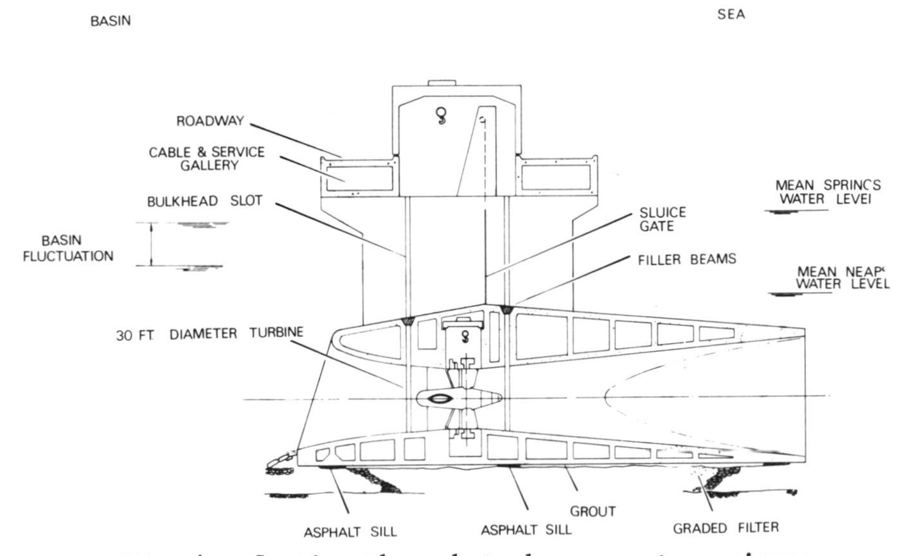

Fig. 4. Section through turbo-generator caisson.

Here, Q_B denotes the rate of flow at any time, counted positive in the seaward direction, and $\Delta\mathcal{J}$ the drop in head across the barrage moving from reservoir to sea. Generation of power with downstream flow is supposed to occur only when the head difference is greater than a and, as indicated above, this gives rise to three different regimes governing the exchange of water between the reservoir and the open sea. The critical head of water, a, is taken as 12 ft. and K, K^1 are constants relating to the system of turbines and sluices for the proposed tidal scheme. Details of this scheme, giving expressions for K and K^1, have been communicated to me by Dr. E. M. Wilson. The following is a summary of the facts I received from him:-

From an energy analysis based on an optimization procedure described by Swales and Wilson (5), and assuming reductions in tidal range from 2.5 ft. on a 36 ft. tide to 1.5 ft. on an 18.5 ft. tide, the optimum number of turbo-generators and refilling sluices likely to produce energy at minimum cost per kWh has been established. There are 120 straight-flow turbines of 30 ft. diameter and 120 venturi sluices 40 ft. square. In addition, 120 power house over-sluices are allowed for, operating at various over-sill heads. The number of over-sluices is the same as the number of turbines because of the station design; the number of venturi sluices is the same purely by coincidence. The physical shapes of the water passages through the straight-flow turbines, the over-sluices and the venturi sluices are shown in Figures 4 and 5.

The scheme is designed for ebb-flow power generation only, such that flow occurs during turbining in a downstream direction under a differential head between 7.5 and 30 ft. and during refilling, in an upstream direction, under a head between 0 and 5 ft. The preceding assumption of a = 12 ft. is a simplification for the tidal computations: the power calculations used optimized starting heads varying from about 10 to 16 ft. and generation was assumed in most cases down to a 7.5 ft. head.

In turbining, flow takes place through the turbines only and the value of K is given by:

$$K = C_1 N_1 A_1 \sqrt{2g} \qquad \ldots\ldots.(2)$$

Refilling takes place through turbines (now idling in reverse), over-sluices and venturi sluices, so that K^1 is given by:

$$K^1 = C_{xy}L + \sqrt{2g}\,(C_2N_2A_2 + C_3N_1A_1) \qquad \ldots\ldots.(3)$$

where g denotes the acceleration due to gravity and

C_1 (a discharge coefficient for turbining) = 0.54
N_1 (the number of turbines) = 120
A_1 (the area of a single turbine water passage) = 225π ft.^{2}
C_2 (venturi sluice discharge coefficient) = 1.60
N_2 (the number of venturi sluices) = 120

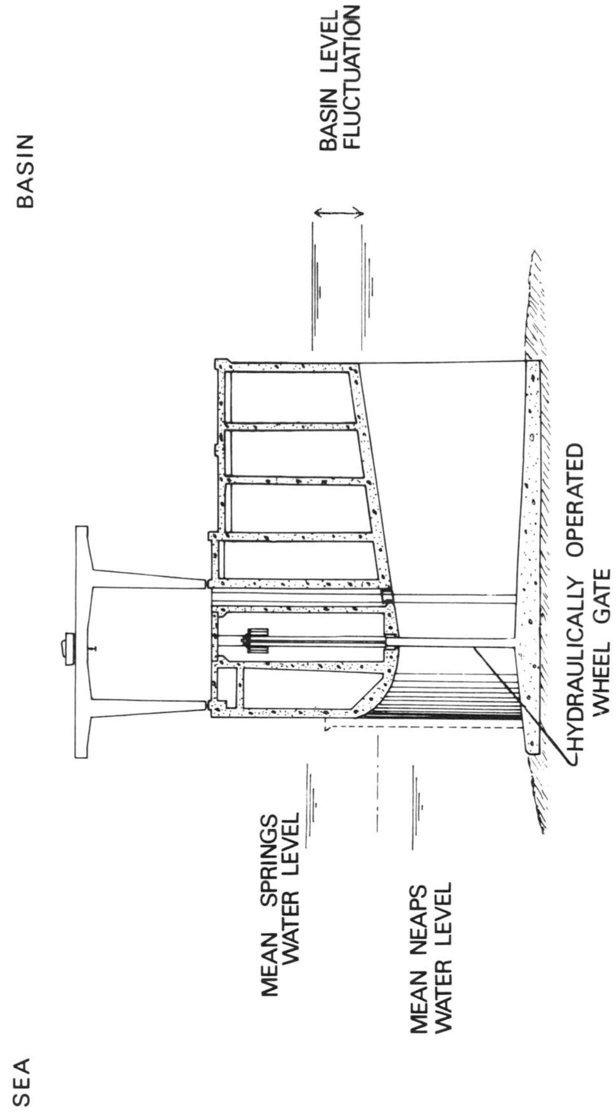

Fig. 5. Refill sluice section.

A_2 (the area of a single venturi sluice) = 1600 ft.^{2}
C_3 (a discharge coefficient for refilling) = 0.84
C_{xy} (see below) = 200 ft.$^{3/2}$/sec.
L (the crest length of all 120 crest sluices) = 3600 ft.

A note is necessary about the term $C_{xy}L$ in (3): the other terms are quite conventional. A formula was used for the rate of flow over the submerged broad-crested over-sluices of the form:

$$Q_f = C_x C_y A^1 \sqrt{2g} \, (|\Delta \mathcal{T}|)^{1/2}$$

where C_x = a discharge coefficient
C_y = a submergence coefficient
A^1 = flow area = DL (depth x crest length).

Since A^1 is not constant this formula is awkward to use and therefore a study was made of all possible discharges through the over-sluices and a simplified "averaged" formula derived:

$$Q_f = C_{xy} L \, (|\Delta \mathcal{T}|)^{1/2}$$

which describes the over-sluice flow without substantial error; L is not combined in the constant factor so that different lengths may be used.

BASIC EQUATIONS

The numerical analysis is based on the differential equation of motion:

$$\frac{\partial u}{\partial t} = -g \frac{\partial \mathcal{T}}{\partial x} - \frac{k u |u|}{h + \mathcal{T}} \qquad \text{.......(4)}$$

and that of continuity:

$$\frac{\partial Q}{\partial x} + b \frac{\partial \mathcal{T}}{\partial t} = 0 \qquad \text{.......(5)}$$

where

$$Q = (A + b\mathcal{T})u \qquad \text{.......(6)}$$

Here, t denotes time,
x distance measured along the medial line of the Channel seaward from section 0,
A area of a vertical cross-section of the Channel when the water surface there is at its mean level,
b breadth of the Channel at the mean level of the water surface,

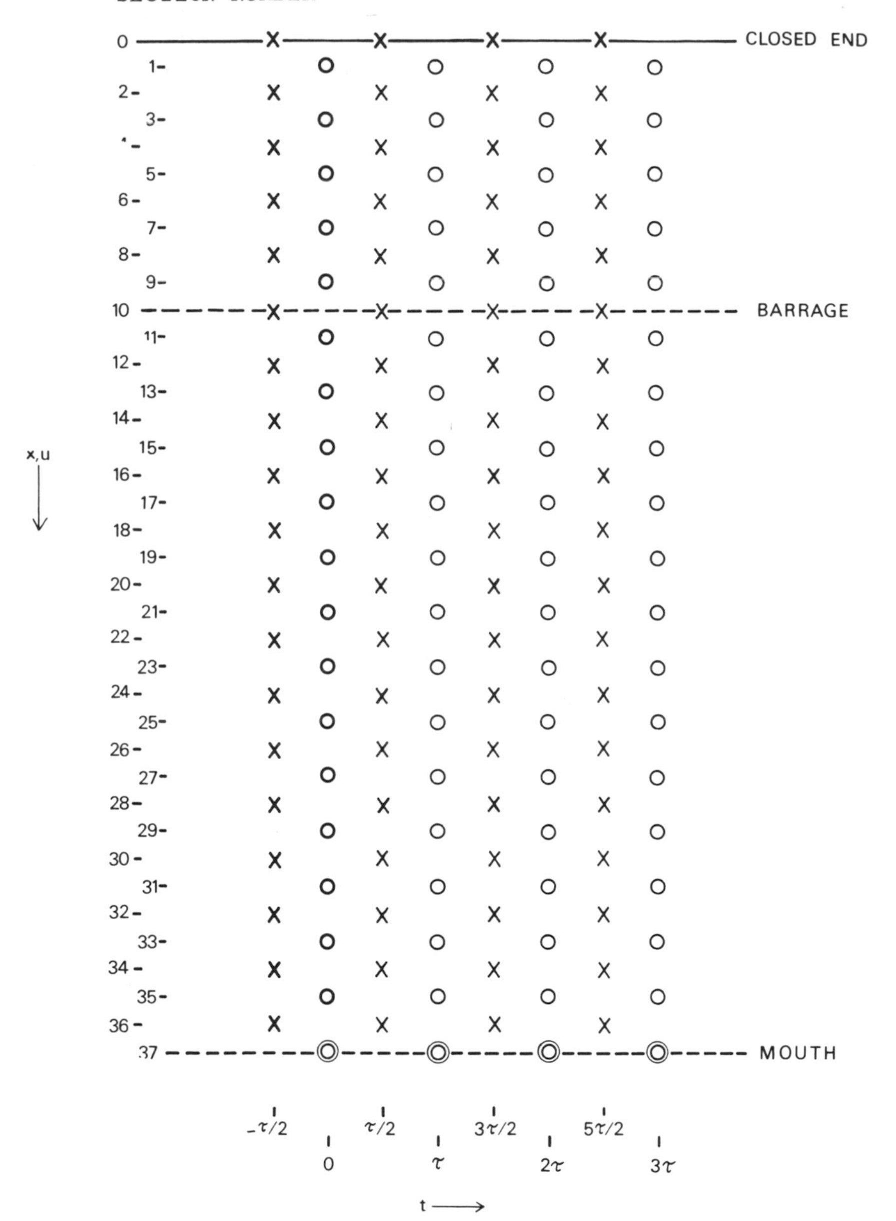

Fig. 6. Computational scheme:
x u point, x zero u
o ζ point, o zero ζ, ◎ prescribed ζ (input at mouth)

h mean value of the depth of water over a cross-section, given by A/b,
$\mathcal{T}$ tidal elevation above the mean level,
u mean value, over a cross-section, of the horizontal tidal current in the direction of increasing x,
Q the total flow over a cross-section,
k a coefficient of friction.

Thus, consideration is given to one-dimensional motion directed along the length of the Channel. Second order accelerations and variations of broadth with changing water level are ignored. A quadratic law of friction is assumed and appears in the final term of equation (4).

A notation is introduced defining $\mathcal{T}$, u, Q (functions of x and t), k, A, h and b (functions of x) at sections 0 to 37 of Figure 1. With τ a fixed time interval and q an integer taking values 0, 1, 2, 3,....., let

$\mathcal{T}_{2r+1,\,q}$ = $\mathcal{T}$ at section 2r + 1 at time $q\tau$,
$u_{2r,\,q-1/2}$ = u at section 2r at time $(q - 1/2)\tau$,
$Q_{2r,\,q-1/2}$ = Q at section 2r at time $(q - 1/2)\tau$,
k_{2r} = k at section 2r,
A_{2r} = A at section 2r,
h_{2r} = h at section 2r,
b_{2r}, b_{2r+1} = b at sections 2r, 2r + 1 respectively,
(r = 0, 1, 2, ..., 18).

Accordingly, u is evaluated at the even-numbered sections at times $t = -\tau/2, \tau/2, 3\tau/2, 5\tau/2, \ldots$ and $\mathcal{T}$ at the odd-numbered sections at times $t = 0, \tau, 2\tau, 3\tau, \ldots$. The pattern formed by these values in space and time is indicated in Figure 6. The tidal computations start with zero values of u and $\mathcal{T}$ at $t = -\tau/2$ and $t = 0$, respectively, and determine in a sequence of forward time steps the u and $\mathcal{T}$ at the higher time levels $(q - 1/2)\tau$, $q\tau$ where q = 1, 2, 3,.... The iterative procedure which effects this progressive buildup of elevations and currents in the Channel is described subsequently. Application of the procedure continues with increasing q until the variations of u and $\mathcal{T}$ differ insignificantly from one tidal cycle to the next. The tidal regime in the Channel is then considered to be established. An integral part of the computational scheme involves equating u to zero at section 0 and $\mathcal{T}$ to prescribed input values at section 37. In addition, the barrage condition given by equation (1) is invoked.

NUMERICAL METHOD

Figure 7 illustrates, using general notation, the two basic steps of the numerical procedure. First, the u-values at $t = (q + 1/2)\tau$ are determined

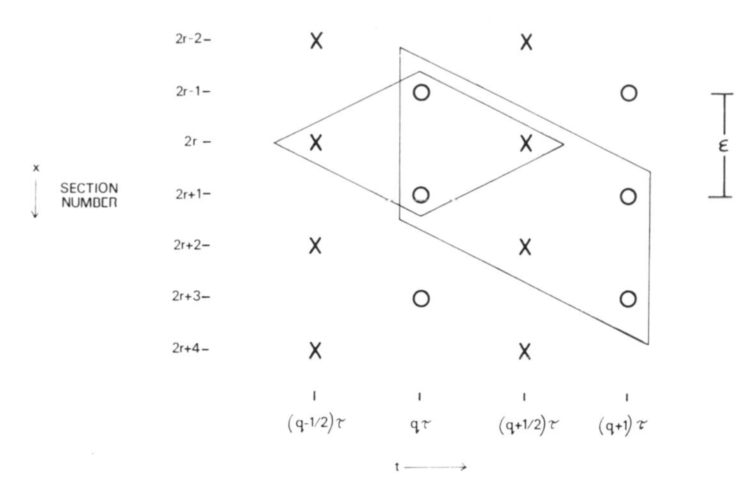

Fig. 7. General notation showing computational steps: x u point, o ζ point

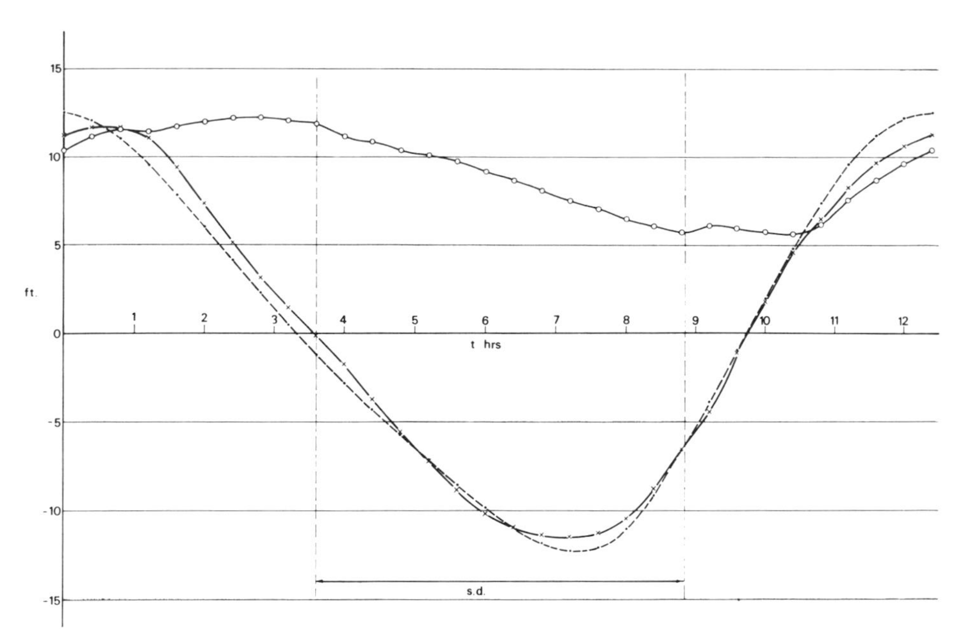

Fig. 8. Tidal curves for M_2:

—o—o—o—o— elevation at section 9 } permeable

—x—x—x—x— elevation at section 11 } barrage

—•—•—•—•— elevation at section 11, impermeable barrage

s.d. = seaward discharge through the turbines

from the u -values at $t = (q - 1/2)\tau$ and the $\mathcal{T}$-values at $t = q\tau$; secondly, the $\mathcal{T}$-values at $t = (q + 1)\tau$ are determined from the $\mathcal{T}$-values at $t = q\tau$ and the u -values at $t = (q + 1/2)\tau$. Repeated application of this routine taking, successively, $q = 0, 1, 2, 3, \ldots$ yields the u and $\mathcal{T}$ through time from their initial values. The routine, derived essentially from finite-difference approximations to the hydrodynamical equations (4) and (5) is now described in detail.

Step 1

Employing central differences in both space and time, equation (4) may be written approximately:

$$(u_{2r,\,q+1/2} - u_{2r,\,q-1/2})/\tau = -g(\mathcal{T}_{2r+1,\,q} - \mathcal{T}_{2r-1,\,q})/\epsilon$$
$$-k_{2r}\, u_{2r,\,q+1/2}|u_{2r,\,q-1/2}|/(h_{2r} + \mathcal{T}_{2r,\,q})$$

where

$$\mathcal{T}_{2r,\,q} = (\mathcal{T}_{2r+1,\,q} + \mathcal{T}_{2r-1,\,q})/2$$

ϵ (= 10.22 km) denotes the distance between alternate sections. It follows that:

$$u_{2r,\,q+1/2} = \frac{u_{2r,\,q-1/2} - g\tau(\mathcal{T}_{2r+1,\,q} - \mathcal{T}_{2r-1,\,q})/\epsilon}{1 + \tau k_{2r}|u_{2r,\,q-1/2}|/(h_{2r} + \mathcal{T}_{2r,\,q})} \qquad \ldots\ldots.(7)$$

which determines the u -values at $t = (q + 1/2)\tau$ from the u -values at $t = (q - 1/2)\tau$ and the $\mathcal{T}$-values at $t = q\tau$. However, special conditions apply at the barrage, section 10: having regard to equation (1) we take

$$Q_{10,\,q+1/2} = \begin{cases} K(\Delta\mathcal{T})^{1/2} & \Delta\mathcal{T} > a \\ 0 & 0 \le \Delta\mathcal{T} \le a \\ -K^{1}(|\Delta\mathcal{T}|)^{1/2} & \Delta\mathcal{T} < 0 \end{cases} \qquad \ldots\ldots.(8.1)$$

with

$$\Delta\mathcal{T} = \mathcal{T}_{9,\,q} - \mathcal{T}_{11,\,q} \qquad \ldots\ldots.(8.2)$$

Hence, equations (7) and (8) yield $u_{2r,\,q+1/2}$ ($r = 18, 17, 16, \ldots \ldots, 6; 4, 3, 2, 1$) and $Q_{10,\,q+1/2}$ in terms of the values of u and $\mathcal{T}$ at the earlier time levels $(q - 1/2)\tau$, $q\tau$.

Step 2

Approximately, equation (5) may be written:

$$(Q_{2r+2,\, q+1/2} - Q_{2r,\, q+1/2})/\epsilon + b_{2r+1}(\mathcal{T}_{2r+1,\, q+1} - \mathcal{T}_{2r+1,\, q})/\tau = 0$$

whence

$$\mathcal{T}_{2r+1,\, q+1} = \mathcal{T}_{2r+1,\, q} - \tau(Q_{2r+2,\, q+1/2} - Q_{2r,\, q+1/2})/\epsilon\, b_{2r+1} \quad ....(9)$$

This relation is applied successively for $r = 17, 16, 15, \ldots, 6$ to determine $\mathcal{T}_{2r+1,\, q+1}$ from the values of $\mathcal{T}$ and Q at the immediately earlier time levels. In its application we take, referring to (6):

$$Q_{2r+2,\, q+1/2} = u_{2r+2,\, q+1/2}\{A_{2r+2} + b_{2r+2}(\mathcal{T}_{2r+3,\, q+1} + \mathcal{T}_{2r+1,\, q})/2\} \quad ....(10.1)$$

and

$$Q_{2r,\, q+1/2} = u_{2r,\, q+1/2}\{A_{2r} + b_{2r}(\mathcal{T}_{2r+3,\, q+1} + \mathcal{T}_{2r-1,\, q})/2\}, \quad ....(10.2)$$

thereby determining a first approximation to $\mathcal{T}_{2r+1,\, q+1}$, $\mathcal{T}^{(1)}_{2r+1,\, q+1}$ say. Then with $Q_{2r+2,\, q+1/2}$ as before, but now taking:

$$Q_{2r,\, q+1/2} = u_{2r,\, q+1/2}\{A_{2r} + b_{2r}(\mathcal{T}^{(1)}_{2r+1,\, q+1} + \mathcal{T}_{2r-1,\, q})/2\}, \quad ....(10.3)$$

(9) yields a second approximation to $\mathcal{T}_{2r+1,\, q+1}$, $\mathcal{T}^{(2)}_{2r+1,\, q+1}$ say. Further approximations are found in the same way until the change from one approximation to the next becomes negligibly small.

The use of (9) is continued, determining in succession:

1. $\mathcal{T}_{11,\, q+1}$ - taking $Q_{12,\, q+1/2}$ as given by (10.1) with $r = 5$, and $Q_{10,\, q+1/2}$ as already found from (8);

2. $\mathcal{T}_{9,\, q+1}$ - knowing $Q_{10,\, q+1/2}$ and taking

$$Q_{8,\, q+1/2} = u_{8,\, q+1/2}\{A_8 + b_8(\mathcal{T}_{7,\, q} + \mathcal{T}_{9,\, q})/2\}$$

to find an approximation $\mathcal{T}^{(1)}_{9,\, q+1}$; then finding a second approximation from (9) using $Q_{8,\, q+1/2}$ as given by (10.3) with $r = 4$; subsequently repeating the use of (9) and (10.3) in this way to yield closer and closer approximations to $\mathcal{T}_{9,\, q+1}$;

3. $\mathcal{T}_{2r+1,\, q+1}$ for $r = 3, 2, 1$ - employing (9), (10.1), (10.2) and (10.3) as already described;

4. $\mathcal{T}_{1,\, q+1}$ - taking $Q_{0,\, q+1/2} = 0$ and $Q_{2,\, q+1/2}$ given by (10.1) with $r = 0$.

The importance of the computational sequence described above lies in the inclusion of the barrage flow $Q_{10,\, q+1/2}$ - calculated in Step 1 and then used in performing Step 2. It should be noted that, according to equations (8.1) and (8.2), the flow through the barrage at time $(q + 1/2)\tau$ is related to the head difference at time $q\tau$ rather than at time $(q + 1/2)\tau$. This approximate formulation was adopted to fit the computational scheme; τ is small in practice and therefore the approximation is acceptable.

THE COMPUTATIONS

The computations were carried out on the KDF9 digital computer at Liverpool University. As program data, the values of b and h for the various sections were taken from the author's earlier paper [1]. A value of 0.0026 was given to the frictional coefficients k_{2r}. To achieve sufficient accuracy, with errors in tidal elevation of less than 0.05 ft. it was found necessary to take $\tau = 0.0125$ hr.; the smallness of this time step was required in order to cope with the rapid changes in flow through the barrage at the commencement of each period of reservoir filling.

Tidal elevations and currents along the length of the Channel were calculated, in the manner already described, for three different cases of tidal input at the mouth (section 37). Prescribing input elevation in the form:

$$\mathcal{T}_M = H \cos(2\pi t/T)$$

the three cases corresponded to:

1. the M_2 tide, taking $H = 7.87$ ft. and $T = 12.4$ hr.,
2. conditions at mean springs, taking $H = 10.4$ ft. and $T = 12.3$ hr.,
3. conditions at mean neaps, taking $H = 4.55$ ft. and $T = 12.5$ hr.

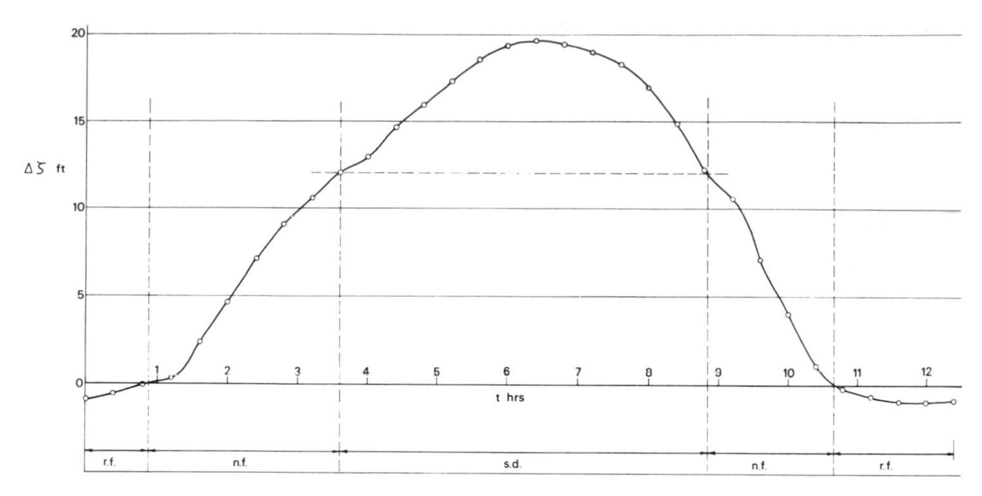

Fig. 9. Head difference across the barrage (M_2):
s.d. = seaward discharge through the turbines
n.f. = no flow through the barrage
r.f. = reservoir filling, upstream flow through the barrage

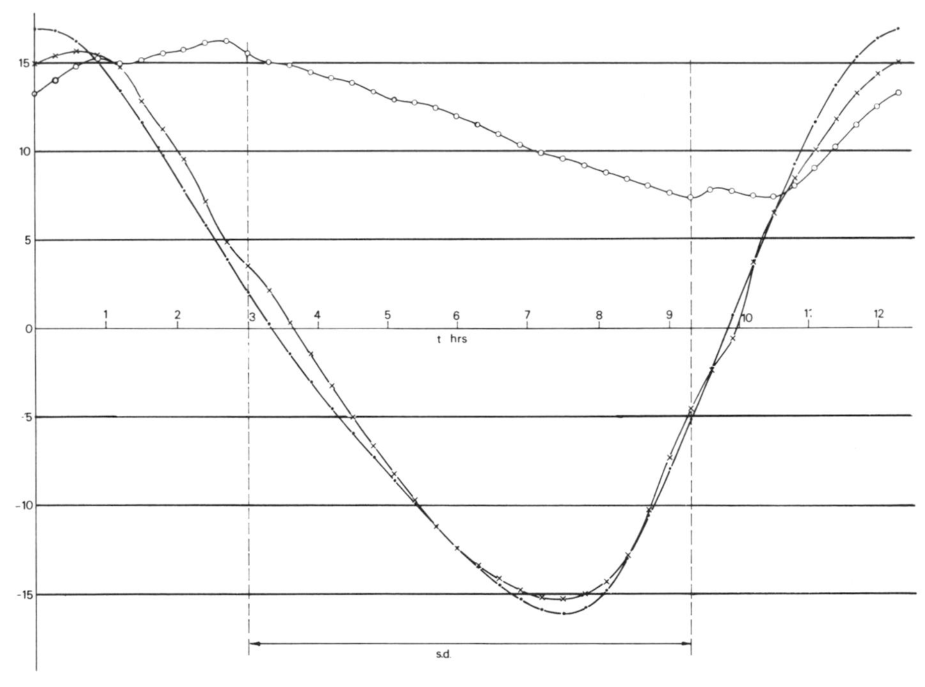

Fig. 10. Tidal curves at mean springs:
o — o — o — o — o elevation at section 9 } permeable
x—x—x—x—x elevation at section 11 } barrage
•—•—•—•—• elevation at section 11, impermeable barrage
s.d. = seaward discharge through the turbines

These values of H and T were derived from Admiralty Tide Tables and refer to the port of Milford Haven which lies at the seaward end of the Channel (Figure 1). Thus, it was assumed that the existing tides at the mouth of the estuary (no barrage) could be used as tidal input to the barraged estuary, i.e., the effect of the barrage on the tides at the mouth was assumed to be negligibly small.

In case (1), the tides generated in response to the M_2 oscillation were built up and established by computations which stepped forward in time through six cycles of M_2; elevations at the odd-numbered sections and currents at the even-numbered sections were printed out at intervals of 0.4 hr. In case (2), computations extending over six cycles of the applied oscillation established the mean spring tides (output every 0.3 hr). In case (3), computations covering eight cycles of the applied oscillation established the mean neap tides (output every 0.5 hr). Results of special interest, obtained from these calculations, are shown in Figures 8 to 13.

Parallel computations to those mentioned above were carried out with an impermeable barrage located at section 10. Results thus obtained are also shown in Figures 8 to 13 for comparison.

DISCUSSION OF THE RESULTS

Some comments are now made on the results of the computations shown in Figures 8 to 13. The figures cover cycles of the average tide (M_2), mean spring tides and mean neap tides. For each regime, tidal curves give the variations of elevation at section 9 (directly upstream of the barrage) and section 11 (directly downstream of the barrage) when the tidal power scheme is operating. The corresponding variations of elevation at section 11 when the permeable water power barrage is replaced by an impermeable barrage are also indicated. Head difference across the barrage, determined as the drop in elevation from section 9 to section 11, is plotted separately for each state of tide.

From the point of view of the generation of power, it is important to note from the curves that the effect of permeability is to reduce the tidal range at section 11 by approximately 1.6, 2.2 and 0.5 ft. for M_2, mean springs and mean neaps respectively. However, the loss in head across the turbines associated with this reduction is counterbalanced by the influence of a seiche-like oscillation in the reservoir set up during refilling. The effect, clearly observable in all three tidal regimes, is a rise in level at section 9 preceding the beginning of seaward discharge through the turbines. The large downstream flow through the turbines appears to retard the reflection of this rise from the barrage so that a higher head than would have been obtained from a reservoir of uniform level occurs over much of the generation period.

The periods of seaward discharge through the turbines, no flow through

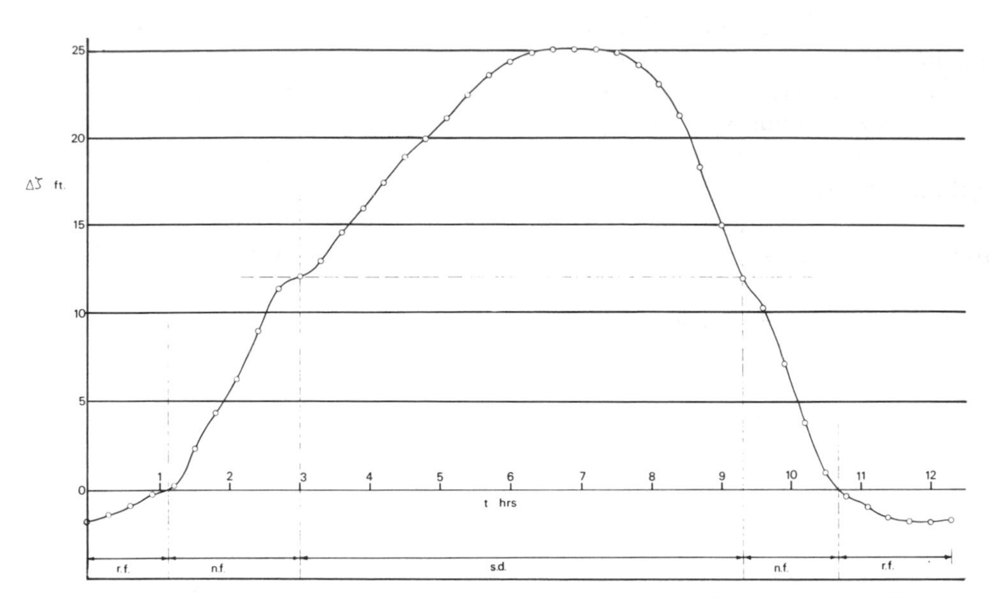

Fig. 11. Head difference across the barrage (mean springs):
s.d. = seaward discharge through the turbines
n.f. = no flow through the barrage
r.f. = reservoir filling, upstream flow through the barrage

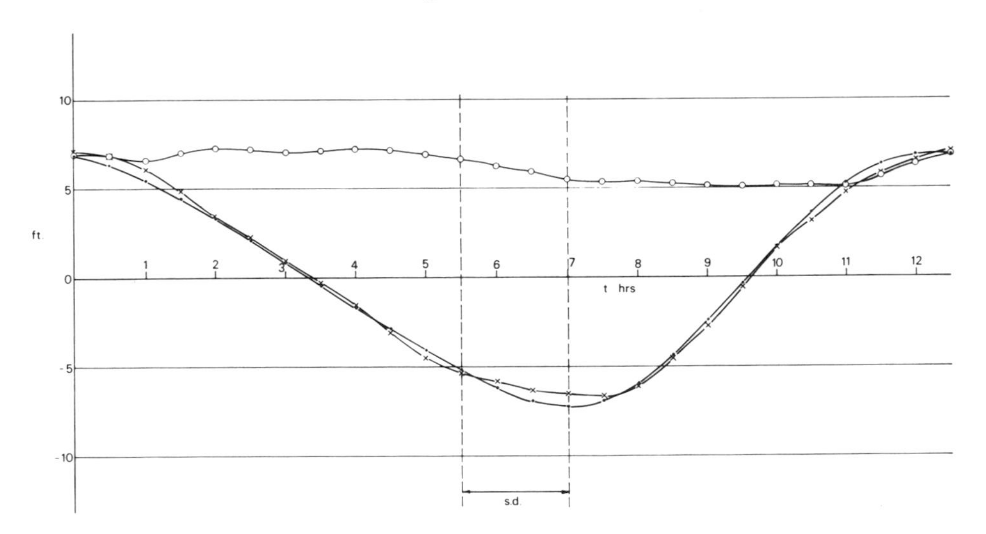

Fig. 12. Tidal curves at mean neaps:
o — o — o — o — o elevation at section 9 } permeable barrage
x — x — x — x — x elevation at section 11 } permeable barrage
• — • — • — • — • elevation at section 11, impermeable barrage
s.d. = seaward discharge through the turbines

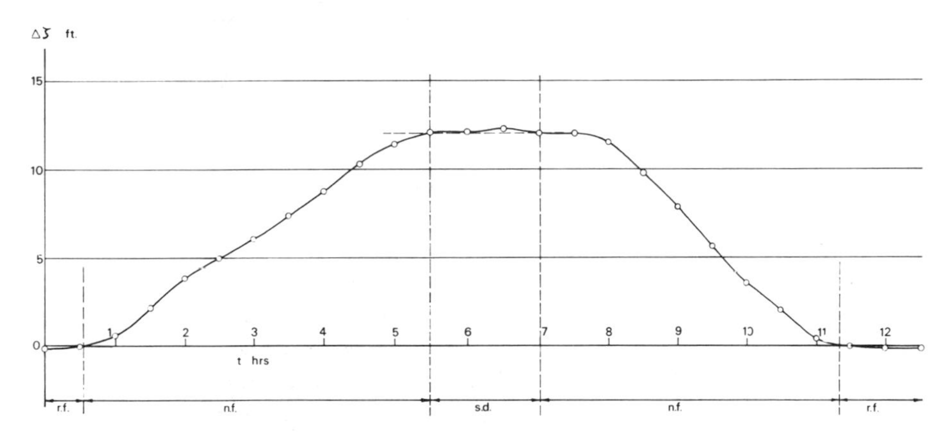

Fig. 13. Head difference across the barrage (mean neaps):
s.d. = seaward discharge through the turbines
n.f. = no flow through the barrage
r.f. = reservoir filling, upstream flow through the barrage

the barrage and upstream flow through the barrage with the reservoir filling are marked in the figures. Clearly, the passage of water through the barrage from sea to reservoir causes the level at section 11 to reach a lower maximum somewhat later. However, the subsequent rate of fall in level at this section is greater than that which occurs with a closed barrage, due probably to the large downstream momentum imparted to the Channel as a result of turbining. Thus, the overall dynamical properties of the reservoir-estuary system play an important part in determining the fluctuations of water level in the vicinity of the barrage.

Impulsive motion which occurs at the commencement and termination of seaward discharge is indicated by characteristic irregularities in the tidal curves -- particularly those for the average tides and the mean spring tides. As might be expected, all the variations in tidal elevation at section 11 are distinctly non-linear, rates of rise on the flood being greater than rates of fall on the ebb.

CONCLUDING REMARKS

The main aim of the paper has been to investigate in a relatively uncomplicated way the likely influence on the tides of a hydro-electric barrage placed across the upper reaches of the Bristol Channel. The work has attempted to seek out trends and determine effects rather than present a comprehensive dynamical account of the Bristol Channel with water power generation. Perhaps the most important feature of the study has been the comparison made between the tides obtained assuming an open barrage representing the action of the power station, and those obtained with a closed barrage located in the same position. The general conclusion must be that the effects of barrage permeability are of considerable importance in the design of the proposed tidal scheme. The price which must be paid for a satisfactory assessment of the energy potential of the Channel is a detailed consideration of the dynamics of the entire estuary with special emphasis on conditions near the barrage and in the reservoir behind it. In this respect a more sophisticated numerical model of the barraged estuary could be conceived than the one presented here, taking into account two-dimensional motion in the horizontal, but this would be a research topic since the numerical methods available to solve the problem have not, to my knowledge, been applied to such a situation.

ACKNOWLEDGMENTS

I wish to thank Dr. E. M. Wilson for his cooperation on the engineering aspects of this paper. It was an enquiry from him, concerning the tidal effects of a permeable as opposed to an impermeable barrage in the Bristol Channel, which led to the commencement of the work.

My thanks are also due to Mr. A. Bamford for assistance in carrying through the computations.

Figures 2 and 3 are reproduced by permission of The Institution of Civil Engineers, being taken from the paper: "Estimated Effects of a Barrage on Tides in the Bristol Channel" in Volume 40 of the Proceedings of The Institution.

REFERENCES

1. Heaps, N. S., "Estimated Effects of a Barrage on Tides in the Bristol Channel", Proc. of Inst. of Civil Engineers, 40, 495 (1968).

2. Proudman, J., "Dynamical Oceanography", Methuen and Co. Ltd., London, 325 (1953).

3. Wilson, E. M., Severn, B., Swales, M. C., and D. Henery, "The Channel Barrage Project", Proc. of 11th Conf. on Coastal Engineering, Vol. 2, Am. Soc. of Civil Engineers, N. Y., 1304 (1969).

4. Rossiter, J. R., and G. W. Lennon, "Computation of Tidal Conditions in the Thames Estuary by the Initial-Value Method", Proc. of the Inst. of Civil Engineers, 31, 25 (1965).

5. Swales, M. C, and E. M. Wilson, "Optimization of Tidal Power Generation", Water Power, 20, 109 (1968).

NOTATION

H_1	amplitude of the M_2 tide
g_1	phase lag of the M_2 tide
σ	speed of the M_2 tide
H	amplitude of the tidal oscillation at the mouth of the Channel
T	tidal period
Q_B	rate of downstream flow through the barrage
$\Delta \mathcal{T}$	drop in head across the barrage, from the reservoir to the sea
a	critical head of water below which there is no turbining
K, K^1	constant coefficients in the barrage formulae
C_1	a discharge coefficient for turbining
N_1	the number of turbines
A_1	the area of a single turbine water passage
C_2	venturi sluice discharge coefficient
N_2	the number of venturi sluices
A_2	the area of a single venturi sluice
C_3	a discharge coefficient for refilling
Q_f	flow rate over the submerged broad-crested over-sluices
L	the crest length of the over-sluices
C_x	a discharge coefficient
C_y	a submergence coefficient
D	depth of flow over the crest sluices

A^1	flow area over the crest sluices
C_{xy}	a constant factor in the formula for Q_f
g	the acceleration due to gravity
t	time
x	distance measured along the medial line of the Channel seaward from section 0
A	area of a vertical cross-section of the Channel when the water surface there is at its mean level
b	breadth of the Channel at the mean level of the water surface
h	mean value of the depth of water over a cross-section
ζ	tidal elevation above the mean level
u	mean value, over a cross-section, of the horizontal tidal current in the direction of increasing x
Q	total flow over a cross-section
k	a coefficient of friction
τ	a fixed time step
q	integer denoting time level
r	integer used in quoting section number
ϵ	distance between alternate sections
ζ_M	tidal elevation at the mouth of the Channel

A MATHEMATICAL MODEL OF THE EFFECT OF A TIDAL BARRIER ON SILTATION IN AN ESTUARY

M. W. Owen and N. V. M. Odd*

INTRODUCTION

The Greater London Council have commissioned the Hydraulics Research Station to carry out a comprehensive investigation of the hydraulic aspects of means of preventing flooding in Central London. The mathematical model of silt movement in the Thames Estuary described in this paper forms part of that investigation. The main purpose of the model was to attempt to reproduce the periodic pattern of suspended silt concentrations, and the location of the main areas of deposition of silt in the estuary: hence to test the effects of silt movement in the estuary of various tidal barriers proposed for flood protection.

As far as is known, this is the first time that such a model has been attempted, and further work is necessary before it can be accepted as completely reliable for investigating detailed changes in an estuary. In particular, fundamental research designed to give a better understanding of the processes of silt movement is continuing at the Hydraulics Research Station [1, 2]. Nevertheless the study has given an overall indication of the changes which could result from various flood protection schemes involving tidal control.

THE THAMES ESTUARY

The main features of the estuary of the River Thames, a map of which is shown in Fig. 1, are its relatively uniform depths (about 25 ft. (7.6 m) at mean tide level) and the exponential variation of cross-sectional area and channel width along its length. From its seaward limit at Southend to its tidal limit at Teddington Weir it is 62 miles (100 km) long, the widths at these limits being 23,000 ft. (7000 m) and 280 ft. (85 m), respectively.

The mean tide range at Southend is 14 ft. (4.3 m), but the rapid decrease

---

*Scientific Officer and Senior Scientific Officer respectively, Ministry of Technology, Hydraulics Research Station, Wallingford, England.

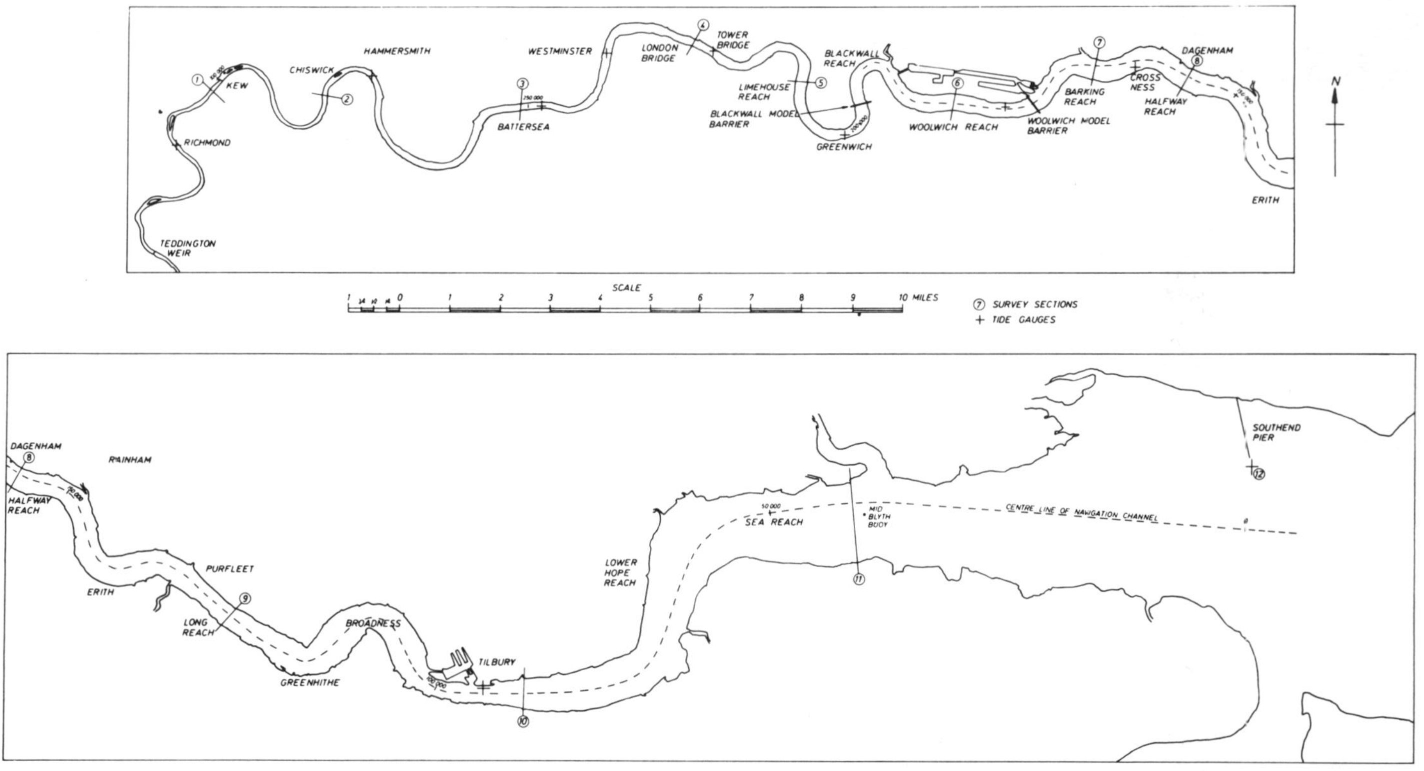

Fig. 1. General map of the Thames Estuary.

in width in the 39 miles (63 km) upstream of Southend outweighs the damping effect of bed friction to amplify the tidal range to a maximum mean value of 18.5 ft. (5.6 m). Further upstream however this is rapidly reduced by friction, giving a mean range of 11.5 ft. (3.5 m) at Teddington. The average freshwater flow over Teddington Weir is 2500 ft.^{3}/s (71 m^3/s).

Saline water intrudes about 40 miles (64 km) upstream of Southend under normal conditions, and is relatively well mixed in depth, with a variation from bed to surface of about 1-2 parts per thousand. The longitudinal density gradient caused by the salinity sets up a net movement of water near the bed landward in the lower reaches and seaward in the upper reaches. The null point or point of zero drift varies according to the freshwater flow, but under normal conditions is about 34 miles (55 km) upstream of Southend.

With one major exception, the estuary has a hard bed made up of gravel, clay and chalk. In the area known as the Mud Reaches, 28-33 miles (45-53 km) upstream of Southend, there are extensive deposits of silt, and dredging has been necessary in the past to maintain the shipping channel. The null point is usually located in this area, and it is generally believed that the siltation there is due mainly to silt transported in the lower layers from both the upstream and downstream directions. A local pocket of silt also occurs in Gravesend Reach, about 15 miles upstream of Southend.

For the greater part of its length, the land levels along the banks of the Thames are lower than the high water levels during normal Spring tides. Extensive embankments and high quays protect about 1 1/4 million people and 65 square miles from regular flooding. These embankments do not however prevent flooding by exceptionally high water levels caused by storm surges originating mainly in the North Sea, such as occurred during January 1953.

The flooding problem is accentuated by the fact that, due to a gradual rise in water level in the North Sea, and a gradual fall in land levels in South East England, mean high water springs levels in the Thames are rising at a rate of about 3 ft/century (0.9 m/century) relative to the land [3]. Although the embankments have been raised from time to time, the stage has now been reached in many places where further raising would endanger their stability, and extensive rebuilding would be necessary. Partly because of this, it has been proposed that flood protection could be provided by a structure placed as far seaward in the estuary as possible which would exclude surge tides from the upper parts of the estuary.

Various types of structure have been proposed including a fixed barrage with locks for shipping or a movable barrier. The main scheme under consideration at present is for a half tide barrier situated at either Woolwich or Blackwall. It was expected that the barrier would be operated during successive tides either throughout the year or during the periods when there is a risk of a North Sea surge. The barrier would be closed when the water level at the site reached mean tide level on the ebb tide, and reopened when the levels on either side of the barrier equalized during the flood tide, the timing of which would depend on the freshwater flow. In the event of a surge forecast, the barrier would not reopen until the next suitable flood tide. The amenity value of maintaining water levels about the present mean tide level at all positions upstream of the barrier site by this method of operation would be considerable.

A comprehensive investigation was necessary to determine the effect that such a barrier would have on the hydraulics of the estuary. Extensive surveys of conditions in the estuary were carried out during Spring and Neap tides during both low and high freshwater flow conditions. Continuous monitoring of the suspended concentrations at four locations in the middle and lower reaches is also under way. A physical model of the estuary was constructed and is being used to investigate changes in tidal propagation, velocities, salinities, localized siltation, etc. At the same time the mathematical model was developed to investigate changes in the silt movement and in the pattern of siltation in the estuary.

DESCRIPTION OF THE MODEL

The term mathematical model is used, in the context of estuaries, to describe the systematic calculation of the motion of a body of matter such as water, salt or silt carried out section by section throughout the estuary at successive time intervals. For the movement of the water the equations are based on the fundamental laws of momentum and continuity. The coefficients in the resulting differential equations are related to the channel geometry and roughness. If these are known accurately little effort is required to prove a bulk flow model of water movement. Usually the channel roughness is not adequately defined, but a few runs with various values of roughness are sufficient to obtain a satisfactory agreement with the field results [4, 5, 6].

For calculating the movement of silt however, where the basic principles involved are still incompletely understood, recourse has to be made to empirical expressions based on experimental results to describe such processes as erosion, deposition, and transport in suspension: the continuity of flow of silt must also be maintained. For this reason, a mathematical model of silt movement has to be proved extensively against surveyed data to establish the validity of the empirical expressions over a wide range of conditions.

In the initial stages of the study the underlying aim was to simplify wherever possible in order to obtain a working model to which modifications and refinements could be added in the light of experience gained. For this reason the Thames was treated as an idealized estuary, rectangular in section, with widths varying exponentially along its length. It was divided into 27 model reaches of equal length 12,000 ft., (3.7 km) extending from Southend Pier to Teddington Weir.

The best fit of the surveyed water surface widths to an exponential curve

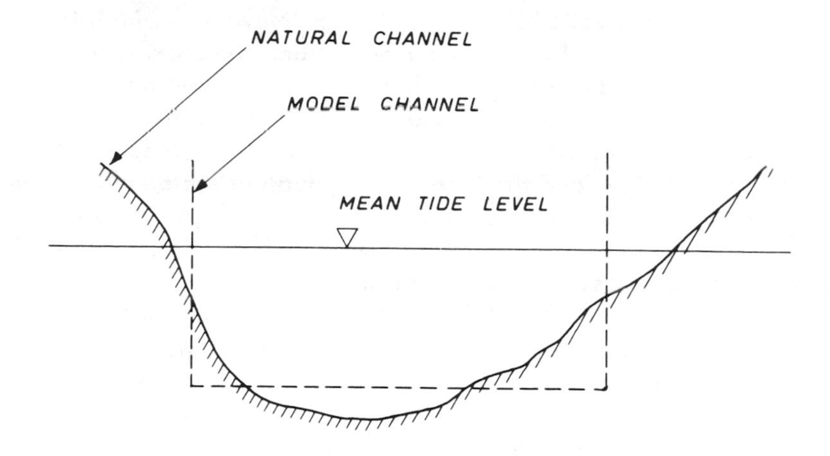

Fig. 2. Schematisation of the channel cross section.

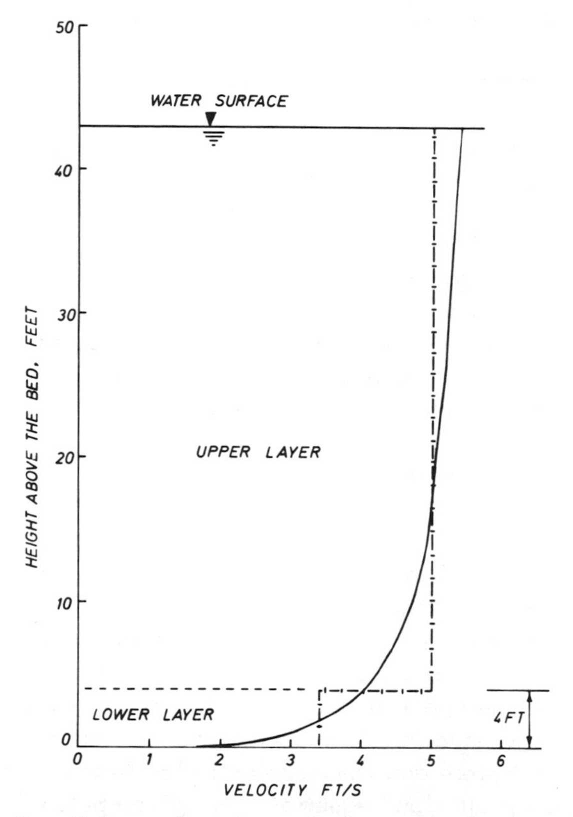

Fig. 3. Schematisation of the velocity profile.

was obtained at a water level of Ordnance Datum (Newlyn), which is approximately mean tide level. For each model reach the cross-sectional area of flow below the Ordnance Datum level was calculated from the surveyed cross-sections, and the model bed elevation then adjusted to give the correct area with the idealized width. Fig. 2 shows a surveyed cross-section, irregular in shape, and the idealized equivalent rectangular model section. Obviously, as the water level changes there will be deviations of the model from the surveyed data, but the geometry of the Thames is such that, with a few exceptions, the simplifications made are reasonable. However, for a distance of about 74,000 ft. (23 km) near the seaward end of the model it was necessary to distinguish between the effective width of flow and the total width: it was assumed that in these reaches the flow took place only in part of the rectangular section, the remainder providing storage volume only. These storage areas were ignored in the calculation of silt movement.

In the natural estuary the flow velocity and the suspended concentration change significantly with depth with the most rapid variation near the bed. In many cases also the mean velocities over a complete tidal cycle -- the 'drift' current -- are in opposite directions at the bed and at the surface. These factors cannot be reproduced in a simple 'bulk flow' model, which assumes uniform distribution of properties throughout the depth and, because they were judged to be important in determining the movement of silt, it was decided to divide the flow into two horizontal layers. The lower one, with a constant thickness of 4 ft. (1.2 m) adjacent to the bed, was considered as a boundary region, where the velocity increases from zero at the bed to almost its main stream value, and the suspended concentration decreases from a very high value near the bed to almost its average value. The remaining depth of flow -- the upper layer -- varied with the tide and could be up to ten times the depth of the lower one. Within each layer it was assumed that all properties were uniformly distributed with depth at their respective average values, and Fig. 3 and 4 show the resulting schematisation of the velocity and suspended concentration profiles respectively.

Water Movement

The movement of silt in an estuary depends to a large extent on the movement of water, and in this particular model it was convenient to calculate the water movement first, and use the output as the input of the silt movement section of the model. In determining the water movement, the method was to calculate the total flow and the flow in the lower layer, and thus deduce the flow in the upper layer. The calculation of the total flow was based on the fundamental equations of momentum and continuity, ignoring second order terms which were considered to be insignificant. The equations, commonly known as the 'bulk flow' equations, are given below using the following notation:

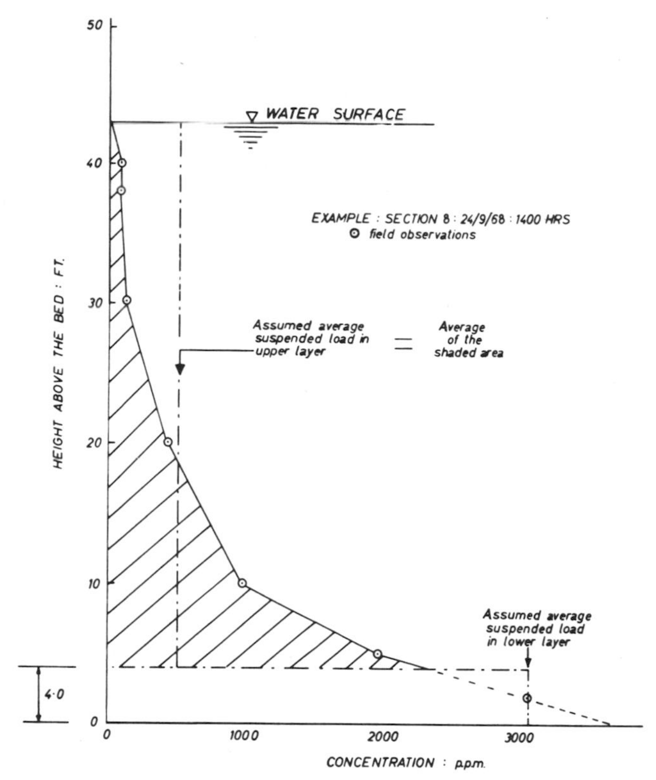

Fig. 4. Schematisation of the suspended concentration profile.

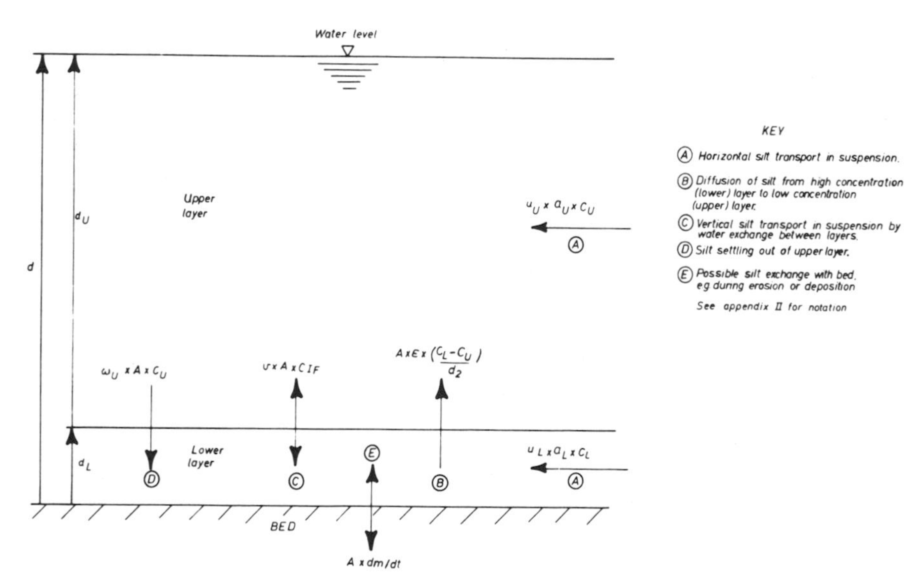

Fig. 5. Processes involved in the movement of silt.

A	plan area of reach	u_*	overall shear velocity
a	cross-sectional area of flow	U	flow velocity along the channel
b	width of channel	v	vertical flow velocity
c	suspended concentration	x	distance along the channel
d	depth of flow	ϵ	kinematic eddy viscosity
g	acceleration due to gravity	λ	friction factor
h	water surface elevation	ρ	density of water - varies with salinity and suspended concentration
k	numerical constant in expression for eddy viscosity. Theoretically 2/15, but value used was 2/150	τ	shear stress acting on sediment on the bed
		τ_c	critical shear stress for erosion
M	constant of proportionality in erosion equation (5)	τ_1	limiting shear stress for deposition
		ω	settling velocity of silt
m	mass of silt		
Q	flow discharge		
t	time		

General suffices

L & U	evaluated for the lower layer and upper layer respectively	e	erosion
		d	deposition
IF	evaluated at the interface between the two layers	m	mixing at interface

Momentum: $$\frac{\partial h}{\partial x} - \frac{\lambda |U| U}{8gd} = \frac{1}{g}\frac{DU}{Dt} \qquad \text{....... (1)}$$

Continuity: $$\frac{\partial Q}{\partial x} + b(x)\frac{\partial h}{\partial t} = 0 \qquad \text{....... (2)}$$

The flow in the lower layer was calculated by similar equations, but additional assumptions were involved:-

1. The friction factor for the lower layer was calculated from the Colebrooke-White equation using a relative roughness and Reynolds number based on the thickness and mean velocity of the lower layer although the shear gradient was based on the total depth of flow.
2. The water surface slope applied equally to both the lower layer and the total flow.
3. Density induced drift currents in the lower layer caused by the longitudinal variation of salinity in the estuary would have a negligible effect on the motion of the water surface or on the total flow calculated by equations (1) and (2).

With these assumptions and the possibility of a vertical water exchange velocity between the two layers, the equations of flow for the lower layer become:

Momentum: $$\frac{\partial h}{\partial x} - \frac{\lambda_L |U_L| U_L}{8gd} + \frac{d}{\rho}\frac{\partial \rho}{\partial x} = \frac{1}{g}\frac{DU_L}{Dt} \quad \ldots\ldots\ (3)$$

Continuity: $$\frac{\partial (a_L U_L)}{\partial X} + bv = 0 \quad \ldots\ldots\ (4)$$

The flow in the upper layer is then calculated to satisfy overall continuity.

Silt Movement

The movement of silt in an estuary can be considered as a cycle of four processes -- erosion, transport in suspension, deposition and bed consolidation. Since each of these is a complex and, in many cases, as yet unknown function of both tidal flow and silt properties, empirical expressions have to be used to describe the relationships.

In the model the equation used to govern erosion was that proposed by Partheniades [7], who described the process as following the expression:

$$\left(\frac{dm}{dt}\right)_e = M \frac{\tau}{\tau_c} - 1 \quad \ldots\ldots\ (5)$$

where M is a constant of proportionality.

Once the material was eroded from the bed it was assumed in the model to be taken immediately into suspension in the lower layer and to move with, and at the same velocity as the water, at the same time gradually diffusing into the upper layer. Within each layer the concentration was assumed to be uniformly distributed in depth although varying continuously with distance along the estuary. The relative distribution of the suspended silt between the two layers depends on the amount of material settling downwards and the net amount moving upwards due to mixing at the interface between the two layers. The net transport of material across the interface is given by the equation:

$$\left(\frac{dm}{dt}\right)_m = v\, c_{IF} - \omega_U c_U + \epsilon\left(\frac{c_L - c_U}{d/2}\right) \quad \ldots\ldots\ (6)$$

The equation used by Krone [8] was used in the model to define the deposition process:-

$$\left(\frac{dm}{dt}\right)_d = - c_L \omega_L \left(1 - \frac{\tau}{\tau_1}\right) \quad \ldots\ldots\ (7)$$

For cohesive sediments, the limiting shear stress for deposition, τ_1, is less than the critical shear for erosion, τ_c, and there thus exists a period during the tide, while $\tau_1 < \tau < \tau_c$ when the bed plays no active part in silt movement within the estuary.

Once material was deposited on the bed, it was assumed to reach its consolidated bed properties immediately so that no account was taken of the effect of age or of bed layers of different erosive strengths. By considering these assumptions and expressions for the various processes involved in silt movement in relation to short elements of the estuary as shown in Fig. 5, the equations of continuity of flow of silt in each layer were formed. Because the silt was assumed to move with the water, these equations could then be combined with the similar continuity equations for the flow of water to give the following differential equations for the changes in concentration in each layer:

Lower layer:

$$d_L \frac{dc_l}{dt} + v \frac{d_L}{d} (c_U - c_L) - \omega_U c_U + ku_* (c_L - c_U) = \frac{dm}{dt} \quad \ldots\ldots\ldots (8)$$

where dc_L/dt is evaluated along the characteristic line $\partial x/\partial t = U_L$

Upper layer:

$$d_U \frac{dc_U}{dt} - v \frac{d_U}{d} (c_U - c_U) + \omega_U c_U - ku_* (c_L - c_U) = 0 \quad \ldots\ldots\ldots (9)$$

where dc_U/dt is evaluated along the characteristic line $\partial x/\partial t = U_U$

The term dm/dt on the right hand side of equation (8), which is the 'driving term' in this pair of equations, is calculated from equation (5) or (7) when either erosion or deposition respectively are occurring. When there is no activity at the bed, its value is zero.

By converting equations (8) and (9) to finite difference form and using the method of characteristics to evaluate the various parameters, they were reduced to two ordinary linear simultaneous equations in terms of the new concentration in each layer. These were then solved numerically for each location and time step considered.

The coefficients in each of these equations are functions of the flow variables, past history of silt movement, and silt properties. The flow variables, such as water depth, bed shear stress, etc., were obtained as output from the water movement part of the model. The past history of silt movement involves a knowledge of the amount of silt on the bed and the suspended concentration in each layer at the previous time intervals, both at the section being considered and at adjacent sections determined by the characteristics lines. The relevant silt properties and a brief description of the methods by which they are evaluated are listed below. A full account of the tests involved is given in Appendix III of Ref. 9.

1. Settling velocity, ω. For a given silt this is a complex function of the concentration, salinity and turbulence level of the flow. The effects of concentration and salinity were measured initially in sedimentation tests in the laboratory. The effect of turbulence cannot be successfully reproduced in the laboratory or measured in the field. However, during the period of the investigation an instrument was developed for another study [10] which enabled settling velocity to be calculated directly from field tests and thus included the effect of turbulence. Results obtained with this instrument were used in later stages of the model.
2. Limiting shear stress, τ_1. This was determined from flume tests. The flume was run at a high discharge to thoroughly mix a charge of suspended silt; then the discharge was gradually reduced step by step until deposition occurred. The bed shear stress at this stage was noted, and the test repeated.
3. Critical shear stress, τ_c. This again was determined from flume tests. A bed was deposited in the flume, either from still or flowing water. Starting from a very low value, the discharge was increased step by step recording at each discharge the shear stress and rate of erosion at the bed. The shear stress at which significant erosion began is the critical value.
4. The constant of proportionality in the erosion equation, M. This is a difficult property to evaluate, but an estimate was obtained from the results of the flume tests [3] for determining the critical shear stress. An accurate estimate would involve a longer series of tests than was practicable to carry out in this instance beginning with a fresh bed for each discharge. Such a series of tests is planned as part of the fundamental research programme on the study of silts and muds which is currently in progress at the Hydraulics Research Station, Wallingford.

SURVEY DATA

Extensive sets of measurements were taken in the estuary during both Spring and Neap tides in September 1968 immediately following a short period of unusually high freshwater flow and again in September/October 1969 following a long period of low freshwater flow. On each occasion simultaneous sets of measurements were made from twelve boats spaced along the estuary, Fig. 1, taking readings where possible at 2, 5, 10, 20, 30 and 40 ft. above the bed and at 5 ft. below the surface. Measurements of flow velocity and direction were taken at 10 minute intervals and of salinity and suspended concentration at 30 minute intervals throughout the tide. The freshwater flow at Teddington Weir and the tide curves at each of the existing tide gauges along the estuary were also obtained on each occasion. It is hoped that eventually the basic data from all these surveys will be published in report form, and will be available from the Hydraulics Research Station.

Because of the relative timing of the surveys and this mathematical model investigation, it was only possible to use the 1968 survey data to verify the model taking the Spring tide of 24th September and the Neap tide of 30th September 1968. Although those surveys followed closely on a short period of very high freshwater flow, the river discharge had returned to normal by the actual date of the first survey. However, during both Spring and Neap surveys the volume of freshwater in the upper tidal reaches of the estuary was greater than usual with the salinity intrusion moved downstream about 8 miles to Woolwich.

Extensive analysis of the survey results was carried out to facilitate rapid comparison with the model results. At each position the tidal and velocity data were subjected to Fourier analysis to obtain the main tidal constituents such as mean tide level, tidal amplitude and phase, and the main velocity constituents such as mean drift velocity, velocity amplitude and phase in the upper and lower layers corresponding to the model layers. The suspended concentration data at each position were analyzed to give the variation throughout the tide of the average suspended concentration in the upper and lower layers and the average value over the complete tide.

PROVING TESTS

A block diagram showing the calculation procedure for each test and the input and output of the two sections of the model is shown in Table 1. The ease with which the model could be divided into the two sections of water movement and silt movement was very convenient for proving tests as the water movement section could be proved independently of, and prior to, the proving of the silt movement section.

The roughness height of the estuary channel was taken as 0.2 ft. (60 mm) throughout the estuary. This gave rise to the same friction factor coefficients as those used in an earlier tidal model of the Thames developed by Rossiter and Lennon [11].

The boundaries of the model were at Teddington Lock and Southend Pier. At Teddington the freshwater discharge was taken as 6000 ft.^{3}/s, (170 m^3/s) or approximately twice the long term average, and the suspended concentrations were taken as constant and equal to the average throughout the tide of the surveyed values. At Southend the observed tide curve was simplified by taking an equivalent sinusoidal tide curve for the reasons discussed previously. The suspended concentrations in each layer were assumed constant and equal to the average throughout the tide of the surveyed values. These end conditions for the suspended load gave rise to a net inflow of material into the estuary of about 43 tons/tide during the Spring tide, and

Table 1
Calculation Procedure

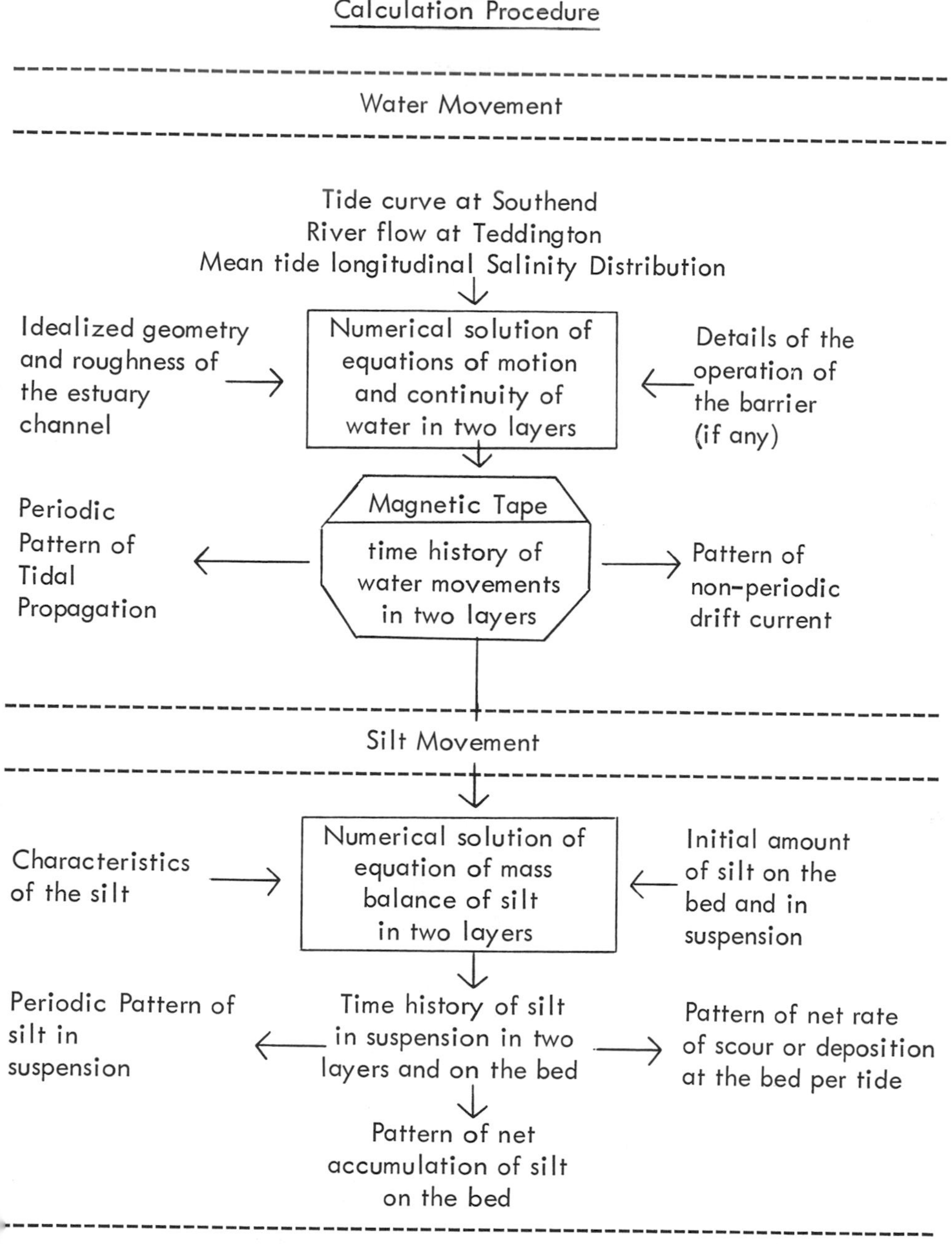

about 26 tons/tide during the Neap tide.

In the space available it will probably be better to consider the Spring tide alone although identical tests were also carried with the Neap tide. There is much more silt movement during a Spring tide, which will give a better idea of the results which can be obtained and the information which can be gained with such a model.

The survey period for the Spring tide of 24th September 1968 extended from Low Water (L.W.) to the following L.W. and the model was therefore run over the same period. To obtain the water movement, the sinusoidal equivalent of the observed tide at Southend and the observed longitudinal salinity distribution at mean tide were introduced into the model together with the instantaneous water surface profile at the L.W. The model then calculated the tidal propagation throughout the estuary, and after two tides settled down to give repeating results. The tidal and velocity data in each layer were then subjected to Fourier analysis in much the same way as the survey data ([9], Appendix IV)) and compared with those results. Fig. 6 shows the comparison for both the tidal and the velocity data giving mean tide level, tidal amplitude (half the tidal range) and phase, drift velocity (or mean velocity over a complete tide), velocity amplitude and phase. In interpreting the velocity data from the model, it must be pointed out that the survey data applies to particular cross-sections, whereas the model results apply to the average cross-section over a 12,000 ft. (3.7 km) reach of river. To obtain a more meaningful comparison, the model results have therefore been adjusted to take account of the difference between the areas of flow of the average model sections and particular survey section before plotting in Fig. 6.

The agreement between the model and survey results in Fig. 6 is good, especially for the tidal data. The velocity data is also in good agreement considering that the survey data applies only to one vertical in the cross-section, whereas the model results give the average over the complete width.

The initial conditions for the tests to simulate the silt movement were taken as the surveyed concentrations in each layer at low tide, and a uniform thickness of potentially active silt spread over the complete bed of the estuary. It was hoped that the model would itself distribute this bed material to form its own "hard", i.e. silt-free and mud reaches. There was no data available as a guide to the amount of active silt present on the bed, but the final value chosen after several trials was 0.26 $lb/ft.^2$ (1.3 kg/m^2) spread over the complete estuary with the exception of about 36,000 ft. (11 km) at the landward end of the model. This gave a total amount of active silt in the system including both bed and suspended material of about 229,000 tons (233,000 tonnes) dry weight of silt. It is important to realize that this is the maximum amount of material actively engaged in the silt movement processes during a Spring/Neap tidal cycle, and does not reflect the total quantity stored in the estuary. There is a considerable source of silt in the 'Mud Reaches', and

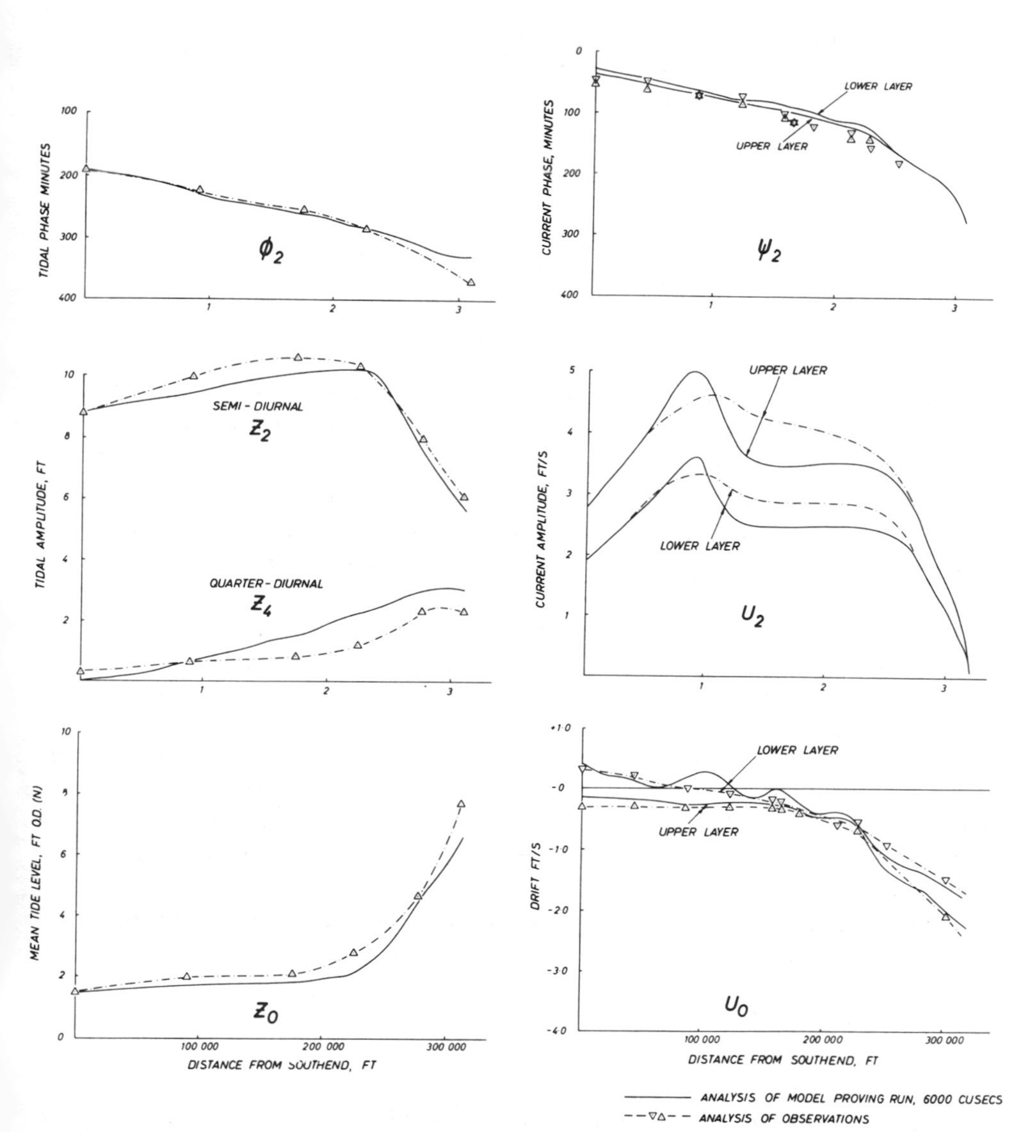

Fig. 6. Main tidal and velocity constituents; Spring tide proving tests.

this area could have been given a limitless supply in the model without significantly affecting the results.

During each test of the silt movement the model was run for four tides to establish uniformity of results, and the fifth tide was then analyzed to check the simulation of the silt concentrations in each layer. After a number of trials with various values of each of the silt constants the results appeared to be in reasonably good agreement with the survey. The initial estimates of the silt constants, based where possible on experimental studies, and the final values chosen as a result of the proving tests were remarkably close. The main changes were to make the critical shear stress for erosion increase with increasing salinity and to reduce the eddy viscosity by a factor of 10 from the usual value of $u_* d/15$. This reduction is believed to be necessary because of the vertical density gradient in the estuary due to the variation of both salinity and suspended concentration from bed to surface.

The main bases of comparison of the model results with the survey results were the variation along the estuary of the average concentrations in each layer throughout the tide and the periodic pattern of concentration in each layer at each of the survey positions. Comparison of the average concentrations in given in Fig. 7, and shows how, in both model and survey results, the peak concentrations occur in the area known as the Mud Reaches. The variation of the concentration in each layer throughout a tide is given for some of the survey sections in Fig. 8. Although the absolute values may not be exactly in agreement, the overall agreement in the shape of the curves is very good considering the complexity of the problem with peak and low concentrations occuring at virtually the correct times. The agreement in the lower layers of sections 7 and 8 is in fact rather better than it looks in the figure. In section 7 for instance, the very high surveyed concentrations between 7 and 9 hours after low water are patently not suspended concentrations. At concentrations about 10,000 - 15,000 mg/1 suspended silt exhibits hindered settling, and behaves more as a fluid mud layer on the bed than as a suspension of silt flocs. Rapid deposition has obviously occurred between 7 and 8 hours after low water around the sampling instrument, which was thus probably sampling the fluid mud layer of the bed. In the model this deposition is reflected in the rapid reduction in suspended concentrations at about this time. The fluid mud layer probably remained until it was completely re-eroded 9 hours after low water, and in the model this is reflected by the rise in suspended concentration.

Although there is virtually no field data available with which to compare it, much useful information can be gained from the model about the part the bed material plays in silt movement throughout the tide. From the relative values of the instantaneous grain shear stress at the bed the critical shear stress for erosion and the limiting shear stress for deposition, a time-distance diagram can be plotted which shows the variation with the state of the tide of the location of zones of potential deposition or erosion within the estuary,

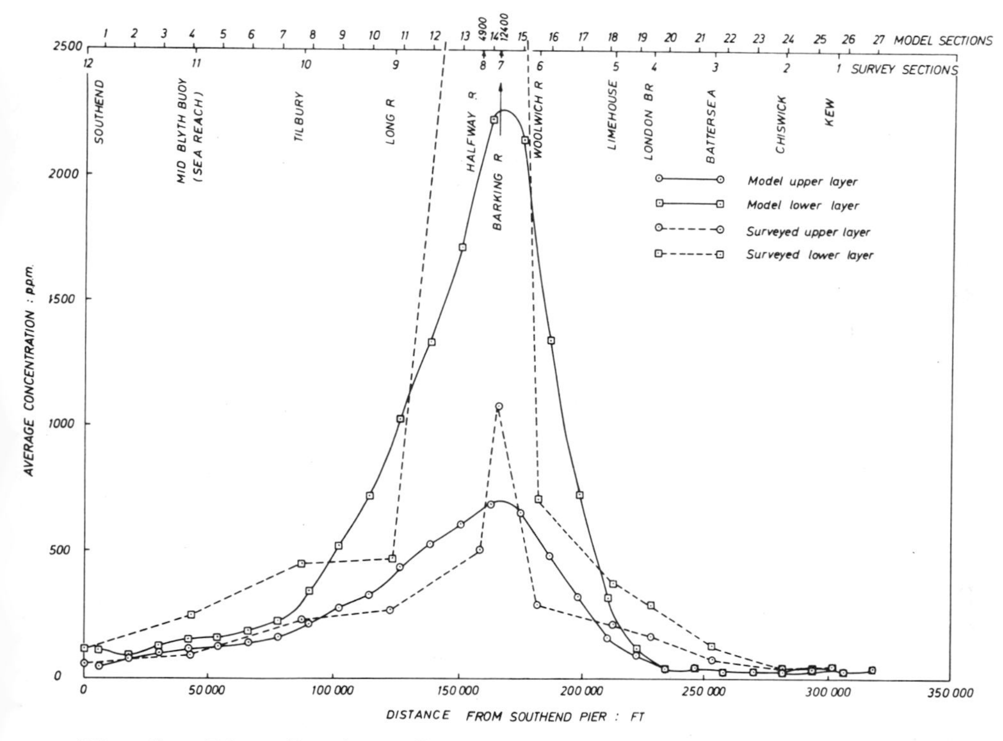

Fig. 7. Distribution of average concentrations; Spring tide proving tests.

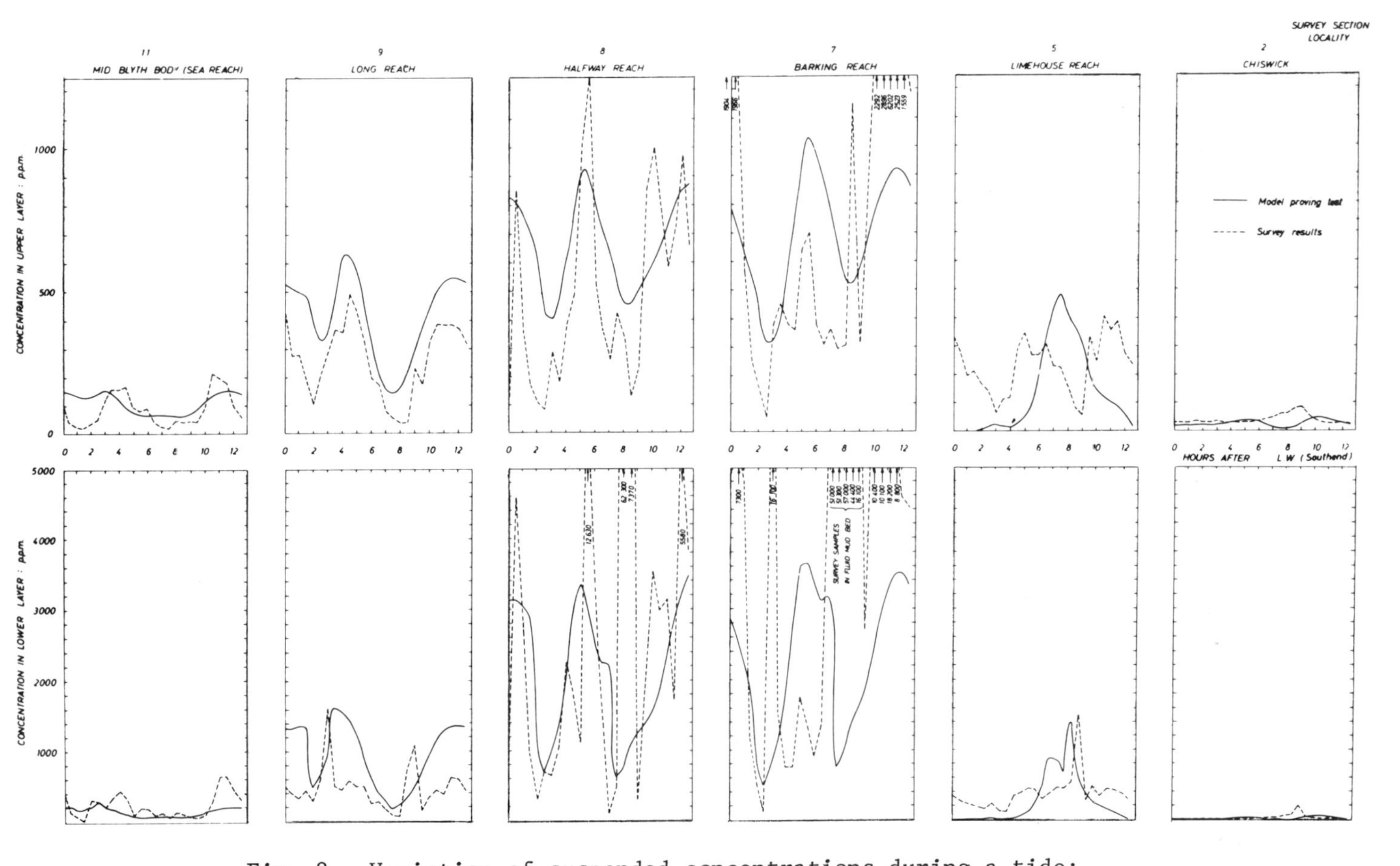

Fig. 8. Variation of suspended concentrations during a tide; Spring tide proving tests.

where any material which may be present will be deposited on or eroded from the bed. Such a diagram for the Spring tide is plotted in Fig. 9. It can be seen that deposition in the estuary can only occur near slack high water or slack low water, and that although erosion can occur during both ebb and flood tides, it is predominant on the flood. This is with the exception of the landward end of the model where erosion occurs mainly with the low water depths during the ebb tide.

By logging the quantities of silt involved in these exchange processes of erosion and deposition, which are given by equations (5) and (7), the net accumulation or removal of silt on the bed at a given location could be determined. This is plotted in Fig. 10 and shows how the model, starting from the initial imposed condition of a uniform distribution of silt over most of the bed of the estuary arranged itself into 'hard' reaches where most of the silt was removed, and a chalk or gravel bed would be expected and mud reaches where silt was deposited as in the so called 'Mud Reaches', which were particularly well defined during the Spring tide tests. The calculation of the net rate of silt exchange at the bed or tide plotted in Fig. 11 confirmed the Mud Reaches as the only area where significant net deposition occurred over a complete tide. The behavior of the bed and the likely rates of siltation at a particular location are probably of more concern than the suspended concentrations in assessing the acceptability of various sites for half-tide barriers, and Figures such as 9, 10 and 11 are essential for this purpose.

The movement of silt during a Neap tide differs only in degree from the Spring tide. There is only about one-tenth of the material in suspension and very little erosion occurs anywhere in the estuary, what little there is taking place mainly during the flood tide. The model results, while confirming the existence of the Mud Reaches and satisfactorily reproducing the variation of average concentrations along the estuary, did not give as good agreement as the Spring tide results for the periodic pattern of concentrations at the survey sections. The fact that very little erosion was present in the model was taken to indicate that conditions were such that the maximum instantaneous shear stresses were very close to the critical shear stress for erosion and small errors in either could have a disproportionately large effect on the suspended concentrations.

During the proving tests the principle of using identical values of the silt constants for both Spring and Neap tides was strictly adhered to unless there was direct evidence to the contrary. Better agreement could probably have been obtained for the Neap tide if this principle had been relaxed, but the only silt constant changed was the settling velocity. Results of field tests (10) showed that the settling velocity varied linearly with concentration during the Spring tide and with the square of concentration during the Neap tide, the difference being attributed to changes in the level of turbulence in the flow. In the model for a particular tide good agreement could also probably have been achieved by adjusting the values of the silt constants independently

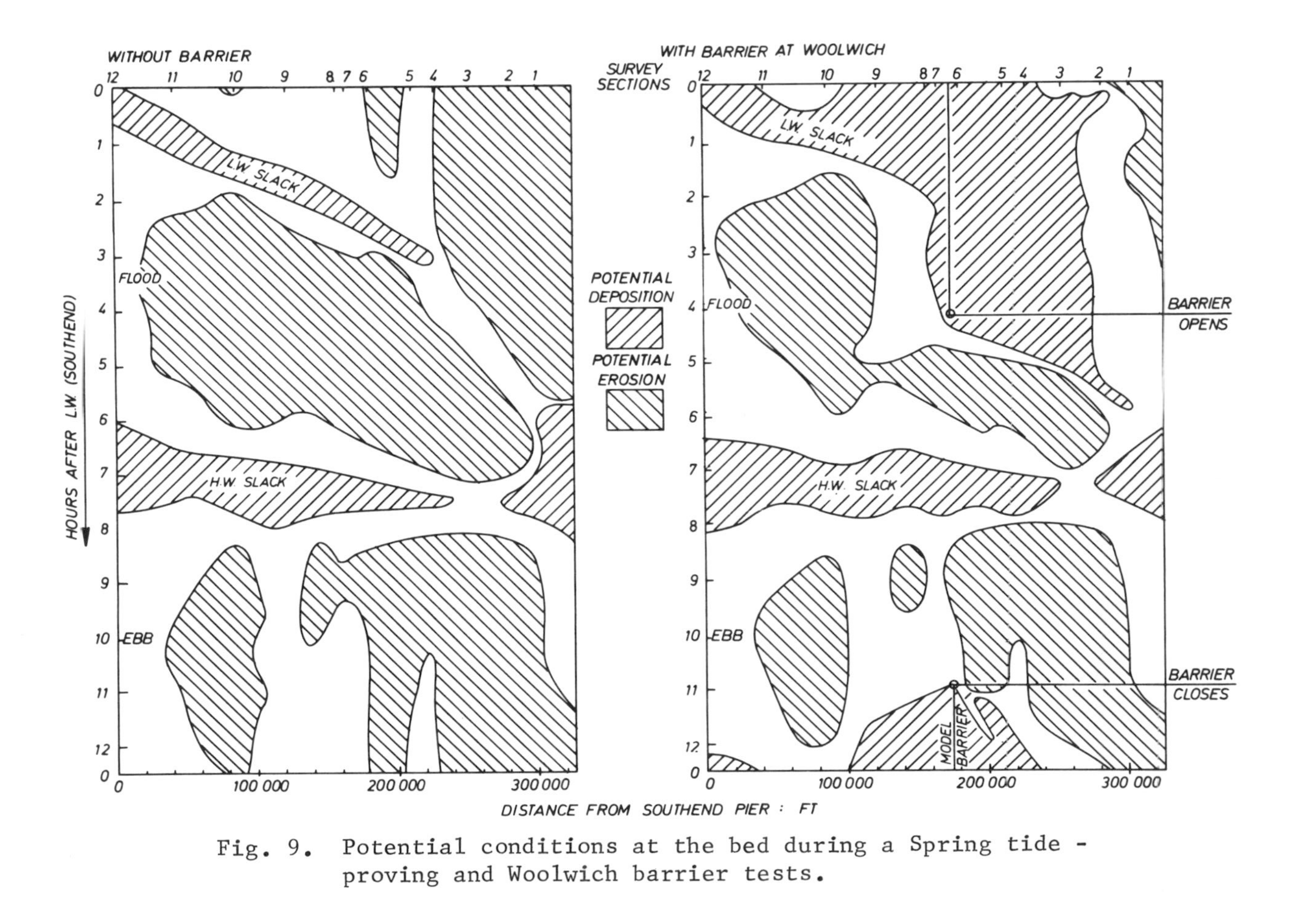

Fig. 9. Potential conditions at the bed during a Spring tide - proving and Woolwich barrier tests.

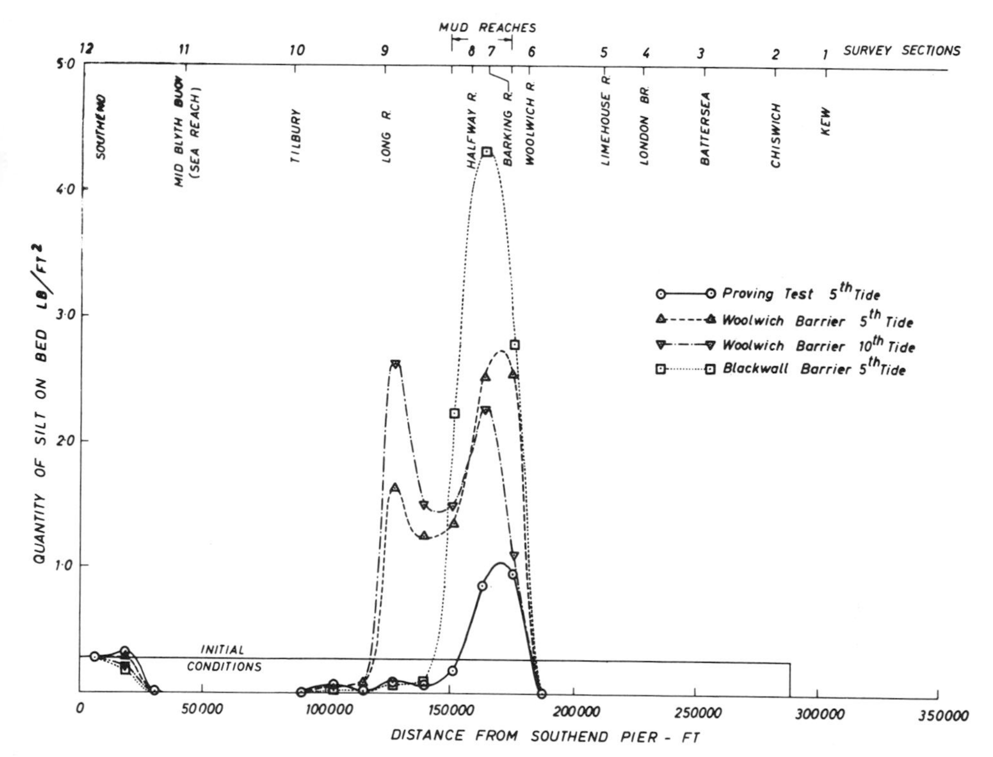

Fig. 10. Distribution of silt on the bed - Spring tide tests.

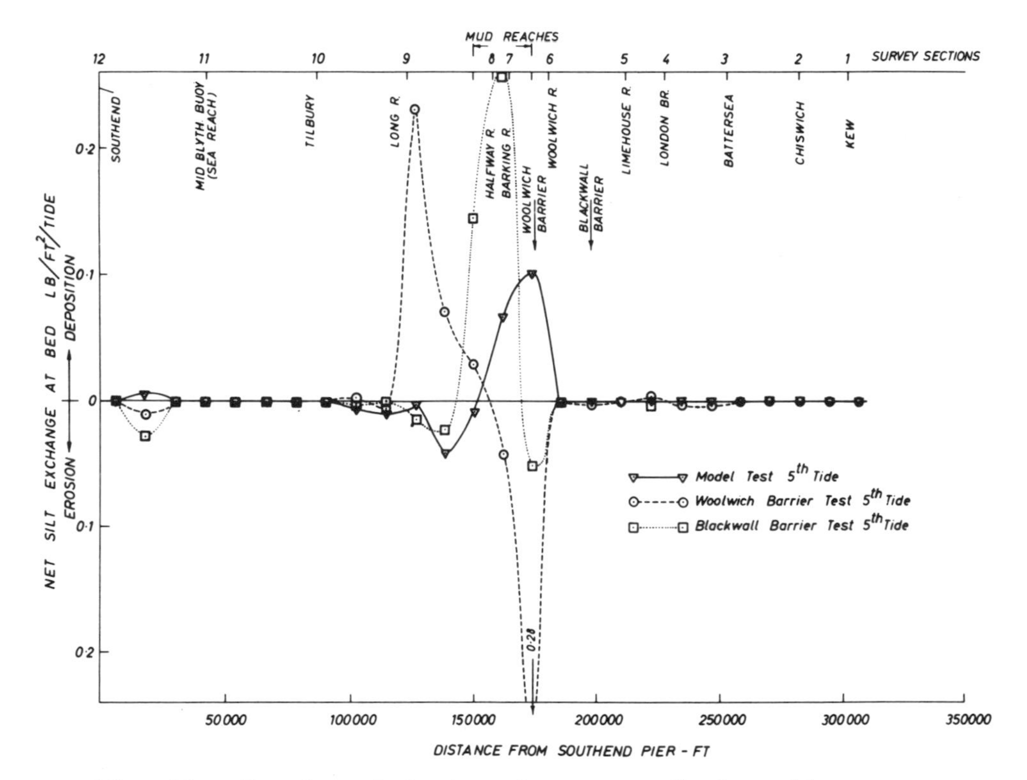

Fig. 11. Net deposition/erosion rate - Spring tide tests.

at each section,resulting in an apparently haphazard distribution along the length of the estuary. In fact the silt constants were fixed along the estuary with the exception of the critical shear stress for erosion which was made to vary with mean salinity according to experience gained in previous investigations of the Thames Estuary (12).

THE EFFECTS OF A HALF TIDE BARRIER

Most of the tests carried out to determine the effect of a half tide barrier were for a site in Woolwich Reach, Fig. 1, which was considered to be the most promising location. Later however, studies of various alternative sites were requested, and one of these in Blackwall Reach, Fig. 1, has also been completed. The others may well be studied in due course.

The proposed half tide barrier in Woolwich Reach was incorporated in the model at the nearest model section. It was programmed to close instantaneously when the water level on the ebb tide reached Ordnance Datum which is approximately mean tide level, and to open again when the water level downstream of the barrier equalled that upstream. The barrier was operated on each successive tide.

As the model does not, at present, reproduce variations in salinity in the estuary, it was necessary to assume a new longitudinal salinity gradient when the barrier was in operation. The mean tide longitudinal salinity distribution was assumed to remain as before, but the landward toe of the salinity intrusion, which extended just upstream of the barrier site during the survey period of September 1968, was adjusted to reach zero at the barrier. Later tests in the physical model of the Thames confirmed that these were reasonable assumptions.

As the barrier is open for about half a tide centered around high water, the initial conditions of suspended concentrations and bed material were taken as those given by the model at high water and the barrier test tides taken from high water to high water, i.e., half a tide out of phase with the proving tests. Most of the tests were run for five tides as before, but the flood of the fourth and ebb of the fifth tide were analyzed as one tide to bring the two series of tests back into phase, and enabling direct comparisons to be made.

The operation of the half tide barrier had very similar effects on tidal propagation during both Spring and Neap tides. Typical tide curves upstream and downstream of the barrier for a Spring tide are shown in Fig. 12. Downstream the high water levels were slightly reduced, the tidal range increased, and the tide advanced in phase, especially at low water which occurred about half an hour earlier than previously. Upstream the high water levels were also slightly reduced, but the main effect was obviously the impounding of the water at or above the previous mean tide level. Typical mean velocity curves for a

Spring tide are shown in Fig. 13 for sites upstream and downstream of the barrier. During the ebb tide the velocities were very little changed until the barrier closed, when the velocities were greatly reduced downstream, and were almost zero upstream. Although the barrier opened again halfway through the flood tide, the peak flood velocities were much lower than previously.

The overall reduction in velocities in the estuary had a marked effect on the suspended concentrations in the estuary. Fig. 14 shows that the average concentrations during a Spring tide were greatly reduced for about 104,000 ft. (32 km) downstream and about 66,000 ft. (20 km) upstream of the barrier site. The reason for this is plain from Fig. 9 showing the location of zones of potential erosion or deposition in the estuary throughout a Spring tide. The barrier tests and proving tests are compared on this figure, and it can be seen that the reduction in velocities has resulted in greatly increased zones of potential deposition, especially while the barrier is closed, and greatly reduced zones of potential erosion. The net effect is that much of the silt previously in suspension is now deposited on the bed resulting in the much reduced suspended concentrations.

Most of the material settled out of suspension deposited soon after the first closure of the barrier during the first ebb tide. At mid ebb tide the suspended concentration approached their peak values, and were highest in the Mud Reaches. Naturally therefore, the quantity of material on the bed in this area rapidly increased as shown in Fig. 10 after five successive Spring tides of barrier operation. This distribution of bed material throughout the estuary was not however a stable condition as the calculation of the net silt exchange per tide given in Fig. 11 showed that the bed material was being rapidly redistributed with material being eroded from the existing Mud Reaches to form new mud reaches about 6 miles downstream. After running the barrier tests for ten tides in all the formation of these new mud reaches showed up clearly on Fig. 10.

While the operation of the barrier during the Neap tide showed most of the features experienced with the Spring tide, with reduced velocities, reduced suspended concentrations, etc., it did not show this redistribution of the bed material, and the silt deposited from suspension remained more or less where it settled in the middle reaches of the estuary.

Although no tests were carried out over a complete Spring/Neap tidal cycle, it was estimated from the results of the discrete Spring and Neap tide tests that the continuous operation of a half tide barrier at Woolwich throughout the year would tend to move the Mud Reaches about 3 or 4 miles (5 or 7km) downstream. This was confirmed by tests in the physical model using crushed perspex grains as a tracer material. From the mathematical model it was also estimated that the rate of movement would be about 1 1/4 million tons dry weight of silt per year, which would continue at a gradually reducing rate until a new regime had been established in the estuary or until all the material in the existing Mud Reaches had been removed. Because of the relatively large

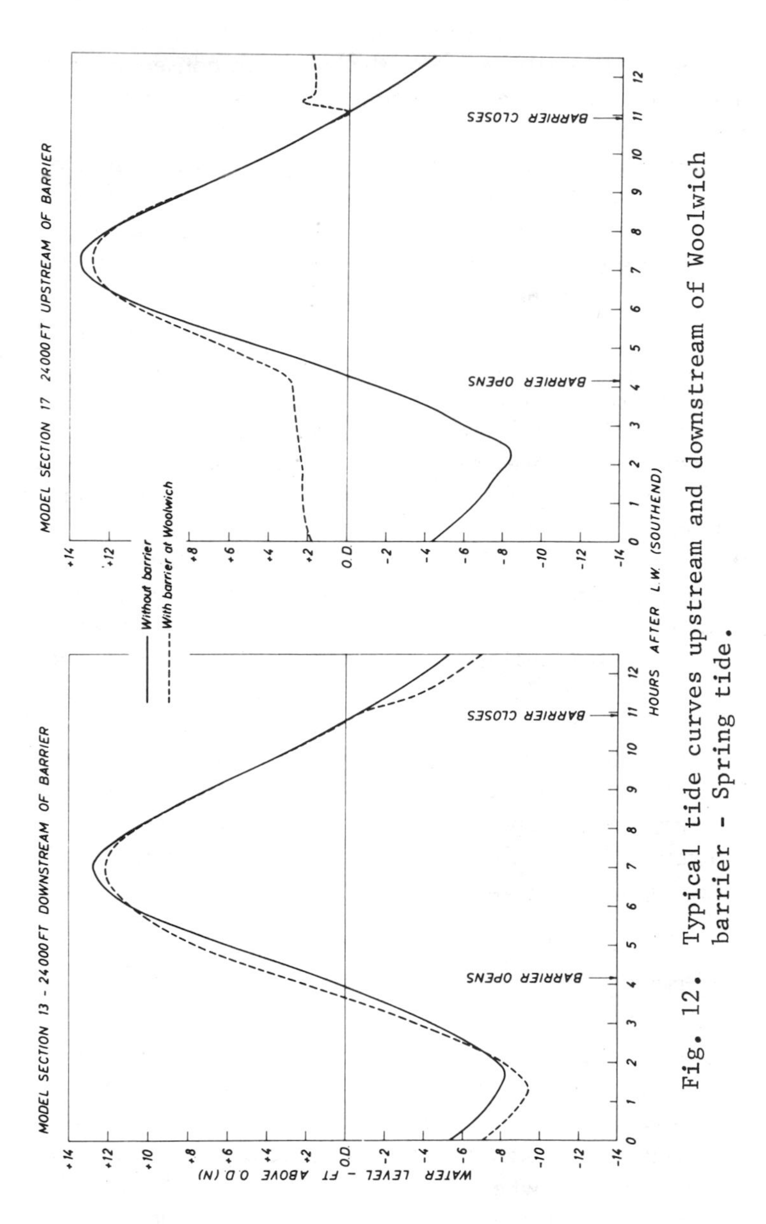

Fig. 12. Typical tide curves upstream and downstream of Woolwich barrier - Spring tide.

amount of dredging which might therefore be necessary to maintain existing navigation channel depths, various other sites for a half tide barrier were considered and one in Blackwall Reach was tested in the model.

The site of the Blackwall barrier was only about 4 miles (7 km) upstream of the Woolwich barrier site so that the changes in tidal volume excluded by the barrier were very little different at either site. In consequence the overall effect of the barrier at Blackwall on tide levels and velocities was very similar to that of the Woolwich barrier. The changes in detail however had a significant effect on silt movement. The moving of the barrier site from Woolwich, which is just at the upstream end of the Mud Reaches, to Blackwall meant that the Mud Reaches were much less affected by the barrier. As with the Woolwich barrier the suspended concentrations in the estuary were greatly reduced by the operation of the Blackwall barrier as a result of the reduced velocities. Again most of the material settled out of suspension was deposited during the first closure of the barrier in the first ebb tide and mainly in the Mud Reaches, and as shown in Fig. 10 remains there after running the model for 5 tides. The net silt exchange per tide, plotted in Fig. 11, shows that there was no substantial redistribution of this material on the bed once it had been deposited, illustrated by the virtual absence of net erosion per tide at any location along the estuary.

From these tests it appeared that from the point of changing existing conditions as little as possible, the Blackwall site seems to offer the better alternative for a half tide barrier since there was very little movement of the Mud Reaches and very little maintenance dredging would be required to retain existing depths in the navigation channel. The deposition of the suspended material during the first tide of operation would amount only to about 50,000 tons dry weight of silt spread over a fairly large area of the middle reaches of the estuary, and would occur only once. The reduction in suspended concentrations as a result of the operation of the barrier at either site could well have beneficial results in reducing the amount of maintenance dredging required in docks impounded with river water.

FUTURE DEVELOPMENT AND POSSIBLE APPLICATIONS

One important development of the model applied to the Thames estuary would be the inclusion of a section to reproduce the movement of salt in the estuary as both the water and silt movement are affected by variations in salinity. This would entail extending the model to a position about 20 miles (32 km) seaward of Southend where the salinity reaches a steady value.

Much of the success obtained in developing a mathematical model of the movement of silt in an estuary has been due to the fact that the Thames is an ideal estuary on which to attempt such a model. Because of its relatively narrow and exponentially varying width, virtually constant depth, good tidal

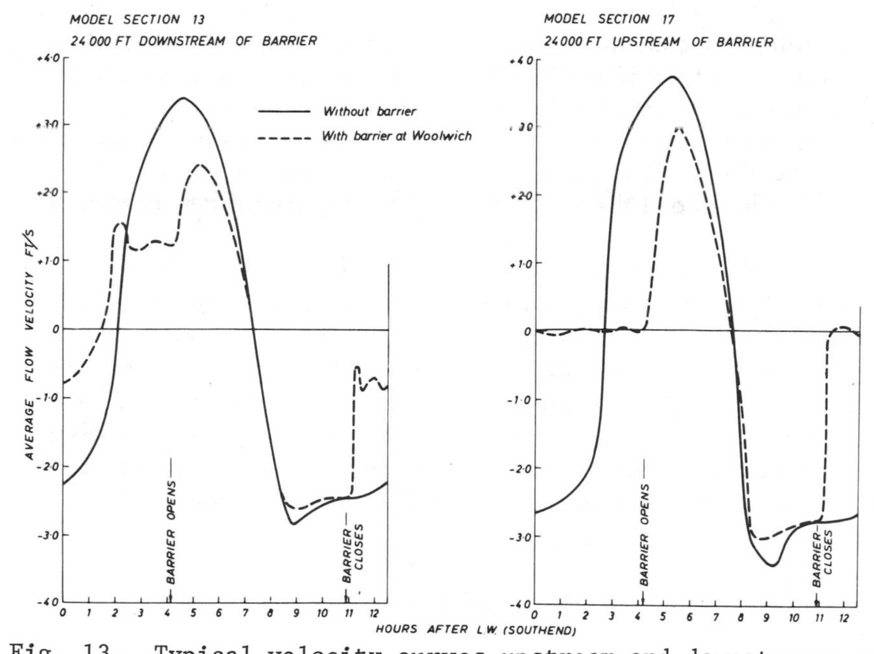

Fig. 13. Typical velocity curves upstream and downstreams of Woolwich barrier - Spring tide.

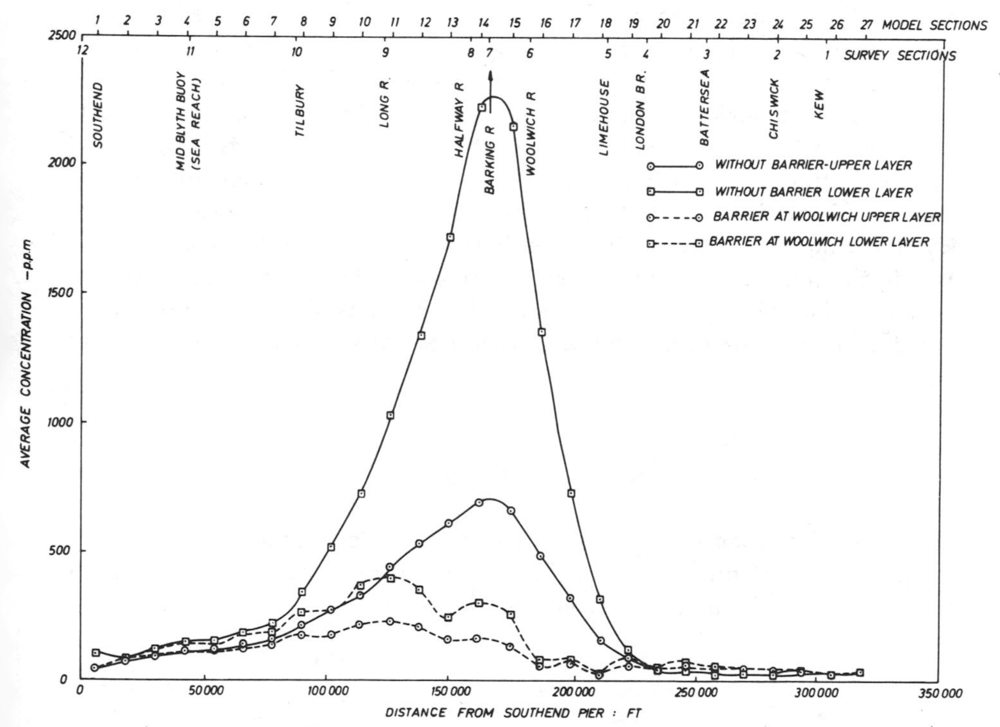

Fig. 14. Distribution of average concentrations - Woolwich barrier tests - Spring tide.

mixing, and tides at the mouth close to sine curves, several simplifications could be made without significantly affecting the accuracy of the model. This enabled most of the effort to be spent on understanding and setting up expressions for the complex processes of silt movement. However, now that the validity of a silt model has been established, most of the simplifications made could be relaxed and the model applied to estuaries of a more complex nature.

This particular model was used to study the effects of a half-tide barrier on silt movement in the estuary, and the barrier was closed instantaneously for simplicity. There is no technical reason however why the barrier could not have been programmed to close over a certain period of time, say over half an hour, which would be more realistic. Further to this it would be a relatively short step to incorporate into the model a fixed barrage complete with tidal sluices and turbines and to study the effect that various modes of operating a tidal power plant would have on silt and water movement.

CONCLUSIONS

1. A mathematical model was established and proved to simulate the movement of water and silt in a somewhat idealized version of about sixty miles of the Thames Estuary. The model successfully reproduced the known regime of siltation during Spring and Neap tides for a constant river flow of about twice the long term average at Teddington.
2. The effects of a movable barrier at either Woolwich or Blackwall on silt movement in the estuary were investigated. From the results it appeared that the better site for a half-tide barrier would be at Blackwall rather than Woolwich since it would only slightly affect the Mud Reaches, and any dredging required to maintain existing depths would be much less than in the new mud reaches caused by a similar barrier at Woolwich.
3. Now that the validity of the silt model has been established, applications to problems involving other types of tide control such as tidal power systems are possible.

ACKNOWLEDGMENT

The work described in this report has been carried out as part of the research programme of the Hydraulics Research Station, and is published with the permission of the Director of Hydraulics Research.

REFERENCES

1. Owen, M. W., "The Properties and Behavior of Muds", Report No. INT 61, Hydraulics Research Station, Wallingford, England (1966).

2. Owen, M. W., "The Settling Velocities of an Estuary Mud", Ibid No. INT 78 (1969).

3. Greater London Council, Thames Flood Prevention, London (1969).

4. Hydraulics Research Station, " The Effects of Proposed Extraction of Water on Siltation in the Nene Estuary", Report No. EX 307, Wallingford, England (1966).

5. Hydraulics Research Station, Derwent Tidal Sluice, Report No. EX 428, Wallingford, England (1969).

6. Hydraulics Research Station, Port Rashid, Dubai Creek Tidal Calculations, Report No. EX 453, Wallingford, England (1969).

7. Partheniades, E., "Erosion and Deposition of Cohesive Soils in Salt Water", Ph.D. Thesis, Univ. of California, Berkeley, California (1962).

8. Krone, R. B., "Flume Studies of the Transport of Sediment in Estuarial Shoaling Processes", Univ. of California, Hyd. Eng. Lab. and Sanct. Eng. Res. Lab. (1962).

9. Hydraulics Research Station, Thames Estuary Flood Prevention Investigation, Report No. EX 479, Wallingford, England (1970).

10. Hydraulics Research Station, Woolwich Ferry Terminals, Report No. EX 467, Wallingford, England (1969).

11. Rossiter, J. R., and G. W. Lennon, "Tidal Conditions in the Thames Estuary by the Initial Value Method", Proc. Instn. Civ. Engrs., 31, 25 (1965).

12. Sir Inglis, C. C., and F. H. Allen, "The Regimen of the Thames as Affected by Currents, Salinities and River Flow", Proc. Instn. Civ. Engrs., 7, 827 (1957).

CORROSION CHARACTERISTICS OF NON-FERROUS METALS IN MARINE APPLICATIONS

M. J. Pryor* and R. V. L. Hall**

INTRODUCTION

This paper is intended to describe the corrosion characteristics of aluminum, copper and titanium base alloys in marine applications. Two types of marine exposure will be considered:

1. Exposure to marine atmospheres in the immediate vicinity of the sea, and
2. Immersion either in static or moving sea water.

The paper is intended as a general guide to the selection of non-ferrous materials that might find use in tidal power technology. Since non-ferrous materials are generally more costly on a per cubic inch basis than low alloy steels, their successful use from the economic standpoint in marine atmospheres or in sea water generally implies that they will be used in the unprotected condition, i.e., that they will not be subjected to painting or to cathodic protection. Even within the three classes of non-ferrous materials discussed in this paper, considerable economic choice exists. The aluminum alloys generally are priced in the cost range of $0.35 to $0.60 per lb. with a density of around 0.1 lb. per cu. in. Copper alloys range from $0.80 to $1.25 per lb. with a density somewhat in excess of 0.3 lb. per cu. in. Titanium and its alloys vary from around $4.00 to $6.00 per lb. with a density of around 0.164 lb. per cu. in.

Selection of materials for corrosion purposes must consider not only the initial cost of the parts which are required to perform a certain structural function but also the corrosion performance in terms of anticipated life together with the potential cost of replacement provided premature failure is experienced. Obviously, it makes no sense to save a few dollars on a small part if the cost of replacing it in operating equipment runs into the vicinity of

---

*Associate Director, Olin Corporation, Metals Research Laboratories, 91 Shelton Avenue, New Haven, Conn., U. S. A.
**Olin Corporation, Metals Research Laboratories, 91 Shelton Avenue, New Haven, Conn., U. S. A.

thousands of dollars.

This paper summarizes quantitative corrosion data collected in the two basic types of environment in long-term simulated service tests together with the mechanical properties and densities of the alloys. These data can provide a general guide to non-ferrous material selection problems in tidal power technology.

CORROSION REACTIONS

In marine environments the three classes of metals considered, aluminum, copper and titanium alloys all corrode with the formation of solid reaction products. Indeed, the composition, morphology and thickness of the reaction product as influenced by the alloy composition will in essence determine the corrosion rates.

In the case of aluminum, the reaction product films formed in a marine environment are duplex in nature. There is an inner thin film of γ-Al_2O_3 which is formed by direct interaction of oxygen and the metal. The additional reaction product film formed in marine atmospheres and in sea water below 70° C comprises bayerite ($\beta Al_2O_3 \cdot 3H_2O$) which forms on top of the γ-Al_2O_3 substrate film which remains intact throughout the corrosion process [1]. The bayerite is formed by an electrochemical reaction which occurs at local anodes and local cathodes on the corroding specimen. The reactions involved are as follows:

anodic $$2Al = 2Al^{+++} + 6e \quad \ldots\ldots\ (1)$$

hydrolysis $$2Al^{+++} + 6H_2O = \beta Al_2O_3 \cdot 3H_2O + 6H^+ \quad \ldots\ldots\ (2)$$

cathodic $$6H^+ + \frac{3}{2} O_2 + 6e = 3H_2O \quad \ldots\ldots\ (3)$$

Total reaction $$2Al + 3H_2O + \frac{3}{2} O_2 = \beta Al_2O_3 \cdot 3H_2O \quad \ldots\ldots\ (4)$$

In the case of copper and its alloys, two anodic reactions appear to take place simultaneously in chloride environments [2]. The first set of reactions leads to cuprous oxide as a direct anodic reaction product as follows:

anodic $2Cu + H_2O = Cu_2O + 2H^+ + 2e$ .... (5)

cathodic $2H^+ + 1/2O_2 + 2e = H_2O$ .... (6)

Total reaction $2Cu + 1/2O_2 = Cu_2O$ .... (7)

The foregoing reactions (5) through (7) may account for between 20 and 50% of the corrosion of copper alloys in marine applications depending upon the nature of the environmental conditions. The remaining part of the corrosion process yields cupric ions as a soluble reaction product by the following reactions:

anodic $Cu = Cu^{++} + 2e$ .... (8)

cathodic $2H^+ + 1/2O_2 + 2e = H_2O$ .... (9)

Total reaction $Cu + 2H^+ + 1/2O_2 = Cu^{++} + H_2O$ .... (10)

Titanium is a passive metal, i.e., the anodic reaction in marine environments leads to the direct formation of a passivating film of TiO_2. The reactions involved are as follows:

anodic $Ti + 2H_2O = TiO_2 + 4H^+ + 4e$ .... (11)

cathodic $4H^+ + O_2 + 4e = 2H_2O$ .... (12)

Total reaction $Ti + O_2 = TiO_2$ .... (13)

CORROSION OF ALUMINUM AND ITS ALLOYS

Aluminum and its alloys share a combination of low cost per pound and low density. In fact, the cost per cubic inch of most aluminum alloys is not much higher than that of 1010 low carbon steel.

Despite the fact that there are dozens of different aluminum alloys, those that find substantial use in marine applications are more limited in number. A selection of the more important marine alloys and their composition is given in Table 1. The typical mechanical properties of the alloys in Table 1 are shown for a limited range of tempers in Table 2.

The alloys in Table 1 cover the range from commercially pure aluminum, (alloy 1100) through the manganese-containing alloys (3003 and 3004) to several

Table 1
Some Aluminum Alloys used in Marine Applications
Chemical Composition*

Alloy	Al	Si	Fe	Cu	Mn	Mg	Cr	Zn	Ti	Others	
										Each	Total
1100	99.9 min.		1.0 Si + Fe	0.20	0.05	-	-	0.10	-	0.05	0.15
3003	Rem.	0.6	0.7	0.20	1.0–1.5	-	-	0.10	-	0.05	0.15
3004	Rem.	0.30	0.7	0.25	1.0–1.5	0.8–1.3	-	0.25	-	0.05	0.15
5052	Rem.		0.45 Si + Fe	0.10	0.10	2.2–2.8	0.15–0.35	0.10	-	0.05	0.15
5154	Rem.		0.45 Si + Fe	0.10	0.10	3.1–3.9	0.15–0.35	0.20	0.20	0.05	0.15
5086	Rem.	0.40	0.50	0.10	0.20–0.7	3.5–4.5	0.05–0.20	0.25	0.15	0.05	0.15
5083	Rem.	0.40	0.40	0.10	0.30–1.0	4.0–4.9	0.05–0.25	0.25	0.15	0.05	0.15
6061	Rem.	0.40–0.8	0.7	0.15–0.40	0.15	0.8–1.2	0.04–0.35	0.25	0.15	0.05	0.15

*Composition in % maximum unless a range is shown.

Table 2
Typical Mechanical Properties of Some Aluminum Alloys used in Marine Applications

Alloy	Temper	T.S. (Ksi)	Typical Mechanical Properties Y.S.* (Ksi)	% El. (in 2")
1100	0	13	5	35
	H14[x]	18	17	9
3303	0	16	6	30
	H14	22	21	8
3004	0	26	10	20
	H34[xx]	35	29	9
5052	0	28	13	25
	H34	38	31	10
5154	0	35	17	27
	H34	42	33	13
5086	0	38	17	22
	H34	47	37	10
5083	0	42	21	22
6061	T6	45	40	12
Alclad**6061	T6	42	37	12

Al-Mg alloys (from 5052 to 5083) which are normally preferred for severe service in marine applications. Also included is one heat-treatable aluminum-magnesium-silicon alloy (6061) of reasonably good corrosion resistance. Other higher strength aluminum alloys such as those based on copper or on zinc, magnesium and copper possess too low a corrosion resistance to be seriously considered for use in marine service in the absence of stringent protective measures.

In this paper the corrosion information that will be presented has been gathered from quantitative tests in a variety of marine atmospheres and in sea water. Such data have the advantage over service experience in that the extent of corrosion and loss of mechanical properties, if any, can be determined in a quantitative fashion.

*0.2% Offset
x Half hard
xx Stabilized after cold rolling
** With 5% per side 7072 containing 1% Zn

Atmospheric Corrosion

The alloys shown in Table 1 exhibit a high degree of resistance to attack by atmospheres immediately adjacent to the sea coast. Classically, exposure testing of this type is conducted on flat panels of significant size mounted at an angle of 45° to the horizontal and facing the ocean. Figure 1 shows a typical exposure rack for marine testing. The intensity of corrosion will vary as a function of distance from the mean tide level and for severe exposure conditions testing at a distance of no greater than 400 ft. from the mean tide level is practised.

It is normal to find in atmospheric testing of aluminum alloys that the groundward surface of the specimen is more severely corroded than the skyward surface. This is because of the comparative absence of washing action by the rain and because of the accumulation of condensate high in chlorides. Accordingly, the corrosion of these two surfaces is often evaluated and reported separately. Corrosion of the skyward and groundward faces of alloy 5086-H34 at Daytona Beach, Florida is shown in Figure 2. This specimen had been exposed for a period of 5 years. It may be readily seen that corrosion is more severe on the groundward surface than on the skyward one. Further, this illustrates a point of major importance with respect to the atmospheric corrosion of aluminum alloys basically suited for marine service. They will all corrode by pitting of very high frequency and of comparatively shallow depth.

Some recent quantitative information on the corrosion characteristics of many of the alloys shown in Table 1 has been developed for marine exposure at Kure Beach, North Carolina and at Point Reyes, California. In this work the extent of corrosion was judged by three criteria:

1. the loss in weight after cleaning reaction products from the panels,
2. the deepest pits that could be found on the specimen, and
3. the loss in ultimate tensile strength due to corrosion.

The results of these seven year tests are summarized in Table 3. This study [3], confirms that comparatively little loss in mechanical properties and little danger of perforation exists with the aluminum alloys described in Tables 1 through 3. This is, of course, not so with the whole range of aluminum alloys because those containing copper show high corrosion rates in chloride-containing atmospheres.

Some information on the kinetics of marine atmospheric attack can be obtained from longer time tests (up to 20 years) that have been conducted on one of the alloys from Table 1. Specifically, the results depicted in Figures 3a and b show the depth of pitting and loss of tensile strength as a function of time for alloy 3003-H14 over a twenty year period at Pt. Judith, R. I. [4]. Figure 3a shows that the pitting is initially relatively rapid but that after about two years further penetration becomes very very limited indeed. This decrease in rate with increasing time has been referred to as "the self-stopping nature of the corrosion of aluminum and its alloys". The effect is due to the accumulation

Fig. 1. Atmospheric corrosion test racks.

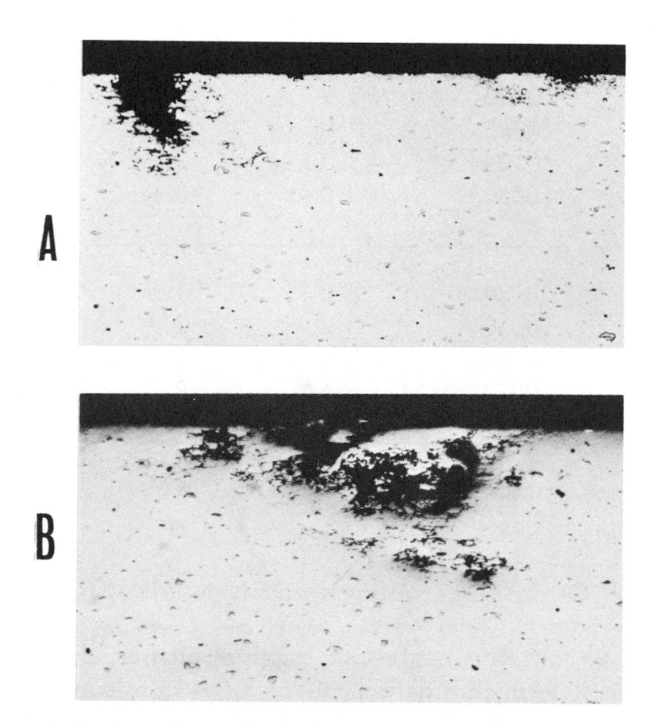

Fig. 2. Typical atmospheric corrosion pitting of aluminum alloy 5086-H34; (a) skyside, (b) groundside, after 5 years at Daytona Beach, Florida.

Table 3
Atmospheric Corrosion of Some Aluminum Alloys after 7 Years of Seacoast Exposure (Kure Beach, N.C., "A"; Pt. Reyes, Calif., "C")

Alloy	Site	Avg. Corr. Rate Mils per yr.	Avg. 4 Deepest Pits Sky	Avg. 4 Deepest Pits Ground	% T.S. Losses by Corrosion
1100-H14	A	.0122	1.5	2.1	0
	C	.0051	0.7	0.4	0
3003-H14	A	.0015	1.7	1.4	0.9
	C	.0055	0.6	1.6	0
3004-H14	A	.0094	1.5	1.7	0.7
5052-H34	A	.0085	1.5	1.7	1.2
	C	.0025	1.0	0.3	0.3
5154-H34	A	.0072	1.8	4.3	0.7
	C	.0023	0.6	2.6	0.5
5083-0	A	.0120	2.2	1.4	1.1
	C	.0063	0.06	NAA	0.5
5086-H34	A	.0104	3.3d	2.7d	1.7
	C	.0030	1.0d	0.9d	0.4
6061-T6	A	.0134	2.3	1.5	0.4
	C	.0060	0.2	2.7	0.2
6061-T6	A	.0113	0.5	1.5	0.7
Clad	C	.0070	NAA	2.0	0.2

Note: NAA = No appreciable attack.
d = Intergranular.

of the bayerite ($\beta Al_2O_3\ 3H_2O$) reaction products described in equations (2) and (4).

Where it is important to minimize the depth of pitting, the concept of "alcladding" is utilized with considerable success. A thin layer of a dissimilar alloy (generally 7072 containing 1% Zn) is roll-clad to one or both sides of the base aluminum alloy. The cladding is selected to act as a built-in sacrificial anode. Once corrosion penetrates to the core-clad interface, the

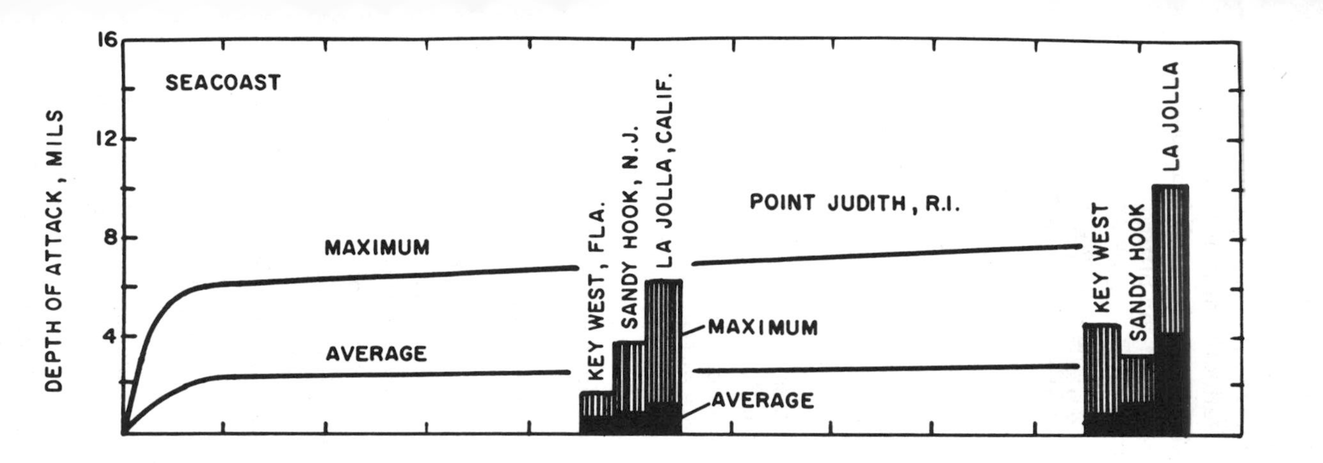

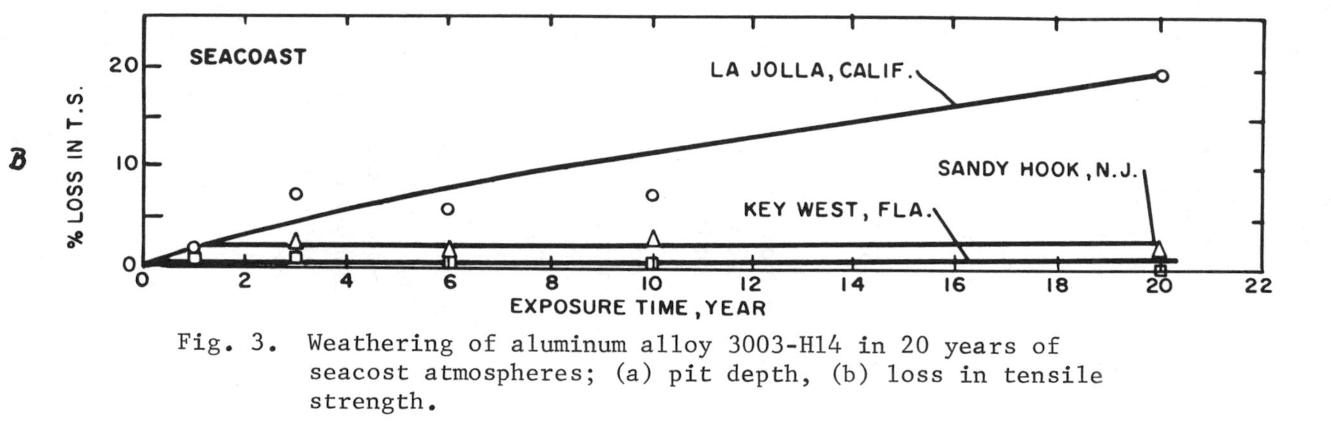

Fig. 3. Weathering of aluminum alloy 3003-H14 in 20 years of seacost atmospheres; (a) pit depth, (b) loss in tensile strength.

sacrificial action of the cladding comes into action causing the corrosion to spread laterally along the core-clad interface rather than to penetrate into the base metal. In a high conductivity electrolyte such as sea water, corrosion can be maintained to the depth of the core-clad interface until as much as 95% of the cladding has been consumed. The effectiveness of this principle is illustrated in Figure 4 which shows the corrosion of bare and alclad 3003 in the same industrial atmosphere [5]. It may be seen that the pitting is much shallower in the alclad 3003 and that it is confined to the depth of the core-clad interface which is readily visible. In contrast, the bare 3003 is much more deeply pitted. It is relatively common amongst the alloys in Table 1 to clad 3003, 3004 and 6061 for corrosion protection. It is not so common to clad the aluminum-magnesium alloys containing from 2.5% magnesium because their level of marine corrosion resistance is high in the unclad condition.

CORROSION OF ALUMINUM AND ITS ALLOYS IN SEA WATER

Corrosion under conditions of immersion in sea water is generally much more severe than in marine atmospheres. Quantitative information on corrosion inquiet sea water can be obtained in one of three ways:

1. in constant immersion at fixed depth below the surface of the ocean; this involves attaching the specimens to a type of raft or float
2. at constant position but totally immersed, and
3. at constant position but immersed only intermittantly.

Of the three foregoing methods, 3 will generally be found to be most corrosive; the majority of information reported in the literature used method 2 above.

Typical fixed position immersion specimens from some of the author's tests in Florida are shown in Figure 5. Immediately obvious is the extensive marine fouling which collects on the specimens in periods as short as three months. The marine growths that accumulate during corrosion testing on most metals (the problem is less severe with rich copper alloys) promotes conditions of uneven aeration and numerous crevices which tend to make the pitting of metals much more severe than in atmospheric attack where these complications do not exist. Information on the weight loss in periods of up to ten years at two locations (Harbor Island, North Carolina and Halifax, Nova Scotia) are shown in Figures 6a and b [6]. These are for alloys 1100-H14, 5052-H34, and 6061-T6. This attack is much more severe than is experienced in the marine atmospheres on comparable alloys. Characteristically, the corrosion is still pitting [7] but of considerably greater depth.

One of the authors analyzed the frequency distribution of individual pit depths in four aluminum alloys, 3003, 5050 (1.4% Mg), 5052 and 6061 after two years complete immersion at fixed depth in sea water at Daytona Beach, Florida. Contrary to what might have been expected, the population

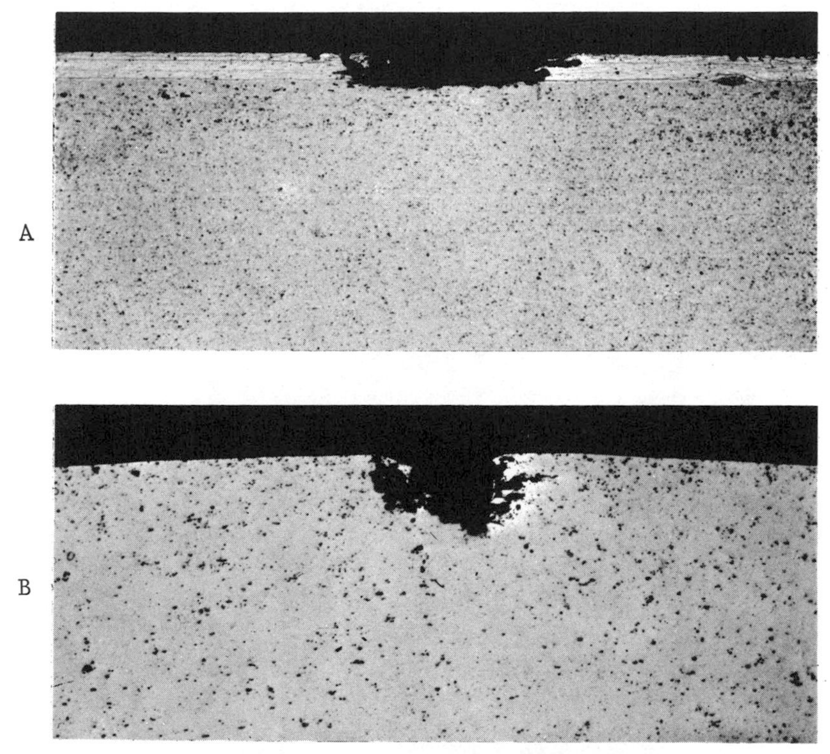

Fig. 4. Pitting of Alclad (a), and bare (b), aluminum alloy 3003 after 10 years in industrial atmosphere.

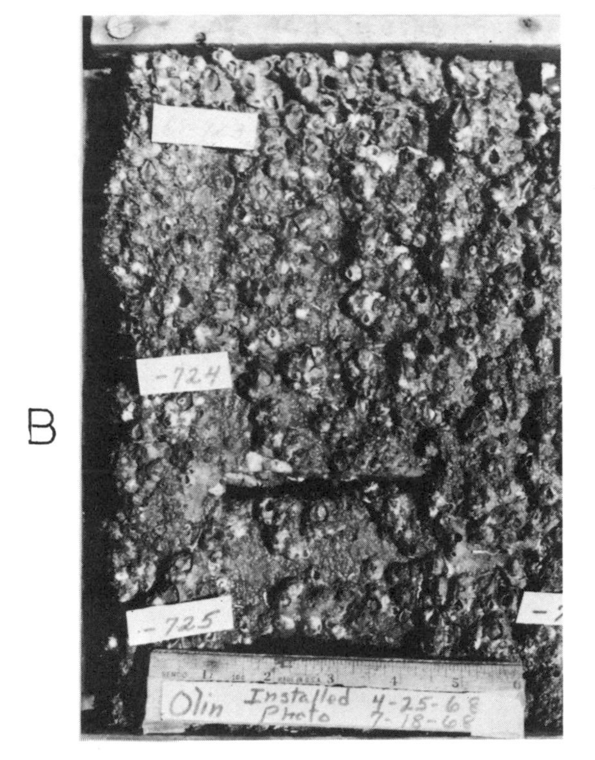

Fig. 5. Overall (a), and close-up (b), of sea water immersion test racks after 3 months - Daytona Beach.

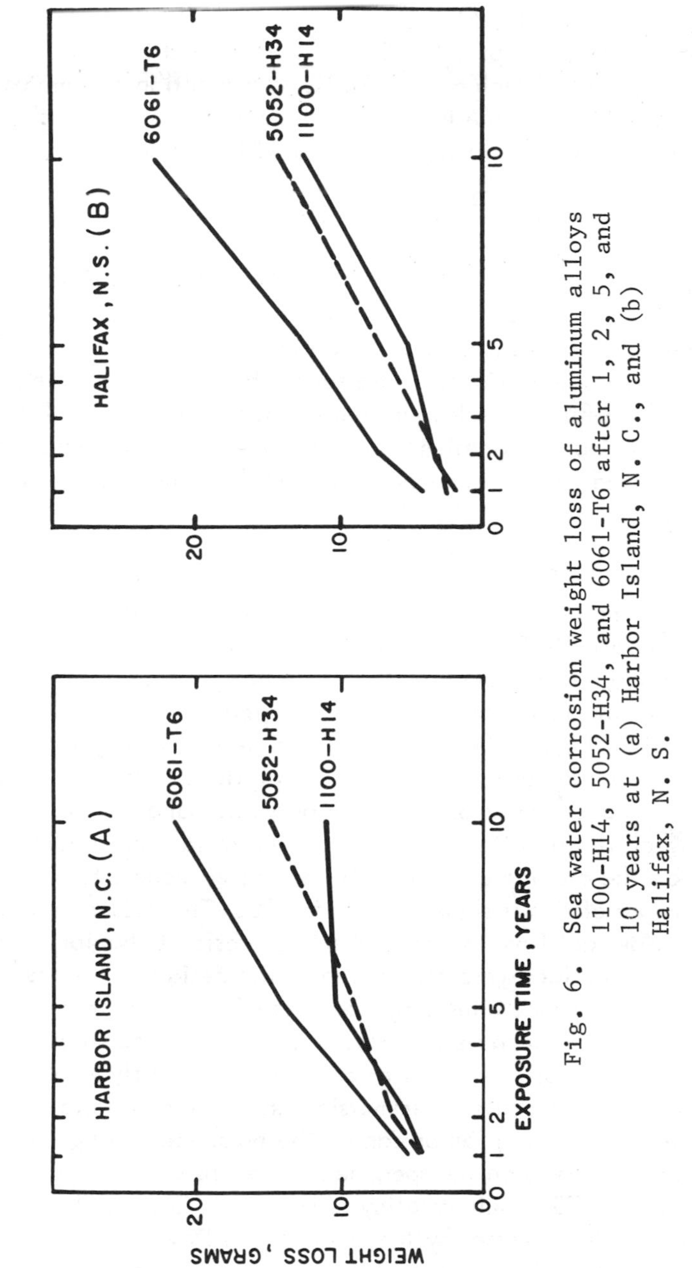

Fig. 6. Sea water corrosion weight loss of aluminum alloys 1100-H14, 5052-H34, and 6061-T6 after 1, 2, 5, and 10 years at (a) Harbor Island, N. C., and (b) Halifax, N. S.

vs depth plots were far from Gaussian as shown in Figure 7. A typical pit, approximately 4 mils deep, in 6061-T4 is shown in Figure 8. Mathematical manipulation showed that the pitting population could be normalized with the square root of the pit depth, Figure 9, thereby permitting the normal statistical values of median pit depth, standard deviation and probable error to be determined. Conduct of such measurements as a function of time permits some degree of predictability to the longer term behavior. As is the case of atmospheric exposure the rate of pitting decreases with increasing time.

Summary of Aluminum Corrosion Tests

The foregoing information shows that there are a number of aluminum alloys with various strength levels that can be used in the unprotected condition in marine atmospheres immediately adjacent to the sea coast. The corrosion under immersion conditions is more severe but here again the nature of the corrosion of aluminum tends to be somewhat self-limiting. The more corrosion resistant aluminum-magnesium alloys can be used in the unprotected condition with good life expectancy if reasonably heavy thicknesses are utilized.

Aggravating Factors

The foregoing quantitative corrosion information was collected on samples that were comparatively free from aggravating factors, at least, insofar as the atmospheric exposures are concerned. For instance, aluminum alloys are basically susceptible to the phenomena of crevice corrosion which is illustrated diagrammatically in Figure 10a. Where a crevice exists oxygen is consumed within the crevice and corrosion products precipitate at its mouth thereby hindering the replacement of oxygen from outside. Under such conditions aggressive attack results with the anode located inside the crevice and cathode located outside the crevice under conditions of free access to oxygen. A practical example of this is shown in Figure 10b. The presence of crevices in equipment will detract from the atmospheric corrosion behavior reported in this paper where crevices were eliminated in the design of the test. The degree to which corrosion damage caused by crevices will detract from the immersion-corrosion results is less because the marine growth, and particularly barnacles, (Figure 5), provides a great number of crevices, and partly explains the difference between the pitting rates in immersion and in atmospheric exposure.

A second aggravating factor can be the presence of large applied or residual tensional stresses on the specimens. For instance, Figure 11 shows a cross-section of a 7075 aluminum alloy (Al-Zn-Mg-Cu) which had been exposed under bending stresses by the sea coast in Daytona Beach, Florida.

The failure is intergranular and is stress corrosion cracking, i.e., a combination of an applied or residual tensile stress with a susceptibility of the grain boundaries in the metal to preferential corrosion. The alloys in

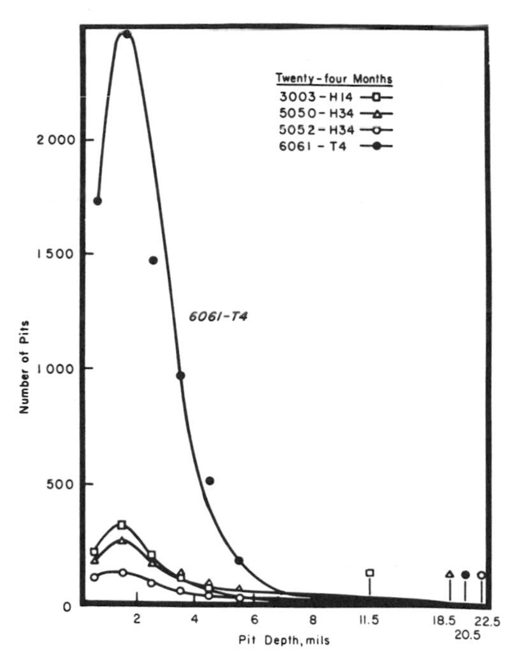

Fig. 7. Frequency distribution curves of pit-depths of aluminum alloys 3003-H14, 5050-H34, 5052-H34, and 6061-T4 after 2-years immersion in Daytona Beach sea water.

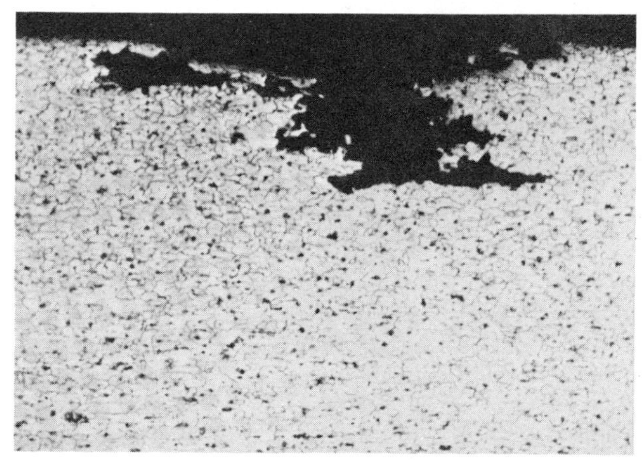

Fig. 8. Typical corrosion pit (.004" deep) in aluminum alloy 6061-T4 after 2 year immersion in sea water at Daytona Beach, Florida.

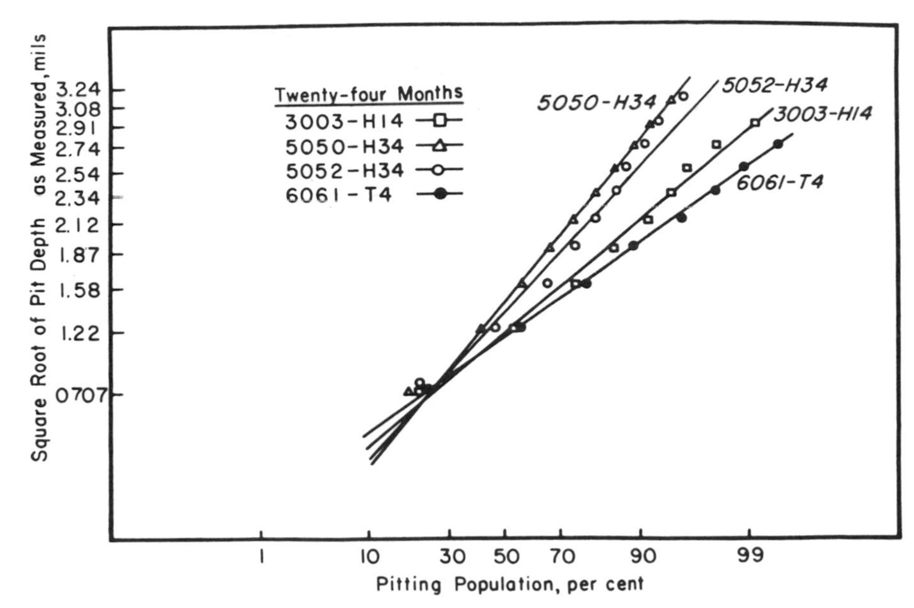

Fig. 9. Plot of square root of pit depth vs. pit population of aluminum alloys 3003-H14, 5050-H34, 5052-H34, and 6061-T4 after 2 years immersion in Daytona Beach sea water.

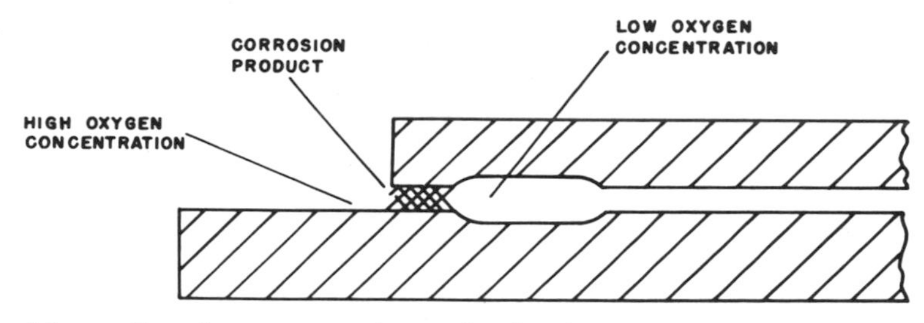

Fig. 10a. Crevice corrosion of aluminum - diagram of crevice corrosion mechanism.

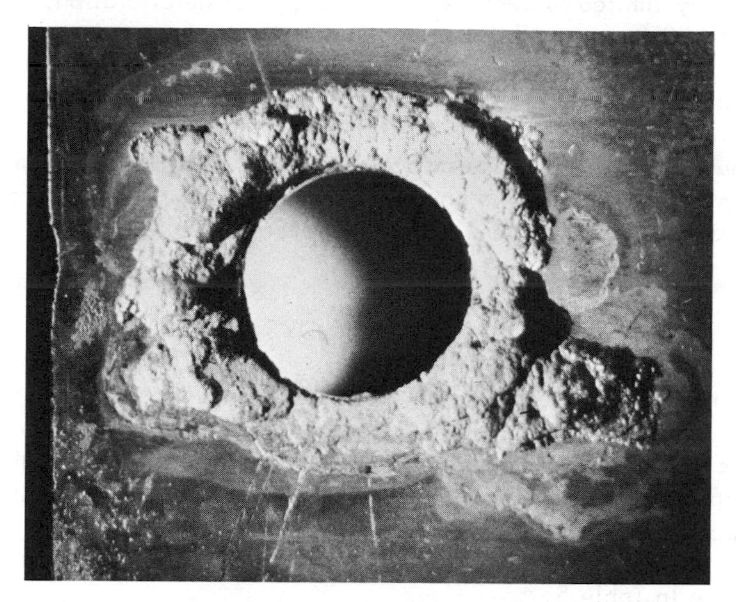

Fig. 10b. Crevice corrosion under washer of aluminum coupon tested in hot 3.4% sodium chloride.

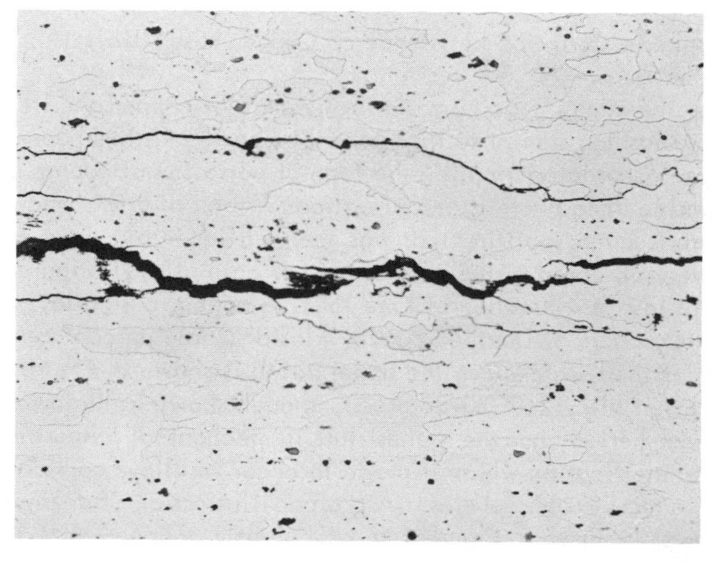

Fig. 11. Stress corrosion cracked aluminum alloy 7075-T6, exposed 5 years in the atmosphere at Daytona Beach, Florida.

Table 1 show very limited susceptibility to this type of deterioration. Only alloys 5086 and 5083 can exhibit a susceptibility to stress corrosion cracking and this susceptibility is only developed on holding for prolonged times within the temperature range of 200–350° F, i.e., a condition not normally encountered in marine service.

Aluminum and the rest of the alloys considered here have problems with respect to compatibility in marine environments. This compatibility problem is considered as a general problem in the final section of this paper.

COPPER AND ITS ALLOYS

A wide range of copper alloys shows satisfactory performance in marine atmospheres. A more limited range of copper alloys shows excellent performance under conditions of immersion in quiet or moving sea water. Table 4 presents the chemical composition of some of the copper alloys that find significant application in marine environments.

The mechanical properties of the foregoing alloys in a limited range of tempers is shown in Table 5.

It may be seen in comparison with Table 2 that the mechanical properties of copper alloys are somewhat superior to those of aluminum alloys. Their higher cost per pound and higher density can be justified particularly under conditions of immersion in sea water by a much higher level of corrosion resistance, particularly under conditions of relatively rapid movement by or through the environment.

Atmospheric Corrosion Resistance of Copper Base Alloys

Testing of the atmospheric corrosion resistance of copper and its alloys is conducted in a fashion similar to that described earlier in this paper for aluminum alloys. Characteristically, the form of corrosion of copper base alloys in the marine atmosphere is one of uniform attack at a fairly low rate. The shallow high frequency pitting that was seen earlier in aluminum alloys is seldom, if ever, observed in the case of copper base alloys. Figure 12 shows the weight loss as a function of time for five copper base alloys exposed to the marine atmosphere at Daytona Beach, Florida. The percentage loss of tensile strength of the specimens in the same period is shown in Figure 13. It should be noted that alloy 260 (70–30 brass), though showing the lowest weight loss in a four-year period, has the highest loss in mechanical properties. This is because of dezincification which is prone to occur in alloys containing more than 20% zinc in marine atmospheres. A typical illustration of dezincification of 70–30 brass is shown in Figure 14a. Obviously, copper-base alloys subject to parting of this type in a marine atmosphere are not recommended for load carrying members. Dezincification is absent in the case of the alloy

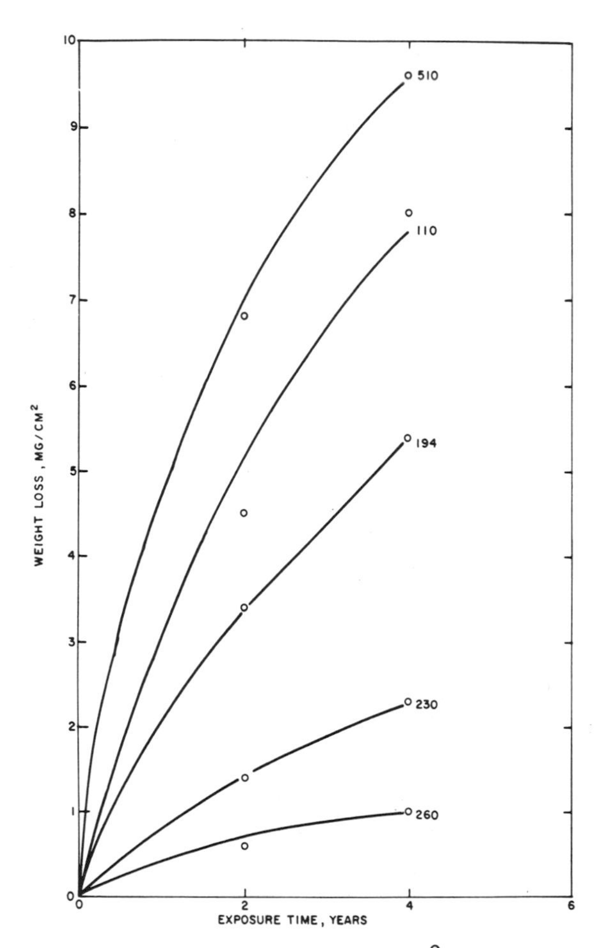

Fig. 12. Weight loss by corrosion (mg/cm^2) of copper alloys 110, 194, 230, 260, and 510 after 2 and 4 years atmospheric exposure at Daytona Beach, Florida.

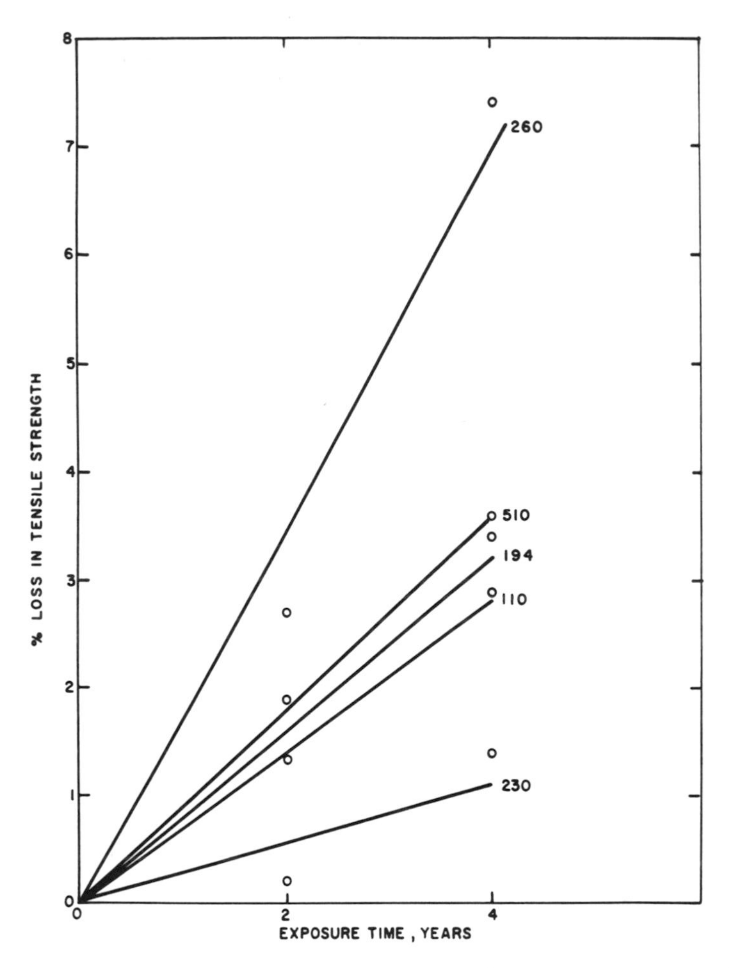

Fig. 13. Percent loss of tensile strength by corrosion of copper alloys 110, 194, 230, 260, and 510 after 2 and 4 years atmospheric exposure at Daytona Beach, Florida.

Table 4
Some Copper Alloys used in Marine Applications
Chemical Compositions*

CDA Alloy	Cu	Zn	Sn	Ni	Mn	Al	Fe	P	As	S.C.C.
110	99.9 Min.	-	-	-	-	-	-	-	-	I
122	99.9 Min.	-	-	-	-	-	-	.15-.4	-	I
194	97.0-97.8	.05-.20	.03	-	-	-	2.1-2.6	.01-.04	-	I
230	84.0-86.0	Rem.	-	-	-	-	.05	-	-	MS
260	68.5-71.5	Rem.	-	-	-	-	.05	-	-	S
443	70-73	Rem.	.8-1.2	-	-	-	.06	-	.02-.10	S
510	95 Nom.	.30	3.5-5.8	-	-	-	.10	.03-.35	-	SS
687	76.0-79.0	Rem.	-	-	-	1.8-2.5	.06	-	.02-.10	S
706	88.7 Nom.	1.0	-	9.0-11.0	1.0	-	.5-2.0	-	-	I
715	70 Nom.	1.0	-	29.0-33.0	1.0	-	.7	-	-	I

Note: SCC - Stress Corrosion Cracking Susceptibility
I - Immune
S - Susceptible
MS - Moderately
SS - Slightly Susceptible

*Composition in % maximum unless a range is shown.

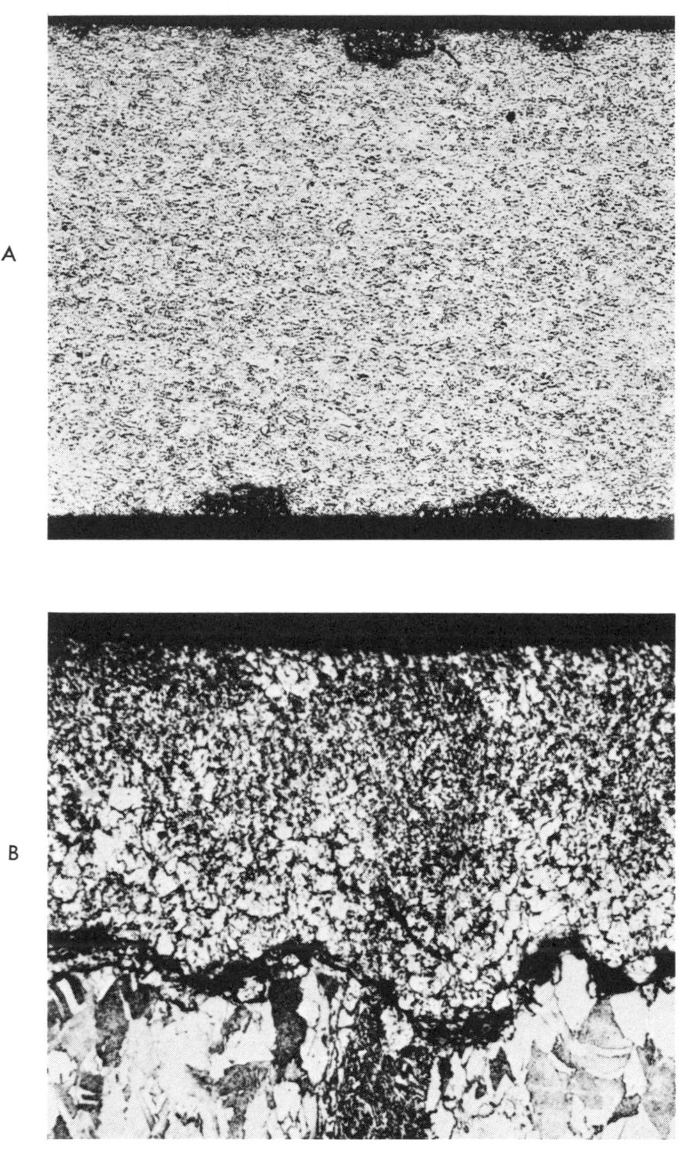

Fig. 14. Dezincification of copper alloy 260. (A) After 4 years in Daytona Beach, Florida atmosphere; (B) welded tube exposed to flowing warm brine.

Table 5
Typical Mechanical Properties of some Copper Alloys used in Marine Applications

CDA Alloy	Form	Temper	T.S. Ksi	Y.S.* Ksi	% Elongation in 2"
110 & 122	Tube	Ann. .025 mm.	34	11	45
		Light Drawn 15%	40	32	25
194	Strip	Half Hard	53-63	48-58**	9.5
230	Tube	Ann. 0.015 mm.	44	18	45
260	Tube	Ann. 0.025 mm.	52	20	55
443	Tube	Ann. 0.025 mm.	53	22	65
510	Strip	Ann. 0.025 mm.	50	21	52
687	Tube	Ann. 0.025 mm.	60	27	55
706	Tube	Ann. 0.025 mm.	44	16	42
		Light Drawn 15%	60	57	10
715	Tube	Ann. 0.025 mm.	60	25	45

Note: Annealed = Ann.

*1/2% Extension.
**.2% Offset.

230, (85–15 brass) and the remainder of the alloys shown in Figures 12 and 13. Longer time atmospheric corrosion data of copper alloys up to and including 20 years duration were collected earlier by Tracy (8). The results for a limited number of alloys are shown in Figure 15a and b. Comparison of Figure 15* with Figure 3 shows that the corrosion rate of copper-base alloys does not fall off as rapidly with time as aluminum base alloys. However, in the absence of parting of selected alloys, the corrosion is generally uniform and so little, if any, significant structural damage other than that due to general thinning results from extremely long term corrosion in this type of atmosphere.

Corrosion under Immersion Conditions

One of the significant features of the use of copper base alloys in sea water is that corrosion resistance is not much lower than experienced under marine atmospheric conditions. Alloys basically suited to service in sea water include inhibited Admiralty Brass, alloy 443, arsenical aluminum-brass, alloy 687 and the two copper-nickel alloys 706 (90–10 copper-nickel) and 715 (70-30 cupro-nickel). These alloys have the capability not only to withstand attack by quiescent sea water but also can withstand attack by a rapidly moving sea water as for instance may be encountered in condenser tube service. The high zinc alloy 443 and 687 contain arsenic additions which effectively prevent dezincification that would otherwise occur in marine environments. Figure 16 shows the weight loss as a function of time in 40° C recirculating sodium chloride solution for alloys 122, 443, 687, 706 and 715. Generally speaking, the form of attack at the nominal velocity of 5 fps in these tests is relatively uniform. The weight loss figures per unit area are somewhat higher than those shown in Figure 12 for atmospheric attack. Many of the corrosion rates appear to be self-limiting rather similar to what was observed earlier in aluminum. This is on account of the gradual formation of comparatively protective films composed primarily of cuprous oxide (Equation 5) formed by direct anodic reaction, with secondary components of cupric hydroxychloride (Cu_2OH_3Cl) and CuO formed by precipitation from solution. Depending upon the alloying additions contained in solid solution in the substrate metal, the cuprous oxide reaction product films can exert a pronounced protective influence against subsequent corrosion by acting as ohmic resistances between the local half cell corrosion reactions.

It is fairly characteristic of copper alloys that in a given environment the corrosion rates will fall as the velocity increases. This will only occur up to a point known as the break-away velocity which is characteristic of each copper alloy. This effect is depicted in Figure 17 for alloys 122, 194, 443, 687 and 706. Information obtained at Harbor Island, North Carolina on the

---

*Detailed data from 1944 and 1945 reports of Subcommittee Vi of B-3 committee.

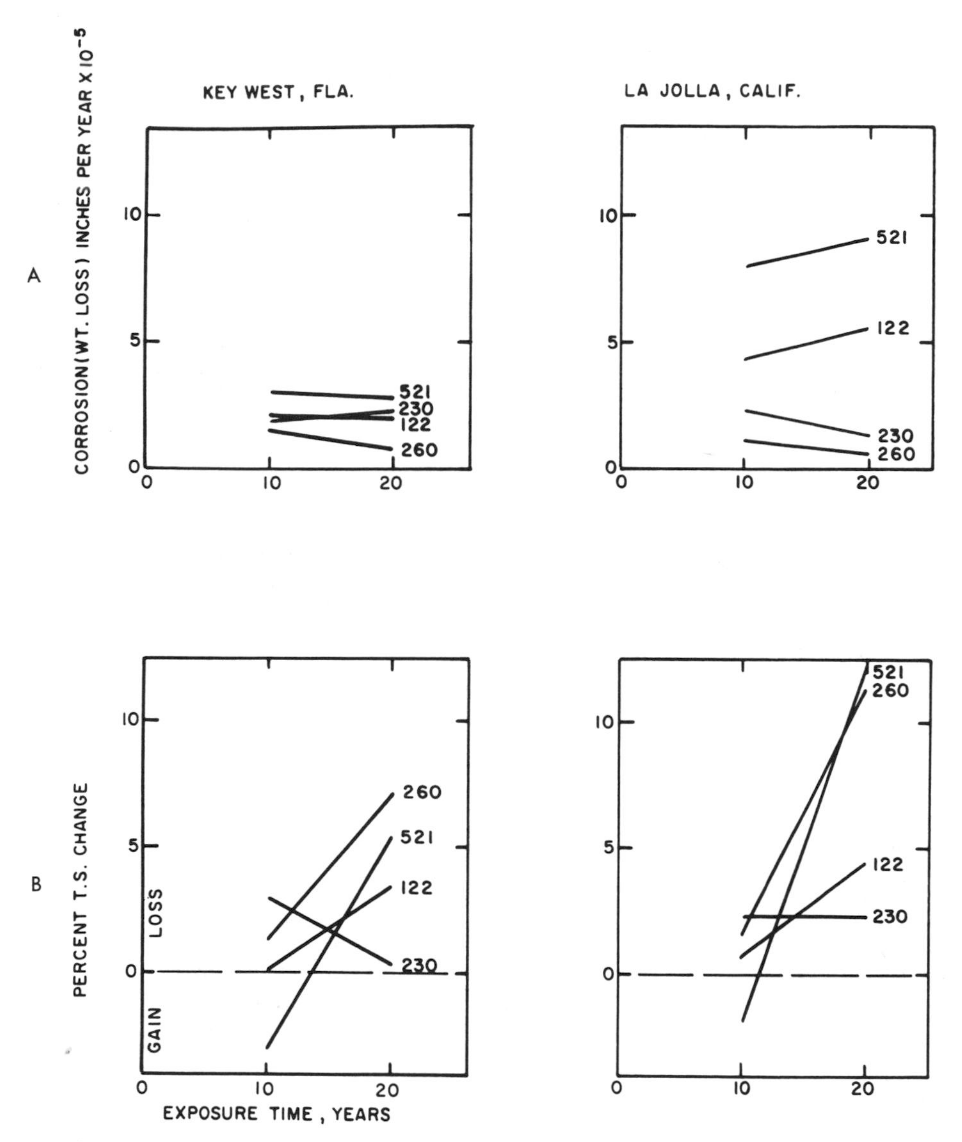

Fig. 15. Atmospheric corrosion of some copper alloys after 10 and 20 years at Key West, Florida and La Jolla, California. (a) Weight loss to corrosion penetration/rate; (b) per cent loss of tensile strength.

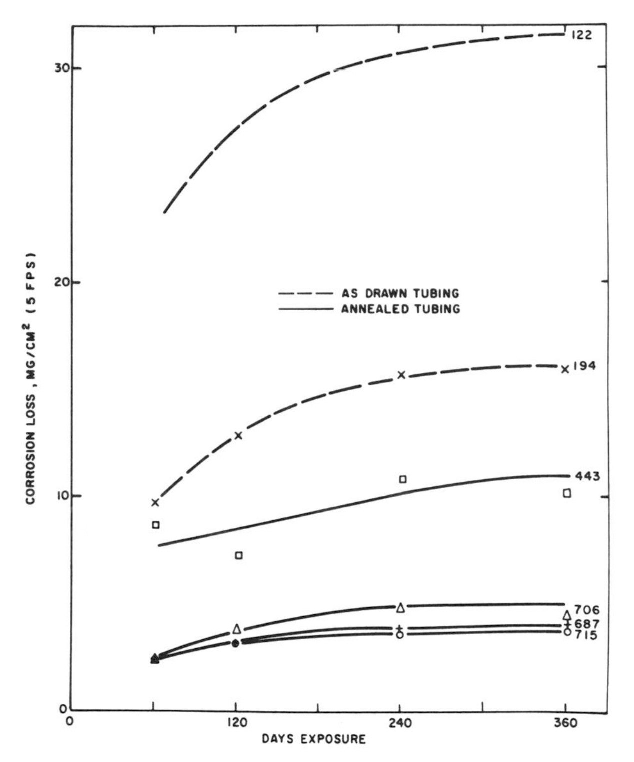

Fig. 16. Corrosion weight loss (mg/cm^2) of copper alloys 122, 194, 443, 687, 706, and 715 in 3.4% sodium chloride at 40°C recirculated at 5 ft. per sec. for 60, 120, 240, and 360 days.

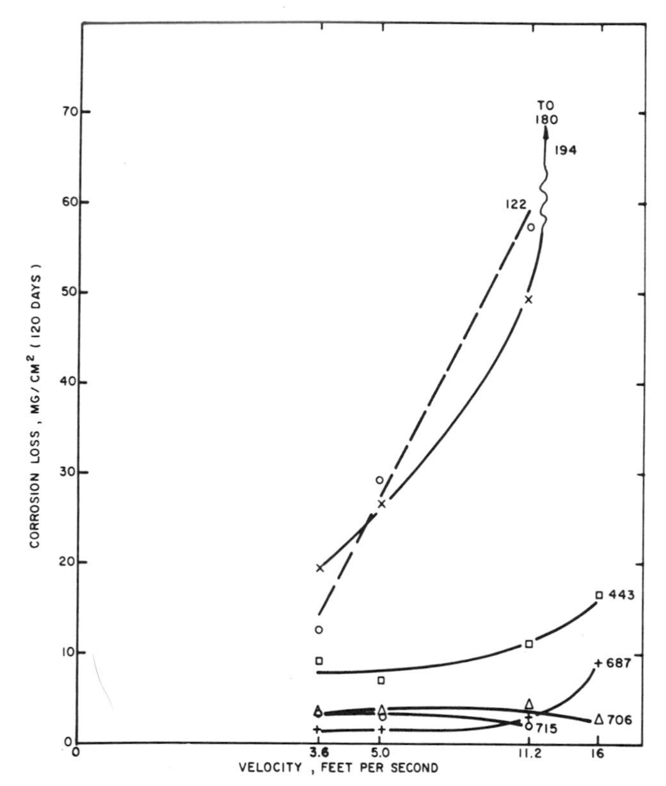

Fig. 17. Corrosion weight loss (mg/cm^2) of copper alloys 122, 194, 443, 687, 706, and 715 in 3.4% sodium chloride at 40°C recirculated for 120 days at velocities of 3.6, 5.0, 11.2, and 16 ft. per sec.

suggested maximum velocity various copper alloys can withstand in unpolluted sea water is shown in Table 6 [9].

Table 6
Velocity Effects on Copper Alloys in Clean Sea Water

Cu Alloy	Suggested Maximum Velocity Ft/Sec
110 & 122	3
443	5
687	8
706	12
715	> 15
400 Monel	>30

Some similar information has also been reported by C. L. Bulow on the extent of impingement attack on the leading edge of a variety of copper alloy specimens exposed to flowing sea water in a trough at an average velocity of 2 to 3 fps at Kure Beach, North Carolina [10]. The results in Table 7 again attest to the superiority of the aluminum-brass alloy 687 and the copper-nickel alloys over the brasses and the coppers which are not well suited for use under conditions of immersion in moving sea water.

Table 7
Sea Water Impingement Attack on Cu Alloys - Kure Beach, N. C.

Alloy	Depth of Impingement* Attack i.p.y.	
	Avg.	Max.
110	0.0118	.0388
122	0.0095	.0303
443	0.0095	.0170
260	0.0103	.0225
687	0.0019	.0039
715	0.0008	.0023

One of the important factors in developing maximum corrosion resistance in the copper-nickel alloys is the inclusion of iron in these alloys. Without the inclusion of this element corrosion rates tend to be seriously degraded.

---

*Velocity from 2 to 3 feet per second.

Typically the benefits of including iron in a variety of copper-nickel alloys can be seen in results depicted in Figure 18 [11]. Considerable argument has attended the reason why iron is so effective in improving the corrosion rate of the copper-nickel alloys. Some, like Stewart and LaQue [12], have felt that iron hydroxides are incorporated in the overall corrosion product as a separate phase and improve its corrosion resistance. However, recent work by one of the authors [13] has shown that the iron is effective in being included in the cuprous oxide where it modifies its defect structure in such a fashion as to eliminate the large number of cation vacancies that normally exist in this structure.

The copper alloys per se are not as susceptible to fouling during immersion in sea water as are aluminum alloys and titanium alloys. Basically, most copper alloys except alloys 715 and higher nickel alloys release sufficient cupric ions into the sea water in the immediate vicinity of the alloy so as to be relatively toxic to the marine organisms. As, however, corrosion resistance increases and less cupric ions are introduced into the environment conventional fouling occurs. In an alloy such as 715 in quiet sea water this leads to undesirable pitting. Accordingly, under conditions where stagnant exposure to sea water is anticipated alloys such as 687 and 706 will often perform better. Of course, under conditions of very rapid solution movement there is much less tendency for marine fouling to build up.

Summary

In summary, copper base alloys with the exception of those uninhibited alloys containing more than 20% zinc show generally good resistance to marine atmospheric corrosion. Corrosion is essentially uniform in nature and does not exhibit the pitting found to be characteristic in aluminum alloys. Under conditions of immersion, a more limited range of copper alloys including 443, 687, 706 and 715 exhibit outstanding corrosion resistance with rather minimal tendency towards pitting and parting. The copper-nickels show the best velocity characteristics but the higher nickel-alloy as 715 shows relatively lower resistance to pitting under quiescent conditions.

Aggravating Factors

Parting is a general term including such factors as dezincification in brasses, dealuminification in aluminum-bronzes and denickelification in cupro-nickels. Figure 14b has shown a typical microsection of a dezincified 260 alloy coupon where the zinc has passed into solution leaving behind a porous sponge comprising the copper that was originally present in the solid solution. On account of this dezincification straight brasses are seldom used in sea water these days. Both inhibited Admiralty alloy 443 and inhibited brass alloy 687 have additions of arsenic to them which is intended to exercise

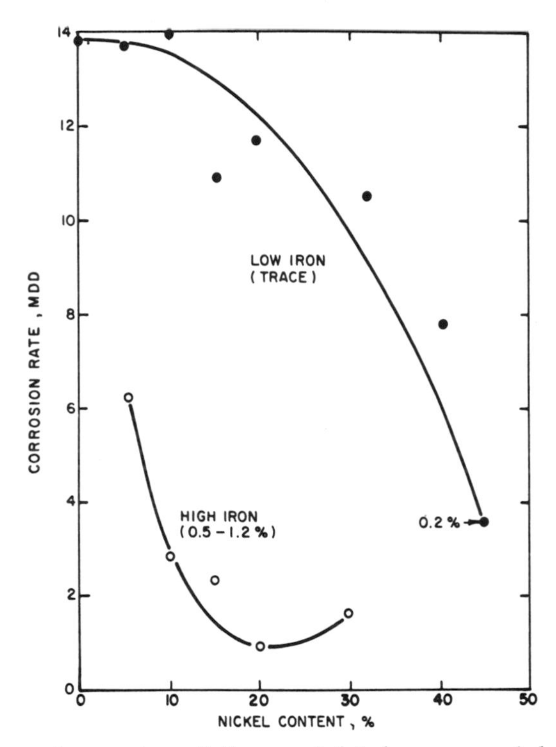

Fig. 18. Corrosion rate of low and high cupro-nickels in flowing sea water at 2 ft. per sec. for 175 days.

a positive control over dezincification. The element in small amounts in the order of 0.1% or less is extremely effective in inhibiting dezincification of brasses. The copper-nickels described in this paper are not particularly prone to denickelification under the type of service conditions described in this manuscript.

Stress Corrosion Cracking

A number of the alloys described in Table 4 are susceptible to stress corrosion cracking in the appropriate environment and under the presence of applied tensional stresses. Figure 19 shows a typical stress corrosion cracking of an intergranular nature in alloys 260 and 230. Of the two alloys, alloy 260 is much more susceptible to this form of attack. Of the alloys described in Table 4, alloy 260 can be extremely susceptible to stress corrosion cracking particularly in the presence of traces of ammonia. Table 4 also contains a statement of the relative degree of susceptibility to stress corrosion cracking of the other alloys considered.

Pitting

Although pitting is not a normal occurrence in copper alloys in marine atmospheric attack it can be sometimes observed under conditions of immersion particularly where the flow rates are relatively low (localized attack associated with a very high flow rate is impingement corrosion). Figure 20 shows a typical example of pitting at moderate velocity of alloy 706 in sodium chloride solution at 40° C. Generally alloy 706 has good resistance to pitting corrosion perhaps somewhat better than alloy 715 under conditions of low velocity. The alloy is less prone to local velocity effects than alloys 443 and 687, both of which can show severe local attack on occasion.

TITANIUM AND ITS ALLOYS

Tables 8 and 9 list the composition and mechanical properties of the range of grades of commercially pure titanium and the titanium alloys. In the so-called commercially pure grades of titanium mechanical properties are controlled by the amounts of impurities such as carbon, nitrogen, hydrogen and particularly iron. Tensile strengths ranging from 35 to 80,000 psi may be obtained primarily by varying the iron content. It may be seen that in the grades of commercially pure titanium mechanical properties are not particularly outstanding. High tensile and yield strength are only obtained in the heat treatable alloys particularly the Ti-6Al-4V alloy. However, alloys of this nature do not tend to find much use in sea water on account of lowered stress corrosion resistance. Accordingly, for most applications one is concerned with

A

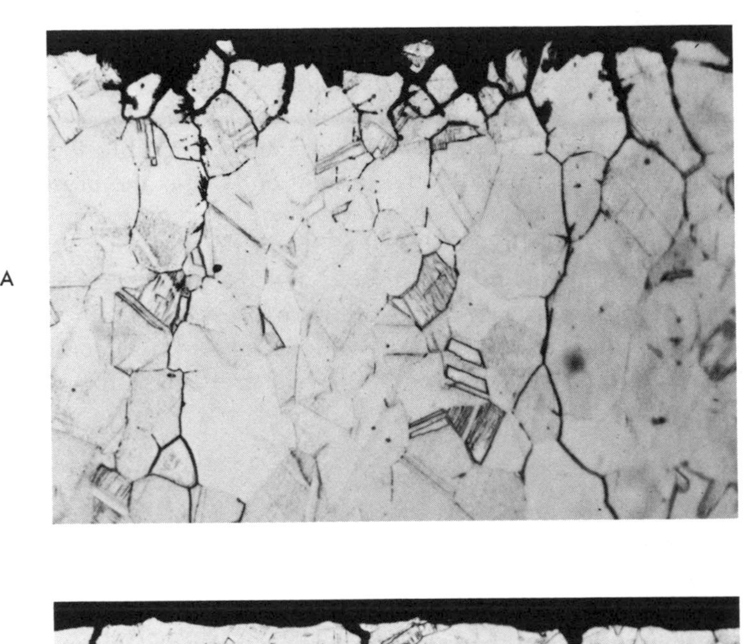

B

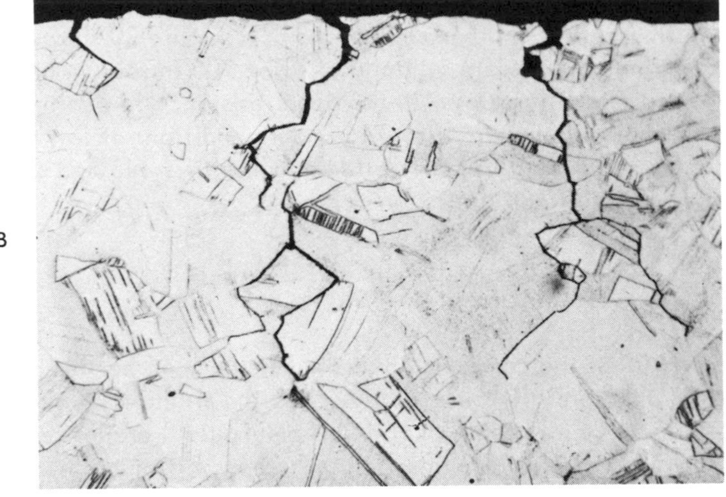

Fig. 19. Stress corrosion cracking (intergranular) of copper alloys 260(a) and 230(b) exposed to ammonia atmosphere. (Photo at 500X Mag.)

Fig. 20. Corrosion pitting of copper alloy 706 in 3.4% sodium chloride at 40°C flowing at 5.0 ft. per sec. for 60 days.

titanium in grades through 35 to 75A.

Table 8
Properties of Commercially Pure Grades of Titanium (14)

Grade	Ti35A	Ti50A	Ti65A	Ti75A
% C max.	0.08	0.08	0.08	0.08
% N max.	0.05	0.05	0.05	0.05
% H max.	0.015	0.015	0.015	0.015
% Fe max.	0.12	0.20	0.25	0.30
UTS, psi, sheet	35,500	50,000	65,000	80,000
0.2% YS, psi	25,000	40,000	55,000	70,000
Elong. in 2", %	25	22	20	15

Table 9
Some Properties of Common Titanium Alloys (14)

Alloy	Ti-0.2 Pd	Ti-5Al-2.5 Sn	Ti-6Al-4V
% C max.	0.08	0.08	0.08
% N max.	0.05	0.05	0.05
% H max.	0.015	0.017	0.015
% Fe max.	0.25	0.50	0.25
% Pd nom.	0.15	-	-
% Al nom.	-	4.0-6.0	5.75-6.75
% Sn nom.	-	2.0-3.0	-
% V nom.	-	-	3.5-3.5
UTS psi	50,000	120,000	130,000
0.2% YS, psi	40,000	115,000	120,000
Elong in 2", %	22	10	10

Strength to weight ratio is reasonably good on account of the lower density of titanium (0.164 lb./cu. in.) compared with copper alloys. However, the high price of the material and its comparatively modest strength dictate that it be used quite selectively for maximum economics.

There is no doubt that the corrosion resistance of the commercially pure grades of titanium is quite outstanding in the marine atmosphere and in quiescent or moving sea water. Although titanium and its alloys can be subject to very severe crevice corrosions this is not usually observed until temperatures are significantly in excess of 250°F. Therefore, in most uses of titanium in the atmosphere and in immersion at normal ambient temperature in sea water, the metal is passive and no significant weight loss or detectable attack is observed. Typical of the high corrosion resistance of commercially pure titanium are the results shown in Table 10 which are from rotating disc experiments at high velocity. It may be seen that the velocity characteristics of the commercially pure titanium are substantially better even than those of the copper alloy 715.

Table 10
Corrosion of Titanium and other Metals in High-Velocity* Sea Water [15]

Material	Weight loss gm	Critical velocity fps	Loss in thickness at outer edge, in.
Titanium**	0.05	>25	0.000
110	22.41	18	0.012
230	9.62	19	0.009
715	18.95	18	0.010

COMPATIBILITY OF METALS IN A NUMBER OF ENVIRONMENTS

To date, we have considered the corrosion characteristics of aluminum, copper and titanium alloys in marine atmospheres and under conditions of immersion in still and moving sea water. Some aggravating factors have been considered, amongst them stress corrosion cracking, pitting and parting. However, in practical equipment there remains one further serious complication. That is the metallic connection of dissimilar metals in sea water or in marine atmospheres. The consequence of this bimetal connection can be disastrous insofar as one member of the couple is concerned. For instance, Table 11 (from "Corrosion and its Prevention at Bimetallic Contacts, H. M. Stationary Office, London 1958, pp. 1.91-1.94 of Vol. 1) presents an unusually complete summary of bimetallic corrosion phenomenology by Evans and Rance [16]. This table shows that basically titanium and its alloys are without serious galvanic corrosion problems, i.e., titanium itself will not suffer increased corrosion on coupling to a wide variety of dissimilar metals. To a major degree this same situation exists with the copper alloys and intensification of corrosion is only obtained on connection to a limited number of metals such as titanium, stainless steels, and, of course, the noble metals.

*(5-in. disk specimens rotated at 1140 rpm for 60 days in sea water at 80° F; peripheral speed, 26 fps)
**Composition: carbon 0.01-0.04%; nitrogen 0.006-0.03%; oxygen 0.10-0.20%.

TABLE 11

Degree of Corrosion at Bimetallic Contacts

(Based on data provided by members of the I.S.M.R.C. Corrosion and Electrodeposition Committee and others, and arranged by Mrs. V. E. Rance)

A = The corrosion of the 'metal considered' is not increased by the 'contact metal.'
B = The corrosion of the 'metal considered' may be slightly increased by the 'contact metal.'
C = The corrosion of the 'metal considered' may be markedly increased by the 'contact metal.'
(Acceleration is likely to occur only when the metal becomes wet by moisture containing an electrolyte, e.g., salt, acid, combustion products. In ships, acceleration may be expected to occur under in-board conditions, since salinity and condensation are frequently present. Under less severe conditions the acceleration may be slight or negligible.)
D = When moisture is present, this combination is inadvisable, even in mild conditions without adequate protective measures.

Metal Considered \ Contact Metal	1 Au, Pt, Rh, Ag	2 Monel Inconel Nickel Mo Alloys	3 Cupronickels, Silver Solder, Aluminium Bronzes, Tin Bronzes, Gunmetals	4 Copper, Brasses, 'Nickel Silvers'	5 Nickel	6 Lead, Tin, and Soft Solders	7 Steel and Cast Iron	8 Cadmium	9 Zinc	10 Mg and Mg Alloys (Chromated)	11 Stainless Steels	12	13	14 Chromium	15 Titanium	16 Al and Al Alloys
											Austenite 18/8Cr/Ni	18/2 Cr/Ni	13% Cr			
1. Gold, platinum, rhodium, silver	-	A	A	A	A	A	A	A	A	A	A	A	A	A	A	A
2. Monel, Inconel, Nickel/molybdenum Alloys	B	-	A	A	A	A	A	A	A	A	A	A	A	A	A	A
3. Cupronickels, Silver Solder, Aluminium Bronzes, Tin Bronzes, Gunmetals	C(k)	B or C	-	A	A	A	A	A	A	A	B or C	B	A	B or C	B or C	A(e)
4. Copper, Brasses, 'Nickel silvers'	C(k)	B or C	B or C	-	B or C	B or C(p)	A	A	A	A	B or C	B or C	A	B or C	B or C	A(e)
5. Nickel	C	B	A	A	-	A	A	A	A	A	B or C	B or C	A	B or C	B or C	A
6. Lead, Tin and Soft Solders	C	B or C(t)	B or C(q)	B or C(q)	B	-	A or C(r)	A	A or C(r)	A	B or C	B or C	B or C	B or C	B or C	A
7. Steel and Cast Iron (a) (f) (w)	C	C	C	C	C(k)	C(k)	-	A(m)	A(m)(1)	A	C	C	C	C(k)	C	B(m)
8. Cadmium (u)	C	C	C	C	C	B	C	-	A	A	C	C	C	C	C	B
9. Zinc (u)	C	C	C	C	C	B	C	B	-	A	C	C	C	C	C	C(j)
10. Magnesium and Magnesium Alloys (chromated) (b) (a)	D	D	D	D	D	C	D	B or C	B or C	-	C	C	C	C	C	B or C(c)
11. Austenitic 18/8	A	A	A	A	A	A	A	A	A	A	(v)	A	A	A	A	A
12. Stainless Steel 18/2 Cr/Ni	C	A or C(s)	A or C(s)	A or C(s)	A	A	A	A	A	A	A	(v)	A	A	(o)	A
13. 13% Cr.	C	C	C	C	B or C	A	A	A	A	A	C	C	(v)	C	C	A
14. Chromium	A	A	A	A	A	A	A	A	A	A	A	A	A	-	A	A
15. Titanium	A	A	A	A	A	A	A	A	A	A	A	A	A	A	-	A
16. Al and Al Alloys (n) (a) (w)	D	C	D(c)	D(e)	C(k)	B or C	B or C	A	A	A(c)(h)	B or C	B or C	B or C	B or C(d)	C	(v)

Notes to Table 11

(a) The exposure of iron, steel, magnesium alloys, and unclad aluminum-copper alloys in an unprotected condition in corrosive environments should be avoided whenever possible even in the absence of bimetallic contact.

(b) The behavior of magnesium alloys in bimetallic contacts is particularly influenced by the environment, depending especially on whether an electrolyte can collect and remain as a bridge across the contact. The behavior indicated in the Table overleaf refers to fairly severe conditions. Under conditions of total immersion or the equivalent, magnesium alloys should be electrically insulated from other metals. In less severe conditions complete insulation is not necessary, but steel, brass and copper parts should be galvanized or cadmium plated and jointing compound (D.T.D. 369A) used during assembly. Under conditions of good ventilation and drainage, contacts classified as D have given satisfactory service, e.g., brass and steel push-fit and cast-in inserts in magnesium castings.

(c) Where contact between magnesium alloys and aluminum alloys is necessary, adverse galvanic effects will be minimized by using aluminum alloys containing little or no copper (0.1% max.).

(d) If in contact with thin (decorative) chromium plate, the symbol C, but with thick plating (as used for wear resistance) the symbol is B.

(e) When contacts between copper or copper-rich materials and aluminum alloys cannot be avoided, a much higher degree of protection against corrosion is obtained by first plating the copper-rich material with tin or nickel and then with cadmium, than by applying a coating of similar thickness.

(f) The corrosion of mild steel may sometimes be increased by coupling with cast iron, especially when the exposed area of the mild steel is small compared with the cast iron.

(g) Instances may arise in which corrosion of copper or brasses may be accelerated by contact with bronzes or gunmetals, e.g., the corrosion of copper seawater-carrying pipe-lines may be accelerated by contact with gunmetal valves, etc.

(h) When magnesium corrodes in sea water or certain other electrolytes, alkali formed at the aluminum cathode may attack the aluminum.

(j) When it is not practicable to use other more suitable methods of protection, e.g., spraying with aluminum, zinc may be useful for the protection of steel in contact with aluminum, despite the accelerated attack upon the coating.

(k) This statement should not necessarily discourage the use of the 'contact metal' as a coating for the 'metal considered', provided that continuity is good; under abrasive conditions, however, even a good coating may become discontinuous.

(l) In most supply waters at temperatures above about 60° C, zinc may accelerate the corrosion of steel.

(m) In these cases the 'contact metal' may provide an excellent protective coating for the 'metal considered', the latter usually being electrochemically protected at gaps in the coating.

(n) When aluminum is alloyed with appreciable amounts of copper it becomes more noble and when alloyed with appreciable amounts of zinc or magnesium it becomes less noble. These remarks apply to bimetallic contacts and not to inherent corrosion resistance. Such effects are mainly of interest when the aluminum alloys are connected with each other.

(o) No data available.

(p) In some immersed conditions, the corrosion of copper or brass may be seriously accelerated at pores or defects in tin coatings.

(q) In some immersed conditions there may be serious acceleration of the corrosion of soldered seams in copper or copper alloys.

(r) When exposed to the atmosphere in contact with steel or galvanized steel, lead can be rapidly corroded with formation of PbO at narrow crevices where the access of air is restricted.

(s) Serious accleration of corrosion of 18/2 stainless steel in contact with copper or nickel alloys may occur at crevices where the oxygen supply is low.

(t) Normally the corrosion of lead/tin soldered seams is not significantly increased by their contact with the nickel-base alloys but under a few immersed conditions the seams may suffer enhanced corrosion.

(u) The corrosion product on zinc is, in certain circumstances, more voluminous and less adherent than that on cadmium. Where this is known to be the case, it should be borne in mind in making a choice between these two metals.

(v) These joints are liable to corrosion in crevices where these are not filled with jointing compound (See para. 8 of the Introduction).

(w) Corrosion products from iron or steel reaching aluminum, or corrosion products from aluminum reaching iron or steel may sometimes cause serious local corrosion through oxygen-screening or in other ways, when the total destruction of metal is finished.

It is beyond the scope of this paper to consider palliative measures for bimetallic corrosion which usually involve protective coatings, insulation, etc. However, it must be pointed out the attractions of aluminum such as low density, low cost and good corrosion resistance cannot be realized without the necessity of expending care and design and selection in preventing bimetallic corrosion.

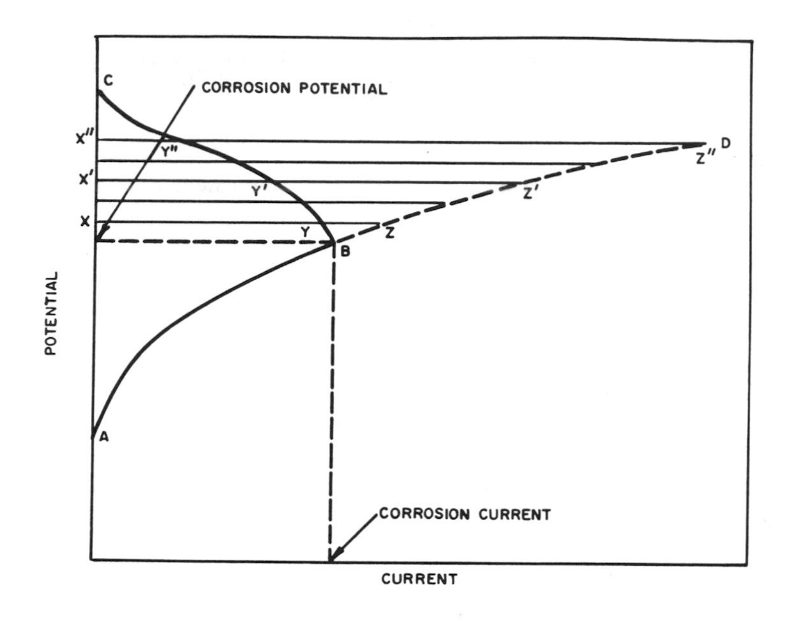

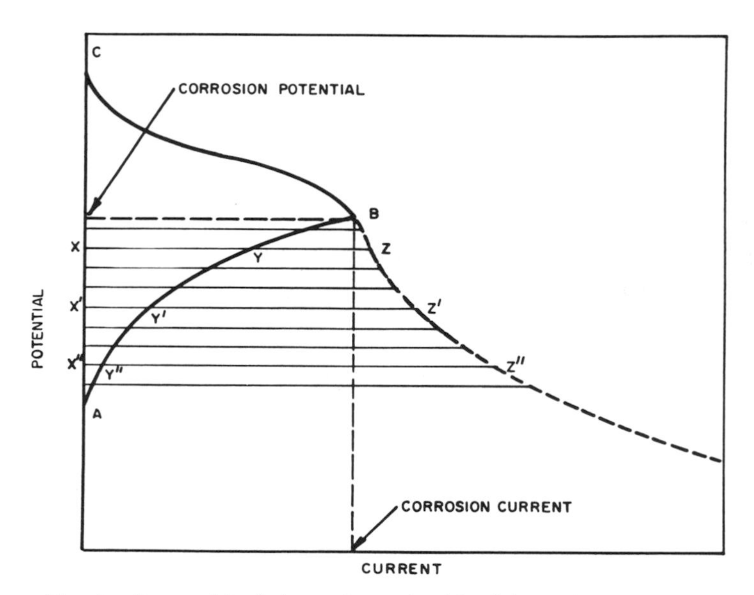

Fig. 21. Typical anodic(a) and cathodic(b) polarization diagrams.

By contrast, galvanic corrosion problems with the very active aluminum alloys are serious when connected to a variety of dissimilar metals in marine environments. This tends to reduce the attractive advantages of unprotected aluminum in practical apparatus. In marine immersion the only essentially compatible metals are cadmium and zinc, thereby indicating that under such conditions appropriate protective measures have to be applied to a multi-metal assembly containing aluminum alloys.

Galvanic corrosion is due to difference in corrosion potentials between the two metals. Generally speaking, the one with the lower corrosion potential will suffer a greater rate of corrosion while that with a higher or more noble potential will be partly or completely protected against corrosion. The rationale behind this is illustrated diagrammatically in Figure 21[17], which shows the effect of anodic and cathodic polarization on corrosion current, i.e., corrosion rate. In Figure 21a which pertains to the noble member of the corrosion couple, it may be seen that as the potential is moved in the active direction, the corrosion current, i.e., rate decreases along the line "BA". By contrast, the active member of the couple in Figure 21b can be seen to suffer increased or greater corrosion as the potential is moved in the noble direction. As this is coupled to a more noble member the corrosion current or corrosion rate of the more active member is given by a continuation of line "AB". Depending upon the slope of the anodic polarization curve, large increases in corrosion rate can be obtained with relatively small changes in potential.

The quantitative effect of such coupling can be shown in Table 12 [18].

Table 12
Weight Loss of Iron for Fe/M Bimetallic Couples in 1 Per Cent NaCl

Metal (M)	Weight Loss of Iron mg	Weight Loss of Metal (M) mg	Difference in Standard Electrode Potentials[4] V*
Copper	183.1	0.0	+0.785
Nickel	181.1	0.2	+0.19
Tin	171.1	2.5	+0.30
Lead	183.2	3.6	+0.31
Aluminum	9.8	105.9	-1.23
Cadmium	0.4	307.9	+0.04
Zinc	0.4	688.0	-0.32
Magnesium	0.0	3104.0	-1.90

*The + sign indicates that iron has a less noble standard electrode potential than the second metal.

REFERENCES

1. Pryor, M. J., "Electrode Reactions on Oxide Covered Aluminum", Z. Electrochem., 62, 782 (1958).

2. North, R. F., and M. J. Pryor, Corrosion Science. In press.

3. McGeary, et al, "Atmospheric Exposure of Non-ferrous Metals and Alloys-Aluminum: Seven-Year Data", Metal Corrosion in the Atmosphere, ASTM STP 435, Am. Soc. for Testing and Materials, 141 (1968).

4. Walton, C. J., and W. King, "Resistance of Aluminum-Base Alloys to 20-Year Atmospheric Exposure", Symposium on Atmospheric Corrosion of Non-Ferrous Metals, ASTM STP 175, 21 (1955).

5. Mattsson, E., and S. Lindgren, "Hard Rolled Aluminum Alloys", ASTM STP 435, 240 (1968).

6. Godard, H. P., et al, "Aluminum", The Corrosion of Light Metals, 3, Wiley, N. Y. (1967).

7. Summerson, T. J., Pryor, M. J., Keir, D. S., and R. J. Hogan, "Pit Depth Measurements as a Means of Evaluating the Corrosion Resistance of Aluminum in Sea Water", Metals, ASTM STP 196, 157 (1956).

8. Tracy, A. W., "Effect of Natural Atmospheres on Copper Alloys: 20-Year Test", Symposium on Atmospheric Corrosion of Non-Ferrous Metals, ASTM STP 175, 67 (1955).

9. LaQue, F. L., and A. H. Tuthill, "Economic Considerations in the Selection of Materials for Marine Applications", The Soc. of Naval Architects and Marine Engineers, New York Metropolitan Sect., 16, Feb. (1961).

10. Bulow, C. L., "Corrosion and Biofouling of Copper-Base Alloys in Sea Water", Trans. Electrochem. Soc., 87, 319 (1945).

11. Tracy, A. W., and R. L. Hungerford, "The Effect of Iron Content of Cupro Nickel on its Corrosion Resistance in Sea Water", Proc. A.S.T.M., 45, 591 (1945).

12. Stewart, W. C., and F. L. LaQue, "Corrosion Resisting Characteristics of Iron Modified 90:10 Cupro Nickel Alloy", Corrosion, 8, 259 (1952).

13. North, R. F., and M. J. Pryor, "The Nature of Protective Films Formed on a Cu-Fe Alloy", Corrosion Science, 9, 509 (1969).

14. Rance, V. E., and U. R. Evans et al, "Corrosion and its Prevention at Bimetallic Contacts", H. M. Stationary Office, London, 1958 - See Vol. I "Corrosion" by L. L. Shreir ed., 1.91-1.94, George Newnes, Ltd. and Wiley, N. Y. (1963).

15. Pryor, M. J., Bimetallic Corrosion, "Corrosion" Vol. I, L. L. Shreir ed., 1.70, George Newnes, Ltd. and Wiley, N. Y. (1963).

16. Bauer, O., and O. Vogel, "The Rusting of Iron in Contact with Other Metals and Alloys", Mitt. kgl. Materialprufsungsamt 36, 114 (1918).

CORROSION BEHAVIOR OF FERROUS ALLOYS IN MARINE ENVIRONMENTS

J. F. McGurn*

INTRODUCTION

Man has for centuries utilized the ocean. It has served as a freeway, source of food, playground, battle-field and, of growing concern in our modern age, as a waste disposal system.

Use of tidal power goes back many years into history. For example, William the Conqueror's Domesday Book of 1085 mentions a tide mill at the entrance to the south coast port of Dover (1). The twentieth century, however, with its advancing technology, provides a modern base for study of tidal sites for electric power generation. Cook Inlet - Alaska, Gulf of San Jose - Argentina, Bay of Fundy - Canada, Passamaquoddy - Canada and United States, Yang-tse-kiang Estuary - China, Bristol Channel - England, Solway Firth - England and Scotland, Rance River - France, The Collier Bay - Western Australia, and in Russia - Kislaya Guba Bay, Mezen and Penzhina Bay, are but a few of the sites that either have been under study or that are producing power today.

Harnessing tidal power presents a challenge to those engaged in ocean engineering. Major objectives are to slow down the effects of corrosion and, in some cases, stop its progress. In order to meet this challenge, selection of materials for either an initial or a replacement installation will be related to such factors as:

1. Unnecessary plant shutdowns with loss of service.
2. Reduction in maintenance costs.
3. Improvement in plant efficiency.
4. Preservation of an acceptable outward appearance.

Ocean engineers must possess a knowledge of corrosive characteristics of various environments, corrosion principles, materials resistant to corrosion, physical and mechanical properties of such materials and how they are fabri-

---

*Market Development Engineer, The International Nickel Company of Canada Limited, Toronto - Dominion Center, Toronto 1, Ontario, Canada.

cated. This knowledge combined with intelligent materials selection will enable ocean engineers to meet the challenge of harnessing tidal power.

THEORETICAL CONSIDERATIONS

Corrosion may be defined as the destruction of a substance, usually a metal or alloy, by chemical or electro-chemical reaction with its environment. In a normal marine environment, this reaction is electro-chemical.

To those not regularly involved with corrosion engineering problems, the response on viewing a piece of steel covered with a layer of rust might be -- "what went wrong? -- why did the steel rust?" When you consider that nature retains most metals in the form of oxides or sulfides, the answer is apparent. In reverting to its more stable condition, it is normal for steel to react with oxygen in the air to produce rust or iron oxide.

A fundamental corrosion cell consists of:

1. an anode, where corrosion is occuring
2. metallic path to conduct electrons from the anode to the cathode
3. a cathode, where a reaction is occurring
4. an electrolytic path connecting the anode area to the cathode area, to complete the electrical circuit.

Such cells lead to corrosion in many forms such as uniform attack, pitting, intergranular corrosion, galvanic or dissimilar metal corrosion, concentration cell, dezincification or parting, stress and erosion corrosion, cavitation and fretting. Explanations for a few of these will be given where necessary in this paper to illustrate various points.

It is estimated that corrosion in all forms costs approximately ten billion dollars a year in North America. Such economic losses may be classified as:

1. direct losses
2. indirect losses.

Direct losses are those costs incurred in the replacement of corroded structures, including labour. Indirect losses are more difficult to assess; however, the following may be considered:

1. plant shutdown
2. loss of product through corroded systems
3. decreased efficiency
4. contamination of raw material and product.

Loss through hidden costs can result in the use of unnecessarily high corrosion factors leading to over-design. Unknown corrosion rates, or uncertain corrosion control methods lead to equipment which is designed with corrosion allowances greater than those required if more complete data had been available. An adequate knowledge of the corrosive environment and its effect on materials, by careful observations in an experimental test facility, can provide data necessary to simplify component design and reduce material and labour costs.

Galvanic Effects

If the effects of a galvanic couple, or dissimilar metal corrosion, are not anticipated, accelerated attack will lead to failure of the less noble or anodic metal. In pumps and valves provision is made for this form of sea water corrosion by designing a component with greater wall thickness. Materials for valve bodies are either identical or anodic so they will protect the more critical valve components. Galvanic compatibilities of valve and pump trim with body materials in sea water, are shown in Figure 1. Similarly, pump bodies are designed to sacrifice themselves to protect the impeller, thus maintaining maximum pumping efficiency for longer periods of time.

To assist ocean engineers to select materials for sea water applications, a listing called the Galvanic Series of Metals in Flowing Sea Water has been compiled. This series is given as Table 1. Those metals at the top left are anodic, which means they will tend to be sacrificed through galvanic corrosion when coupled to metals further down in the series. The further the metals are apart the greater will be the rate of attack. Metal combinations as galvanically close together as practical will prove effective in slowing down this form of corrosion.

Table 1
Galvanic Series of Metals in Flowing Sea Water (5)

Anodic or Least Noble

Zinc and Aluminum Base Alloys	Nickel Base Alloys
Mild Steel and Cast Iron	Titanium Base Alloys
Austenitic Stainless Steel-Active	Austenitic Stainless Steel - Passive
Ni-Resist*	Hastelloy C**
Copper Base Alloys	Monel* Nickel Copper Alloy
	Graphite
	Cathodic or Most Noble

Maximum corrosive effects occur when the surface area of the more noble metal is greater than that of the less noble component. This is particularly true for fasteners, and to assist designers in selecting materials for fasteners reference may be made to Figure 2.

Where practical, two metals may be effectively separated through the

---

*Inco Registered Trademark.
**Trademark Union Carbide Corporation.

BODY MATERIAL	TRIM		
	Brass or Bronze	Nickel-Copper Alloy 400	Type 316
Cast Iron	Protected	Protected	Protected
Austenitic Nickel Cast Iron	Protected	Protected	Protected
M or G Bronze 70/30 Copper-Nickel Alloy	May Vary (1)	Protected	Protected
Nickel-Copper Alloy 400	Unsatisfactory	Neutral	May Vary (2)
Alloy 20	Unsatisfactory	May Vary	May Vary

(1) Bronze trim commonly used. Trim may become anodic to body if velocity and turbulence keep stable protective film from forming on seat.

(2) Type 316 is so close to nickel-copper alloy 400 in potential that it does not receive enough cathodic protection to prevent it from pitting under low velocity and crevice conditions.

Fig. 1. Galvanic compatibility of valve and pump trim with body materials in sea water[14].

BASE METAL	FASTENER							
	Aluminum (1)	Carbon Steel	Silicon Bronze	Nickel	Nickel Chromium Alloy	Type 304	Nickel Copper Alloy 400	Type 316
Aluminum	Neutral	(2) Comp.	(2) Unsatis-factory	(2) Comp.	Comp.	Comp.	(2) Comp.	Comp.
Steel and Cast Iron	N.C.	Neutral	Comp.	Comp.	Comp.	Comp.	Comp.	Comp.
Austenitic Nickel Cast Iron	N.C.	N.C.	Comp.	Comp.	Comp.	Comp.	Comp.	Comp.
Copper	N.C.	N.C.	Comp.	Comp.	(3) Comp.	(3) Comp.	Comp.	(3) Comp.
70/30 Copper-Nickel Alloy	N.C.	N.C.	N.C.	Comp.	(3) Comp.	(3) Comp.	Comp.	(3) Comp.
Nickel	N.C.	N.C.	N.C.	Neutral	(3) Comp.	(3) Comp.	Comp.	(3) Comp.
Type 304	N.C.	N.C.	N.C.	N.C.	May (4) Vary	(3) Neutral	Comp.	(4) Comp.
Nickel-Copper Alloy 400	N.C.	N.C.	N.C.	N.C.	May (4) Vary	May (4) Vary	Neutral	May (4) Vary
Type 316	N.C.	N.C.	N.C.	N.C.	May (4) Vary	May (4) Vary	May (4) Vary	(4) Neutral

(1) Anodizing would change ratings as fastener.
(2) Fasteners are compatible and protected but may lead to enlargement of bolt hole in aluminum plate.
(3) Cathodic protection afforded fastener by the base metal may not be enough to prevent crevice corrosion of fastener particularly under head of bolt fasteners.
(4) May suffer crevice corrosion, under head of bolt fasteners.
Note: Comp. = Compatible, Protected. N.C. = Not Compatible, Preferentially Corroded.

Fig. 2. Galvanic compatibility of fastener and base metal combinations in sea water[14].

use of insulators. If this cannot be done, it may be necessary to coat both metals. It is important to remember if only the less noble is coated, any break or imperfection in the coating will concentrate severe attack on a relatively small area whereas a similar discontinuity in the coating of the more noble metal will have little effect. Further data, with area effects being taken into account, may be found in UHLIG's Corrosion Handbook (2).

The corrosion resistance of stainless steel is attributed to the formation of a protective oxide surface film. This passive film is formed whenever oxygen, even in trace amounts, is present in the environment. If a small area of the film is broken in service and insufficient oxygen is available to allow it to reform, such as in a crevice, an anodic site is established. Corrosion can now progress because the passive oxide acts cathodically, causing attack at the anodic site usually in the form of a deep pit. In sea water with velocities greater than approximately 5 feet per second, stainless steel remains passive; however, if crevices are present or velocities fall below 5 feet per second and non-uniform marine fouling occurs, anodic sites can be created. Further discussion on this form of corrosion will be found in the section under Marine Environment -- effects of velocity.

Selective Corrosion

Selective corrosion, selective leaching, dezincification, or the preferred term "parting", are used to describe the removal of one or more components of a solid-solution alloy. One such material to be affected by this form of corrosion is cast iron. Carbon in the casting acts as the cathodic member of the cell while iron is anodic and corrodes. As this parting corrosion or graphitization continues, a porous layer of carbon remains on the surface. When the environment contains suspended matter, such as sand, or if there is mechanical abrasion, the carbon layer will be removed and further parting takes place until failure occurs. However, if the porous carbon layer remains in place, it can stimulate galvanic attack on adjacent materials since graphite is at the noble end of the galvanic series. Table 2 summarizes ways of dealing with this type of corrosion.

Ni-Resist* austenitic nickel cast iron is being used in increasing quantities for various components in pumps, valves and fittings. The long-term low corrosion rates of these alloys result from the greater nobility of the alloyed matrix, which decreases the selective corrosion effect between the graphite and the matrix. Typical compositions and mechanical properties of Ni-Resist* irons are shown in Table 3 (3).

---

*Inco Registered Trademark.

Table 2
Selective Corrosion [5]

	Graphitization	Intergranular Corrosion of Austenitic Stainless Steels
Susceptible Component	Cast Iron Ductile Iron	Heat of welding or slow cooling of casting leads to selective attack of stainless steel in sea water but has little effect in marine atmosphere.
Solution to Problem	Use austenitic nickel cast iron	1. Anneal. 2. Use low carbon grades -- 304L, 316L, CF-4, CF-4M. 3. Use stabilized grade -- 347 or 321. 4. Avoid welding after annealing of susceptible grades.

Stress Corrosion Cracking

Stress corrosion cracking results from the combined effects of stress and corrosion. The stress is tensile in nature; in other words it tends to pull the metal apart and may be due to either static or cyclic forces. Where cyclic forces are involved the term corrosion fatigue is applied. Stress corrosion cracking can be initiated by tensile stress which may be internal or residual in nature, or externally applied. Residual stress may result from cold forming, unequal cooling rates during heat treatment or welding operations during manufacture or fabrication. Externally applied loads, which may be encountered during the life of a component, are usually allowed for in the initial design so as to maintain permissible stress levels in service. If, over a period of time, conditions are such that the component is subjected to high loads, the problem of stress corrosion cracking will be increased. Stress cracking may also be experienced if applied and residual stresses combine to produce an excessively high value. Should these loads be cyclic in nature failure will occur, in time, from metal fatigue.

Figure 3 is the result of work by Brown [1] in which yield strengths are related to stress corrosion or hydrogen embrittlement for susceptible high strength steel alloys.

It is generally accepted that in sea water above a temperature approxi-

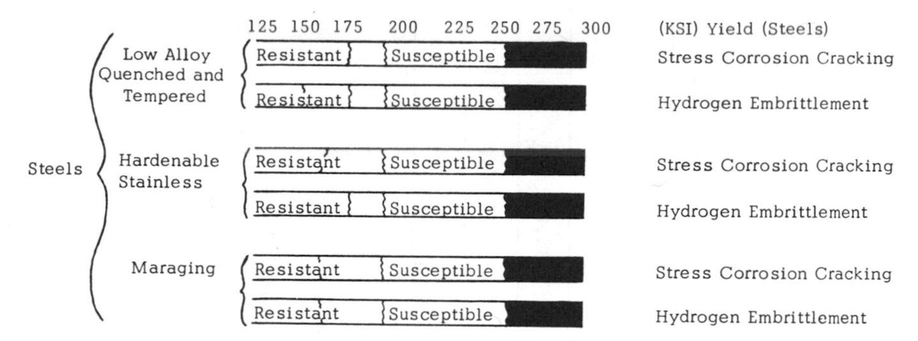

Fig. 3. Stress limits imposed by stress corrosion or hydrogen embrittlement in sea water[14].

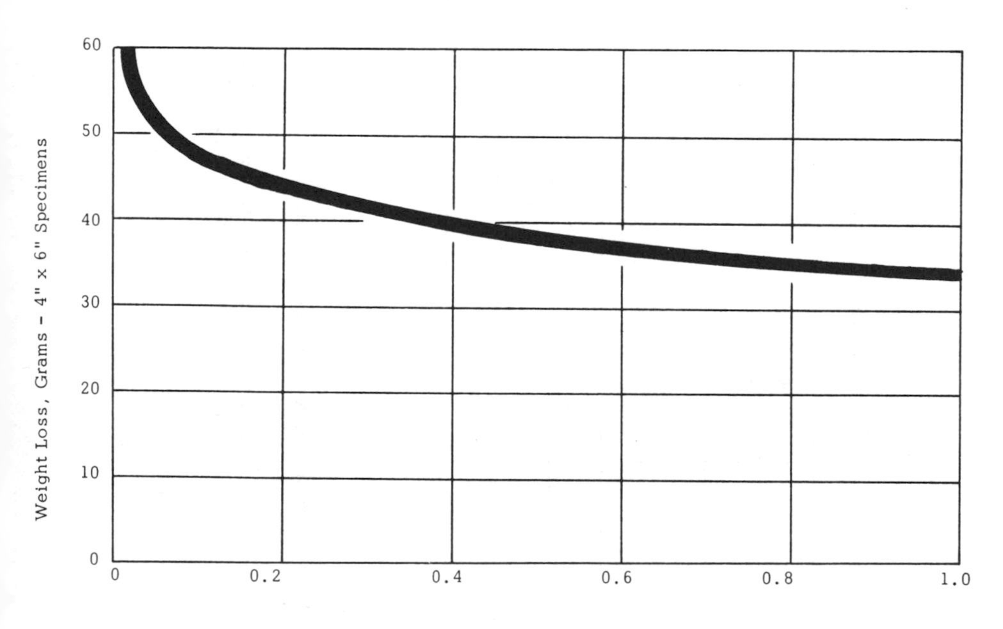

Fig. 4. Effect of copper content on corrosion of steel in marine atmosphere at Kure Beach, N. C. 90 months[5].

Table 3 Ni-Resist* Austenitic Cast Iron(3)

A. Typical Chemical Composition

Name	Type 1[1]	Type 1b	Type 2	Type 2b	Type 3	Type 4	Type 5
Total Carbon	3.00 Max.	3.00 Max.	3.00 Max.	3.00 Max.	2.60 Max.	2.60 Max.	2.40 Max.
Silicon	2.00	2.00	2.00	2.00	2.00	5.50	1.50
Manganese	1.25	1.25	1.25	1.25	0.50	0.50	0.50
Nickel	15.50	15.5	20.00	20.00	30.00	30.00	35.00
Copper	6.50	6.50	0.50 Max.	0.50 Max.	0.50 Max.	0.50 Max.	0.50 Max.
Chromium	2.00	3.00	2.00	4.5	3.00	5.00	.10 Max.[2]

[1]Where the presence of copper offers increased corrosion resistance, Type 1 is preferred.

[2]Where higher hardness, greater strength and added heat resistance are desired, the chromium may be 2.5-3.0% at the expense of increased expansivity.

* Inco Registered Trademark.

continued

Table 3 (continued) B. Typical Mechanical Properties

	Type 1	Type 1b	Type 2	Type 2b	Type 3	Type 4	Type 5
Tensile Strength (psi x 10^3)	25-30	25-35	25-30	25-35	25-35	25-35	20-25
Compressive Strength (psi x 10^3)	100-200	-	100-120	130-160	100-130	80	80-100
Torsional Strength (psi x 10^3)	35-40	-	35-40	45-60	35-45	29	30-35
Torsional Modulus (psi x 10^6)	4.5	-	4.5	5.5	5.0	4.0	4.5
Modulus Elasticity (psi x 10^6) (at 25% of Tensile Strength)	12-14	14-16	15-16.2	15-16.5	15-15.5	15	10.5
Permanent Set Point (psi)	3000	-	3000	-	-	-	-
Transverse Properties (18"): load-(thousand pounds) deflection-inches	 2.0-2.2 .3-.6	 - -	 2.0-2.2 .3-.6	 2.4-2.8 .2-.4	 2.0-2.4 .5-1.0	 1.8 .3-.6	 1.8-2.0 .5-1.0
Hardness, Brinell	130-170	150-210	125-170	170-250	120-160	150-210	100-125
Toughness by Impact (foot-pounds [1])	100	80	100	60	150	80	150

[1] 1.2 arbitration bar unnotched -- struck 3" above supports. (Gray Iron shows 25-35 ft.-lbs.)

mating 150° F, particularly where the metal surface may become salt encrusted, the common austenitic stainless steel grades such as Types 304 and 316 become quite susceptible to stress corrosion cracking. Under ambient temperature conditions such as those found in coastal architectural, and topside marine hardware applications, stress corrosion cracking of Type 304 stainless steel is uncommon. It is also infrequent in pump impellers, propellers and other castings made of CF-4, CF-8, and CF-8M, as well as wrought austenitic stainless steel valve trim, under marine operating conditions.

Fouling

As marine organisms such as barnacles grow, a portion of the surface to which they are attached becomes sealed off from the environment. This may lead to concentration cell corrosion which causes pits to form. In piping systems, fouling by organisms can be such that flow rates are reduced or completely stopped. Table 4 rates fouling resistance of various materials in quiet sea water. It should be noted that, if sea water velocities are above 3 feet per second, organisms will have difficulty in attaching themselves and fouling is minimized. However, if marine organisms become attached during periods of low velocity, and the operating velocity is high, the food supply is rapidly replenished and accelerated growth takes place. As shown in Table 4, ferrous alloys do not possess the necessary resistance to fouling, however, those alloys containing copper in increasing percentages offer better protection. The anti-fouling characteristic of high copper alloys is attributed to the toxicity of the corrosion products to the organisms at corrosion rates which provide at least 5 mg. of copper per sq. dm. a day [5]. This is equivalent to uniform attack at a rate of about one mil. per year.

MARINE ENVIRONMENTS

Of the commonly occurring natural environments the marine atmosphere and sea water are the most corrosive. Therefore, proper application and evaluation of the performance of metals in or near the ocean is of prime importance. Although the mineral content of the major oceans is remarkably uniform, these waters are not simple sodium chloride solutions. They contain many other inorganic salts as well as dissolved gases, suspended solids, organic matter and biological creatures. Researchers may reproduce the mineral composition of sea water for laboratory work but they cannot reproduce the living nature of the ocean; therefore, engineers must go to the sea to solve corrosion problems [6].

Table 4
Fouling Resistance–Quiet Sea Water (14)

Above 3 ft. per sec. continuous velocity (about 1.8 knots) fouling organisms have increasing difficulty in attaching themselves and clinging to the surface, unless already attached securely.

Arbitrary Rating Scale of Fouling Resistance		Materials
90–100	Best	Copper 90/10 copper-nickel alloy.
70–90	Good	Brass and bronze.
50	Fair	70/30 copper-nickel alloy, aluminum bronzes, zinc.
10	Very Slight	Monel* Nickel Copper Alloy 400.
0	Least	Carbon and low alloy steels, stainless steels, nickel-chromium-high molybdenum alloys, titanium.

Atmosphere

One of the basic materials of construction is carbon steel; low cost, strength and availability suggest that it will continue to receive wide use. However, unprotected steel surfaces, subject to exposure in marine and other atmospheres, will soon become completely covered by rust. Such ease of attack and the extensive use of steel account for the greatest destruction of any metal on a tonnage basis. Copson (7) has proposed that the corrosion rate of steel in the atmosphere is directly related to the rate at which a ferrous corrosion product is leached from the barrier film of rust. When one of the products of corrosion is soluble a fully protective film is almost impossible to form.

Improvements in the resistance of carbon steel to corrosion by salt-containing atmospheres is achieved by alloying with elements such as copper, nickel, chromium and phosphorous. Individual beneficial effects of the first three are shown in Figures 4, 5 and 6. Ferrous materials containing these elements when exposed to air, form a corrosion product which is less permeable, more adherent and protective than in the case of carbon steel. The value of such protective films is shown in the time-corrosion curves of Figure 7 where the curve for alloy steel has a relatively low slope. Steels containing these

*Inco Registered Trademark.

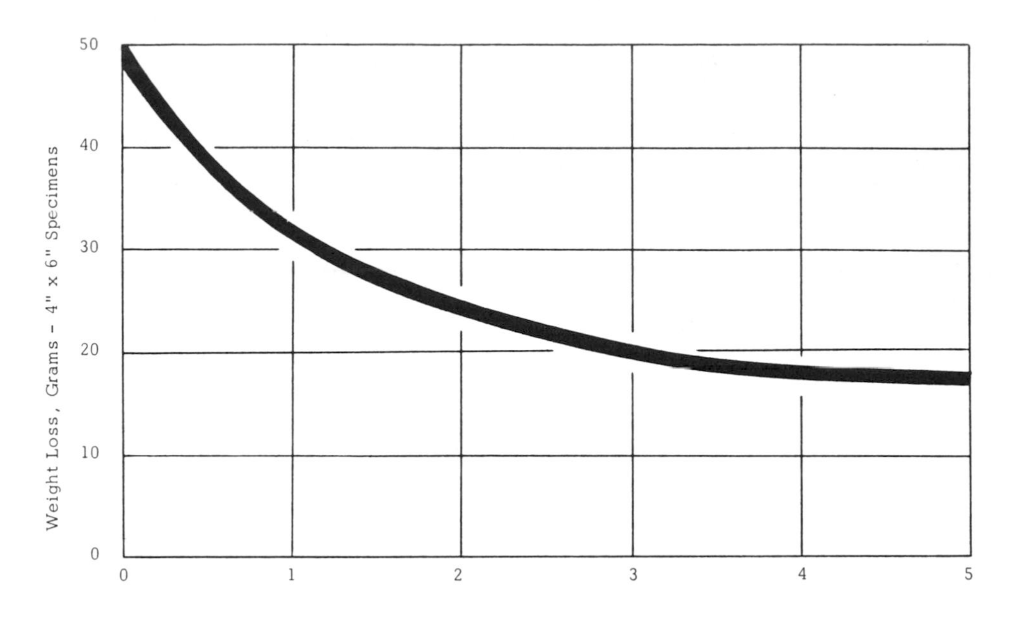

Fig. 5. Effect of nickel content on corrosion of steel in marine atmosphere at Kure Beach, N. C. for 90 months[5].

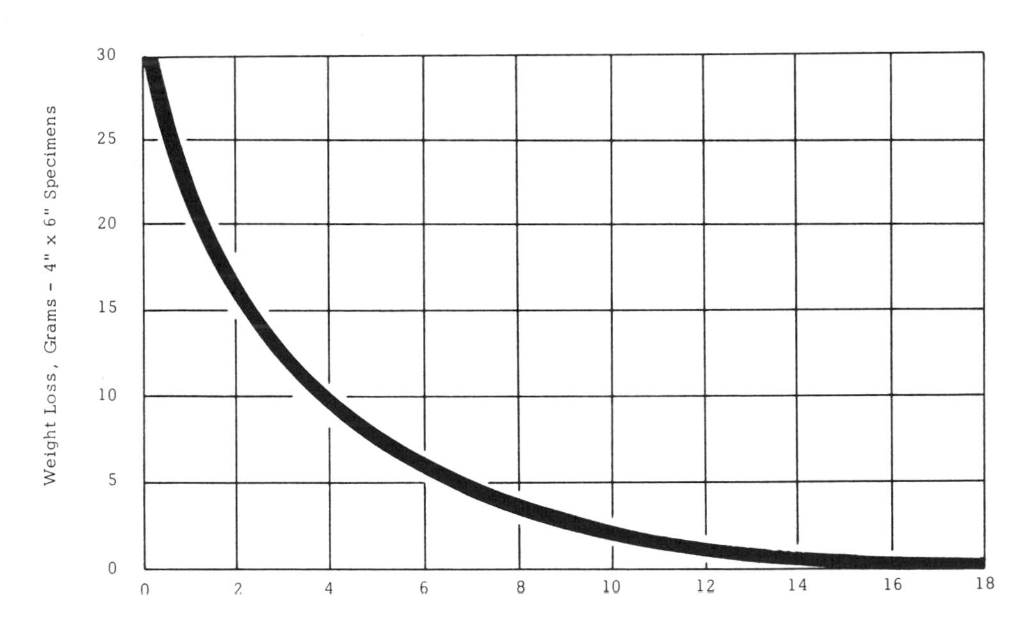

Fig. 6. Effect of chromium content on corrosion of steel in marine atmosphere at Kure Beach, N. C. for 90 months[5].

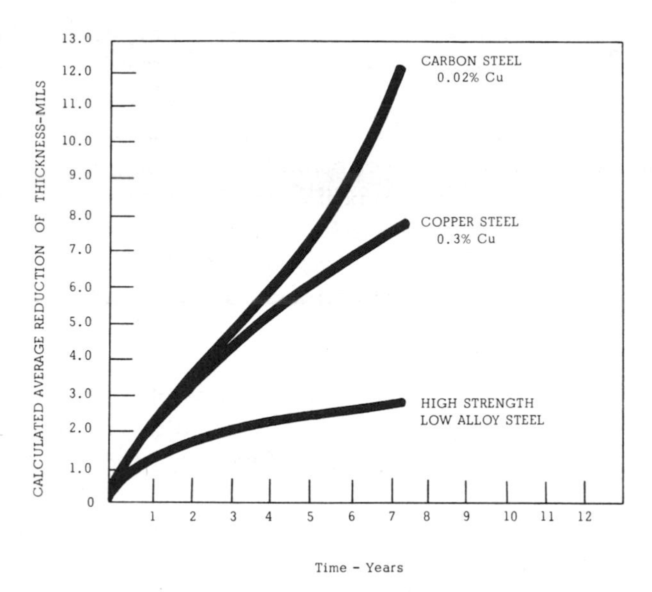

Fig. 7. Time-corrosion curves of steels in marine atmosphere at Kure Beach, N. C.[5]

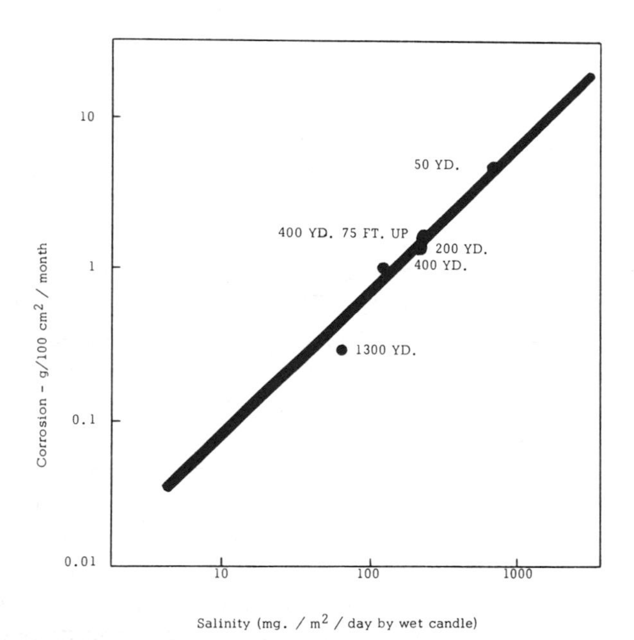

Fig. 8. Relation between corrosion of mild steel, salinity, and distance from ocean[5].

elements are commonly known as high-strength low-alloy steels (HSLA) and, as shown, have two to four times better corrosion resistance than carbon steel in a marine atmosphere. Their improved mechanical properties can lead to designs with thinner cross-sections and substantial reductions in weight.

The resistance of materials to corrosion by salt-containing atmospheres is evaluated by exposing metal panels positioned in specially constructed racks. Material composition, atmospheric conditions such as salt content, temperature and humidity in addition to degree of shelter from sun, wind and rain, and distance from the ocean, are all recorded to assist in the final analysis.

The effect of distance from the ocean is shown in a study made by Ambler and Bain on mild steel in the tropical environment of Nigeria [8]. A summary of their results in Figure 8 indicates there was a tenfold reduction in corrosion within the first 1,000 yards from the ocean. Similarly, Inco's Kure Beach, North Carolina test station revealed typical corrosion rates for steel, on a lot situated 80 feet from the ocean, to be twelve times greater than rates observed on specimens located on the 800 foot lot.

At the above test sites, the temperature range was from 76°F to 92°F in Nigeria accompanied by high humidity and 18° to 92°F at Kure Beach. Typical corrosion rates for steel approximately 50 yards from the ocean at the two locations were 30 mils per year and 15 mils per year respectively. Effects of elevation and distance from the ocean, illustrated in Figure 9, indicates that corrosion reached a maximum at an elevation of about 25 feet, while a short distance inland corrosion increased directly with elevation to a height of 50 feet. HSLA steels under the above conditions have superior corrosion resistance.

The use of protective coatings such as paints, enamels, zinc or aluminum, including sprayed zinc or aluminum with appropriate sealers, will assist in slowing down corrosion of carbon steel. The beneficial effect of alloying with respect to the performance of paint, is shown in Figure 10.

Corrosive conditions which are not satisfied by HSLA steels require the higher alloy content of stainless steels. These are classified according to their predominating microstructure as ferritic, martensitic or austenitic. The austenitic grades are usually preferred in many applications for their good corrosion resistance, particularly where durability, ease of maintenance and appearance are important factors. The protective oxide film necessary for corrosion resistance is readily formed on stainless steels when the surface is exposed to air.

Table 5 lists the composition of the various wrought austenitic stainless steels and their cast alloy counterpart. The more widely-used types and their tensile properties, in the annealed condition, are also shown. These steels possess the ability to retain a substantially unchanged appearance after long exposure to the atmosphere under many conditions. In marine atmospheres the more commonly used 18 chromium, 8 nickel (Type 304) stainless steel may

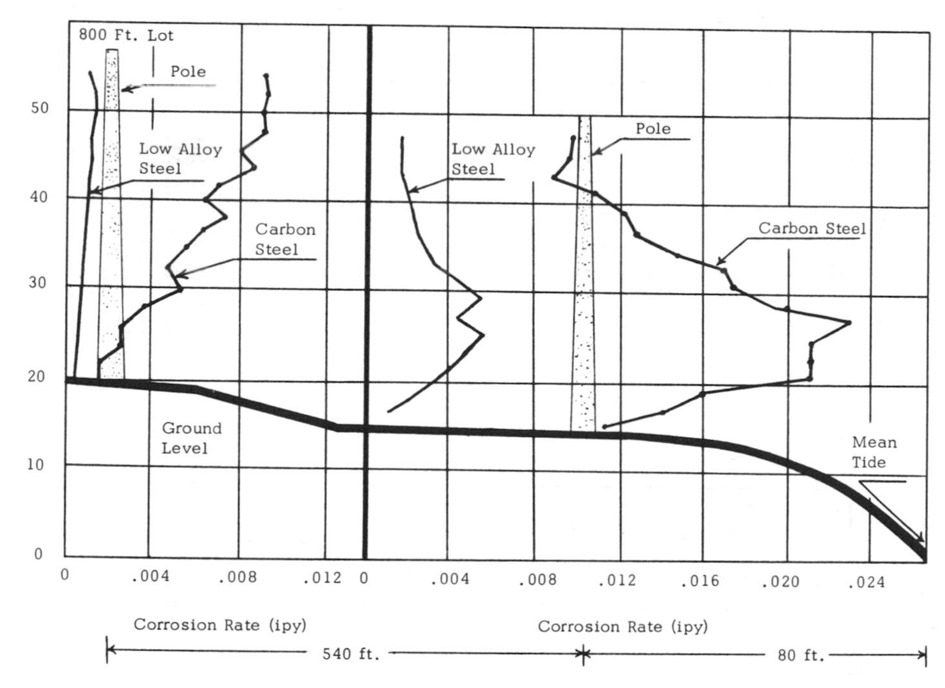

Fig. 9. Effect of height and location on the corrosion of steel exposed in the atmosphere at Kure Beach for 2.1 years(5).

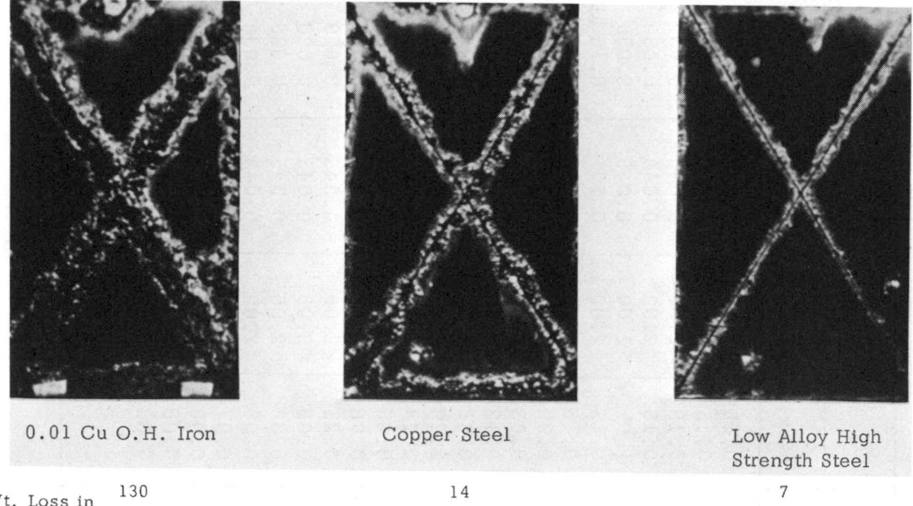

Wt. Loss in Grams of Unprotected Panels in 18 Month: 130 | 14 | 7

Fig. 10. Effect of composition on corrosion resistance of steels and performance of protective coating(5).

Table 5. A. Nominal Composition for Wrought Chromium-Nickel Stainless Steel[9]

AISI Type	Composition - %							
	C Max.	Mn Max.	P Max.	S Max.	Si Max.	Cr	Ni	Mo	Other
201	0.15	6.50	0.060	0.030	1.00	17	4.5	-	N 0.25 Max.
202	0.15	8.00	0.060	0.030	1.00	18	5	-	N 0.25 Max.
301	0.15	2.00	0.045	0.030	1.00	17	7	-	-
302	0.15	2.00	0.045	0.030	1.00	18	9	-	-
302B	0.15	2.00	0.045	0.030	2.50	18	9	-	-
303	0.15	2.00	0.20	0.15 Min.	1.00	18	9	0.060 Max.	-
303Se	0.15	2.00	0.20	0.06	1.00	18	9	-	Se 0.15 Min.
304	0.08	2.00	0.045	0.030	1.00	19	10	-	-
304L	0.03	2.00	0.045	0.030	1.00	19	10	-	-
305	0.12	2.00	0.045	0.030	1.00	18	11	-	-
308	0.08	2.00	0.045	0.030	1.00	19	11	-	-
309	0.20	2.00	0.045	0.030	1.00	23	13	-	-
309S	0.08	2.00	0.045	0.030	1.00	23	13	-	-
310	0.25	2.00	0.045	0.030	1.50	25	20	-	-
310S	0.08	2.00	0.045	0.030	1.50	25	20	-	-
314	0.25	2.00	0.045	0.030	2.25	24	20	-	-
316	0.08	2.00	0.045	0.030	1.00	17	12	2.50	-
316L	0.03	2.00	0.045	0.030	1.00	17	12	2.50	-
317	0.08	2.00	0.045	0.030	1.00	19	13	3.50	-
D319	0.07	2.00	0.045	0.030	1.00	18	13	2.50	-
321	0.08	2.00	0.045	0.030	1.00	18	10	-	Ti 5 x C Min.
347	0.08	2.00	0.045	0.030	1.00	18	11	-	Cb-Ta 10 x C Min.
348	0.08	2.00	0.045	0.030	1.00	18	11	-	Cb-Ta 10 x C Min.; Ta 0.10 Max.; Co 0.20 Max.

continued

Table 5 (continued) B. Nominal Composition for Cast Chromium-Nickel Stainless Steel

Cast Alloy Designation	Wrought Alloy Type [1]	Composition - %								
		C max	Mn max	P max	S max	Si max	Cr	Ni	Mo	Other
CD-4MCu	-	0.04	1.00	0.04	0.04	1.00	26	5.5	2.0	Cu 3.0
CE-30	-	0.30	1.50	0.04	0.04	2.00	28	9.5	-	-
CF-3	304L	0.03	1.50	0.04	0.04	2.00	19	10	-	-
CF-8	304	0.08	1.50	0.04	0.04	2.00	20	9.5	-	-
CF-20	302	0.20	1.50	0.04	0.04	2.00	20	9.5	-	-
CF-3M	316L	0.03	1.50	0.04	0.04	1.50	19	11	2.5	-
CF-8M	316	0.08	1.50	0.04	0.04	1.50	20	10.5	2.5	-
CF-12M	316	0.12	1.50	0.04	0.04	1.50	20	10.5	2.5	-
CF-8C	347	0.08	1.50	0.04	0.04	2.00	20	10.5	-	Cb 8 x C min, 1.0 max. or Cb-Ta 10 x C min., 1.35 max.
CF-16F	303	0.16	1.50	0.17	0.04	2.00	20	10.5	1.5 max.	Se 0.27
CG-8M	317	0.08	1.50	0.04	0.04	1.50	20	11	3.5	-
CH-20	309	0.20	1.50	0.04	0.04	2.00	24	13.5	-	-
CK-20	310	0.20	1.50	0.04	0.04	2.00	25	20.5	-	-
CN-7M	-	0.07	1.50	0.04	0.04	1.50	20	29	2.5	Cu 3.5

continued

[1] Wrought alloy type numbers are included only for the convenience of those who wish to determine corresponding wrought and cast grades. The chemical composition ranges of the wrought materials differ from those of the cast grades.

Table 5 (continued) C. Typical Tensile Properties of Annealed Stainless Steel Sheet and Plate[3]

Type	Yield Strength - p.s.i.		Ultimate Strength - p.s.i.		% Elongation in 2"	
	Sheet	Plate	Sheet	Plate	Sheet	Plate
301	40,000	40,000	110,000	105,000	60	55
304	35,000	30,000	85,000	85,000	50	60
304L	28,000	28,000	75,000	75,000	50	50
316	40,000	35,000	90,000	85,000	50	55
316L	32,000	32,000	75,000	75,000	50	50

Typical Tensile Properties of Annealed Stainless Steel Castings[3]

ACI Designation	Yield Strength - p.s.i.	Ultimate Strength - p.s.i.	% Elongation in 2"
CF-3	32,000	72,000	70
CF-8	37,000	77,000	55
CF-3M	40,000	83,000	50
CF-8M	42,000	80,000	50

develop superficial staining in the form of scattered patches of faint rust film, with little evidence of attack beneath the film. This discolouration develops during the first few months of exposure after which it does not appear to progress much further. Rust films or stains have been removed easily after as long as fifteen years, to reveal a bright surface which has suffered very little attack. If periodic cleaning is not practical an alternative choice would be a molybdenum-containing Type 316 stainless steel which has greater resistance to staining. Although intensity of staining is markedly diminished as distance from the ocean increases, none of the austenitic stainless steels suffer significant structural damage after exposure of 1 and 15 years as recorded in Tables 6 and 7 respectively [(9)]. The surface finish is also an important factor. Atmospheric corrosion tests carried out in marine sites at Biarritz, France and Walvis Bay, South Africa suggest that panels with a number of 3 or 2 B finish will exhibit a faint rust film while electro-polished material revealed, for the most part, only a thin transparent film [(10)].

The designer should be aware that other materials such as aluminum, copper or steel coupled or attached to stainless steel may suffer from galvanic corrosion. This form of attack can occur when dissimilar metals are electrically connected or coupled to each other and continually wetted by the salt spray. The voltage or potential difference of the couple permits an electrical current to flow which causes the less resistant or anodic metal to corrode or sacrifice itself while the cathodic stainless steel is protected. Table 8 shows the result of tests in which Types 304 and 316 stainless steels were coupled to a number of other metals. The tests were carried out for a period of five years and show the ratios of corrosion rates of coupled to uncoupled specimens [(9)]. The effects of galvanic corrosion promoted by salt spray on exposed surfaces would be decreased by the action of sun and rain. The rain washes the salt from the metal surface and prevents salt accumulation while sunlight assists by providing a drying action. In locations where metals are sheltered the beneficial effects of sun and rain will not be realized and some corrosion may be expected.

Splash Zone

Maximum rates of corrosion occur in the splash zone because the rust films which form have little or no opportunity to dry and develop the protective characteristics necessary to lower rates of attack, and to the high content of dissolved oxygen in splashing sea water. This effect compares with the continuously submerged surface, where a combination of corrosion products and marine growths retard diffusion of oxygen to the corroding surfaces.

Upgrading to alloy steels will assist in overcoming corrosion in the splash zone. For example, HSLA steel containing 0.5% nickel, 0.5% copper and 0.12% phosphorous has shown five times better corrosion resistance than carbon steel [(11)]. Figure 11 compares the distribution of corrosion for AISI

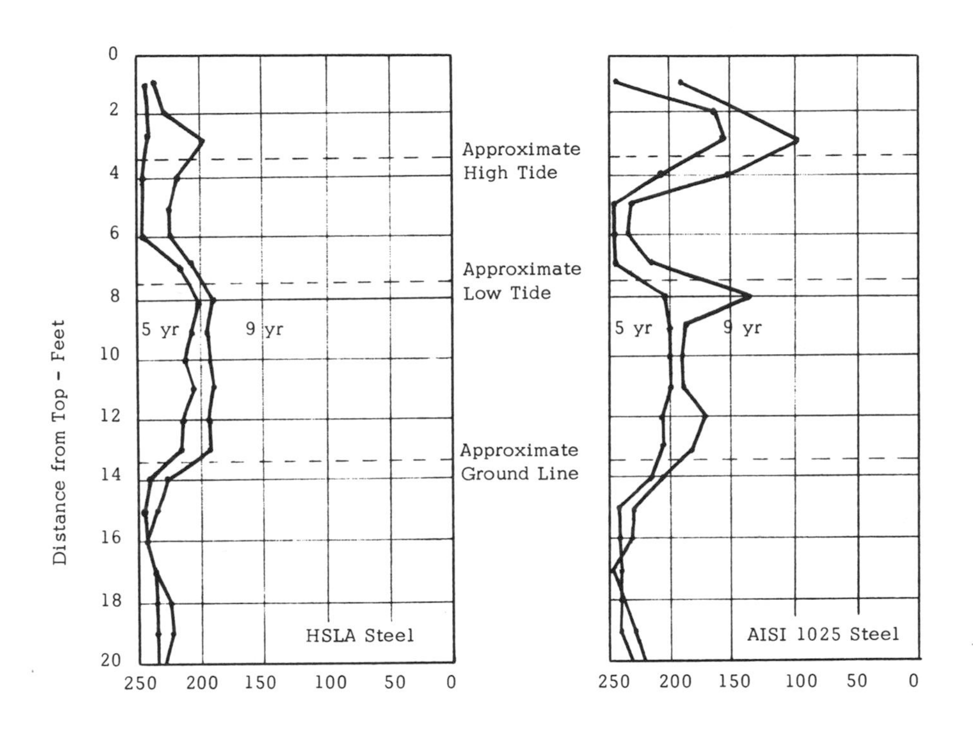

Fig. 11. Distribution of corrosion for HSLA steel and AISI 1025 plain carbon steel[6].

Table 6. Appearance of Austenitic Stainless Steels After Exposure in a Marine Atmosphere for One Year, 800 ft. from Ocean, Kure Beach, N. C.[9]

AISI Type	Appearance
301	Scattered faint rusting, several well developed rust spots.
302	Scattered general rusty discoloration over entire surface.
304	Scattered faint rusting.
308	Scattered faint rusting, about the same as Type 304.
309	Scattered faint rusting with several well developed rust spots but less than with Type 304.
310	Scattered faint rusting, about the same as on Type 316 and 317.
316	Scattered faint rusting, much less than on Type 304.
317	Scattered faint rusting, much less than on Type 304.
321	Scattered faint rusting with several well developed rust spots.
347	Scattered faint rusting with several well developed rust spots.

International Nickel Company data.

Table 7. Performance of Stainless Steels in a Marine Atmosphere 15 Years, 800 ft. from Ocean, Kure Beach, N. C.[9]

AISI Type	Average Corrosion Rate, mpy	Average Depth of Pits, mils	Appearance*
301	<.001	1.6	Light rust and rust stain on 20% of surface.
302	<.001	1.2	Spotted with rust stain on 10% of surface.
304	<.001	1.1	Spotted with slight rust stain on 15% of surface.
308	<.001	1.6	Spotted by rust stain on 25% of surface.
309	<.001	1.1	Spotted by slight rust stain on 25% of surface
310	<.001	0.4	Spotted by slight rust stain on 20% of surface.
316	<.001	1.0	Extremely slight rust stain on 15% of surface.
317	<.001	1.1	Extremely slight rust stain on 20% of surface.
321	<.001	2.6	Spotted with slight rust stain on 15% of surface.
347	.001	3.4	Spotted with moderate rust stain on 20% of surface.

* All stains easily removed to reveal bright surface

International Nickel Company data

Table 8 Galvanic Corrosion of Several Materials When Coupled to Types 304 and 316 Stainless Steels[9]

(Ratios of Corrosion Rates of Coupled to Uncoupled Specimens)

Material	Exposed to Atmospheres at:							
	New York, N.Y.		Altoona, Pa.		State College, Pa.		Kure Beach, N.C.	
	Coupled to:		Coupled to:		Coupled to:		Coupled to:	
	304	316	304	316	304	316	304	316
Aluminum	3.0	3.9	0.75	0.75	2.5	1.5	1.1	0.75
Aluminum Alloy 2024	4.5	4.6	2.6	2.9	2.1	1.0	2.3	2.0
Aluminum Alloy 5053	1.8	1.8	2.5	3.8	2.0	1.7	5.2	4.8
Copper	1.9	1.9	-	2.1	2.2	2.3	1.5	1.9
Architectural Bronze	1.5	1.5	1.4	1.4	1.8	1.8	1.4	1.5
Lead	1.8	2.1	2.3	1.7	2.5	2.6	2.1	2.2
Zinc	2.2	2.2	2.5	2.5	2.2	2.2	1.8	2.0
Monel* Nickel Copper	1.5	1.5	1.7	1.6	2.3	1.7	1.9	1.9
Mild Steel	1.4	1.3	2.0	1.2	1.5	1.3	2.2	2.2

*Inco Registered Trademark

1025 plain carbon steel and HSLA steel piling which is partially immersed in sea water. The splash zone, approximately 3 feet from the top, is shown to be more corrosive to carbon steel than to HSLA steel.

Protective coatings may be used to advantage; however, the demanding requirements of surface preparation could lead to unfavourable results, particularly during attempts to recoat installed structures. Non-ferrous metal sheathing can also provide an effective barrier to corrosion in the splash zone. An off-shore drilling platform, for example, has nickel-copper alloy sheathing which is providing satisfactory corrosion protection after a period approximating 20 years.

Tidal Zone

Metal surfaces in the tidal zone become wetted by highly aerated sea water as the tide rises while the surfaces below low tide remain in contact with practically oxygen-free water. This differential aeration can form a cathodic area in the tidal zone with respect to the submerged anodic surfaces. An internally generated current results which is sufficient to protect the metal in the tidal zone. Additional protection for iron and steel may be realized by using submerged sacrificial anodes which are made of magnesium or zinc.

Referring again to Figure 11 and the tidal zone area, it is observed that benefits can be realized in the selection of HSLA steel over plain carbon steel. The use of coatings will offer additional protection while nickel-copper alloy suggested for the splash zone will provide a more effective measure of corrosion resistance.

Submerged Zone

Attack within the submerged zone is governed by the rate of diffusion of oxygen through layers of rust or other oxide films and marine organisms. Corrosion rates approximate 5 mils per year for carbon steel, however, pollution may produce greater rates of attack. Accelerated corrosion, in the form of generally rounded or wide flat-bottom pits, may be at rates several times that of uniform attack or general wastage. The pitting is due to anodic points which form because of differences in the metal surface or the environment, or through metallurgical or mechanical differences in the metal itself such as impurities. For example, differences in the metal surface could result from improper cleaning or incomplete removal of mill scale on a hot-rolled steel product. This form of corrosion is often difficult to detect because, as often happens, pits are relatively small and usually covered by rust. Pitting is one of the most destructive forms of corrosion and causes more unexpected failures than any other form because of its localized intense type of attack.

Materials selection for this zone demands particular attention because of the large quantities of metal subject to salt water exposure. Ocean engineers look to proper design, metal alloying and cathodic protection, to mention a few, to assist in solving corrosion problems. Proper design will go a long way to minimize corrosion and extend useful life by eliminating turbulence from improper construction and joining techniques. In most fabricated structures welded joints have preference over those which are riveted in order to avoid crevices which can lead to concentration-cell corrosion. This form of corrosion is due to differences in the environment and can be of two common types -- ion cell and oxygen cell. The mechanisms of corrosion within or around crevices are much the same as those involved in pitting. Consequently, susceptibility to crevice corrosion parallels susceptibility to pitting as indicated by Figure 12 and summarized in Table 9. The order of merit in this table is from left to right by classes and top to bottom in each class.

It is worth remembering that all corrosion data are useful for guidance but are never substitutes for actual experience based directly on specific service conditions. Such data will be useful to indicate first approximations of durability and performance. Further reference should be made to information from manufacturers, published literature and, preferably, to test data in the environment in question.

Proper design of components is important to avoid highly localized stresses leading to stress-corrosion cracking. Internal stresses act together with corrosion to form cracks through grain boundaries (intergranular) or across grains (transgranular) resulting in failure. Stress-relieving heat treatments will decrease chances of this form of corrosion occurring. Excessive stresses produced during equipment operation may also result in this form of attack, but knowledge of the stresses which may be encountered in operation can be offset by providing greater section thickness.

Selection and performance of materials for marine service are based on corrosion resistance and mechanical strength -- the two principal properties necessary to provide long life. Copper, nickel, and chromium when added to plain carbon steel as a single alloying element, or combinations of these and other elements, provide a list of ferrous alloys to meet many of the demands of ocean engineers.

Sacrificial zinc or magnesium anodes, or impressed direct current are two methods commonly used to obtain cathodic protection. The latter technique is a more scientific method, utilizing an external power source and inert materials for anodes such as high-silicon iron, platinum, graphite, and platinized titanium or tantalum. In some cases impressed currents are supplemented by the use of protective coatings but, when this is done, adjustments in the current level are necessary to compensate for coating deterioration or its accidental removal.

Two additional methods are available to assist those charged with selecting corrosion resistant materials. The first is alloy cladding. This consists of bonding a material such as stainless steel to one or both sides of a carbon or low alloy steel. The principal objective of such a procedure is to provide,

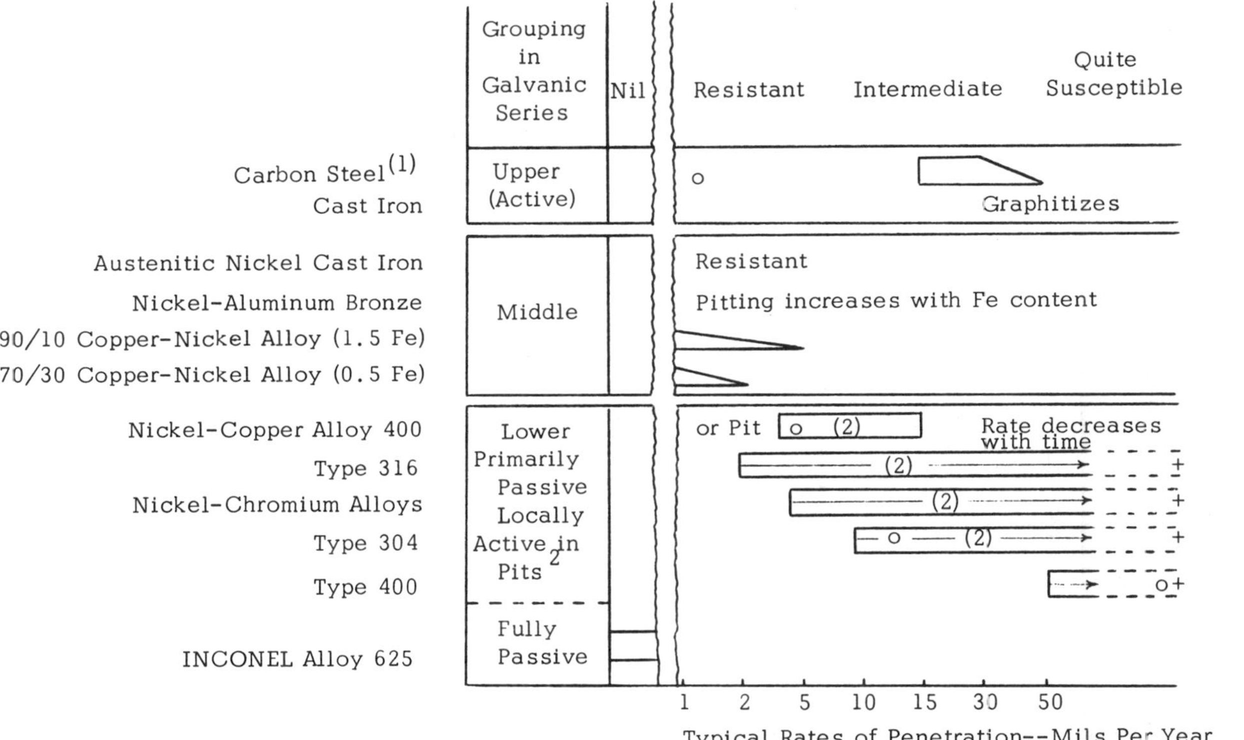

° Data from results of early tests at depths of 2,300 to 5,600 feet.

(1) Shallow round bottom pits.

(2) As velocity increases above 3 fps pitting decreases. When continuously exposed to 5 ft. per sec. and higher velocities these metals, except Type 400 series, tend to remain passive without any pitting over the full surface in the absence of crevices.

Fig. 12. Rates of pitting of metals immersed in quiet sea water - velocity less than 2 ft. per sec.[14]

Table 9. Tolerance for Crevices Immersed in Quiet Sea Water[5]

(Velocity less than 2 feet per second)

Inert	Useful Resistance			
Best	Good	Neutral	Less	Crevices Tend to Initiate Deep Pitting
HASTELLOY*	90/10 copper-nickel alloy (1.5 Fe)	Austenitic Nickel Cast Iron	Incoloy*** Alloy 800	Type 316
Alloy C	70/30 copper-nickel alloy (0.5 Fe)	Cast iron	Alloy 825	Nickel-chromium alloys
Titanium?**	Bronze	Carbon Steel	Alloy 20	Type 304
Inconel*** alloy 625	Brass		Monel*** nickel-copper alloy Copper	Type 400 series

* Trademark Union Carbide Corporation
** Susceptible in hot sea water
*** Inco Registered Trademark

at low cost, the desirable properties of stainless steel and the backing steel for applications where full-gauge alloy construction is not required. While stainless cladding furnishes the necessary resistance to corrosion, abrasion, or oxidation, the backing steel contributes structural strength and improves fabricability and thermal conductivity of the composite.

The second method, which could be considered a form of cladding, involves placing a continuous layer or layers of overlapping weld beads of stainless steel on a carbon steel slab or fabricated shape. Although the weld-deposited slab can be rolled to yield clad steel plates, it is more common to weld-overlay in excess of the required cladding thickness, and machine to obtain the desired surface finish and final thickness [12].

Effects of Velocity

Velocity, as Copson has shown, will affect the corrosion rate of steel [13]. The effects of velocity are as follows:

1. Slight motion makes the environment more uniform thereby lowering the possibility of local attack.
2. Increased motion thins quiescent layers so that corrosives can more readily attack the metal surfaces.
3. Higher velocities may change the nature of the corrosion products and barrier film.

Figure 13 shows typical corrosion rates of common marine materials immersed in quiet sea water defined as velocities less than 2 feet per second. It can be observed that ferrous materials, such as carbon steel, corrode in a uniform manner compared to stainless steel Types 304 and 316, which are rated inert except for deep pitting. It is quite common in quiet sea water, in the presence of marine organisms or other deposits, for 90% to 99% of the surface of these relatively inert alloys to remain virtually unaffected except for a few deep pits. In some cases these pits penetrate the full wall thickness. For most materials listed in Figure 13, average corrosion rates are less than 3 mils per year, which ensures long service if metal loss is uniform and not localized [14].

With motion increasing in the range 3 to 6 feet per second, fouling diminishes, pitting of the more noble alloys slows down and even ceases. As velocities continue to increase the corrosion barrier film is stripped from one after another of the steel and copper base alloys, while the stainless steels and nickel base materials remain inert. Figure 14 shows the effect of sea water velocity on corrosion of steel at ambient temperature and suggests that, once the barrier film is fully stripped away, carbon steel corrodes at several times the rate observed in quiet sea water. Conditions that require materials to withstand high velocities and periods of stagnant water may be selected by referring to Figures 12 and 15 and Table 9.

The passive oxide barrier film that forms on more inert materials, such

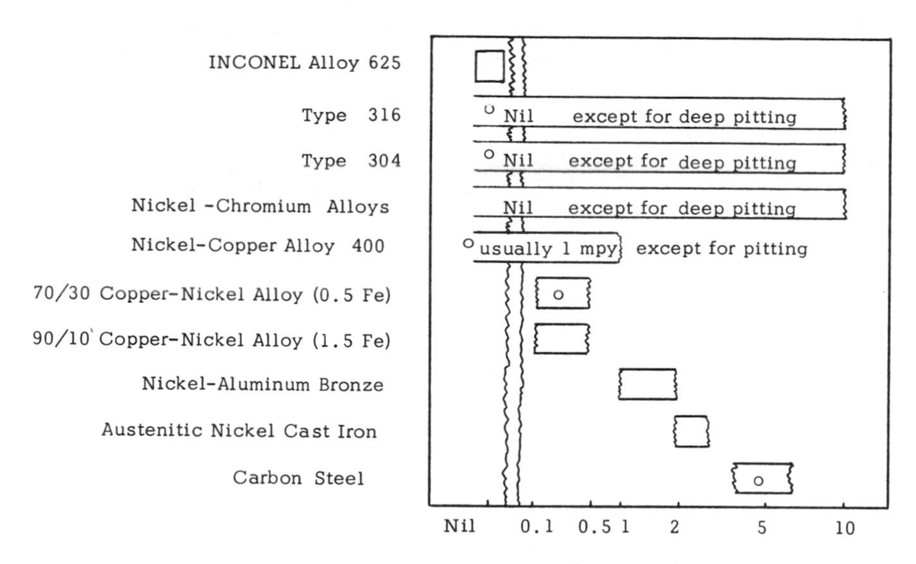

Fig. 13. Rates of general wasting of metals immersed in quiet sea water - velocity less than 2 ft. per sec.[14]

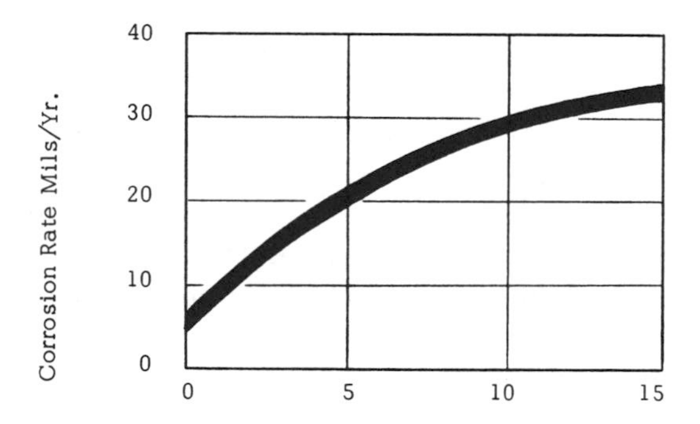

Fig. 14. Effect of sea water velocity on corrosion of steel at ambient temperature[14].

as stainless steel, will be adequately maintained by flowing sea water which contains 6 to 10 parts per million oxygen at temperatures from 40°F to 90°F. In order to prevent deposits in the form of debris, silt and fouling organisms from collecting, a minimum velocity of 5 feet per second has been suggested [2]. Table 10 indicates the effect of low sea water velocity on pitting of welded Types 316 and 310 stainless steels. At extremely high water velocities in the order of 140 feet per second, 30 day tests of Types 304 and 316 stainless steels indicated corrosion rates of about 0.2 mils per year [15].

Table 10
Effect of Sea Water Velocity on Pitting of
Welded Types 316 and 310 Stainless Steels [15]
(1257 days in test)

Material	In Flume at 4 ft./sec.			In Basin at 0 ft./sec.		
	Number of Pits	Maximum Depth, mils	Average Depth,* mils	Number of Pits	Maximum Depth, mils	Average Depth,* mils
316 Plate	0	0	0	87	78	38
Weld	0	0	0	47	130	76
310 Plate	0	0	0	19	110	38
Weld	0	0	0	23	250**	120

EQUIPMENT

Problems of Fabrication

Where fabrication techniques call for welding, selection of suitable material will be important. Welding an 18 chromium - 8 nickel (Type 304) structure can lead to the formation of a "weld decay" band which corrodes preferentially with respect to the balance of the structure when immersed in sea water. This weld decay band is a zone of sensitized alloy in which heat from the welding operation has caused carbon to migrate to grain boundaries precipitating out as chromium carbide. As a result, the sensitized area possesses a lower chromium content than the matrix and becomes anodic so that intergranular corrosion takes place. This form of corrosion can be controlled by three generally used methods:

1. lowering the carbon content of the alloy to less than 0.03%

---

*Average of 10 deepest pits.
**Perforated from one side (thickness of specimen = 0.25 in.).

2. employing post-weld heat treatment (solution quenching)
3. stabilizing the alloy through additions of columbium or titanium.

Lowering the carbon content will reduce the degree of chromium carbide formation at the grain boundaries. Wrought alloys such as 304L and 316L are those containing low carbon levels, the complementary cast alloys being CF-3 and CF-3M respectively.

Post-weld heat treatments which follow forming operations to stress relieve and to obtain maximum corrosion resistance are usually difficult to carry out on fabricated structures. For instance, if thin and thick sections are incorporated in the same structure, the heat treatment may cause distortion on cooling due to stresses produced by temperature differences between sections. It is important that maximum cooling of a heat-treated structure be carried out to prevent reprecipitation of carbides and sensitization leading to intergranular corrosion. Table 11 shows the type of stress-relieving heat treatments which are used for austenitic stainless steels [16]. Because of the varying degrees of stress relief required, the number of grades of stainless steel, and the large number of service requirements, many different treatments are available for optimum results.

Problems associated with heat treatment can be overcome by selecting a stainless steel with a stabilized composition such as Types 321 and 347. These alloys are manufactured with additions of titanium and columbium respectively. To obtain maximum corrosion resistance they are stabilized by annealing to inactivate the carbon through formation of carbides of titanium and columbium and, in so doing, prevent chromium depletion.

Pumps

For pumps and other high velocity applications where there is severe turbulence, metals fall into two different groups:

1. those that are velocity limited, such as steel and copper base materials
2. those that are not velocity limited, such as stainless steels and many nickel base alloys.

Metals that are not velocity limited suffer virtually no metal loss from velocity effects or turbulence, short of cavitation conditions or from sand erosion. Cavitation or cavitation-erosion is caused by the formation and collapse of vapour bubbles in a liquid at a solid-liquid interface. For example, bubbles may occur on the trailing edge of an impeller at the suction side, where pressures are decreased and remain so until the impeller blade passes the outlet of the pump where the bubbles collapse. The forces involved in this action produce pressures as high as 60,000 p.s.i. and cause metal surfaces to cold work. Through a combination of recorded experience and results of accelerated cavitation-erosion tests in sea water, a number of materials have been found which show resistance to this form of attack. Table 12 shows the rating of a number of these materials with a rating of 1 providing the best

Table 11. Suggested Stress-Relieving Treatments for Austenitic Stainless Steels[16]

Anticipated Service Environment, or Other Reason for Treatment	Suggested Thermal Treatment* (entered in order of decreasing preference)		
	Extra-low-carbon grades, such as 304L and 316L	Stabilized grades, such as 318, 321, and 347	Unstabilized grades, such as 304 and 316
Severe Stress Corrosion	A, B	B, A	(a)
Moderate Stress Corrosion	A, B, C	B, A, C	C (a)
Mild Stress Corrosion	A, B,C, E, F	B, A, C, E, F	C, F
Remove peak stresses only	F	F	F
No stress corrosion	None required	None required	None required
Intergranular corrosion	A, C (b)	A, C, B (b)	C
Stress relief after severe forming	A, C	A, C	C
Relief between forming operations	A, B, C	B, A, C	C (c)
Structural soundness (d)	A, C, B	A, C, B	C
Dimensional stability	G	G	G

* Key to letter designations of treatments:

A - Anneal at 1950 to 2050 F; slow cool
B - Stress relieve at 1650 F; slow cool.
C - Anneal at 1950 to 2050 F; quench (e)
D - Stress relieve at 1650 F; quench (e)
E - Stress relieve at 900 to 1200 F; slow cool.
F - Stress relieve at below 900 F; slow cool.
G - Stress relieve at 400 to 900 F; slow cool.
(Usual time, 4 hours per inch of section)

(a) To allow the optimum stress-relieving treatment, the use of stabilized or extra-low-carbon grades is recommended.

(b) In most instances, no heat treatment is required, but where fabrication procedures may have sensitized the stainless steel the heat treatments noted may be employed.

(c) Treatment A, B or D also may be used, if followed by treatment C when forming is completed.

(d) Where severe fabricating stresses coupled with high service may cause cracking. Also after welding heavy sections.

(e) Or cool rapidly.

Table 12. Order of Resistance to Cavitation Erosion in Sea Water(5)

Rating	Alloy or Class	Example of Class
1	Stellite	
2	Age Hardened Nickel Chromium Alloys	Inconel* Nickel Chromium Alloy 718
2	Precipitation Hardened Stainless Steels	17-4 PH
2	Austenitic Stainless Steels	18 Cr - 8 Ni
2	Matrix Stiffened Nickel Chromium Alloys	Inconel* Nickel Chromium Alloy 625
2	Titanium Alloys	
3	Nickel Chromium Alloys	Inconel* Nickel Chromium Alloy 600
3	Age Hardened Nickel Copper Alloy	Monel* Alloy K500
4	Nickel Copper Alloy	Monel* Alloy 400
5	Nickel Aluminum Bronze	Ni-Bral*
6	Copper Nickel Alloy	70% Cu - 30% Ni
7	Manganese Bronze	
8	Tin Bronze	Composition-G (10% Sn)
9	Valve Bronze	Composition-M (6% Sn)
10	Austenitic Flake Graphite Cast Iron 3% Cr	Ni-Resist* Type 3
11	Austenitic Spheroidal Graphite Cast Iron	Ductile Ni-Resist*
12	Heat Treated Alloy Steel	Hy-80
13	Alloyed Ductile Cast Iron	1-1/2% Ni - 0.5% Mo
14	Ductile Cast Iron	
15	Alloyed Flake Graphite Cast Iron	1-1/2% Ni- 0.5% Mo
16	Cast Carbon Steel	
17	Flake Graphite Cast Iron	
18	Aluminum Alloys	(6061-T-6 and 356-T-6)

* Inco Registered Trademark

resistance. Improved pump design and smoother finishes on internal components will assist in overcoming cavitation problems.

A selection of materials for sea water pumps is shown in Table 13 [17]. The high velocity and turbulence associated with pumps preclude the use of carbon and low alloy steels for this application because the less adherent barrier films on these materials offer little protection.

Pipe and Tube

Effects of local turbulence in pipe and tube are taken into account in Figure 16.

Pipelines and other components handling moving sea water suffer from erosion-corrosion or impingement attack. Water impinging on the protective film on a metal surface erodes it and, if the rate of erosion exceeds the rate of repair, the film will be ruptured and the water then corrodes the exposed area of metal surface. This attack occurs when a critical water velocity is exceeded. Ferrous alloys such as stainless steel Types 304 and 316, provide the necessary resistance to velocity effects as long as velocities are maintained above a flow rate approximating 5 feet per second. Below this level, pitting and fouling become problems which have to be considered. For condensers or coolers, such as those used to cool alternators, alloys such as 70/30 or 90/10 copper-nickel would provide the necessary corrosion protection.

TIDAL POWER PROJECTS

Passamaquoddy

Various forms of corrosion and their effect on ferrous alloys were studied under this project. Although the project was not carried through to completion, the report on the study dated October, 1959 [18] is instructive in recommending various ferrous materials for certain major components. Turbine blades, hub, discharge ring, speed ring wearing surfaces adjacent to the wickets both top and bottom, the top and bottom plates of the wicket gates, and the top four feet of the draft tube liner, would be 18% nickel, 8% chromium, 3% molybdenum corrosion resistant steel, designated as AISI Type 316. Nickel-copper alloys were selected for the sleeve on wicket gate shaft extensions as well as the bushings. A similar alloy was used as sleeve material under the packing gland area. The turbine runner cone, head covers, speed ring, draft tube liner, wickets and stay vanes were to be fabricated from structural-grade carbon steel. All exposed carbon steel parts were to be painted with a standard 4-coat vinyl system, and further protected by an impressed-voltage, cathodic-protection system. Anodes recommended were cold-rolled

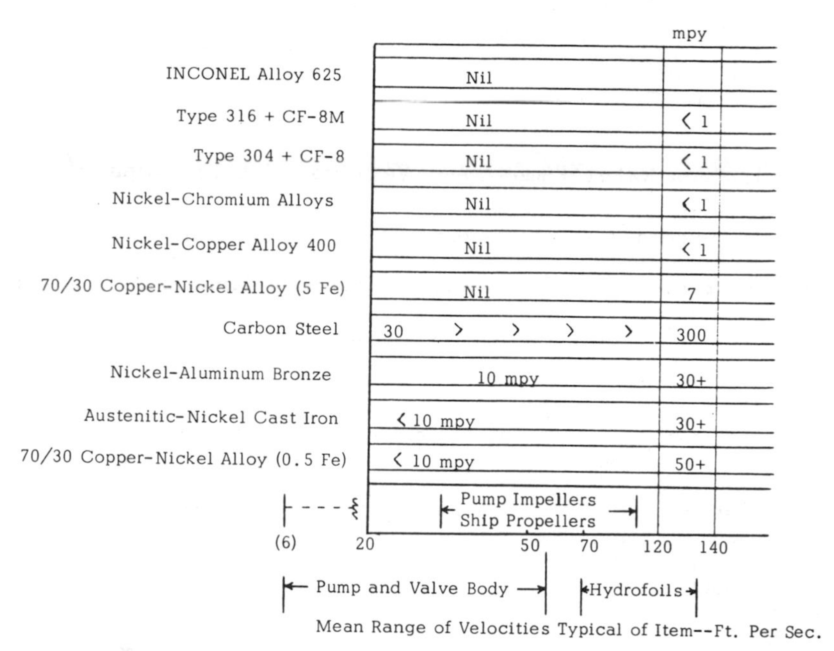

Fig. 15. Effects of high velocities on corrosion of metals in sea water[14].

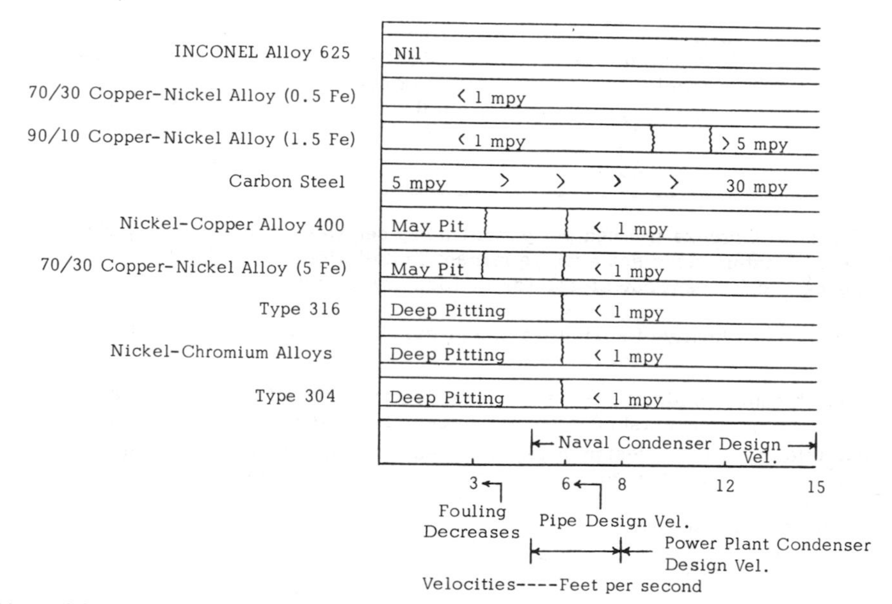

Fig. 16. Effects of velocity of sea water on corrosion of metals in pipe and tube service[14].

Table 13. Sea Water Pumps A. Vertical Wet Pit Type(17)

	Basic Minimum Cost Construction[a]	Preferred	Most Durable
Class of Service	Frequent, but predictable replacement	High service factor low maintenance infrequent maintenance	Minimum out of service time minimum maintenance
Body			
Suction Bell	1. Cast Iron 2. 2% Ni Cast Iron	Ni-Resist*[b] Type 1 or Type 11	Ni-Resist* Type 1 or Type 11
Shroud Liner or Case in Way of Impeller	1. Cast Iron 2. 2% Ni Cast Iron 3. Cement or epoxy coated cast iron 4. Tin or Al bronze	CF-3M (316L)	CF-3M (316L)
Case with Straightening Vanes (Diffuser)	1. Cast iron 2. 2% Ni Cast Iron 3. Cement or epoxy lined cast iron 4. Tin or Al bronze	Ni-Resist* Type 1b[c]	CF-3M (316L)
Discharge	1. Cast Iron 2. 2% Ni Cast Iron 3. Carbon Steel	1. Cast Iron 2. 2% Ni Cast Iron 3. Cement or epoxy or steel	1. 90-10 Cu-Ni 2. Ni-Resist* Type 1 or 11 3. Cement or epoxy lined cast iron or steel

(a) Common cast iron bronze trimmed pumps will pump sea water and are widely used for temporary installations and even permanent installations where high maintenance and frequent out-of-service periods are tolerable.

(b) Ni-Resist refers to family of austenitic nickel cast irons. See ASTM Specification A-436.

(c) The higher chromium grade of Ni-Resist, Type 1b, is preferred for this part of the body to provide greater resistance to turbulence. Type 1b will have a greater shrinkage rate, therefore, design parameters require adjustment to allow for this effect. If designs cannot be adjusted, upgrade to CF-3M (316L).

* Inco Registered Trademark

continued

Table 13 (continued) A. Vertical Wet Pit Type

	Basic Minimum Cost Construction	Preferred	Most Durable
Class of Service	Frequent, but predictable replacement	High service factor low maintenance infrequent maintenance	Minimum out of service time minimum maintenance
Impeller	1. Tin bronze (G or M) 2. Aluminum Bronze[d]	1. CF-8M (316) 2. 29 Cr- 11 Ni 3. CF-3 (304L) 4. Ni-Al Bronze (d)	1. CF-8M (316) 2. 29 Cr - 11 Ni 3. See (e) 4. See (f)
Wear Rings	1. Tin Bronze (G or M) 2. Aluminum Bronze (d) 3. Ni-Resist* Type 1 or Type 11	1. See Wear Ring comment (g)	1. See wear ring comment (g) 2. Type 316 3. Alloy 20
Shaft	1. Type 410 2. Precipitation hardening stainless steel	1. Monel* Alloy K500 2. Monel* Alloy 400 3. Type 316	1. Monel* Alloy K500 2. Monel* Alloy 400 3. Type 316
Bolting	1. Silicon Bronze	1. Monel* Alloy 400	1. Monel* Alloy 400

(d) Aluminum bronze dezincifies rapidly even in quiet sea water with little visible indication of attack until dezincified shell collapses. The use of grades with 4% minimum nickel avoids dezincification. Check with supplier before welded repairs are made for even with the 4% nickel grade, dezincification in the heat-affected zone of weld may be encountered.

(e) CF-3 has given excellent performance as workboat propellers with favourable but limited use in pump impellers. The molybdenum containing grade has better resistance to pitting in sea water and provides a better margin of safety during down periods.

(f) Nickel-aluminum bronze has given excellent service as ship propellers with favourable but also limited use in pump impellers. The stainless alloys have an advantage under the most severe velocity conditions, but in many designs nickel-aluminum bronze could approach the durability of stainless and it is more resistant to pitting during down periods.

(g) Wear rings need special consideration to limit the internal recirculation of water from the high pressure inlet side of the pump. Material guides for sea water pumps, which are resistant to highly turbulent sea water, may be found in Figure 15.

* Inco Registered Trademark

continued

Table 13 (continued) B. Volute Dry Pit Type

Body	Wear Ring	Shaft	Impeller	Comments
Common[a]				
Cast Iron or Nickel Cast Iron	G or M Bronze	Steel	G or M	Lower service factor. Frequent but scheduled replacement.
Shipboard				
1. 70-30 Copper-Nickel (weldable) 2. G or M. Bronze	Monel* Alloy 506 (H Monel*)	1. Monel* Alloy K500 2. Monel* Alloy 400	1. Monel* Alloy 410 (A Monel) 2. Monel* Alloy 411 (E Monel)	Bronze cases with Monel* Alloy 410 may be considered standard construction for better practice on merchant ships. The Navy has shifted from bronze to 70-30 Cu-Ni cases in order to facilitate repair of battle damage by welding.
Coastal Plants				
Ni-Resist* (austenitic Ni Cast Iron) Type 1, Type 1b or Type 11	See Wear Ring Comment[b]	1. Monel* Alloy 400 2. Monel Alloy K500 3. Type 316	CF-8M (316) or Worthite** or 29 Cr - 11 Ni	This construction provides cathodic protection to the impeller and shaft that tends to minimize pitting and crevice corrosion.

(a) Common Cast iron bronze trimmed pumps will pump sea water, and are widely used for temporary installations and even permanent installations where high maintenance and frequent out-of-service periods are tolerable.

(b) Wear rings need special consideration to limit the internal recirculation of water from the high pressure inlet side of the pump. Material guides for sea water pumps, which are resistant to highly turbulent sea water, may be found in Figure 15.

* Inco Registered Trademark
** Worthington Corporation Trademark

Table 13 (continued) B. Volute Dry Pit Type

Special Sea Water Pump Constructions

Body	Wear Ring	Shaft	Impeller	Comments
CF-3M (316L)	Type 316	Type 316	CF-8M (316	All CF-3M (316L) sea water pumps are subject to crevice corrosion and pitting unless considerable care is taken to open the pumps and wash down with fresh water whenever they are removed from service.
Worthite**	Type 316	Alloy 20	Worthite**	Higher alloyed pumps such as Worthite** and Alloy 20 are substantially more resistant to crevice corrosion and pitting. Limited experience indicates it may not be necessary to clean or wash down with fresh water as frequently as with the all CF-8M (316) construction.
CN 7M	Alloy 20	Alloy 20	CN 7M	Higher alloyed pumps such as Worthite** and Alloy 20 are substantially more resistant to crevice corrosion and pitting. Limited experience indicates it may not be necessary to clean or wash down with fresh water as frequently as with the all CF-8M (316) construction.

** Worthington Corporation Trademark

steel rods, 1-1/2 inches in diameter located on the scroll walls upstream from the wickets. In addition, platinum-plated anodes might be placed in the draft tube liner.

The report on Passamaquoddy (18) outlines the following general areas which required investigation to minimize corrosion:

1. selection of materials
2. protective coatings
3. cathodic protection.

Materials selection is probably the most significant area, as the judicious selection of correct materials may often eliminate the need to proceed to more complex and costly methods of construction.

The Passamaquoddy report lists the following materials as appropriate for use in tidal power projects with salient points for each added as brief summaries:

1. Selection of materials
   (a) Structural-grade carbon steel
      - poor resistance to corrosion
      - must be used in conjunction with protective coatings or cathodic protection systems
      - least expensive
      - universally available
      - no problems in manufacture or fabrication.
   (b) Structural-grade low-alloy steel
      - does not provide a marked improvement in corrosion resistance over carbon steel and it is also more costly
      - its greater allowable stress may be advantageous in permitting smaller cross-sections and decreasing structural weight.
   (c) Cast steel
      - considered in the same category as (a) and (b).
   (d) Cast Iron
      - resistance to corrosion erratic and unpredictable due to selective or parting corrosion
      - may find usage in pipes where protection could be afforded by an interior coating, and where flow conditions are not apt to abrade the coating.
   (e) Alloy iron (Type I Ni-Resist*)
      - more expensive than either (c) or (d)
      - resistant to corrosion and abrasion
      - greater strength than (c) or (d).
   (f) Wrought iron
      - no great difference in the rate of corrosion between steel and

---

*Inco Registered Trademark.

wrought iron submerged in sea water.

(g) Manganese bronze
- widespread use for ship propellers
- more expensive with lower strength than cast steel, however, corrosion resistance is greater
- during periods of non-rotation this propeller alloy has the ability to restore a highly resistant skin which acts as a barrier to corrosion during operation.

(h) Nickel-aluminum bronze
- more expensive than (g)
- as propeller material its life expectancy is 15-25 years (based on 5 years experience when report was written)
- has greater erosion-corrosion resistance than (g).

(i) Stainless steel
(1) Type 316
- best performance to all types of sea water corrosion
- subject to pitting at low velocities and to crevice corrosion, however, assurance of freedom from this type of corrosion is realized when cathodic protection is used.

(2) Type 400
- corrosion resistance slightly less than that of Type 316
- serious drawback if used with cathodic protection systems because current densities less than those necessary to achieve minimal protection of steel will cause formation of surface blisters.

2. Protective coatings
(a) Paints
- pigmented lead-in-oil types are being supplanted by synthetic paints or the vinyl and phenol types
- vinyl-type coatings based upon vinyl chloride-acetate copolymers appear to be best suited for underwater exposure
- vinyl coatings dry quickly, are easily repaired, and have good bond characteristics when overlapping previously applied coats.

(b) Metals
(1) Cladding
- corrosion resistant material hot-rolled onto carbon steel; readily adaptable to flat areas
- complex shapes clad using corrosion resistant weld-overlay techniques; method is costly.

(2) Metallizing
- corrosion resistant metal is applied as a hot atomized spray
- technique is fast but it is difficult to obtain a dense, impervious coating.

(3) Galvanizing
- zinc coating is anodic to steel and rapidly corrodes.

(c) Bitumens (generally coal tars or asphalts)
- inexpensive method of protecting steel structures from corrosion
- have poor resistance to abrasion, impact, high velocity and freezing
- newer materials such as tar epoxies (synthetic plastics in a coal tar base) are usually applied cold; catalytic setting; have better abrasion and mechanical properties than coal tars
- may be considered in conjunction with cathodic protection systems.

(d) Elastomers (includes rubber and rubber-like compounds both natural natural and synthetic)
- materials are inert, protect through mechanical action by excluding moisture
- have been used in pipes and pumps handling sea water at velocities up to 15 feet per second with excellent results
- polymerized chloroprene has better resistance to oils, sunlight, aging and heat than rubber.

(e) Vitreous material
- fused glass or porcelain would have good corrosion resistance but mechanical weakness detracts from its use.

(f) Inorganic coatings
- natural oxide compounds formed on aluminum, or chemically developed chromate films on magnesium and zinc, offer some protection to corrosion
- a metal may be protected by a calcareous deposit which often occurs on the metal surface when cathodic protection is used. The film, precipitated from sea water by the impressed current, is largely alkaline in nature. This deposit may lead to destruction of some types of paint coatings and, for this reason, must be closely controlled. The main advantage of coatings of this type is that they permit a reduction in the protective current once the deposit has been established.

3. Cathodic protection
- corrosion of metal in sea water is electro-chemical in nature, involving flow of current between anode and cathode
- cathodic protection inhibits corrosion by reversing naturally occurring currents thereby protecting the metal by making it a cathode instead of an anode
- carbon steel may be protected by using sacrificial anodes made of metals such as cadmium, zinc, aluminum or magnesium all of which are more anodic than carbon steel
- impressed-voltage cathodic protection systems overcome deficiencies of galvanic systems, however, they have a higher initial cost. They permit adjustment of the protective current, which is found to be relatively

high during start-up, and remains so until the cathode polarizes or a calcareous deposit forms, after which the current can be reduced. An impressed-current system, using inert anodes such as platinized steel, will last indefinitely whereas galvanic systems employing magnesium or zinc will require replacement of these materials on a periodic basis.

The three areas outlined above indicate the in-depth studies which are required to reach the conclusions necessary in the selection of materials by ocean engineers concerned with tidal power projects. Specification of corrosion resistant alloys such as Type 316 stainless steel, nickel containing cast irons and bronze alloys, together with protective coatings, or cathodic systems for carbon steel, enable the construction of major components which permit efficient plant operation.

Rance River

In 1967 France placed a tidal power station in operation which consists of 24 submerged, horizontal 4-blade Kaplan turbines each of which drives a 10 MW generator. The project represents approximately 20 years of planning, about 6 years of actual construction and, of equal importance, some three years of experience in corrosion control of a power generating facility operating in sea water.

Preliminary corrosion control investigations were made at atmospheric, tidal zone and submerged test sites. Starting in 1955, long-term immersion and atmospheric tests were carried out on scale models designed to provide sea water flow conditions at expected velocities. Although the Rance report [19] does not list all the systems examined, it does state that the following materials were investigated: 15 grades of stainless steel, 4 grades of nickel - aluminum bronze and 150 types of paint coatings and primers. Important materials-selection data were obtained from the study.

Stainless steel with a martensitic structure and a composition of 17% chromium, 4% nickel, 1.5% molybdenum and 0.1% carbon was selected for:

- Kaplan wheel hubs and blades for 12 of the 24 turbines
- the shell casing of the device for controlling the blades
- turbine cone
- inside flange of the distributor
- guide vanes.

Martensitic stainless steels were used for turbines because of greater strength and outstanding resistance to cavitation after heat treatment.

Stainless steel with an austenitic structure and a composition of 18% chromium, 11% nickel, 2.5 molybdenum and 0.03% carbon (Type 316L) was used for:

- the outside flange of the distributor and cover of the hydraulic duct
- inside sleeve of the Kaplan wheel
- wearing part of the Kaplan wheel joint

- mobile labyrinth of the Kaplan wheel (which prevents the flow of sea water inside the shell)
- all pipe subject to contact with sea water

Two main reasons were cited for selecting austenitic stainless steel:

- the general corrosion resistance of this steel is greater than other grades of stainless
- heat-treatment following welding operations is unnecessary.

Nickel-aluminum bronze with a composition of 9% aluminum, 5% nickel, 4% iron, 1% manganese, balance copper, was used for the remaining 12 turbine blades. This alloy exhibits resistance to cavitation, pitting and corrosion fatigue and is easily repaired when damaged.

The report [19] from which the foregoing information on the Rance project was extracted continues by stating the above metals were selected after studying the effects of various forms of corrosion, such as pitting, intergranular (where heat-treatment could not be used because of size limitations), cavitation and stress. In addition, particular attention was paid to erosion-corrosion because of suspended sand in the sea water.

Certain features of the design specifications for the turbines were satisfied by either martensitic stainless steel or nickel-aluminum bronze. Corrosion test data did not suggest one alloy would perform better than the other so half of the 24 turbine blades were made with martensitic stainless steel and the balance with nickel-aluminum bronze.

Plain carbon steel used for components submerged in the ocean were protected by vinyl paints. Where carbon steel and stainless steel are coupled together, galvanic corrosion is prevented by also using these types of paints.

Prior to plant start-up, problems were encountered when the turbines were left in stagnant sea water for 5 months. Midway through this period an examination of the turbine blades revealed that the martensitic stainless steel suffered from heavy fouling and considerable pitting with marked undercutting. Turbine blades made from nickel-aluminum bronze exhibited some selective corrosion (dealuminumification) but no fouling. A cathodic protection system was subsequently installed which produces sufficient chlorine to prevent growth of fouling marine organisms. As of January this year, the turbine blades have shown no further corrosion and no trace of cavitation or erosion.

Austenitic stainless steel, used for the hydraulic duct, is painted and revealed only isolated traces of shallow pitting. In order to allow for the changes in diameter in this area the sections are joined by flanges. Some crevice corrosion was observed at the joints which are now painted to reduce this hazard.

The plain carbon steel has been painted and further protected by an impressed current cathodic system. Moderate corrosion has been observed where the paint system was damaged.

The nickel-aluminum bronze has shown satisfactory corrosion resistance in sea water [20]. Where greater strength and hardness are necessary, martensitic stainless steel could be specified along with cathodic protection. If cathodic protection cannot be used because its beneficial effects are masked or shielded by other components, the higher alloy austenitic stainless steel should be employed.

The importance of the Rance power project cannot be over emphasized. Some problems particularly related to corrosion have been successfully overcome by operating personnel. Other challenges will arise in the future. The Rance experience will act as a guide to assist others to combat the effects of corrosion in future tidal power projects.

Water Control Systems

Stoplogs, bulkhead gates and trashracks for the Passamaquoddy project [18] were to be structural grade carbon steel. Trashracks would have been protected against corrosion by a coal-tar enamel-primer coat and an impressed-voltage, cathodic-protection, system. Stoplogs and bulkhead gates were to receive a vinyl-type coat followed by four double-spray coats of red lead vinyl primer.

Because of the overhaul schedule for the Passamaquoddy project, rotation of the gates would be such that one set would be held in reserve under atmospheric conditions. This periodic exposure of the gates, to air and sunlight, would kill marine growth. Stoplogs would be used infrequently so that they would also be exposed to marine atmospheric conditions much of the time.

No mention is made in the Passamaquoddy report regarding use of high-strength low-alloy steels to conserve weight and decrease section size of water control systems. Similarly, the use of ferrous alloys such as stainless steel has been omitted.

The Rance tidal power station report [19] states that the Rance water control system uses stainless steel for rails and rollers but the actual type is not given. No mention is made of material selection for gates; however, structural grade carbon steel cathodically protected, would be the probable choice.

SUMMARY

Theoretical aspects of corrosion and the behavior of ferrous alloys in marine environments have been reviewed; and examples of selected materials for the proposed Passamaquoddy and existing Rance Tidal Power Project have been outlined. Sound engineering judgment can lead to the economical and practical use of ferrous alloys in marine environments only if corrosion is recognized as a major, if not a controlling factor, in design.

It may not always be possible to eliminate sea water corrosion, although it can frequently be controlled to within acceptable limits by the proper selection and intelligent use of materials available today in the field of ocean engineering.

ACKNOWLEDGMENT

The author gratefully acknowledges the assistance of his associates in the preparation of this paper and wishes to thank The International Nickel Company of Canada, Limited for permission to publish it.

REFERENCES

1. "Moon Power", Ontario Hydro News (1969).

2. Uhlig, H. H., Corrosion Handbook, John Wiley and Sons, Inc., New York (1948).

3. "Nickel-Containing Materials in Gates for Water Control Systems", Inco, Canada, A-488.

4. Brown, B. F., and L. S. Birnham, "Corrosion Control for Structural Metals in the Marine Environment", U.S. Naval Research Laboratory, 6167, Ed. W. S. Pellini (1964).

5. LaQue, F. L., "Materials Selection for Ocean Engineering", 1966 Statewide Lecture on Ocean Engineering, University of California (1966).

6. Covert, R. A., Todd, B., and B. A. Weldon, "Nickel-Containing Materials for Ocean Engineering, IMAS, London (1969).

7. Copson, H. R., "A Theory of the Mechanism of Rusting of Low Alloy Steels in the Atmosphere", Proc. A.S.T.M., 45 (1945).

8. Ambler, H. R., and A. A. J. Bain, J. App. Chem., 5 (1955).

9. "Corrosion Resistance of the Austenitic Chromium-Nickel Stainless Steels in Atmospheric Environments", Inco, Canada, A-318.

10. Evans, T. E., "Atmospheric Corrosion Behavior of Stainless Steels and Nickel Alloys", 4th Int. Congress on Metallic Corrosion, Amsterdam (1969). (In press)

11. Law, R. J., and F. L. LaQue, "Behavior of Materials in Marine Environment", Canadian Forces Corrosion Prevention Conference (1966).

12. "Stainless Clad Steels", Inco, Canada, A-316.

13. Copson, H. R., "Effects of Velocity on Corrosion", Corrosion, 16 (1962).

14. "Guidelines for Selection of Marine Materials", Inco, Canada, A-404.

15. "Corrosion Resistance of the Austenitic Chromium - Nickel Stainless Steels in Marine Environments", Inco, Canada, A-319.

16. Metals Handbook, 2, Am. Soc. for Metals, Metals Park, Ohio (1964).

17. "Sea Water Pumps - A Guide for Selection of Component Materials", Inco, N.Y., M-110.

18. "Tidal Power Plant and Corrosion Prevention", International Joint Commission on the Passamaquoddy Tidal Power Project, App. 8, Washington, D. C. (1959).

19. "The Role of Nickel in the Tidal Power Station of the Rance", Documentation Centre of Société le Nickel, Paris (1966).

20. Private Communication, B. A. Weldon, The International Nickel Company, England.

PERFORMANCE OF CONCRETE IN A MARINE ENVIRONMENT

C. R. Wilder*

INTRODUCTION

Concrete is a versatile engineering material, probably more widely used than any other construction material. Its versatility stems from our ability to tailor it to provide specific properties within a fairly wide range and to mold it into almost any desired shape and accounts for its wide acceptance and use. It can be mixed at the construction site or at a central location and transported to the site, or it can be used to precast structural elements at a central "factory" location for later erection at a specific site.

DEFINITION OF CONCRETE

Concrete is a mixture of portland cement, water, fine aggregate and coarse aggregate. Sometimes a chemical admixture is used to produce a specific effect, for example an air-entraining admixture to provide a controlled amount of intentionally entrained air in the concrete. In every case the portland cement paste is the binder which coats all of the aggregate particles and fills the spaces between them. Hydration of the cement binds all of these particles into a solid mass.

The quality of the concrete is influenced most markedly by the quality of the cement paste or binder. Cement paste quality is a direct function of the relative amounts of cement and water making up the paste. Therefore we use the water-cement ratio as an index in discussing paste quality. Low water-cement ratios produce more desirable cement pastes than high water-cement ratios. As the water-cement ratio is decreased the paste becomes denser, stronger, more impermeable to the passage of fluids and gases, and less subject to volume change on wetting and drying.

---

*Manager, Portland Cement Association, Public Works & Transportation Section, Skokie, Illinois.

Concrete made with natural sand and gravel or crushed aggregate will have proportions within the range shown in Fig. 1. The "lean" mix concrete would be suitable for buried footings or foundations in a non-aggressive environment. The "rich" mix would be used for concrete which is exposed to the environment, such as bridge decks or thin sections in contact with fresh water.

While the cement paste, or binder, probably is the most important single factor affecting the qualities of the concrete, these qualities also are dependent on the properties of the other components, the manner in which the concrete is mixed, and the subsequent treatment of the resulting mixture, particularly during its early life while it is being cured.

COMPONENT MATERIALS

Portland Cements

Portland cement is required to meet the specifications of the Canadian Standards Association (CSA), the American Society for Testing and Materials (ASTM), various government agencies, and other specifiers. The following are the five basic types of portland cement, as covered in the ASTM and CSA specifications, with a brief description of the primary use for which each was designed. In some areas these various cements also are available with an interground air-entraining addition as one means for providing intentionally entrained air in concrete.

ASTM Type I, CSA Normal. This type is a general purpose cement, suitable for use when the special properties of the other types are not required. This would be where the concrete is not subject to specific exposures such as sulfate attack from soil or water, or to an objectionable temperature rise because of the heat generated by hydration.

ASTM Type II. Type II cement is used where precaution against moderate sulfate attack is important, as in drainage structures exposed to sulfate concentrations higher than normal but not unusually severe. Type II will usually generate less heat at a slower rate than Type I or Normal cement, and therefore may be used in structures of considerable mass, such as large piers, heavy abutments, and heavy retaining walls. Its use is particularly beneficial when the concrete is placed in warm weather.

ASTM Type III, CSA High-Early-Strength. This cement provides high strengths at an early age, and is used when forms are to be removed as soon as possible or when the structure must be put into service quickly. In cold weather its use permits a reduction in the controlled curing period. Although richer mixes of Type I or Normal cement may be used to gain high early strength, Type III or High-Early-Strength cement may provide it more satisfactorily and/or more economically.

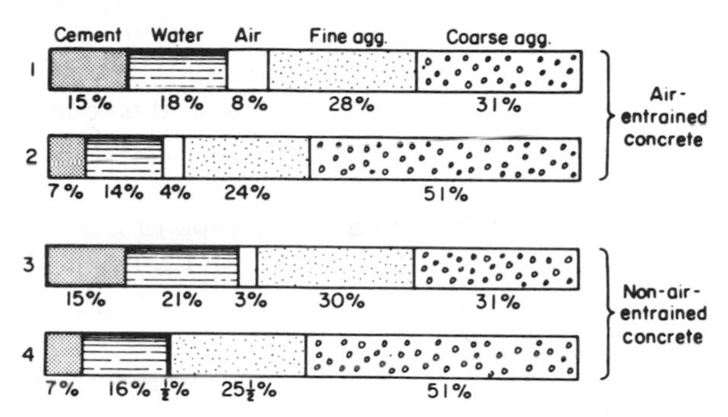

Fig. 1. Range in proportions of materials used in concrete. Bars 1 and 3 represent rich mixes with small aggregates. Bars 2 and 4 represent lean mixes with large aggregates.

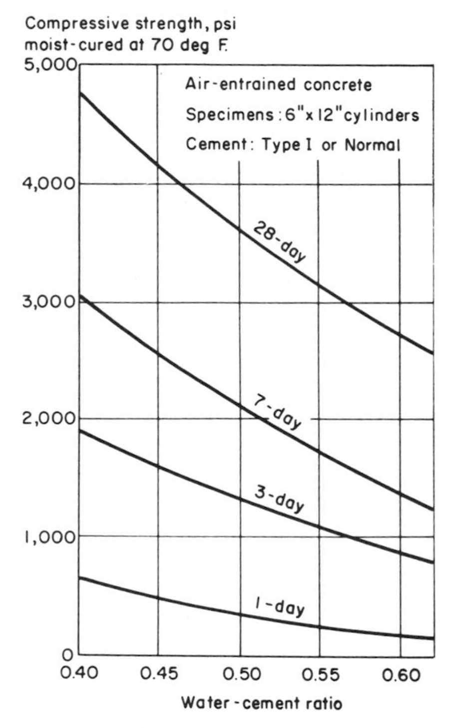

Fig. 2. Typical age-strength relationships of air-entrained concrete based on compression tests of 6 x 12 in. cylinders. Note that strength increases as curing period increases and water-cement ratio decreases.

ASTM Type IV. Type IV cement develops strength at a slower rate than Type I or Normal cement, and is intended for use in massive concrete structures, such as large gravity dams where the rate and amount of heat generated must be minimized. Since 1953 very little Type IV cement has been used.

ASTM Type V, CSA Sulfate-Resisting. This type of cement is used only in concrete exposed to severe sulfate attack, and is used principally where soils or groundwater have a high sulfate content. It gains strength more slowly than Type I or Normal cement.

Four principal compounds are found in portland cement. Typical calculated compound compositions for the ASTM and CSA portland cement types are shown in Table 1. The four column headings are abbreviations for tricalcium silicate, dicalcium silicate, tricalcium aluminate, and tetracalcium aluminoferrite, respectively. Tricalcium silicate hardens rapidly and is largely responsible for initial set and early strength. Dicalcium silicate hardens slowly and contributes to strength increases at ages beyond about one week. Tricalcium aluminate contributes largely to early strength development and liberates a large amount of heat during the first few days of hardening. Tetracalcium aluminoferrite hydrates moderately rapidly and contributes moderately to strength. The particular influence of tricalcium aluminate on concrete durability will be discussed later.

Table 1
Typical Calculated Compound Composition and Fineness of Portland Cements

Type of portland cement		Compound composition, percent*				Fineness, sq. cm. per g.**
ASTM	CSA	C_3S	C_2S	C_3A	C_4AF	
I	Normal	50	24	11	8	1800
II		42	33	5	13	1800
III	High-Early-Strength	60	13	9	8	2600
IV		26	50	5	12	1900
V	Sulfate-Resisting	40	40	4	9	1900

Aggregates

As shown in Fig. 1, aggregates occupy 60 to 80 percent of the volume

*The compound compositions shown are typical. Deviations from these values do not indicate unsatisfactory performance. For specification limits see ASTM C150 or CSA A5.

**Fineness as determined by Wagner turbidimeter test.

of concrete so their characteristics certainly have an effect on the properties of the concrete. Aggregates for good, high quality concrete should consist of clean, hard, strong, and durable particles free of chemicals, coatings of clay, or other fine materials that might affect bonding of the cement paste to the aggregates and hydration of the cement.

Important properties which can be measured include: resistance to abrasion, resistance to freezing and thawing, compressive strength, chemical stability, particle shape and surface texture, absorption, specific gravity, and gradation. The importance to be attached to any of these properties depends upon the use for which the concrete is intended.

Particle shape affects primarily the economy of the concrete mix, while all the other properties named directly affect the quality of the concrete. Specifications for aggregates also limit the amount of materials considered to be deleterious such as organic impurities, silt, clay, coal, lignite, and certain other lightweight and soft particles.

Mixing Water

Fresh, potable mixing water is best for use in concrete. It is well to avoid the use of water which contains appreciable amounts of acids, alkalies, oils, or organic materials, especially sugar. Water containing less than 2000 parts per million of total dissolved solids generally is satisfactory for making concrete. Seawater, which contains up to 35,000 ppm, or 3.5 percent,of salt, is generally suitable for use in making unreinforced concrete. It should be noted that while concrete made with seawater may have higher early strength than if fresh water were used, its strength at later ages, after 28 days, often is lower. This strength reduction may be allowed for by reducing the water-cement ratio.

Seawater also may be used for making reinforced concrete, although this may increase the risk of corrosion. This risk is reduced if the reinforcement has sufficient cover -- at least 3 in. -- and if the concrete is watertight and contains an adequate amount of entrained air. Reinforced concrete structures made with seawater and exposed to the marine environment should have a water-cement ratio of not more than 0.45. Seawater should not be used for making prestressed concrete in which the prestressing steel is in contact with the concrete as would be the case with pretensioning.

Admixtures

Admixtures may be used to produce specific effects in concrete such as the entrainment of air. The principal reason for using intentionally entrained air is to improve the concrete's resistance to freezing and thawing. Additional reasons for using air entrainment will be presented later.

Other admixtures such as water-reducing agents, retarders, or accelerators may be used to produce the specific effects implied by their names. The ACI report, "Admixtures for Concrete", is a good summary of present day knowledge and practice on the use of admixtures of all kinds.

Proportioning Concrete Mixtures

The term "proportioning concrete mixtures" includes determination of the most economical and practical combination of concrete ingredients that is workable in a freshly-mixed state and that will develop the required qualities when it is hardened. Generally speaking, a properly proportioned mixture is one which achieves three objectives:

1. Adequate workability in the freshly-mixed state
2. The required qualities of the hardened concrete
3. Economy

Workability

Workability of freshly-mixed concrete is the property that determines the amount of effort required to fully consolidate the concrete. It is attained by proper adjustment of the relative amounts of fine and coarse aggregates and the total amount of aggregate used in the selected mixture. There is no quantitative measure for workability. The judgment of adequate workability is an art as well as a science, and a little experience makes such a judgment relatively simple.

Concrete Quality

Concrete quality including adequate strength, watertightness, and resistance to freezing and thawing or seawater exposure is dependent on cement content and water-cement ratio, entrained air and proper curing.

Economy

Economy is achieved primarily by minimizing the amount of cement required without sacrificing workability and quality. Since quality is largely dependent on the water-cement ratio, minimizing the water requirement results in a parallel reduction in cement requirement. This can be accomplished by using:

1. the stiffest practicable mixture
2. the largest practicable maximum size aggregate
3. the optimum ratio of fine to coarse aggregate.

Concrete mixtures should be of a consistency and workability suitable for the conditions of the job which include not only consideration of form dimensions and reinforcement but also equipment available for placing and compacting.

CONSTRUCTION PROCEDURES

Even when the greatest of care is exercised in selecting proper concrete materials and proportions the quality of the resultant concrete is influenced by mixing and placing procedures and other construction practices. Each step in handling, transporting, and placing concrete should be carefully controlled to maintain uniformity within a batch and from batch to batch so that the completed work is consistent throughout.

To produce concrete of uniform quality the ingredients for each batch must be measured accurately. Most specifications rightly require that batching be done by weight rather than by volume. Use of a weight batching system provides greater accuracy, simplicity and flexibility. Flexibility is necessary because changes in aggregate moisture content require frequent adjustments in quantities of water and aggregates. If the moisture content of the fine aggregate changes from six percent to eight percent, for example, the water-cement ratio of a typical mix would be increased from perhaps 0.45 to 0.48 unless an adjustment is made in the amount of mixing water added. This could have a material effect on the desired properties of the concrete. Water of course can be measured accurately by either volume or weight.

Concrete is mixed, handled and transported by many methods. On many large jobs concrete is mixed on the job site in a stationary mixer, but in the United States about two-thirds of the concrete currently being used is ready-mixed concrete. Regardless of mixing, handling and placing techniques used, it is essential to avoid separation of the coarse aggregate from the mortar or of water from the other ingredients.

CONCRETE PROPERTIES PARTICULARLY IMPORTANT IN SEA-WATER EXPOSURES

All of the desirable concrete properties mentioned are of importance to concrete which is to be exposed to seawater. In addition, seawater exposure involves wetting the concrete by a solution of salts, principally magnesium sulfate and sodium chloride. The sulfate combines with the aluminates to form ettringite, a form of calcium sulfo-aluminate, which can produce disruptive pressures within the concrete; chlorides may promote corrosion of reinforcement; alkalies present may participate in the alkali-aggregate reaction which is not uncommon in some parts of the United States and here in Nova Scotia. Thus concrete exposed to seawater should be made with cement of adequately controlled aluminate content and with non-reactive aggregate; embedded steel should be well covered by concrete of low permeability; and good construction practices should be followed.

Strength and Impermeability

As discussed, strength and impermeability are dependent on the water-cement ratio and adequate curing. Typical water-cement ratio to strength relationships are shown in Fig. 2. Note the substantial advantage of increasing the period of curing from 7 to 28 days. With 7 days of curing, an excellent air-entrained concrete (w/c = 0.45) will have a strength of less than 3000 psi, while if the same concrete is given 28 days of moist curing its strength should be in excess of 4000 psi.

Curing simply means providing an adequate supply of water to permit hydration of the cement in the mixture. General curing requirements are shown in Table 2 and methods of curing in Table 3. Curing may be provided by applying moisture, as in one of the six methods listed, or by preventing evaporation of the water included in the mix. If a curing membrane is used it must be continuous and must be maintained throughout the entire curing period. It is important that the concrete never be permitted to dry out during the curing period.

Table 2
General Curing Requirements

Time	3-7 days (minimum)
Temperature	50° - 100° (may be higher)
Moisture	Saturated at all times

Table 3
Methods of Curing

Seal in Mixing Water
- Waterproof paper (reinforced)
- Polyethylene sheets
- Spray curing compound

Supply Additional Water
- Ponding
- Soaker hose
- Wet sand
- Wet burlap
- Immersion (precast)
- Steam curing (saturated air)

The importance of curing to strength development is illustrated in Fig. 3. Note that strength development continues so long as moisture is made available. If curing is interrupted and then resumed after an air drying period, the strength will again increase. However, such concrete will never

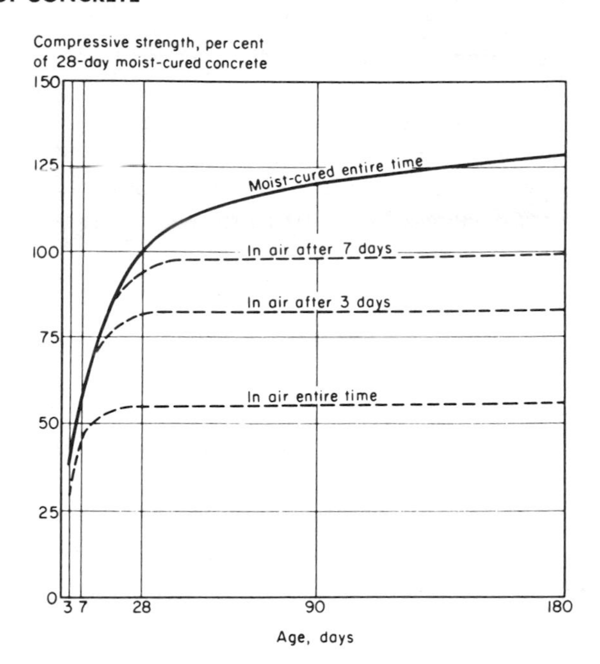

Fig. 3. Strength of concrete continues to increase as long as moisture is present for hydration of cement.

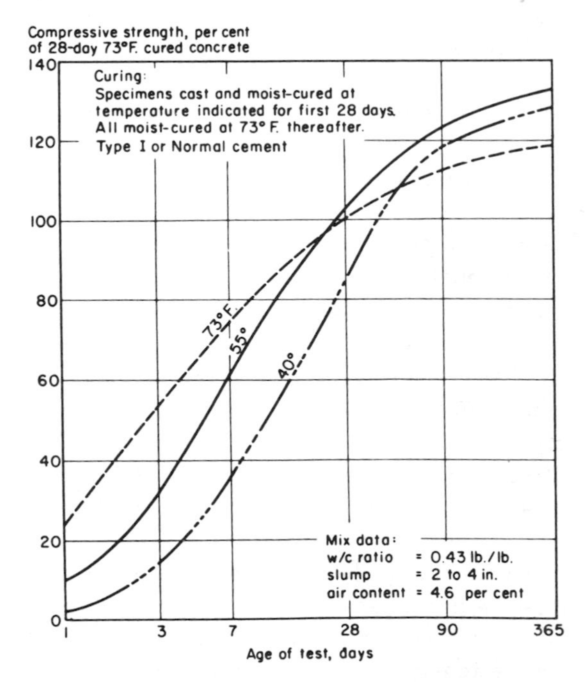

Fig. 4. Effect of temperature on concrete compressive strength at various ages.

be as strong as concrete which is adequately moist cured from the time it is cast.

Since hydration is a chemical process, it proceeds more rapidly as curing temperatures are increased. As seen in Fig. 4, concrete mixed, placed and cured moist at higher temperatures has higher early strength, but more moderate temperatures generally result in somewhat higher strength at later ages.

Low Permeability

Low permeability is an important requirement for concrete subjected to hydraulic pressures and for reinforced concrete which is exposed to chloride solutions such as seawater. As shown in Fig. 5, both water-cement ratio and curing have important influences on permeability. Note that the low water-cement ratio mortar showed no leakage when tested after 7 days of moist curing while the high water-cement ratio concrete was somewhat permeable even after extended moist curing.

Durability

All concrete is expected to resist a wide variety of exposure conditions, some more severe than others. Of immediate interest to us is the ability of concrete to resist all of the potentially destructive elements to which ocean structures are exposed. These include freezing and thawing which is of course dependent on geographical location and exposure, various forms of chemical attack by seawater, and corrosion of the reinforcing steel.

Freezing and Thawing

When water freezes, it increases in volume by about 9 percent. Thus any water trapped within concrete is a potentially destructive influence, since concrete is not sufficiently elastic to stand such internal pressures. Fortunately, freezing of water in concrete is progressive rather than instantaneous which means that there is time for unfrozen water displaced by the growing mass of ice to move ahead of the advancing ice front. To avoid distress in concrete, a relief zone must be reached before the pressure generated by the ice exceeds the tensile strength of the concrete. Factors of importance are the amount of free water available and its freezing rate, the permeability of the concrete paste, and the distance to relief zones. These factors help us design concrete that will satisfactorily withstand freezing and thawing.

The concrete must be impermeable to prevent or minimize the entrance of water from external sources, which would increase the freezable water content of not only the cement paste but also the aggregate. Concrete also must be strong enough to resist those tensile forces generated by the freezing

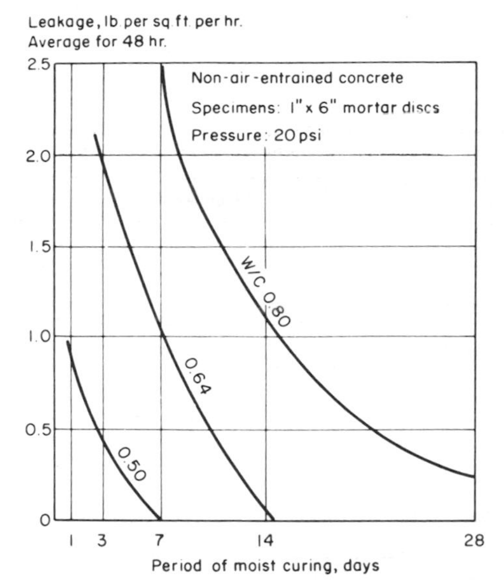

Fig. 5. Effect of water-cement ratio and curing on watertightness. Note that leakage is reduced as the water-cement ratio is decreased and the curing period increased.

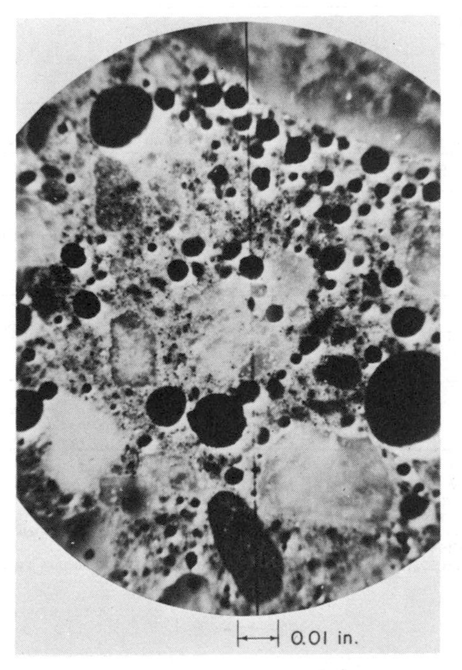

Fig. 6. Polished section of air-entrained concrete as seen through a microscope. The solid black circles are entrained air voids.

Table 4 **Recommended Maximum Water-Cement Ratios (lb. per lb. of cement) for Plastic Mixes for Various Types of Structures and Exposure Conditions**

Type of structure		Exposure conditions					
		MILD—Temperature rarely below freezing, infrequent snow or frost. Air entrainment recommended.		SEVERE—Wide range of temperature, long periods of freezing, or frequent alternations of freezing and thawing. All concrete must be air-entrained.		In sea water, within the range of fluctuating sea water level or exposed to spray, or in contact with high sulfate soils or waters. All concrete must be air entrained.†	
		In air, or continuously submerged in fresh water.	In fresh water within the range of fluctuating water level, or exposed to spray.	In air, or continuously submerged in fresh water.	In fresh water within the range of fluctuating water level, or exposed to spray.		
Irrigation canal, stormwater conveyance, or flood control canal lining.	unreinforced	0.62	0.58	0.53	0.53	0.44	Function or location
	reinforced	0.62	0.58	0.53	0.53	0.44	Function or location
Water or sewage treatment units: tank walls, tank floor slabs or roofs.		0.53	0.53	0.49	0.49	0.44	Function or location
Concrete deposited by tremie in water.		0.44*	0.44*	0.44*	0.44*	0.44*	Function or location
Concrete that will be protected from weather by enclosure or back-fill, but which may be exposed to freezing and thawing for several months a year before such protection.		**	N. A.	0.53	N. A.	N. A.	Function or location
Concrete protected from weather: interior of buildings, concrete below grade.		**	N. A.	**	N. A.	**	Function or location
Thin sections: railings, curbs, sills, ledges, ornamental concrete, piles.	1″-2″ cover	0.49	0.49	0.44	0.44	0.44	Dimension
	2″-3″ cover	0.53	0.53	0.49	0.49	0.44	Dimension
Moderate sections: retaining walls, piers, abutments, beams, girders, siphons, tunnel linings.	2″-3″ cover over reinforcing	**	0.53	0.49	0.49	0.44	Dimension
Mass portions of concrete (exterior), minimum thickness of these portions, 2 ft.		**	0.53	0.53	0.49	0.44	Dimension
Mass portions of concrete (interior): dams, gravity walls.		*Water-cement ratio should be selected on basis of strength, workability, thermal properties, and volume-change requirements that should be established individually for such structures.*					Dimension

N.A. Not applicable.
*Plus air entrainment for workability.
**Water-cement ratio should be selected on basis of strength and workability requirements.
†Sulfate-resisting cement must be used (see Table 3 also):
For 150-1,000 ppm sulfate (as SO_4), use Type II or a portland cement with a C_3A content of less than 8 percent; portland blast-furnace-slag cement, Type IS (MS) or Type ISA (MS), may be used.
For over 1,000 ppm sulfate (as SO_4), use Type V or a portland cement with a C_3A content of less than 5 percent.

of the water which cannot be eliminated. Low water-cement ratio is important to provide both adequate strength and impermeability. Observation of concrete in service over the years has led to the development and general acceptance of recommended maximum water-cement ratios applicable for various exposures, as shown in Table 4. Note that the differences are dependent not only on location of the concrete, in seawater or fresh water, but also on thickness of the member.

Low water-cement ratios and adequate curing produce concrete which is relatively impermeable and has a relatively small amount of freezable water. Low permeability results in minimizing penetration of freezable water into the paste. It is necessary to avoid the buildup of destructive hydraulic pressures during freezing, and this can be done by the use of intentionally entrained air. Both laboratory and field performance studies show conclusively that air entrained concrete is much more resistant to freezing and thawing than ordinary, non-air-entrained concrete.

Fig. 6 is a photomicrograph of a section of hardened concrete showing intentionally entrained air voids. To effectively reduce the path length which unfrozen water must travel the air voids must consist of numerous small bubbles. The average maximum distance from any point in the cement paste to the nearest air void must be less than 0.01 in. This is called the void spacing factor.

The volume of entrained air required, and this is what is generally measured in the field, is dependent on the mortar content of the concrete which varies with the maximum size of coarse aggregate. Optimum air contents for different maximum sizes of coarse aggregate are shown in Table 5.

Table 5
Recommended Air Contents for Concretes Subject to Severe Exposure Conditions

Maximum-size Coarse Aggregate, in.	Air Content, percent by volume*
3/8	7.5 ± 1
3/4	6 ± 1
1-1/2	5 ± 1
3	3.5 ± 1
6	3 ± 1

Both laboratory studies and field observations have demonstrated the

*The air content of the mortar fraction of the concrete should be about 9 percent.

remarkable increase in resistance to freezing and thawing accomplished by entrained air. The results of some laboratory tests are shown in Fig. 7, the curve relating the number of cycles of freezing and thawing required to reduce the dynamic modulus of elasticity 50 percent with the air content in percent. A comparison of field performance is shown in these photographs (Fig. 8) of two concrete boxes which had been exposed to northern Illinois winters for 16 years. The concrete mixes were identical in every respect except that the box on the right was made with air entrained concrete while that on the left was not air entrained.

Table 6 lists a number of other advantages of air entrainment. Note particularly the improvement in watertightness and resistance to sulfate attack, both of which result primarily from reduced water content in the mix and more uniform consolidation during placing.

Table 6
Advantages of Air-Entrained Concrete

Hardened Concrete
- Increases freeze-thaw resistance 10-20 times over regular concrete
- Improves resistance to salt action (scaling)
- Improves resistance to sulfate action
- Improves watertightness

Fresh Concrete
- Improves workability
- Reduces segregation
- Reduces bleeding
- Finishes sooner
- Reduces required sand content (about 100 lb./cu. yd.)
- Reduces water requirement (25-40 lb./cu. yd.)

Resistance to Chemical Attack

While concrete is subject to attack by quite a variety of chemicals, we are particularly concerned with those aspects of chemical attack which may occur in ocean structures. The principal elements in solution in seawater are chlorine, sodium, magnesium, sulfur, calcium and potassium. The dissolved salts of major concern to concrete are the chlorides, sulfates and alkalies. There is general agreement that the compounds of most concern are the sulfates, and that the reaction which is potentially most damaging to concrete is that between the sulfates in the seawater and the hydration products of the tricalcium aluminate component in the cement paste.

Without going too deeply into the chemical aspects of all this the destructive action is due primarily to the formation of ettringite, a high-sulfate form of calcium sulfoaluminate. This crystalline reaction product is

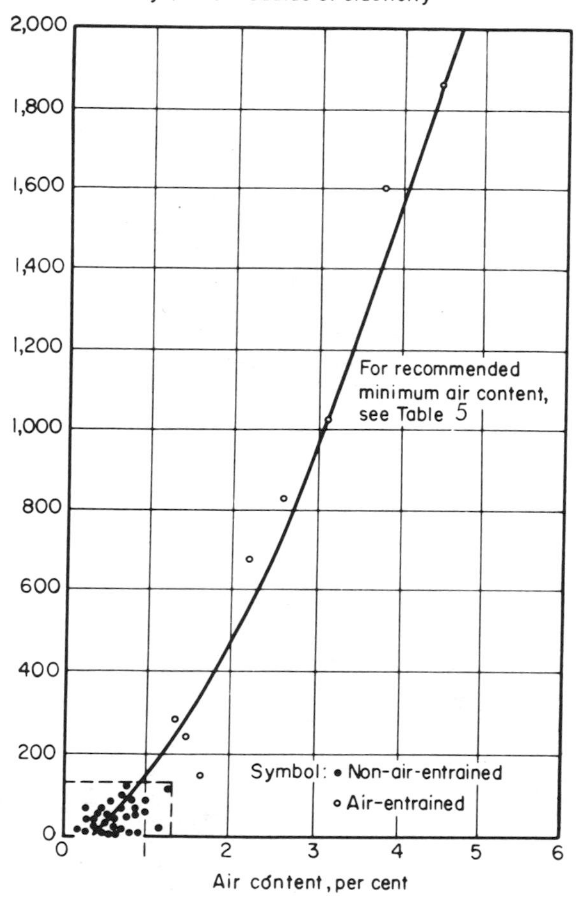

Fig. 7. Effect of entrained air on the resistance of various concretes to freezing and thawing in laboratory tests. Note the improvement in durability with increased air contents.

Fig. 8. (left)

Fig. 8. (right)

Fig. 8. Effect of severe weather on non-air-entrained (left) and air-entrained concretes after nearly 20 years of exposure. Except for entrained air, the two box-type specimens were constructed similarly with lean-mix, high slump concrete. The test plot at Naperville, Ill., was part of the long-time study of cement performance in concrete.

larger in volume than the original aluminate constituent, and expansion and consequent weakening of the entire system results when it is formed. It is fortunate that the chlorides present in seawater act to retard or inhibit the action of the sulfates.

It is appropriate at this stage to describe briefly the results of the Portland Cement Association's Long-Time Study of Cement Performance in Concrete, as it relates to concrete performance in sulfate and seawater exposures.

In this program 27 different portland cements were studied, some of each of the five ASTM types in concretes of different proportions in a variety of exposure conditions. Fig. 9 shows the locations of the test plots. Those of interest are the exposures to sulfate soils and to seawater, specifically the Cape Cod Canal, St. Augustine and Newport Beach test piles, the Los Angeles Harbor test rack, and the sulfate resistance test plot in Sacramento, which is shown in Fig. 10. Observation of these specimens is continuing. Results so far show the importance not only of the chemical composition of the cement used, that is the type of cement, but also of the cement content of the concrete. High cement content pastes are low water-cement ratio pastes and hence are less permeable to the sulfate solutions.

Fig. 11, based on observations of these sulfate-exposure specimens, leads to several conclusions. The seven bag mix concretes, representing water-cement ratios of about 0.45, were more resistant to sulfate attack than the leaner mixes, regardless of cement type. While the type of cement, based on C_3A content, is important in the leaner mixes, its importance decreases as the cement content is increased. This is attributable to changes in water-cement ratio, as shown in Fig. 12 which also shows the beneficial effect of air entrainment permitting a reduction in water content, thus further reducing the water-cement ratio.

Three exposure sites in the Long-Time Study involve concrete pilings in seawater. Those piles in the Cape Cod Canal, obviously in the freezing zone, have demonstrated the value of air entrainment in resisting damage to freezing. The St. Augustine installation has demonstrated the importance of cement type in resisting attack by the sulfates in seawater. The only piles which have shown evidences of sulfate attack were those made with a cement containing 12.2 percent tricalcium aluminate. None of the other piling, all made with cements having lower C_3A contents, showed any evidence of such distress after 25 years of exposure.

Comparing the performance of identical concretes in sulfate soil and seawater exposures, it is apparent that chemical attack by the sulfates in seawater is considerably slower than in high-sulfate soils because of the inhibiting effect of the chlorides present in seawater. It is largely on the strength of this evidence that for concrete in seawater it is recommended that the cement should contain no more than 8 percent tricalcium aluminate, the maximum for ASTM Type II cement, in contrast with the 5 percent maximum

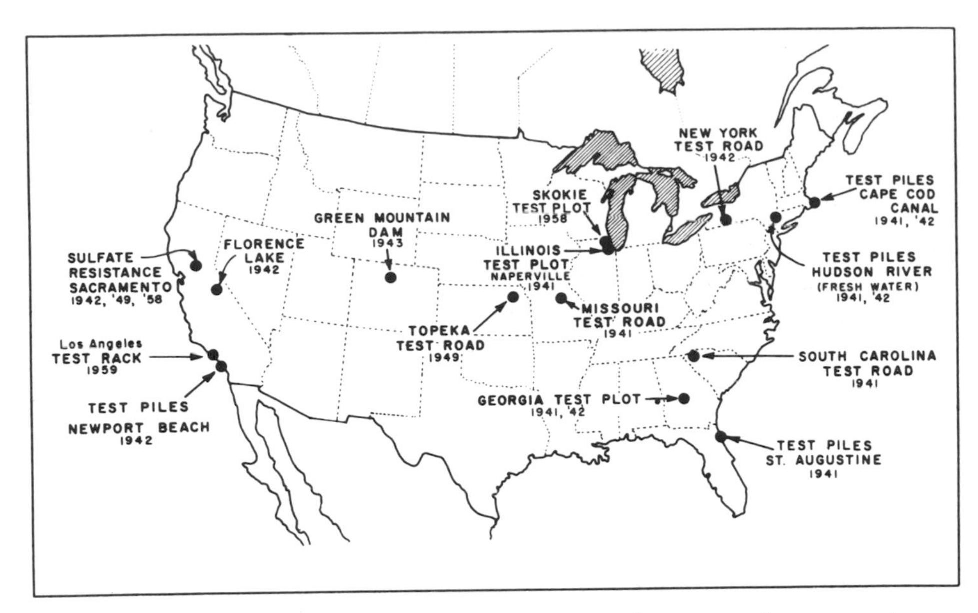

Fig. 9. Long-time study of cement performance in concrete. Locations of test projects and installation dates.

Fig. 10. Concrete specimens exposed to sulfate soil in Sacramento, Calif.

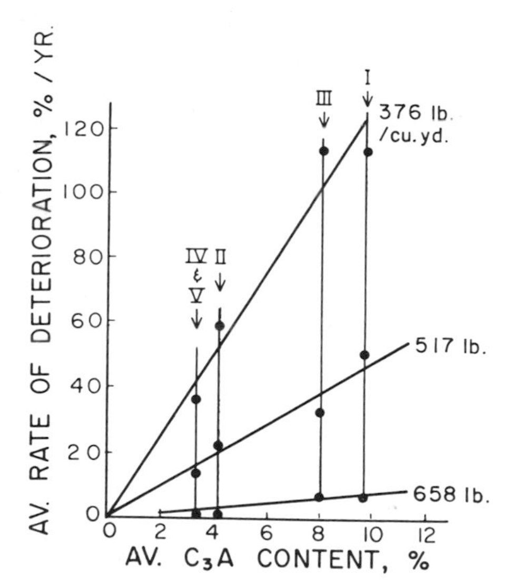

Fig. 11. Effect of cement content and average C_3A content on deterioration of concrete specimens in Sacramento sulfate soil test plots.

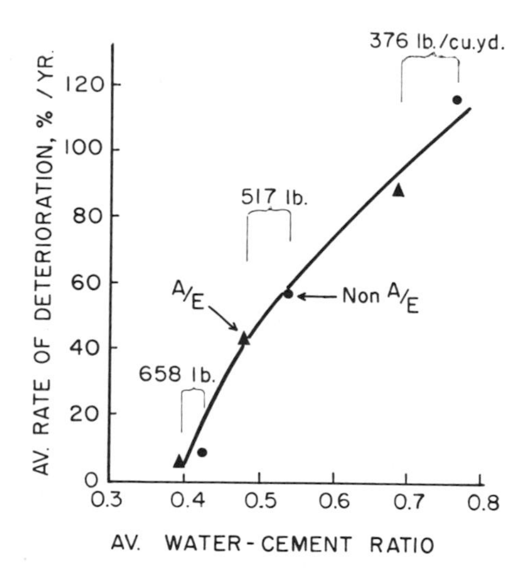

Fig. 12. Effect of water-cement ratio and air-entrainment on resistance of concrete to sulfate deterioration.

recommended for sulfate exposure. It is essential that for good performance the other important practices for making quality concrete -- low water-cement ratio, proper handling, and adequate curing -- be followed.

One other aspect of chemical attack deals with the possibility of a reaction between the alkali ions, sodium and potassium, in seawater and reactive silica in the aggregate. As early as 1937 it was noted that some of the concrete seawalls in Southern California were being adversely affected by the alkali-silica reaction. Some Nova Scotia aggregates are known to be reactive. An obvious solution to this problem is to use an aggregate which will not react with the alkalies in either the cement or the seawater. The cement-alkali reaction can be prevented by using low-alkali cement.

The foregoing gives confidence that concrete adequately durable for seawater exposures can be made if the following principles are followed: low water-cement ratio; entrained air; sound aggregates; satisfactory mixing, placing, and consolidation; adequate curing.

One major cause of deterioration remains and that is the problem of corrosion of steel reinforcement. Corrosion of steel in the presence of moisture is almost always an electrochemical action, which is not limited to concrete in seawater. Since seawater contains such a large quantity of chlorides which greatly enhance the electrochemical reaction reinforced concrete exposed to seawater is particularly vulnerable to this type of attack.

Concrete generally provides an ideal environment for the protection of reinforcing steel. Although even the best cement paste is somewhat porous and can hold both oxygen and moisture, which would normally be expected to promote rusting, the moisture present is highly alkaline. This solution provides a thin, tightly adherent, protective film over the surface of the encased reinforcing steel. So long as this film remains in contact with the steel, corrosion cannot take place. The protective alkaline film can be impaired by reductions in pH resulting from carbonation of the concrete by leaching of the alkalies from the concrete or by a sufficient concentration of chloride ions in solution adjacent to the film.

Carbonation can result from the reaction of carbon dioxide in the air with the calcium hydroxide in solution. It is fortunate, however, that with a low water-cement ratio paste and with adequate concrete cover over the reinforcing steel, penetration of carbonation is exceedingly slow. Under such conditions leaching of alkalies also is very unlikely. Even if the film is destroyed, free oxygen must be available before corrosion can take place. For those portions of the structure completely submerged, migration of dissolved air through water-filled pores is extremely slow and corrosion of steel is rarely a problem.

With respect to chloride penetration, there is some evidence that Type I Portland cement, with a relatively high tricalcium aluminate content, is more protective against corrosion than is Type V, sulfate resistant cement. The ability of cement to remove chlorides from solution is directly related to the

C_3A content of the cement, because of the formation of calcium chloroaluminate from the chlorides, both the amount formed and the amount of chlorides being neutralized depending on the C_3A content of the cement. This conclusion is confirmed by the performance of the reinforced concrete piling tested in the ocean at St. Augustine, Fla. The open vertical bars in Fig. 13 indicate lineal feet of corrosion cracking as a function of the C_3A content of the cement. The solid dots, most of them along the zero line, indicate the degree of sulfate deterioration of the concrete. It is worth noting that the only piles showing evidences of sulfate attack are those made with a cement containing 12.2 percent tricalcium aluminate.

In normal reinforced concrete tensile cracks which develop under working load are small but nevertheless provide paths by which chlorides and oxygen can get to the steel. In prestressed concrete the amount of cracking under service is considerably less, and those cracks that do develop due to overloads are generally kept tightly closed by the prestressing force. However, some special attention should be given to the problem of corrosion of steel in prestressed concrete.

Two principal methods of prestressing are in use: post-tensioning and pre-tensioning. In post-tensioning the reinforcement is not bonded to the concrete. The tensioning force is applied and the load transferred to the concrete after it has hardened. In pre-tensioning the concrete is cast around reinforcement which has been stretched between end anchorages. After hardening the reinforcement is cut and the force transferred to the concrete through the bond. The high tensile strength reinforcement used with pre-tensioning is generally small diameter wire used singly or in strands. The consequences of significant corrosion, particularly of the pitting variety, are more serious than for similar occurrences on the larger diameter reinforcing bars used in reinforced concrete construction or with some systems of post-tensioning. Since the presence of chloride in the concrete is of critical importance with respect to corrosion, neither seawater nor admixtures containing calcium chloride should be used in making prestressed concrete.

The performance of prestressed concrete in a chloride environment has been generally good. This is due in part to the fact that uniformity of mixing, placing and compaction of prestressed concrete is generally observed more closely than for cast-in-place reinforced concrete. As a result, the prestressing tendons generally are uniformly and completely embedded in high quality concrete paste with a minimum of voids adjacent to the steel which could serve as areas of moisture and oxygen collection.

PERFORMANCE OF CONCRETE IN OCEAN STRUCTURES

A number of inspections have been made of reinforced concrete coastal structures. In a 1941 comprehensive survey of shore structures along the

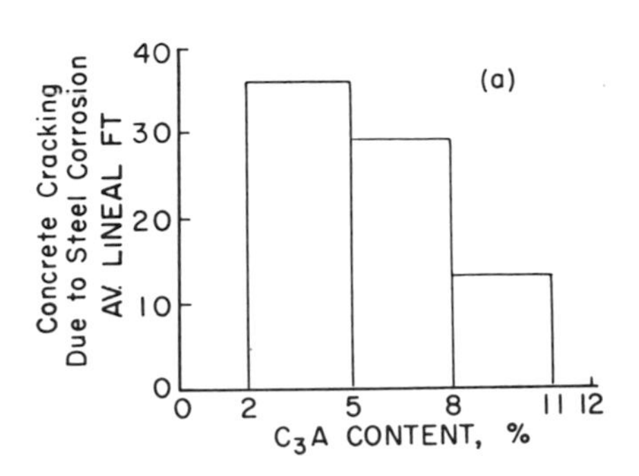

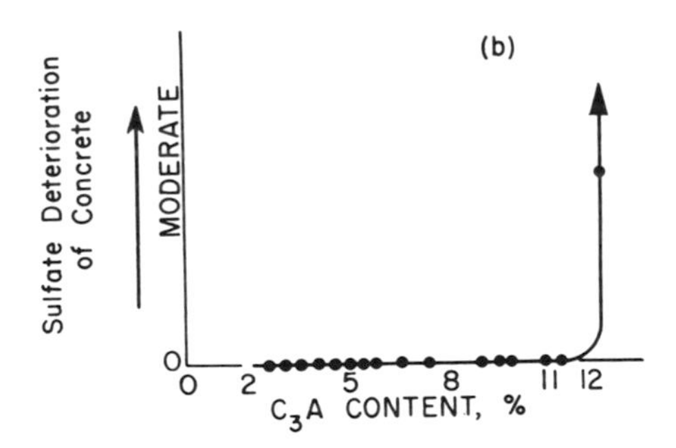

Fig. 13. Effect of C_3A contents on concrete cracking due to corrosion of reinforcement and concrete deterioration due to sulfate attack. Concrete piling exposed to sea water at St. Augustine, Florida.

Pacific coast of both the United States and Canada, the late Homer Hadley, a Seattle, Wash. consulting engineer, found no evidence of sulfate attack from seawater nor any evidence of deterioration due to freezing and thawing. He concluded that the primary problem was erosion and abrasion of surfaces due to mechanical wear, and that the second most common problem was rusting of reinforcement.

In 1956, the British magazine, "Concrete and Constructional Engineering", described the performance of the first reinforced concrete marine structure built in Great Britain. This was a jetty at Southampton built in 1899 as a reinforced concrete deck supported on concrete piles. An inspection in 1956 showed the jetty to be in excellent condition with no signs of deterioration.

Also in the 1950's, Bryant Mather of the U.S. Army Corps of Engineers made an extensive review of the performance of concrete in coastal areas of the United States, including some naval installations in the Philippine and other Pacific Islands. Mather concluded; "The service records indicate that good concrete can endure for a period of 50 years and more without excessive maintenance requirements. The records indicate the necessity of an adequate cover (preferably 3 in. or more) of dense, waterproof concrete over reinforcing steel. The combined use of aggregates permeated with salts, seawater for mixing and rusted reinforcement produce concretes subject to rapid disintegration." Obviously no one intends to build that kind of concrete.

CONCLUSIONS

This presentation has provided a description of concrete as a material suitable for many types of structures exposed to a variety of environments, especially seawater. The technology of concrete is such that component materials can be selected and combined to provide a high level of performance under a wide variety of conditions.

In conclusion, it is emphasized that while knowledge of the performance of concrete in a marine environment is by no means complete, enough is known about its characteristics to have confidence that marine structures can be built to give at least 50 and probably 100 years or more of trouble free service.

BIBLIOGRAPHY

Design and Control of Concrete Mixtures, Portland Cement Assn., Skokie, Ill., U.S.A. (1968).

Symposium on Concrete Construction in Aqueous Environments, Publication SP-8, American Concrete Institute, Detroit, Mich., U.S.A. (1964).

Concrete for Hydraulic Structures, Portland Cement Assn., Skokie, Ill., U.S.A. (1969).

Verbeck, G. J., "Field and Laboratory Studies of Sulphate Resistance of Concrete", Performance of Concrete, pp 113-124, University Press, Toronto, Canada (1968).

Hansen, W. C., Twenty-Year Report on the Long-Time Study of Cement Performance in Concrete, Bulletin 175, Portland Cement Association, Skokie, Ill., U.S.A (1965).

Hadley, H. M., "Concrete in Sea Water: A Revised Viewpoint Needed", Trans., A.S.C.E., 107, 345 (1942).

Wentworth-Shields, F. E., "Contribution to the Fiftieth Anniversary Number", Concrete and Constructional Eng., 51, 25 (1956).

Mather, B., "Factors Affecting Durability of Concrete in Coastal Structures", Technical Memorandum No. 96, Beach Erosion Board, Washington, D.C., U.S.A. (1957).

SOME NEGLECTED ECONOMIC ASPECTS OF POWER PRODUCTION WITH REFERENCE TO TIDAL POWER

A. N. T. Varzeliotis*

INTRODUCTION

The search for the perpetual motion machine has been long and futile. Even today efforts continue and it appears that people will never cease to seek this machine despite the fact the Antoine Lavoisier has demonstrated to the satisfaction of the scientific community that "you can't get something for nothing".

As man could not wait for the coming of the perpetual motion machine, he searched for, and found, other means of avoiding manual work. As a matter of fact,man has increased his needs for goods to such a great extent that even if he wanted to work manually he would be unable to satisfy his highly inflated appetite for "economic goods". Consequently, man depends on other than manual sources of energy to perform his ever-increasing workload.

Man has tapped sources of energy such as animals, coal, hydrocarbons, waterfalls, uranium, and some of his own brothers whom he has converted into slaves. For the satisfaction of his future needs man is looking into utilizing, on a commercial scale, such sources as solar energy and the heat of the earth's molten interior.

From a broad point of view, the sources of energy which serve man may be grouped in two categories:

1. Non-renewable, and
2. Renewable sources.

In the first category are coal, petroleum and other hydro-carbons, and minerals. These sources of energy, although known to exist in large quantities, are limited in extent and once used cannot be replaced. In the second category, the energy of falling water, solar energy, and aeolic (wind) energy, stand virtually alone at the present time.

---

*Engineering Division, Inland Waters Branch, Dept. of Energy, Mines & Resources, 404-1001 W. Pender St., Vancouver 1, B. C., Canada.

It is worth noting here that the non-renewable sources of energy are more than energy sources for they can, and do, satisfy other needs of man. For example, petroleum is currently used in the making of clothing, paints, plastics,and an array of other commodities,and research continues to develop new products. Uranium serves such fields as medicine and agriculture. On the other hand, when the energy of water is utilized for power production its form does not change, it is only its position in space which changes. Our familiar "hydrologic cycle" perpetuates itself and the water is "re-positioned", to be used again. Hence, hydro-electric power utilizes a renewable energy source.

The hydrologic cycle is not the only natural phenomenon which involves the lifting of water. Tides also effect the raising of huge water volumes. The raised tidal waters possess potential energy relative to the low sea and may be tapped by man to satisfy his needs for work. Being of periodic occurrence, tides are a renewable energy source.

Most of the energy consumed by contemporary societies is in the form of electricity, which can be produced from any of the energy sources available. All that matters at the consumption end of the power line is that enough electricity is delivered when it is needed, the source utilized for its production being of no direct concern to the consumer. Hence, the different sources of energy are in direct competition for the markets. The competitive position of a source of energy is determined by a variety of factors, such as technical feasibility for its development, international border tariffs, location of the source relative to the markets and even by the economic philosophy of the planners.

We may note now and bear in mind that:

1. power production has to be planned,
2. power plants have a rather long useful life, and
3. in Canada, power production is, for all practical purposes, a public venture.

The facts that power production must be planned long before the demand occurs and that the power plants have a long lifespan imply that the need for power and its cost have to be estimated well in advance of production, sale and consumption. As we very well know, predicting the future is not an exact science. However, predictions have to be made and power planning engineers and economists must use the best techniques available.

The fact that power production and planning are done by, or on behalf of, governments has added various and far reaching dimensions which need not be considered if power were to be developed by private enterprise. For example, the French Government decorates some hydro-electric powerhouses with artwork to attract the art-loving tourist to spend money in the village near the plant. Private enterprise would not likely spend funds on artwork because the economy of the village would be viewed as being beyond the direct concern and scope of a private power company. Thus, the planning of power production by governments is a process involving the evaluation of a great number of variables such as transportation, pollution, irrigation, industry, relocation of people, recreation, fisheries, etc. Consequently, planning is not necessarily aimed at the cheapest power production as such, but towards the best "benefit to cost ratio" of a project. Both the costs and the benefits can be numerous and some of them may be quite intangible.

Because some of the benefit or cost components described are intangible, they are often either totally ignored, or paid lip service in the evaluation of alternatives when planning for power production. For example, the cost of air pollution is difficult to express in dollars, hence it is usually either ignored or simply mentioned in a couple of lines in the feasibility reports.

Other benefit or cost components are evaluated in what may be called "ways of convenience". Such evaluations leave much to be desired, may produce erroneous results,and may shift the choice of development to other than the optimum source.

The following discussion covers some of the neglected or mistreated aspects which affect the selection of power developments. It leads to the conclusion that,should these aspects be evaluated properly, tidal power may prove to be economically feasible or at least less costly than it is believed to be.

ASPECTS RELEVANT TO EVALUATION OF ENERGY SOURCE DEVELOPMENT

Fossil Fuelled Thermal Power

Regardless of whether the fuel is coal, or petroleum, or even natural gas, the fact remains that the producer of thermal power has to buy fuel regularly throughout the life of the power plant. This raises the odds against making a good guess on the future cost of power, since the price of fuel may change frequently because of absolutely unpredictable factors such as war, discovery of new fuel deposits, technological changes which may produce extremely demanding alternative uses for the fuel, etc. Uncertainty in estimating costs calls for a contingency allowance, or insurance premiums, which should be expressed in dollars. Invariably this is not done.

Although the supply of fuels is large in comparison to our present needs, the fact that fuels exist in limited quantities raises in the minds of some people questions of both a philosophic and an economic nature. For example, people question the amount of the country's resources which should be consumed by any given generation of citizens. Others say "fear not the future" and use as much as needed because science will provide future generations with alternatives. It is likely that energy alternatives will be forthcoming, but what about the alternative for other uses of oil, known and undiscovered?

The next question is: What is the price of coal or oil? Or, to put it in more specific terms, what is the "resource value" component of the price?

Oil is found in many countries. Thus its price is determined by the market forces on an international level. A country whose only resource is oil, if pressed by competition, may not put any resource value on it and may even forego taxing the oil producers so that production may continue to provide jobs for citizens. This may compel other countries to do the same. Consequently, oil may appear cheap, simply because the resource component may not have been added to the price. Still, there is a value on the in-the-ground oil for its quantity is limited. Should, then, a country which has been forced by others to waive the "resource value" use the market price of the oil to compare thermal energy to other sources, or should a resource value be placed on it while renewable sources of energy are still available? For example, for equal all-inclusive cost, any sensible man will choose hydro over thermal power because he can have both the in-the-ground oil and the power at the same time without additional cost. This raises the all-important question: how much more expensive than thermal must hydro power be before thermal power should be developed?

Further, thermal plants are a major source of pollution, which appears to be one of the greatest problems to be considered in the near future. In our ever-increasingly polluted world how can a dollar value be attached to the pollution caused by a thermal plant? And if its present cost cannot be estimated how can the pollution cost for every year of the anticipated life of the plant be determined?

On the pro side of thermal power is the fact that thermal power plants do not occupy substantial amounts of land and that they can be built near the market. An additional advantage is that future generations who will inherit the plant ready made will pay something for the power by buying fuel instead of the present generation shouldering all the costs.

Nuclear Power

The problems associated with thermal power, as far as the fuel is concerned,are, to varying degrees, present here as well.

For nuclear power the problem of pollution arises from the disposal of the radio-active waste. In the U.S.A. the waste is placed in steel containers and dumped into the sea while in Canada the waste is buried. I do not believe that either of these methods constitute an acceptable solution to the problem. Will we ever solve it? Maybe we will, but then how can we predict its cost? Is it fair to base our estimate of power cost on the cost of the disposal methods we use today? What about the possibility of pollution when the containers disintegrate? The "thermal pollution", i.e. the heat added to coastal or river waters, upsets the natural temperature of the environment and this too can be of adverse consequences.

Another problem which is bound to arise is the dismantling and disposal of obsolete nuclear power plants. This problem has not yet presented itself; however, it will not be long before it appears.

Despite all the precautions taken and the safety factors built in the nuclear plant, the danger of accident is always present. This worries the population of the areas surrounding such plants. More than that, people inhabiting areas near the sites of nuclear reactors have been reluctant to believe the engineers' assurance that radiation is no problem.

Then, it should be mentioned that, at the present state of the art, nuclear power plants suffer a high rate of "outage" from service due to malfunctioning and failures. Thus, nuclear power has the added drawback of being somewhat unreliable. Naturally, if we were more knowledgeable we could attach a dollar value to these hazards and uncertainties and this would increase the cost of nuclear power

Finally, the cost of research and development of nuclear power has been and is vast. It should not be ignored in the costing of plant.

River Hydro

"Hydro" has been the best all-round source of power for man. This source is renewable. However, the sites where such sources may be developed are limited in number and thus become more scarce as we continue to develop new sites.

The cost of developing water power varies greatly between alternative sites. The reasons are well known and need not be related here. However, one of the cost components is the value of land flooded by storage reservoirs. This cost is accounted for in the evaluation; nevertheless the way it is evaluated is, to say the least, unrealistic.

What is the value of the land to be flooded by the backwaters of a power dam? Usually the answer is provided by real estate agents working in the area of the development. The price paid for land acquisition is usually lumped with the other capital expenditures and thus the future changes of land value are absolutely ignored in the planning of the project. Is this practice correct? I doubt it, as I doubt the soundness of the "present day value of the land".

To start with, many "classical" economists would argue that in order to arrive at the land component of the power cost, it is proper to consider the land as being rented to the project with the rent being free to follow the market trends. Alternatively, at any given time in the plant's useful life, the land component of the power cost should be based on the "replacement value" of the flooded land.

Next the price of land can be considered. For this, let it be assumed that a hydro power development will affect farmlands. What determines the market value of farmland? Obviously, the income which can be derived by selling its products. The price of the farm product, be it meat, grain or anything else, is determined by the forces of demand and supply which have to be balanced in the market place. Demand and supply set the price and by doing so determine the magnitude of future production which will arrive at the market to become "supply".

Now let us see what "demand" is. One of the definitions of the term given by the dictionary* is: "the desire to purchase and possess coupled with (money) power". As we are informed by our friendly neighbourhood communication media, substantially more than half of the world's population lives in poverty. Even in our own rich country a very substantial number of people live in relative poverty. Evidently the poor are poor because they cannot "couple with power" their desire to purchase the basics, for those who cannot afford luxuries but can afford the basics are not thought of as being poor. Thus some people do not convert their "desire to purchase" or all of their "desires to purchase" into "demand" and therefore do not influence the pattern and magnitude of planning for future production. Consequently, the market doesn't force the cultivation of the soil to the extent that it would if the poor had buying "power". This puts certain agricultural lands below the economic productivity level, forces them out of production, and lowers the value of the better farmlands.

The fact that we are not happy living in the company of poverty, be it in our own country or elsewhere, adds dimensions to the argument and thus it may be concluded that the market price of land is below its true value.

It is unnecessary to comment on the seriousness of the above aspect!

The next point to be considered is the assertion that the flooded land of a hydro project can be reclaimed and restored to agriculture on short notice. This may appear reasonable justification for the low relative value we attach to the land when planning river hydro. In fact it is not. In order to return the land to agriculture, the value of farmland has to increase enough to cover the cost of (a) removing the dam, or part of it; (b) taking care of the economic value of the remaining life of the plant and providing power from another source; and (c) paying for the disturbance of the economic and social patterns which have been established around the reservoir. There is no likelihood that such a reversal of the present day relative values of farmland and power will occur within the lifespan of a hydro power plant.

If the land to be flooded is woodland the economy is deprived of a renewable timber resource. The fact that pollution threatens to destroy life on the planet adds dimensions to the loss of oxygen-producing woodlands.

Frequently the lands to be flooded are inhabited to some degree. Provision of dollar compensation and relocation have not been found adequate to make up for the disruption to the lives of those who have to be moved.

Another negative aspect of the flooding of land is the possibility of

---

*Random House Dictionary, First Edition, Random House, New York.

mineral wealth being hidden underwater.

In order to clarify a misconception it is noted that the cost of hydro is not necessarily passed on to the future generations which will partly enjoy the benefits of the projects. Lending to future generations is really meaningless for we will not be alive to collect. This is specially true when the financing is done within the country, instead of in foreign money markets where payment is, strictly speaking, made in "goods".

The effect of river hydro on migratory fish must also be considered. The art of fish passing facilities has been advanced but not to the point where the problem can be dismissed. Power dams do disrupt the ecology of marine life.

Now we may turn to the advantages of river hydro. The main component of the power cost is the expenditure for the building of the plant, because maintenance and operation are relatively inexpensive and fuel does not have to be purchased. On the basis that these projects are financed over long periods of time and on the assumption that money will continue to lose its purchasing value, this aspect contributes materially to the desirability of hydro over thermal in the conventional method of cost comparison. It also means that the cost of power, at least its upper value, can be easily foreseen at the planning stage, except for the flooded land value which was discussed above. Thus there is little risk involved in estimating the future price of hydro power relative to that of thermal power.

On the subject of reliability, hydro power has surpassed practically all other forms of power. The outage of hydro plants has been very low, and maintenance has been minimal.

It is rarely, if ever, that the benefits of a hydro development occur only in the form of power. Roads built to the site are charged to the power and benefit areas along their route. They also benefit the area surrounding the reservoir. The reservoir will normally provide improved water transportation facilities and may provide sport and even commercial fishing if the flooded lands are properly cleared and if polluting waste is excluded. Further, the reservoir provides the people of the area with valuable recreation in the form of sailing, boating, and swimming. Consequently, it may boost the economy of the area and indeed benefit the country's foreign exchange position with tourists from other countries coming to visit the plant and seek recreation around it.

Improved flood control is usually a very important benefit to be derived from developing hydro-electric power. Further, the land downstream from a dam may become more valuable because the dam may provide irrigation water as well as power, because the navigability of the river downstream from the dam may be improved, and finally because power-consuming industries may be attracted.

Tidal Power

The drawbacks of river hydro are not only absent from a tidal power development; some of them shift to the side of benefits.

Like hydro, tidal power utilizes a renewable source of energy. Tidal bays suitable for power development are not to be found in abundance; however, quite a few countries have potential tidal power sites. Canada has been blessed with one of the best tidal bays, the Bay of Fundy.

Tidal power is more dependable than river hydro because tides are, for all practical purposes, unaffected by weather, unlike river hydro power which is affected by the familiar "dry and wet years" caused by variations in the hydrologic cycle. Hence a tidal bay may be developed to its full potential.

Tidal power development involves the creation of storage reservoirs, or, as they are better known, "tidal power basins". As these basins are cut off from the sea no additional land has to be flooded and added to the already vast water-covered surface of our planet.

On the other hand, a tidal power development results in reclamation of land from the sea. This is so regardless of the scheme of development, for inside a tidal power basin the tidal range is reduced. Land reclamation can, however, become significant and even substantial when the development creates a "Low Basin".

Areas surrounding High Basins may benefit from improved navigation routes and from stabilization of docking conditions as low tides are excluded.

While hydro dams often carry roads, these roads usually connect places which have been separated by construction of the dam and creation of the reservoir. That is, they often rectify a situation of their own creation. Tidal barrages may carry roads connecting areas which have always been separated by water. Thus they may create new links for people living on opposite sides of the basin. The "bridging" of bays provided by tidal barrages is a net benefit and can be quite substantial, especially to the post-development economy of the area.

The fishing industry in the area of a tidal power development may be adversely affected; however, there is always the possibility that, if proper care is taken, it can be improved or changed rather than harmed. I think it is safe to say that we know very little about the marine ecological effects that may result from a large tidal project.

Tidal plants do not force people out of their homes, nor do they pollute the air, the land or the water.

Tides fascinate people; I do not know why. Construction of large scale projects also attracts people. Because tidal power involves heavy construction and because people are fascinated by tides and attracted to the seashore, tidal projects are bound to attract substantially larger numbers of tourists than other projects. This may be a substantial benefit to the area, attributable directly to the project.

Discussed next are some substantial benefits which would accrue from development of a tidal power project in the near future. These are particular benefits due to the timing of the development.

La Rance is a pioneer plant. It is the only plant of its kind constructed, and it has not provided definite answers to the problems related to tidal power plant design, construction and operation. After all, despite the tremendous number of river hydro plants constructed, we are still learning about hydro development by building and by doing research. There is therefore experience to be gained in design, methods of construction, materials of construction, gates, mechanical and electrical equipment and, finally, in operation of tidal plants. Development of the next tidal power project will provide a great deal of knowledge which will assist others in developing their tidal power potential. Because the "state of the art" is not highly developed at the present time, knowledge acquired becomes an exportable commodity of great value.

It should be noted that the experience to be acquired from a tidal power development is not only applicable to tidal power plant construction but can be invaluable in a variety of heavy marine construction such as harbours, causeways, etc. Hence the export of knowledge and products could be quite a substantial benefit.

CONCLUSIONS

The ever-increasing demand for power, scale economics and the nature of the most common energy sources combine to dictate large-scale power projects. Such projects usually have economic and social implications which extend well beyond those directly connected to power per se.

There is a choice between energy sources available to those planning for power production. Each source has a number of particular factors which affect the economics of its development. Some of these factors either go unrecognized or are paid lip service because it is difficult to express their dollar value. Of course, inability to evaluate or unwillingness to include certain cost-benefit components do not make them irrelevant. They are there and their effect is bound to be felt. Even the degree of certainty of evaluation is different for different sources of power. This should be taken into account.

Unless all the significant factors which contribute to the benefits to be derived and to the costs to be encountered in the development of each energy source under comparison are taken into account, with their proper weight, the comparison can be unfair and the selection made, most probably, can be misleading.

The basic assumptions used have a decisive bearing on the selection of the energy source to be developed. In addition, the thoroughness of the economic approach to the evaluation and comparison of energy sources affects the outcome of the comparison. Hence, the fairness of the evaluation to be made depends on how realistic are the criteria applied and on how well the subjects have been examined.

Familiarity breeds partiality. Some sources of energy have been utilized for long periods and thus they have made some loyal friends among the planners. In addition, the ability to utilize energy sources is not equally advanced for all sources.

Feasibility studies have shown tidal power to be in an unfavourable economic position relative to other sources of power. One of the main causes is the effect of the interest rates used in the evaluation. I suggest, however, that there are other reasons, some of which have been discussed in this paper.

In summary it appears that we must:

1. View power development from a very broad point of view. When planning a large power project, the effect and products of other interests must be recognized and their costs must be considered. Research and development costs should also be taken into account when sources of power are compared.
2. Recognize the fact that we have to live with inflation, even if at a controlled rate. We must also recognize that when relatively large amounts of money are involved the decision on where and how this money could be invested cannot be left entirely to a process which resembles auctioning the interest rate for short term returns. This should be reflected in setting a reasonable interest rate that should apply to power development. It is not accidental that the United States applies a very low rate of interest to the development of its water resources.
3. Undertake research to clarify and evaluate the economic and social aspects of large scale projects such as those of power development.

ENVIRONMENTAL EFFECTS OF TIDAL POWER DEVELOPMENT

D. H. Waller*

INTRODUCTION

If increasing samples of environmental crises that confront this earth can be said to have any redeeming features, perhaps one of them is the prospect that man will at last realize that he can never conquer nature. We can at least hope that the persistent demonstration of the unintended side effects of man's efforts to mould the earth to his designs will bring him to the realization that he will succeed in "mastering" nature only to the extent that he understands natural processes and is able to use his understanding to predict the consequences of his actions.

The blame for failing to predict, or for ignoring, the unwanted extra effects of technological development has been laid by some on the engineer. Although "The work of the engineer is to utilize the raw materials and energy of the universe for the benefit of mankind"[1], the engineering profession is being accused, from within and without, of lack of social awareness, and "the single-minded engineering mentality" [2] has been castigated as the source of most of our environmental problems.

It is true that most contemporary environmental problems result from products and structures that were designed and produced by engineers, and most engineers will probably concede that many of the structures and the products have been produced with little consideration for their environmental effects. They will point out, however, that engineers have operated in a political and economic climate in which the effectiveness of a project or a product has been largely judged in accordance with standards of performance that have been based solely on economic efficiency, excluding consideration of consequences or effects that were unrelated to project objectives.

Perhaps the major reason why decisions have been left to engineers, or

---

*Assistant Director, Atlantic Industrial Research Institute, 1334 Barrington St., Halifax, Nova Scotia, Canada.

why engineers' recommendations have been adopted, is that where decisions are based on the most economical method of achieving a single specific project objective the engineer, having examined alternatives in the light of their ability to achieve the objective, is the logical one to select the most appropriate solution.

Decisions made by engineers in these circumstances have been to the benefit of mankind only to the extent that the objectives of the projects' sponsors happened to coincide with those of society at large. It is not surprising that when unconsidered or unanticipated effects arise from these decisions the engineer is blamed, despite the fact that the decisions he has made were based on the criteria that were generally acceptable to society when the decisions were made.

The attitudes of society are changing rapidly: the public now wants and expects that considerations other than project efficiency should be considered, and it is expected that consideration will be extended to effects that cannot necessarily be evaluated in monetary terms. A society that was unaware until recently of adverse environmental effects of technological change, or considered those effects unimportant in the context of an apparently limitless environment and inexhaustible resources, has begun to recognize that neither the physical resources contained in our environment, nor the capability of that environment to resist stress without failure, are unlimited. And a society that -- in a period of developing technology -- regarded each development as necessary and inevitable, and the undesirable side effects as the price of progress, is recognizing that in many areas technology has brought us to the point where choices are available: "It used to be true that most things that were technologically possible were done. ..... Certainly, in the future, this cannot and must not be so. As our ability to do all kinds of things, and the scale of them, increase -- for the scale is planetary for so many things today -- we must try to realize a smaller and smaller fraction of the things that we can do. Therefore, an essential element of engineering from now on must be the element of choice." [3] It is not essential for the survival or the progress of mankind that the SST be flown or that a sea-level canal be constructed at Panama, or that tidal power be developed.

These changing public attitudes imply a different role for the engineer in the decision-making process, a role that recognizes that decisions must be based on considerations that may extend far beyond the efficiency of the project as described in terms of its primary objective.

Recent studies of tidal power development provide examples of major projects where consideration has been given to environmental effects. The review that follows is based very largely on the reports of two of these: the 1959 International Passamaquoddy study [4] and the 1969 Bay of Fundy study [5]. The terms of reference of both the Passamaquoddy study and the Bay of Fundy study include explicit consideration of environmental effects. The governments of the United States and Canada in 1956 directed the International Joint

Commission to determine the cost of developing the tidal power potential of Passamaquoddy Bay and to determine whether that cost would allow power to be produced at an economically feasible price; it also requested the Commission "to determine the effects, beneficial or otherwise, which such a power project might have on the local and national economies in the United States and Canada, and to this end, to study specifically the effects which the construction, maintenance and operation of the tidal power structures might have upon the fisheries in the area" (4). In 1966 the Atlantic Tidal Power Programming Board was established by the governments of Canada, New Brunswick and Nova Scotia to carry out and report on initial studies and surveys of the physical and economic potential of the development of electric power from the tides of the Bay of Fundy and its transmission to markets in Canada and the United States. The Board was also directed "to assess the effects of the power schemes on other interests, such as fisheries, navigation and transportation as well as on the tidal regimes of the Bay of Fundy" (5).

Information about the environmental effects of tidal power developments is presented below by first considering water levels and predicated oceanographic changes, and then revising the environmental changes that are expected to follow from these changes.

WATER LEVEL VARIATIONS AND PHYSICAL EFFECTS

The double-basin scheme proposed for the 1959 Passamaquoddy Project would result in water levels in the high basin that would be permanently higher than the mean ocean level, and levels in the low basin that would be permanently lower than the mean ocean level. If a single-basin, single-effect scheme was operated in Minas Basin -- the most economical site considered in the 1969 Fundy Report -- the water level in the Basin would vary from one foot above to twenty-five feet above mean sea level, compared with an extreme variation of natural levels from twenty-six feet below to twenty-eight feet above mean sea level. If the same Basin was operated in a double-effect mode the expected variations would be from twenty-nine feet below to thirty feet above mean sea level. A double-effect scheme, while operating over a tidal range that was only slightly greater than the natural range, could -- depending on the mode of operation selected in accordance with power demands at a particular time -- produce irregular fluctuations in level. During some periods natural level variations would occur because power was not required; at other periods the level would be held at a high or a low position in anticipation of peak-load power requirements while at other times the plant would be operated in a manner corresponding to the single-effect mode.

The Passamaquoddy scheme would not measurably change the resonant system of tides in the Bay of Fundy, and the only effect to be expected on tidal range would be an increase of up to one foot at the head of the Bay (4, 6).

A mathematical model simulating tidal conditions in the Bay of Fundy indicated that a tidal power development in Minas Basin would reduce the tidal range at various points in the Bay of Fundy outside the Basin by up to five percent (5).

In general, the physical changes brought about by construction of a tidal power plant would appear to be: the physical obstruction created by the dams, altered mean levels and level ranges in the impoundments, change in current patterns, reduced velocities, reduced tidal exchange (flushing), less vertical mixing, lower rates of oxygen up-take, greater stratification, higher summer temperatures, winter icing, reduced salinity in surface layers, reduced dissolved oxygen in deeper layers, and erosion and siltation effects.

The Passamaquoddy studies (4) indicated that currents within the high pool and the low pool, and in a semi-enclosed area outside of the Passamaquoddy impoundments, would be markedly altered; outside of these areas only slight alterations in tidal stream direction and speeds were anticipated. Tidal inflow to the pools would be reduced and dilution of fresh water would therefore be decreased. Velocities in the tidal pools would be reduced during periods when gates were closed with an accompanying reduction in vertical mixing. The consequent stratification would give rise to greater seasonal variations in surface water characteristics in both pools: summer surface temperatures would be increased, ice formation would occur in winter, and surface salinities would be reduced. In the deeper waters temperatures and salinity changes would be minor. Oxygen concentrations in the deeper layers, which are nearly at saturation levels due to vigorous tidal mixing, would decrease but would be unlikely to fall below 50 percent saturation. Erosion and siltation were not matters of serious concern in the Passamaquoddy report, presumably because river inflows are relatively small and of low turbidity, and the shoreline materials within the proposed impoundments would not be subject to significant erosion. The possibility of some silting as a result of reduced turbulence was recognized in connection with effects on fisheries in the impoundments.

The Bay of Fundy study indicated that tidal amplitude in the embranchments immediately below the tidal power sites would be reduced by up to 15 percent, and that reduced amplitude and changed phase might be expected to reduce velocities of tidal streams by up to 20 percent. The reduced velocities could result in changes in turbulence, sedimentation regimes, sediment deposition and scouring within the basin, but the extent of such changes is described as unpredictable (5).

Siltation and erosion were important concerns in the Bay of Fundy investigations because the embranchments of the Bay contain large amounts of sediment, which is mobile under the influence of strong tidal currents, and because observation at the sites of existing tidal dams -- built for marshland protection -- indicate that substantial deposits of sediment can be expected to accumulate on the seaward side of such structures. Preliminary investigations in the Bay of Fundy indicated that construction of tidal power dams would reduce turbulence and turbidity, resulting in a slow increase in sediment upstream of the dams. Where the dams were operated in the single-effect mode, upstream sediment accumulation would be aggravated because in order to avoid abrasion only the clearest water would be allowed to leave the basin through the turbines. Other effects of the tidal power developments would be: changes in currents and water levels, resulting in altered sedimentation and erosion patterns; sedimentation and erosion resulting when equilibrium conditions in bays, river mouths, and tidal reaches of rivers are disturbed; and effects on coastline erosion. The latter effects would depend on the natural rate of erosion of sea cliffs, on protection from storm effects provided by the tidal power structures, and on altered current patterns and tidal ranges resulting from tidal power development.

ENVIRONMENTAL EFFECTS

For purposes of this discussion environmental effects are taken to include those results of physical changes brought about by a tidal power project that do not directly effect the efficiency or cost of the project with respect to the production of power.

Fisheries

In the Passamaquoddy study attention was directed to the effects of the project on inorganic nutrients and on zooplankton concentrations that might affect fish stocks as well as to the effects of altered physical factors that could affect fish. "Anticipated reductions in erosion, scouring, and turbulence together with higher water temperatures resulting from impoundment and the less rapid transportation of soluble compounds and mineral elements carried into the area and by fresh water runoff should enhance the supply of fish- and shellfish-supporting nutrients except for those species which would be denied access by engineer structures" [8]. Existing zooplankton populations in the area proposed for the Passamaquoddy high pool were considerably lower than in the ocean or in the low pool areas; it was predicted that flow of ocean water into the high pool and back to the ocean via the low pool would tend to equalize zooplankton concentrations.

Precise effects on individual fish species are perhaps irrelevant to this discussion, but it is interesting to note the variety of ways in which the altered Passamaquoddy conditions were expected to affect different types of fish. Herring, which breed outside the present area and only enter to feed during periods of migration, would freely pass the tidal inlet gates into the high pool, would be unaffected by the turbines in their passage to the low pool, and would return to the ocean in regular numbers. Ground fish, such as cod, haddock, and pollock, which breed outside the project area and migrate into and out of the area each year, would be unable to enter the low pool except via the high pool and turbines, and the numbers entering via the gates would be seriously reduced in comparison with existing stocks. Winter flounder, a ground fish that breeds in the project area, would not be affected. Growth of lobster larvae would be favoured by higher temperatures. Clams would be seriously affected: in the upper pool an entirely new and steeper shoreline would be unproductive for ten years and when re-established would be restricted to a smaller area; in the lower pool about one-half of the clam beds would be lost due to lower water levels, but would be re-established by the end of the ten-year period. Scallops would be reduced by silting, but in the main deep water channels where velocities would remain sufficiently high, they would increase because of reduced tidal exchange which would limit migration from the breeding area. Possible effects on predators of scallops are noted: reduction in the number of bottom feeding fish, and possible increase in the number of starfish. Anadromous fish, such as salmon, alewives, smelt and trout, were expected to increase because of increased temperatures and reduction of tidal amplitude, provided fishways were installed to facilitate their passage into the lower pool and past the power house into the upper pool. A significant adverse effect of the power project would be increased numbers of shipworms, which would be encouraged by warmer temperatures, reduced turbidity, and retention due to reduced water exchange. The cost of protection of wooden boats, gear and structures ".... could easily equal or exceed the value of all fisheries from the Quoddy region" (9).

The value of commercial fish landings in areas likely to be affected by the Bay of Fundy tidal power developments was estimated, and the probable effects on fisheries were assessed (5). (On-going Federal Department of Fisheries studies, in areas that include the tidal power sites, will provide further information on these points.) No adverse effects on fisheries in areas outside the impoundments are anticipated; clam bed productivity could be improved by reduced silt loadings. Commercial fisheries inside the impoundment area are relatively unimportant except for a small part of the clam fisheries, which would be affected by changed tidal levels and silt accumulations. It is not expected that anadromous fish would be seriously affected. Factors that might influence these fish would be effects of turbidity changes on food organisms, effects of flow patterns on fish movements, and the possibility that salmon smolts passing out through the turbines might encounter concentrations of predators such as striped bass.

Wilson (10) in discussing the English single pool Severn proposal, notes that the passage of ascending adult fish and descending salmon smolts might be adversely affected but comments that biologists are reluctant to do more than speculate, and concludes that -- based on the passage of fish through faster turbines in Scotland -- the project should have little effect on the fishery. Bernstein (11) noted that economics of tidal power can be improved by using the basins for oyster culture, and that experiments in this direction were being conducted in the Netherlands. Oyster production was discussed in connection with the Passamaquoddy project, but insufficient experience was available to permit a prediction to be made (8). Bernstein also described a proposal to use the power house of the proposed Russian single pool Kislayna scheme as a herring trap by opening and subsequently closing all gates during the passage of a school of herrings.

Navigation

Tidal power developments may restrict navigation -- unless appropriate corrective measures are taken -- because of barriers presented by power dams, because of reduced low water levels, or because of the effects of silting. Most major tidal power schemes have incorporated locks to allow ships to pass the dam structures. In two of the three sites considered by the 1969 Fundy study, commercial navigation is considered to be practically non-existent, and no locks are provided; the third site at Shepody Bay would include a gated navigation channel to provide access to the City of Moncton (5).

Reduced tidal ranges on the seaward side of the proposed Minas Basin tidal power site would make it necessary to deepen access channels in order to permit vessels to enter several ports at high tides, and the costs of compensatory dredging were included in project estimates (5). The 1959 Shepody-Cumberland scheme (12) would require canals in several locations to provide continued access to harbours.

Pollution

Tidal power development does not, of itself, represent a pollution source, but creation of an impoundment could aggravate an existing problem or create a problem where one did not previously exist. Factors that would increase pollution potential would be decreased dilution due to reduced tidal exchange, increased temperatures, and reduced dissolved oxygen concentrations due to decreased re-aeration rates. A pollutant that, in the absence of an impoundment, was removed without nuisance by dilution and tidal exchange, might well be a problem if its concentration was higher and less dissolved oxygen was available to meet the demands of decomposing micro-organisms. The problem would be worse if settleable material was deposited, because of low velocities, in stratified deep water with an initially low dissolved oxygen concentration. Benthic organisms could be smothered and the water made intolerable for fish.

The Passamaquoddy report (4) anticipated no pollution problems because of the small quantity of pollution entering the project area. In the case of the 1945 Petitcodiac tidal power scheme, where sewage from the city of Moncton was discharged untreated into the river, the only practical solution was provision of complete sewage treatment in association with tidal power development, because ". . . . the Petitcodiac River, at present a muddy, fast-flowing stream, will be transformed into a more-or-less clear water lake, and during periods of drawdown there will be a tendency for the current to become sluggish, with a decided probability that sewage will alternately move up and down stream, perhaps several times before ultimately being discharged into Shepody Bay" (13).

The 1969 Fundy report notes that in deep tidal impoundments on Prince Edward Island inadequate flushing has resulted in seasonal stagnation and pollution. This report also draws attention to the need to consider the possibility of pollution that would be associated with an increase in population of the areas around the impoundments (5).

Other Effects

The effects of altered water levels on water-front properties were taken into account by the 1959 Passamaquoddy report: docks and piers in the low pool area that were built for use only at high tide would have to be extended; lobster ponds in the high pool would have to be relocated, and herring weirs would be useless (4). Ice formation would affect fish weirs, and would restrict winter lobster fishing (14).

Because of the potential effects of water level changes due to the proposed Bay of Fundy tidal power structures, data were obtained on elevations of all important structures -- such as dikes, wharves, roads, transportation structures, and sewers -- that might be affected, and these elevations were considered in selecting maximum and minimum operating levels in the proposed tidal basins (5). The levels selected were expected to produce slight adverse effects on marshland drainage, but these effects were considered to be more than offset by the decrease in severe storm damage that would accrue from protection of dikes that the tidal power structures would provide.

It was estimated that 800,000 tourists would be attracted to the Passamaquoddy area each year; the same report noted that in 1956, 735,000 people visited Acadian National Park in Maine (4). The Fundy report states that a tidal power development should produce a significant increase in total tourist visits, but admits that the total traffic from this source would be impossible to estimate (5). The Fundy report notes that the recreational value of the tidal impoundments might be enhanced if the mode of operation was such that the tidal range was reduced.

On the basis of a preliminary assessment the Fundy report considered that the tidal power developments could be designed to incorporate a paved highway, and thereby shorten distances between major centres. The estimated savings in road-user costs could be used to justify the additional expenditures necessary to provide the roadways and approach roads (5). Both the Rance scheme and the proposed Severn scheme utilize the top of their dams to provide direct highway links across the estuaries.

DISCUSSION

It may be concluded from both the Passamaquoddy study and the Bay of Fundy study that the economic importance of those effects that are not related directly to tidal power production is small when compared with the large capital expenditures that these projects involve. In the case of the proposed Passamaquoddy Project, the total value of lost commercial fisheries, and of measures which would be provided to preserve fisheries, was too small to be included in cost-benefit considerations. Tourist benefits, although somewhat larger, were excluded from the cost-benefit analyses (4). The Fundy report concluded that "benefits that might accrue from a tidal power project in areas such as transportation, regional development, tourism and recreation would not significantly affect the competitive position of tidal power" (5). Both studies did include some corrective measures such as navigation locks and dredging in their financial analyses.

The reports of tidal power studies indicate that the environmental effects that have received attention are those that have recognizable economic implications. As awareness of the more subtle effects of environmental change increases, future studies of such projects may have to include consideration of effects that are less easily identified and more difficult to evaluate. Biological studies could be extended to include not only fisheries, but also what might be described as the totality of ecological effects.

Present knowledge does not suggest a serious likelihood that further studies would alter conclusions about the relatively small environmental effects of tidal power developments. However, a conclusion regarding the unimportance of such effects should be based on consideration of the results of explicit study of these effects, and not on generalizations about their apparent importance. Examples of effects that might deserve consideration include the effects of altered tidal regimes on bottom organisms, other than shellfish, in the inter-tidal zone, direct or indirect effects on wild fowl, and eutrophication effects.

The ecologists' concern for such effects is expressed in terms of food chains, biogeochemical cycles, and nutrient budgets. Many changes in the numbers of wildlife species have been the result, not of direct environmental insult, or of predators, but of changes that occur far down the food chain: changes to organisms that are apparently unimportant. E. P. Odum says, "If nothing else, ecology teaches us not to make snap judgments as to whether a thing or an organism is 'useful' or 'harmful'" (15). Ecologists are also worried about changes in swamps, wetlands, and estuaries in terms of their role in the biogeochemical cycles that make life on earth possible. Although there is no special reason for concern that changes due to tidal power development in areas of the sea-bottom exposed during tidal cycles are significant in this context, this point should not be left unexamined. A tidal power development may be expected to create more quiescent conditions in the basin upstream of the dam resulting in increased transparency of the water due to sedimentation of suspended matter; and the possibility of increased light penetration, together with the possibility of increasing concentrations of nutrient substances that are trapped in the basin because of reduced tidal flushing action, suggests the possibility of increased productivity of these waters.

A major obstacle in the way of effective consideration of the results of engineering developments is the absence of an ability to completely predict ecological effects. The ecologist can predict a direction, i.e., the nature of the effect that an environmental change may produce, but present knowledge is not such that for large systems the rate of change, or the new equilibrium condition, can be predicted with certainty. The fact that the ecologist cannot necessarily say with certainty that any effect of environmental change, whether good or bad, will be significant in magnitude suggests two things: that the ecologist should be rendered every encouragement and support in his continuing efforts to sharpen his predictive tools; and that in the absence of certainty about the consequences of our actions, we should be prudent about the decisions we adopt.

Uncertainty about ecological effects, and the haunting knowledge of recent examples of unforeseen and unwanted spinoffs from technological changes, argue for careful consideration of such effects in the case of large scale, capital intensive, and essentially irreversible schemes such as tidal power developments. In the absence of certain knowledge of effects, the course of action that preserves the option to reverse a decision in the light of new knowledge has an inherent advantage over a course of action that is irreversible. Making of an irrevocable decision carries with it a heavy responsibility to assure that every effort has been made to anticipate all effects of that action including effects that may be subtle or remote.

One important and difficult aspect of the problem of evaluating effects of technological change is the absence of effective mechanisms to account for and to control effects that may be unimportant in the context of a single project but are, in the aggregate, matters of serious concern. It is probable that if adverse ecological effects were found to be associated with tidal power developments, most of them would be effects that were not unique to such developments: a reduction in the number of some form of wildlife would be apparent only on a local scale; and any effect on biogeochemical cycles would only be apparent as part of a continental or world-wide system. It might well be argued then that these effects do not matter in terms of the evaluation of a single project. But such effects, unnoticed or unrecorded, may in the aggregate amount to the elimination of whole species or threaten the continuity of biogeochemical cycles that are essential to life.

Deevy [16] has pointed out that nearly one-half of the pre-colonial area of swamps, marshlands, and estuaries within the continental United States -- essential in their role in recycling sulfur by reduction in anaerobic bottom muds -- have been lost to development. As the area available to fulfill this function is decreased, the importance of the area that remains is increased, and if encroachment continues, there is danger that an essential part of an essential cycle will be blocked.

Such incremental effects may not always be adverse. Utilization of tidal energy as a power source might not, in the context of the magnitude of all other demands on fossil fuels, represent a significant reduction in the rate of depletion of the reserves of these fuels; nevertheless, it does imply some conservation of fossil fuel resources, and the nature and magnitude of this saving should be acknowledged. These kinds of effects present particular difficulties in relation to economic analyses: what value should be placed on the incremental contribution that a single project makes to the creation or solution of a problem of continental or global magnitude?

The technical feasibility of tidal power development can be investigated and demonstrated without reference to alternative power sources. But the desirability of tidal power development can only be determined from a comparison with alternative sources of power. Brooks and Bowers, in summarizing the views of a panel of the National Academy of Sciences, state, "Since the progress of science in recent years has greatly broadened the spectrum of technological possibilities, the nation is in a position to choose among many technological paths to a given objective. Thus two important aspects of technology assessment are the evaluation of alternate means to the same end and a comparison of their social and economic costs". (17)

The Fundy report, in concluding that economic development of tidal power in the Bay of Fundy is not feasible under prevailing circumstances, indicated that one of the circumstances that would justify further examination of the project was a significant increase in the costs of alternative sources of power, and it pointed to the control of environmental pollution as an important contribution to such costs.(5)

Fossil-fuel and nuclear power plants represent sources of heated cooling water that may create or aggravate pollution problems and ecological disturbances in receiving waters, and fossil-fuel plants also provide sources of air pollutants in the form of both gases and particulate matter. Pollution control for such plants could be significant in a comparison of the costs of these power sources with the cost of tidal power. The Fundy report compares an annual cost for tidal power from a single-effect plant of 5.6 mills per kWh produced with an incremental energy cost of 3 mills per kWh for a fossil-fuel thermal plant and 1 mill per kWh for a nuclear plant, i.e., a difference of 2.6 to 4.6 mills per kWh. A recent review of power plant pollution control costs indicated that the costs of air and water (thermal) pollution control for a fossil-fuel plant would be in the order of 0.8 to 1.9 mills per kWh, and that water (thermal) pollution control for a nuclear plant would cost 0.6 to 1.6 mills per kWh.[18]

These pollution control costs bear no necessary relationship to the actual cost represented by the environmental damage that would result in the absence of pollution control measures. Nevertheless, these increased costs--of control measures, or of adopting pollution-free alternatives--represent the minimum value that the public must place on environmental quality before it will accept the increased prices or taxes that it must pay to avoid or eliminate pollution.

Nuclear plants represent additional causes of environmental concern. Ecologist Lemont C. Cole has expressed one of these: "Before the controlled release of atomic energy the total amount of radioactive material under human control consisted of about ten grams of radium or ten curies of radioactivity. Probably a billion times this amount of radioactivity has already been disseminated into the environment, and we are not really yet into the atomic age. A plant of modest size (by present dreams) is being constructed on the shores of Lake Ontario near Oswego, New York, which will, by the company's own estimate, release to the atomosphere 130 curies per day. Knowing that exposure to radioactivity shortens life, causes malignancies, and can produce genetic effects that can damage future generations, have we cause for complacency?"[19] This effect, like some of the other incremental effects identified earlier, is difficult to quantify or to evaluate in relation to a single power project. The second cause for concern is associated with the danger of nuclear reactor accidents. The order of magnitude of this hazard may be identified by an estimate (in 1961) that the risk of death associated with nuclear reactor accidents "would be for the large power reactors of the order of 1 per $\$10^9$ of investment in contrast, for example, to the death of one passenger per $\$10^7$ approximately of investment in scheduled airlines in North America"[20]. The importance that the public attaches to these risks may, of course, greatly exceed the mathmetical risk, and where an alternative to nuclear power exists fear of these hazards could weigh heavily in favour of that alternative. If the consequences of our actions could be predicted with certainty and if, having predicted these consequences, it was possible to assign economic values to them knowing that such values were universally accepted by the community at large, the role of the engineer in the decision-making process could remain unaltered: the costs and benefits of each project could be evaluated and balanced against those of alternatives, and a single decision would emerge. But our predictive tools are not perfect. One can forecast only the direction, but not the magnitude of many of the effects from environmental changes, and the possibility exists that unanticipated effects may later appear. Effective methods are not available for assigning values to all of the effects that can be identified; the difficulty of quantifying the value of incremental effects has already been mentioned. Such difficulties are compounded by the fact that the value that society attaches to environmental quality changes, both as the amount of unaltered environment becomes scarce relative to the demands that are made upon it, and as the awareness of hazards to the environment is heightened. These difficulties are compounded further by the fact that each element of society values environmental quality according to standards that reflect its own objectives and aspirations.

In the absence of the ability to reliably and completely predict and to evaluate consequences of engineering activity, the choice between alternatives must be based on value judgments. Such judgments properly belong in the political sphere. The responsibility for assessing public attitudes and for rationalizing conflicting objectives properly belongs to the politician, not the engineer. It is when the engineer presumes to make such judgments on behalf of society that he exposes himself to criticism, because the engineer has no method of prior assessment or of after-the-fact evaluation of society's values, nor has he a mechanism for making his basis of assessment of choice known to the public. A political system, on the other hand, does provide such mechanisms, and those who make decisions on behalf of society are exposed to public scrutiny and rebuke.

The role of the engineer must increasingly be one in which his concerns will be 1) to identify all of the physical consequences of his works; 2) to utilize the advice and opinion of physical and social scientists in the effort to identify and, to as great an extent as possible, quantify the environmental and social effects of those physical changes; and 3) to present, for discussion and decision in the political sphere, alternative methods of achieving desired public goals, together with an evaluation of the costs and the implications of each alternative.*

The ultimate decision about the future of tidal power will be a political decision. The Bay of Fundy report, for example, has provided the foundation -- in the form of an engineering feasibility study, an economic evaluation, and an assessment of implications in comparison with alternatives -- that can form the basis for the ultimate political decision about the development of the tidal power potential of the Bay of Fundy. That political decision will, of course, involve consideration of altered economic factors such as interest rates; but it will depend to a very great extent on the value judgments that are made by politicians. They must decide when the point has been reached at which society is prepared to accept the additional costs of tidal power in order to avoid the effects of pollution due to alternative power sources.

---

*I realize that projects that are the concern and responsibility of private agencies or industries provide no mechanisms for the consideration of non-economic consequences of engineering projects. Because social costs, i.e.,external diseconomies, are excluded from the ordinary decision-making processes of competitive business, mechanisms of public regulation and control must be employed to deal with environmental effects of a wide variety of structures and products, such as biocides, automobiles, air and water pollutants, and local development activities.

Although that time has not yet arrived, there is little doubt that the increasing demands of an exploding population on a finite, fragile, and already abused environment will ultimately call into use any alternative that promises that the quality of man's environment does not have to be sacrificed as the price of technological advancement.

REFERENCES

1. Parr, J. G., The Social Impact of Engineers, Engineering Journal, December (1968).

2. Commoner, Barry, cf Crosby Field, ASHRAE Journal, 45, October (1968).

3. Gell-Mann, Murray, Science, 166, 723 (1969).

4. Report of the International Passamaquoddy Tidal Power Project, October (1959).

5. "Feasibility of Tidal Power Development in the Bay of Fundy", Atlantic Tidal Power Programming Board, October (1969).

6. Appendix 1 of Reference (4), (1959).

7. Trites, R. W., ibid Chapter 7.

8. Dow, R. L., ibid Appendix III, Chapter 6.

9. Medcof, J. C., ibid Appendix II, Chapter 11.

10. Wilson et al, "The Bristol Channel Barrage Project", 11th Conference on Coastal Engineering, London (1968).

11. Bernstein, L. B., "Tidal Energy for Electric Power Plants", Translated by the Israel Program for Scientific Translations, Jerusalem (1965).

12. Foundation Engineering Co. (Can), "Cumberland Basin - Shepody Bay" (1959).

13. Acres Ltd., "Report on Tidal Power, Petitcodiac and Memramcook Estuaries" (1945).

14. International Passamaquoddy Fisheries Board Report to International Joint Commission, October (1959).

15. Odum, E. P., "Fundamentals of Ecology", 2nd ed., W. B. Saunders, London (1959).

16. Deevy, E. S., "In Defense of Mud", National Water Commission, Nov. (1969).

17. Brooks, H., and R. Bowers, "The Assessment of Technology", Scientific American, 222, 2 (1970).

18. Anon, "Cut Pollution at What Price?", Electrical World (1970).

19. Cole, L. C., "Our Man-Made Environmental Crisis", Exploding Humanity -- the Crisis of Numbers, University of Toronto, October (1968).

20. Lawrence, G. C., "Reactor Siting in Canada", AECL Report 1375, October (1961).

INDEX